Lecture Notes in Computer Science 9788

Commenced Publication in 1973
Founding and Former Series Editors:
Gerhard Goos, Juris Hartmanis, and Jan van Leeuwen

Osvaldo Gervasi · Beniamino Murgante
Sanjay Misra · Ana Maria A.C. Rocha
Carmelo M. Torre · David Taniar
Bernady O. Apduhan · Elena Stankova
Shangguang Wang (Eds.)

Computational Science and Its Applications – ICCSA 2016

16th International Conference
Beijing, China, July 4–7, 2016
Proceedings, Part III

 Springer

Editors

Osvaldo Gervasi
University of Perugia
Perugia
Italy

Beniamino Murgante
University of Basilicata
Potenza
Italy

Sanjay Misra
Covenant University
Ota
Nigeria

Ana Maria A.C. Rocha
University of Minho
Braga
Portugal

Carmelo M. Torre
Polytechnic University
Bari
Italy

David Taniar
Monash University
Clayton, VIC
Australia

Bernady O. Apduhan
Kyushu Sangyo University
Fukuoka
Japan

Elena Stankova
Saint Petersburg State University
Saint Petersburg
Russia

Shangguang Wang
Beijing University of Posts
    and Telecommunications
Beijing
China

ISSN 0302-9743   ISSN 1611-3349   (electronic)
Lecture Notes in Computer Science
ISBN 978-3-319-42110-0   ISBN 978-3-319-42111-7   (eBook)
DOI 10.1007/978-3-319-42111-7

Library of Congress Control Number: 2016944355

LNCS Sublibrary: SL1 – Theoretical Computer Science and General Issues

Printed on acid-free paper

This Springer imprint is published by Springer Nature
The registered company is Springer International Publishing AG Switzerland

Preface

These multi-volume proceedings (LNCS volumes 9786, 9787, 9788, 9789, and 9790) consist of the peer-reviewed papers from the 2016 International Conference on Computational Science and Its Applications (ICCSA 2016) held in Beijing, China, during July 4–7, 2016.

ICCSA 2016 was a successful event in the series of conferences, previously held in Banff, Canada (2015), Guimares, Portugal (2014), Ho Chi Minh City, Vietnam (2013), Salvador, Brazil (2012), Santander, Spain (2011), Fukuoka, Japan (2010), Suwon, South Korea (2009), Perugia, Italy (2008), Kuala Lumpur, Malaysia (2007), Glasgow, UK (2006), Singapore (2005), Assisi, Italy (2004), Montreal, Canada (2003), (as ICCS) Amsterdam, The Netherlands (2002), and San Francisco, USA (2001).

Computational science is a main pillar of most present research as well as industrial and commercial activities and it plays a unique role in exploiting ICT innovative technologies. The ICCSA conference series has been providing a venue to researchers and industry practitioners to discuss new ideas, to share complex problems and their solutions, and to shape new trends in computational science.

Apart from the general tracks, ICCSA 2016 also included 33 international workshops, in various areas of computational sciences, ranging from computational science technologies to specific areas of computational sciences, such as computer graphics and virtual reality. The program also featured three keynote speeches and two tutorials.

The success of the ICCSA conference series, in general, and ICCSA 2016, in particular, is due to the support of many people: authors, presenters, participants, keynote speakers, session chairs, Organizing Committee members, student volunteers, Program Committee members, Steering Committee members, and many people in other various roles. We would like to thank them all.

We would also like to thank our sponsors, in particular NVidia and Springer for their very important support and for making the Best Paper Award ceremony so impressive.

We would also like to thank Springer for their continuous support in publishing the ICCSA conference proceedings.

July 2016

Shangguang Wang
Osvaldo Gervasi
Bernady O. Apduhan

Organization

ICCSA 2016 was organized by Beijing University of Post and Telecommunication (China), University of Perugia (Italy), Monash University (Australia), Kyushu Sangyo University (Japan), University of Basilicata (Italy), University of Minho, (Portugal), and the State Key Laboratory of Networking and Switching Technology (China).

Honorary General Chairs

Junliang Chen	Beijing University of Posts and Telecommunications, China
Antonio Laganà	University of Perugia, Italy
Norio Shiratori	Tohoku University, Japan
Kenneth C.J. Tan	Sardina Systems, Estonia

General Chairs

Shangguang Wang	Beijing University of Posts and Telecommunications, China
Osvaldo Gervasi	University of Perugia, Italy
Bernady O. Apduhan	Kyushu Sangyo University, Japan

Program Committee Chairs

Sen Su	Beijing University of Posts and Telecommunications, China
Beniamino Murgante	University of Basilicata, Italy
Ana Maria A.C. Rocha	University of Minho, Portugal
David Taniar	Monash University, Australia

International Advisory Committee

Jemal Abawajy	Deakin University, Australia
Dharma P. Agarwal	University of Cincinnati, USA
Marina L. Gavrilova	University of Calgary, Canada
Claudia Bauzer Medeiros	University of Campinas, Brazil
Manfred M. Fisher	Vienna University of Economics and Business, Austria
Yee Leung	Chinese University of Hong Kong, SAR China

International Liaison Chairs

Ana Carla P. Bitencourt	Universidade Federal do Reconcavo da Bahia, Brazil
Alfredo Cuzzocrea	ICAR-CNR and University of Calabria, Italy
Maria Irene Falcão	University of Minho, Portugal

Robert C.H. Hsu	Chung Hua University, Taiwan
Tai-Hoon Kim	Hannam University, Korea
Sanjay Misra	University of Minna, Nigeria
Takashi Naka	Kyushu Sangyo University, Japan
Rafael D.C. Santos	National Institute for Space Research, Brazil
Maribel Yasmina Santos	University of Minho, Portugal

Workshop and Session Organizing Chairs

Beniamino Murgante	University of Basilicata, Italy
Sanjay Misra	Covenant University, Nigeria
Jorge Gustavo Rocha	University of Minho, Portugal

Award Chair

Wenny Rahayu	La Trobe University, Australia

Publicity Committee Chair

Zibing Zheng	Sun Yat-Sen University, China
Mingdong Tang	Hunan University of Science and Technology, China
Yutao Ma	Wuhan University, China
Ao Zhou	Beijing University of Posts and Telecommunications, China
Ruisheng Shi	Beijing University of Posts and Telecommunications, China

Workshop Organizers

Agricultural and Environment Information and Decision Support Systems (AEIDSS 2016)

Sandro Bimonte	IRSTEA, France
André Miralles	IRSTEA, France
Thérèse Libourel	LIRMM, France
François Pinet	IRSTEA, France

Advances in Information Systems and Technologies for Emergency Preparedness and Risk Assessment (ASTER 2016)

Maurizio Pollino	ENEA, Italy
Marco Vona	University of Basilicata, Italy
Beniamino Murgante	University of Basilicata, Italy

Advances in Web-Based Learning (AWBL 2016)

Mustafa Murat Inceoglu	Ege University, Turkey

Bio- and Neuro-Inspired Computing and Applications (BIOCA 2016)

Nadia Nedjah   State University of Rio de Janeiro, Brazil
Luiza de Macedo Mourell   State University of Rio de Janeiro, Brazil

Computer-Aided Modeling, Simulation, and Analysis (CAMSA 2016)

Jie Shen   University of Michigan, USA and Jilin University,
                            China
Hao Chenina   Shanghai University of Engineering Science, China
Xiaoqiang Liun   Donghua University, China
Weichun Shi   Shanghai Maritime University, China
Yujie Liu   Southeast Jiaotong University, China

Computational and Applied Statistics (CAS 2016)

Ana Cristina Braga   University of Minho, Portugal
Ana Paula Costa Conceicao   University of Minho, Portugal
  Amorim

Computational Geometry and Security Applications (CGSA 2016)

Marina L. Gavrilova   University of Calgary, Canada

Computational Algorithms and Sustainable Assessment (CLASS 2016)

Antonino Marvuglia   Public Research Centre Henri Tudor, Luxembourg
Mikhail Kanevski   Université de Lausanne, Switzerland
Beniamino Murgante   University of Basilicata, Italy

Chemistry and Materials Sciences and Technologies (CMST 2016)

Antonio Laganà   University of Perugia, Italy
Noelia Faginas Lago   University of Perugia, Italy
Leonardo Pacifici   University of Perugia, Italy

Computational Optimization and Applications (COA 2016)

Ana Maria Rocha   University of Minho, Portugal
Humberto Rocha   University of Coimbra, Portugal

Cities, Technologies, and Planning (CTP 2016)

Giuseppe Borruso   University of Trieste, Italy
Beniamino Murgante   University of Basilicata, Italy

Databases and Computerized Information Retrieval Systems (DCIRS 2016)

Sultan Alamri   College of Computing and Informatics, SEU,
                            Saudi Arabia
Adil Fahad   Albaha University, Saudi Arabia
Abdullah Alamri   Jeddah University, Saudi Arabia

Data Science for Intelligent Decision Support (DS4IDS 2016)

Filipe Portela	University of Minho, Portugal
Manuel Filipe Santos	University of Minho, Portugal

Econometrics and Multidimensional Evaluation in the Urban Environment (EMEUE 2016)

Carmelo M. Torre	Polytechnic of Bari, Italy
Maria Cerreta	University of Naples Federico II, Italy
Paola Perchinunno	University of Bari, Italy
Simona Panaro	University of Naples Federico II, Italy
Raffaele Attardi	University of Naples Federico II, Italy

Future Computing Systems, Technologies, and Applications (FISTA 2016)

Bernady O. Apduhan	Kyushu Sangyo University, Japan
Rafael Santos	National Institute for Space Research, Brazil
Jianhua Ma	Hosei University, Japan
Qun Jin	Waseda University, Japan

Geographical Analysis, Urban Modeling, Spatial Statistics (GEO-AND-MOD 2016)

Giuseppe Borruso	University of Trieste, Italy
Beniamino Murgante	University of Basilicata, Italy
Hartmut Asche	University of Potsdam, Germany

GPU Technologies (GPUTech 2016)

Gervasi Osvaldo	University of Perugia, Italy
Sergio Tasso	University of Perugia, Italy
Flavio Vella	University of Rome La Sapienza, Italy

ICT and Remote Sensing for Environmental and Risk Monitoring (RS-Env 2016)

Rosa Lasaponara	Institute of Methodologies for Environmental Analysis, National Research Council, Italy
Weigu Song	University of Science and Technology of China, China
Eufemia Tarantino	Polytechnic of Bari, Italy
Bernd Fichtelmann	DLR, Germany

7th International Symposium on Software Quality (ISSQ 2016)

Sanjay Misra	Covenant University, Nigeria

International Workshop on Biomathematics, Bioinformatics, and Biostatisticss (IBBB 2016)

Unal Ufuktepe	American University of the Middle East, Kuwait

Land Use Monitoring for Soil Consumption Reduction (LUMS 2016)

Carmelo M. Torre	Polytechnic of Bari, Italy
Alessandro Bonifazi	Polytechnic of Bari, Italy
Valentina Sannicandro	University of Naples Federico II, Italy
Massimiliano Bencardino	University of Salerno, Italy
Gianluca di Cugno	Polytechnic of Bari, Italy
Beniamino Murgante	University of Basilicata, Italy

Mobile Communications (MC 2016)

Hyunseung Choo   Sungkyunkwan University, Korea

Mobile Computing, Sensing, and Actuation for Cyber Physical Systems (MSA4IoT 2016)

Saad Qaisar	NUST School of Electrical Engineering and Computer Science, Pakistan
Moonseong Kim	Korean Intellectual Property Office, Korea

Quantum Mechanics: Computational Strategies and Applications (QM-CSA 2016)

Mirco Ragni	Universidad Federal de Bahia, Brazil
Ana Carla Peixoto Bitencourt	Universidade Estadual de Feira de Santana, Brazil
Vincenzo Aquilanti	University of Perugia, Italy
Andrea Lombardi	University of Perugia, Italy
Federico Palazzetti	University of Perugia, Italy

Remote Sensing for Cultural Heritage: Documentation, Management, and Monitoring (RSCH 2016)

Rosa Lasaponara	IRMMA, CNR, Italy
Nicola Masini	IBAM, CNR, Italy Zhengzhou Base, International Center on Space Technologies for Natural and Cultural Heritage, China
Chen Fulong	Institute of Remote Sensing and Digital Earth, Chinese Academy of Sciences, China

Scientific Computing Infrastructure (SCI 2016)

Elena Stankova	Saint Petersburg State University, Russia
Vladimir Korkhov	Saint Petersburg State University, Russia
Alexander Bogdanov	Saint Petersburg State University, Russia

Software Engineering Processes and Applications (SEPA 2016)

Sanjay Misra   Covenant University, Nigeria

Social Networks Research and Applications (SNRA 2016)

Eric Pardede	La Trobe University, Australia
Wenny Rahayu	La Trobe University, Australia
David Taniar	Monash University, Australia

Sustainability Performance Assessment: Models, Approaches, and Applications Toward Interdisciplinarity and Integrated Solutions (SPA 2016)

Francesco Scorza	University of Basilicata, Italy
Valentin Grecu	Lucia Blaga University on Sibiu, Romania

Tools and Techniques in Software Development Processes (TTSDP 2016)

Sanjay Misra	Covenant University, Nigeria

Volunteered Geographic Information: From Open Street Map to Participation (VGI 2016)

Claudia Ceppi	University of Basilicata, Italy
Beniamino Murgante	University of Basilicata, Italy
Francesco Mancini	University of Modena and Reggio Emilia, Italy
Giuseppe Borruso	University of Trieste, Italy

Virtual Reality and Its Applications (VRA 2016)

Osvaldo Gervasi	University of Perugia, Italy
Lucio Depaolis	University of Salento, Italy

Web-Based Collective Evolutionary Systems: Models, Measures, Applications (WCES 2016)

Alfredo Milani	University of Perugia, Italy
Valentina Franzoni	University of Rome La Sapienza, Italy
Yuanxi Li	Hong Kong Baptist University, Hong Kong, SAR China
Clement Leung	United International College, Zhuhai, China
Rajdeep Niyogi	Indian Institute of Technology, Roorkee, India

Program Committee

Jemal Abawajy	Deakin University, Australia
Kenny Adamson	University of Ulster, UK
Hartmut Asche	University of Potsdam, Germany
Michela Bertolotto	University College Dublin, Ireland
Sandro Bimonte	CEMAGREF, TSCF, France
Rod Blais	University of Calgary, Canada
Ivan Blečić	University of Sassari, Italy
Giuseppe Borruso	University of Trieste, Italy
Yves Caniou	Lyon University, France

Maurizio Pollino	Italian National Agency for New Technologies, Energy and Sustainable Economic Development, Italy
Alenka Poplin	University of Hamburg, Germany
Vidyasagar Potdar	Curtin University of Technology, Australia
David C. Prosperi	Florida Atlantic University, USA
Maria Emilia F. Queiroz Athayde	University of Minho, Portugal
Wenny Rahayu	La Trobe University, Australia
Jerzy Respondek	Silesian University of Technology, Poland
Ana Maria A.C. Rocha	University of Minho, Portugal
Maria Clara Rocha	ESTES Coimbra, Portugal
Humberto Rocha	INESC-Coimbra, Portugal
Alexey Rodionov	Institute of Computational Mathematics and Mathematical Geophysics, Russia
Jon Rokne	University of Calgary, Canada
Octavio Roncero	CSIC, Spain
Maytham Safar	Kuwait University, Kuwait
Chiara Saracino	A.O. Ospedale Niguarda Ca' Granda - Milano, Italy
Haiduke Sarafian	The Pennsylvania State University, USA
Jie Shen	University of Michigan, USA
Qi Shi	Liverpool John Moores University, UK
Dale Shires	U.S. Army Research Laboratory, USA
Takuo Suganuma	Tohoku University, Japan
Sergio Tasso	University of Perugia, Italy
Parimala Thulasiraman	University of Manitoba, Canada
Carmelo M. Torre	Polytechnic of Bari, Italy
Giuseppe A. Trunfio	University of Sassari, Italy
Unal Ufuktepe	American University of the Middle East, Kuwait
Toshihiro Uchibayashi	Kyushu Sangyo University, Japan
Mario Valle	Swiss National Supercomputing Centre, Switzerland
Pablo Vanegas	University of Cuenca, Equador
Piero Giorgio Verdini	INFN Pisa and CERN, Italy
Marco Vizzari	University of Perugia, Italy
Koichi Wada	University of Tsukuba, Japan
Krzysztof Walkowiak	Wroclaw University of Technology, Poland
Robert Weibel	University of Zurich, Switzerland
Roland Wismüller	Universität Siegen, Germany
Mudasser Wyne	SOET National University, USA
Chung-Huang Yang	National Kaohsiung Normal University, Taiwan
Xin-She Yang	National Physical Laboratory, UK
Salim Zabir	France Telecom Japan Co., Japan
Haifeng Zhao	University of California, Davis, USA
Kewen Zhao	University of Qiongzhou, China
Albert Y. Zomaya	University of Sydney, Australia

Reviewers

Abawajy, Jemal	Deakin University, Australia
Abuhelaleh, Mohammed	Univeristy of Bridgeport, USA
Acharjee, Shukla	Dibrugarh University, India
Andrianov, Sergei Nikolaevich	Universitetskii prospekt, Russia
Aguilar, José Alfonso	Universidad Autónoma de Sinaloa, Mexico
Ahmed, Faisal	University of Calgary, Canada
Alberti, Margarita	University of Barcelona, Spain
Amato, Alba	Seconda Universit degli Studi di Napoli, Italy
Amorim, Ana Paula	University of Minho, Portugal
Apduhan, Bernady	Kyushu Sangyo University, Japan
Aquilanti, Vincenzo	University of Perugia, Italy
Asche, Hartmut	Posdam University, Germany
Athayde Maria, Emlia Feijão Queiroz	University of Minho, Portugal
Attardi, Raffaele	University of Napoli Federico II, Italy
Azam, Samiul	United International University, Bangladesh
Azevedo, Ana	Athabasca University, USA
Badard, Thierry	Laval University, Canada
Baioletti, Marco	University of Perugia, Italy
Bartoli, Daniele	University of Perugia, Italy
Bentayeb, Fadila	Université Lyon, France
Bilan, Zhu	Tokyo University of Agriculture and Technology, Japan
Bimonte, Sandro	IRSTEA, France
Blecic, Ivan	Università di Cagliari, Italy
Bogdanov, Alexander	Saint Petersburg State University, Russia
Borruso, Giuseppe	University of Trieste, Italy
Bostenaru, Maria	"Ion Mincu" University of Architecture and Urbanism, Romania
Braga Ana, Cristina	University of Minho, Portugal
Canora, Filomena	University of Basilicata, Italy
Cardoso, Rui	Institute of Telecommunications, Portugal
Ceppi, Claudia	Polytechnic of Bari, Italy
Cerreta, Maria	University Federico II of Naples, Italy
Choo, Hyunseung	Sungkyunkwan University, South Korea
Coletti, Cecilia	University of Chieti, Italy
Correia, Elisete	University of Trás-Os-Montes e Alto Douro, Portugal
Correia Florbela Maria, da Cruz Domingues	Instituto Politécnico de Viana do Castelo, Portugal
Costa, Fernanda	University of Minho, Portugal
Crasso, Marco	National Scientific and Technical Research Council, Argentina
Crawford, Broderick	Universidad Catolica de Valparaiso, Chile

Cuzzocrea, Alfredo	University of Trieste, Italy
Cutini, Valerio	University of Pisa, Italy
Danese, Maria	IBAM, CNR, Italy
Decker, Hendrik	Instituto Tecnológico de Informática, Spain
Degtyarev, Alexander	Saint Petersburg State University, Russia
Demartini, Gianluca	University of Sheffield, UK
Di Leo, Margherita	JRC, European Commission, Belgium
Dias, Joana	University of Coimbra, Portugal
Dilo, Arta	University of Twente, The Netherlands
Dorazio, Laurent	ISIMA, France
Duarte, Júlio	University of Minho, Portugal
El-Zawawy, Mohamed A.	Cairo University, Egypt
Escalona, Maria-Jose	University of Seville, Spain
Falcinelli, Stefano	University of Perugia, Italy
Fernandes, Florbela	Escola Superior de Tecnologia e Gest ão de Bragança, Portugal
Florence, Le Ber	ENGEES, France
Freitas Adelaide, de Fátima Baptista Valente	University of Aveiro, Portugal
Frunzete, Madalin	Polytechnic University of Bucharest, Romania
Gankevich, Ivan	Saint Petersburg State University, Russia
Garau, Chiara	University of Cagliari, Italy
Garcia, Ernesto	University of the Basque Country, Spain
Gavrilova, Marina	University of Calgary, Canada
Gensel, Jerome	IMAG, France
Gervasi, Osvaldo	University of Perugia, Italy
Gizzi, Fabrizio	National Research Council, Italy
Gorbachev, Yuriy	Geolink Technologies, Russia
Grilli, Luca	University of Perugia, Italy
Guerra, Eduardo	National Institute for Space Research, Brazil
Hanzl, Malgorzata	University of Lodz, Poland
Hegedus, Peter	University of Szeged, Hungary
Herawan, Tutut	University of Malaya, Malaysia
Hu, Ya-Han	National Chung Cheng University, Taiwan
Ibrahim, Michael	Cairo University, Egipt
Ifrim, Georgiana	Insight, Ireland
Irrazábal, Emanuel	Universidad Nacional del Nordeste, Argentina
Janana, Loureio	University of Mato Grosso do Sul, Brazil
Jaiswal, Shruti	Delhi Technological University, India
Johnson, Franklin	Universidad de Playa Ancha, Chile
Karimipour, Farid	Vienna University of Technology, Austria
Kapcak, Sinan	American University of the Middle East in Kuwait, Kuwait
Kiki Maulana, Adhinugraha	Telkom University, Indonesia
Kim, Moonseong	KIPO, South Korea
Kobusińska, Anna	Poznan University of Technology, Poland

Korkhov, Vladimir   Saint Petersburg State University, Russia
Koutsomitropoulos,   University of Patras, Greece
  Dimitrios A.
Krishna Kumar, Chaturvedi   Indian Agricultural Statistics Research Institute
                                   (IASRI), India
Kulabukhova, Nataliia   Saint Petersburg State University, Russia
Kumar, Dileep   SR Engineering College, India
Laganà, Antonio   University of Perugia, Italy
Lai, Sen-Tarng   Shih Chien University, Taiwan
Lanza, Viviana   Lombardy Regional Institute for Research, Italy
Lasaponara, Rosa   National Research Council, Italy
Lazzari, Maurizio   National Research Council, Italy
Le Duc, Tai   Sungkyunkwan University, South Korea
Le Duc, Thang   Sungkyunkwan University, South Korea
Lee, KangWoo   Sungkyunkwan University, South Korea
Leung, Clement   United International College, Zhuhai, China
Libourel, Thérèse   LIRMM, France
Lourenço, Vanda Marisa   University Nova de Lisboa, Portugal
Machado, Jose   University of Minho, Portugal
Magni, Riccardo   Pragma Engineering srl, Italy
Mancini Francesco   University of Modena and Reggio Emilia, Italy
Manfreda, Salvatore   University of Basilicata, Italy
Manganelli, Benedetto   Università degli studi della Basilicata, Italy
Marghany, Maged   Universiti Teknologi Malaysia, Malaysia
Marinho, Euler   Federal University of Minas Gerais, Brazil
Martellozzo, Federico   University of Rome "La Sapienza", Italy
Marvuglia, Antonino   Public Research Centre Henri Tudor, Luxembourg
Mateos, Cristian   Universidad Nacional del Centro, Argentina
Matsatsinis, Nikolaos   Technical University of Crete, Greece
Messina, Fabrizio   University of Catania, Italy
Millham, Richard   Durban University of Technoloy, South Africa
Milani, Alfredo   University of Perugia, Italy
Misra, Sanjay   Covenant University, Nigeria
Modica, Giuseppe   Università Mediterranea di Reggio Calabria, Italy
Mohd Helmy, Abd Wahab   Universiti Tun Hussein Onn Malaysia, Malaysia
Murgante, Beniamino   University of Basilicata, Italy
Nagy, Csaba   University of Szeged, Hungary
Napolitano, Maurizio   Center for Information and Communication
                                   Technology, Italy
Natário, Isabel Cristina   University Nova de Lisboa, Portugal
  Maciel
Navarrete Gutierrez, Tomas   Luxembourg Institute of Science and Technology,
                                   Luxembourg
Nedjah, Nadia   State University of Rio de Janeiro, Brazil
Nguyen, Tien Dzung   Sungkyunkwan University, South Korea
Niyogi, Rajdeep   Indian Institute of Technology Roorkee, India

Oliveira, Irene	University of Trás-Os-Montes e Alto Douro, Portugal
Panetta, J.B.	Tecnologia Geofísica Petróleo Brasileiro SA, PETROBRAS, Brazil
Papa, Enrica	University of Amsterdam, The Netherlands
Papathanasiou, Jason	University of Macedonia, Greece
Pardede, Eric	La Trobe University, Australia
Pascale, Stefania	University of Basilicata, Italy
Paul, Padma Polash	University of Calgary, Canada
Perchinunno, Paola	University of Bari, Italy
Pereira, Oscar	Universidade de Aveiro, Portugal
Pham, Quoc Trung	HCMC University of Technology, Vietnam
Pinet, Francois	IRSTEA, France
Pirani, Fernando	University of Perugia, Italy
Pollino, Maurizio	ENEA, Italy
Pusatli, Tolga	Cankaya University, Turkey
Qaisar, Saad	NURST, Pakistan
Qian, Junyan	Guilin University of Electronic Technology, China
Raffaeta, Alessandra	University of Venice, Italy
Ragni, Mirco	Universidade Estadual de Feira de Santana, Brazil
Rahman, Wasiur	Technical University Darmstadt, Germany
Rampino, Sergio	Scuola Normale di Pisa, Italy
Rahayu, Wenny	La Trobe University, Australia
Ravat, Franck	IRIT, France
Raza, Syed Muhammad	Sungkyunkwan University, South Korea
Roccatello, Eduard	3DGIS, Italy
Rocha, Ana Maria	University of Minho, Portugal
Rocha, Humberto	University of Coimbra, Portugal
Rocha, Jorge	University of Minho, Portugal
Rocha, Maria Clara	ESTES Coimbra, Portugal
Romano, Bernardino	University of l'Aquila, Italy
Sannicandro, Valentina	Polytechnic of Bari, Italy
Santiago Júnior, Valdivino	Instituto Nacional de Pesquisas Espaciais, Brazil
Sarafian, Haiduke	Pennsylvania State University, USA
Schneider, Michel	ISIMA, France
Selmaoui, Nazha	University of New Caledonia, New Caledonia
Scerri, Simon	University of Bonn, Germany
Shakhov, Vladimir	Institute of Computational Mathematics and Mathematical Geophysics, Russia
Shen, Jie	University of Michigan, USA
Silva-Fortes, Carina	ESTeSL-IPL, Portugal
Singh, Upasana	University of Kwa Zulu-Natal, South Africa
Skarga-Bandurova, Inna	Technological Institute of East Ukrainian National University, Ukraine
Soares, Michel	Federal University of Sergipe, Brazil
Souza, Eric	Universidade Nova de Lisboa, Portugal
Stankova, Elena	Saint Petersburg State University, Russia

Stalidis, George	TEI of Thessaloniki, Greece
Taniar, David	Monash University, Australia
Tasso, Sergio	University of Perugia, Italy
Telmo, Pinto	University of Minho, Portugal
Tengku, Adil	La Trobe University, Australia
Thorat, Pankaj	Sungkyunkwan University, South Korea
Tiago Garcia, de Senna Carneiro	Federal University of Ouro Preto, Brazil
Tilio, Lucia	University of Basilicata, Italy
Torre, Carmelo Maria	Polytechnic of Bari, Italy
Tripathi, Ashish	MNNIT Allahabad, India
Tripp, Barba	Carolina, Universidad Autnoma de Sinaloa, Mexico
Trunfio, Giuseppe A.	University of Sassari, Italy
Upadhyay, Ashish	Indian Institute of Public Health-Gandhinagar, India
Valuev, Ilya	Russian Academy of Sciences, Russia
Varella, Evangelia	Aristotle University of Thessaloniki, Greece
Vasyunin, Dmitry	University of Amsterdam, The Netherlans
Vijaykumar, Nandamudi	INPE, Brazil
Villalba, Maite	Universidad Europea de Madrid, Spain
Walkowiak, Krzysztof	Wroclav University of Technology, Poland
Wanderley, Fernando	FCT/UNL, Portugal
Wei Hoo, Chong	Motorola, USA
Xia, Feng	Dalian University of Technology (DUT), China
Yamauchi, Toshihiro	Okayama University, Japan
Yeoum, Sanggil	Sungkyunkwan University, South Korea
Yirsaw, Ayalew	University of Botswana, Bostwana
Yujie, Liu	Southeast Jiaotong University, China
Zafer, Agacik	American University of the Middle East in Kuwait, Kuwait
Zalyubovskiy, Vyacheslav	Russian Academy of Sciences, Russia
Zeile, Peter	Technische Universitat Kaiserslautern, Germany
Žemlička, Michal	Charles University, Czech Republic
Zivkovic, Ljiljana	Republic Agency for Spatial Planning, Belgrade
Zunino, Alejandro	Universidad Nacional del Centro, Argentina

Sponsoring Organizations

ICCSA 2016 would not have been possible without the tremendous support of many organizations and institutions, for which all organizers and participants of ICCSA 2016 express their sincere gratitude:

Springer International Publishing AG, Switzerland
(http://www.springer.com)

NVidia Co., USA
(http://www.nvidia.com)

Beijing University of Post and Telecommunication, China
(http://english.bupt.edu.cn/)

State Key Laboratory of Networking and Switching Technology, China

University of Perugia, Italy
(http://www.unipg.it)

University of Basilicata, Italy
(http://www.unibas.it)

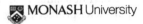

Monash University, Australia
(http://monash.edu)

Kyushu Sangyo University, Japan
(www.kyusan-u.ac.jp)

Universidade do Minho, Portugal
(http://www.uminho.pt)

Contents – Part III

Protein Ligand Docking Using Simulated Jumping

Sally Chen Woon Peh[1(✉)] and Jer Lang Hong[2]

[1] School of Biosciences, Taylor's University, Subang Jaya, Malaysia
chenwoonpeh@yahoo.com
[2] School of Computing and IT, Taylor's University, Subang Jaya, Malaysia
jerlang.hong@taylors.edu.my

Abstract. Molecular docking is an essential topic of study as it is crucial in numerous biological processes such as signal transduction and gene expression. Computational efforts to predict ligand docking is preferable to costly x-ray crystallography and Nuclear Magnetic Resonance (NMR) yet technology today remains incompetent in exploring vast search spaces for optimal solutions. To create efficient and effective algorithms, research has led to De novo drug design: a technique to extract novel chemical structures from protein banks is largely evolutionary in nature, and has found measurable success in optimal solution searching. A study by Shara Amin in 1999 in her novel method: simulated jumping (SJ) has achieved promising results when tested on combinatorial optimization problems such as the Quadratic Assignment and Asymmetric Travelling Salesman problems. Following her success with SJ, we aim to incorporate SJ into protein ligand docking, another optimization problem.

Keywords: Bioinformatics · Simulated jumping · Protein ligand docking · Drug design · Meta-heuristics

1 Introduction

The field of computational docking has increasingly matured: from the study of native conformation to the analysis of all variables that partake in the scoring function. From genomics to proteomics, computational efforts are now more focused on the exploration of protein structure dynamics for drug design (Fig. 1). With the progress of CADD, X-ray crystallography and NMR providing more than 100,000 coordinates of proteins and nucleic acids, the database of available structures are ever increasing. But it was not until the mid-1980 s that the docking problem achieved great progress [2, 3]. DOCK was the first revolutionary program created and has since been incorporated in many algorithms such as the Genetic Algorithm for sampling of ligand conformations [4].

The simulation of the binding association between molecules starts from the identification of protein surfaces, which can be described using mathematical models i.e. geometrical shape or grid based. To model the interaction between two molecules is in itself a worldly task due to the many degrees of freedom (i.e. hydrophobic, van der Waals, electrostatic forces) involved. Moreover, given that molecular structure is complex and non-static most of the time, the solution must be catered to a protein frame that is either static or flexible which further contributes to complexity of the problem.

© Springer International Publishing Switzerland 2016
O. Gervasi et al. (Eds.): ICCSA 2016, Part III, LNCS 9788, pp. 1–10, 2016.
DOI: 10.1007/978-3-319-42111-7_1

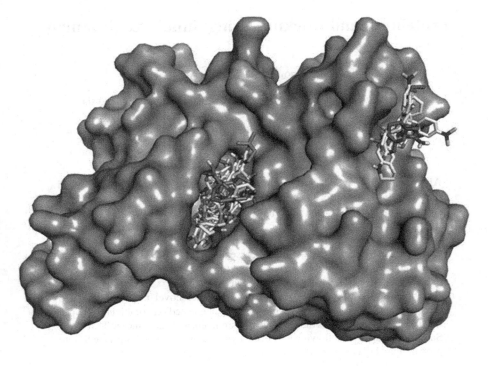

Fig. 1. Protein ligand docking example

A search algorithm should not be thorough but practical due to the size of search spaces. To methodically search through possible binding modes will require an impractically long period of time; yet too little computational expense will result in a small area of space searched. Hence, a balance between both variables is essential. Apart from the searching algorithm, the scoring function is equally as important. Consisting of mathematical methods for prediction of binding affinities, scoring functions have received no lesser attention [5, 6] to its role in determining binding strength. The lower the binding energy needed, the better the docking process.

Docking processes are considered successful if RMSD is less than 2 Å. For a good performance, speed and accuracy are key points of consideration. Common docking processes that focus on these two items are AutoDock [7], DOCK, GOLD [8], DARWIN [9] and DIVALI [10]; with FlexX [11], QXP [12] and Hammerhead [13] also well known for its efficiency.

Studies have shown that Computer Aided Drug Design (CADD) [14, 15] has been the most useful macromolecular crystallography application thus far with promising results in molecular drug design. CADD techniques such as Structure based design (SBD) and De Novo drug design [16] both result in novel drug designs designed in silico and later tested in vitro for feasibility. With LEGEND [17] as the pioneer de novo drug design tool, de novo designs have since been incorporated into the docking problem, with similar algorithms: LigMerge [18], SPROUT [19], Pro_Ligand [20] creating new molecules from SBD techniques.

Over the years, meta-heuristics have been used in conjunction with these tools (i.e. Flux, Lea3D etc.) for improved results. These literatures [21, 22] review the application or hybridization between meta-heuristics and docking tools. The following Sect. 2 details relevant meta-heuristics that have been implemented successfully thus far. Section 3 centers on SJ and its applications, while Sects. 5 and 6 cover its application in ligand docking.

2   Related Work

2.1   Meta Heuristic Techniques

2.1.1   Genetic Algorithm
Genetic Algorithm (GA) is a state-of-the art meta heuristic technique and has shown promising results in solving many complicated problems [23–27]. Its concepts are based on the human genetic structure, whereby a population is repeatedly generated using crossover and mutation operators until a stopping criterion is met. It makes an assumption that crossover operators will yield better offspring generations, hence producing a better result overall which will eventually lead to an optimal solution. To prevent the algorithm from obtaining local optima, mutation operators are utilized (gene structure altered) so that the algorithm will search through the search space from different perspectives.

2.1.2   Simulated Annealing
Simulated Annealing or the Metropolis method [28, 29] is used for conformational and positional searching which works based on annealing concepts. This notion is derived from the slow but steady cooling of glass at various temperatures via non-linear cooling methods. In SA, trial molecules perform a random walk around the target protein. This simulation involves a change in each degrees of freedom of any particular substrate. Trial molecules perform a random walk around the static target protein, in which small random displacement in each of degrees of freedom of substrate is performed at each step of simulation. Energy is then evaluated at each new position and conformation. If the new energy is lower, the step is accepted.

2.1.3   Tabu Search
Tabu Search [30] works on the principle of searching for the optimal solution by ensuring (via a memory storage) that all search spaces are not revisited. This is done by checking on neighbors of any particular search area with a Tabu List: a storage list comprising of previous solution routes. Any moves that lead to previous searched space is prohibited, hence its name. Tabu Search also improves on the local search by accepting solutions that are worse than the average, but only in circumstances where there are no better solutions.

2.1.4   Ant Colony System

The Ant Colony System [31] was developed in 1996 and is based on ants' behavior when forging a path in search for food. Whenever an ant follows a path to a source, it drops pheromone at every interval of the path. These serve to guide other ants in the search for food sources. Consequently, more pheromones are dropped as more ants travel any particular route. It is expected that as time goes by, the shortest path (most optimal) will contain the highest amount of pheromones due to shorter and hence more frequent travel time.

2.1.5   Memetic Algorithm

Memetic Algorithm [32] works by incorporating local search to the existing Genetic Algorithm. The principle is that if a local search is embedded at certain stages of Genetic Algorithm, the accuracy in obtaining the optimal solution is higher. It uses population based approach as a platform where each individual is capable of performing a local refinement via local search. Competitive (selection of better individual solutions) and cooperative (mating or crossover operators) approaches facilitate interactions between solutions to result in an overall better result due to diversity in the population.

3   Simulated Jumping

3.1   Overview

Simulated jumping (SJ) is similar to SA in the way that both are based on cooling and heating. They differ in the speed of the cooling and heating process, and that makes all the difference.

SA is one of a few meta-heuristic which enables the escape of an algorithm from local minima. All generated solutions underwent slow cooling at each temperature for the system to orderly restructure itself in ground state. This process is crucial for an efficient and effective performance of the SA algorithm, and has inspired many to further analyses the effects of initial temperatures and discovers more cooling methods [33]. Hybrids are occasionally utilized as cooling strategies to stabilize the system, and one example is the SA/TS system used in the capacitated cluster problem [34].

While SA places emphasis on the steady state of the cooling and heating system, SJ takes advantage of the turbulent nature stemming from rapid heating and cooling to obtain high probabilities for symmetry in energy states. Also seen as 'jumping', the search space is continuously heated and cooled in different areas to homogenously shake the system to achieve an evenly distributed equilibrium state. This is due to the existence of large energy barriers between domain walls which effectively seals of ground states, 'jumping' helps the algorithm cross that barrier by breaking through via rapid heating and cooling. Author and pioneer of SJ, Shara Amin pioneered SJ with the assumption that all NP hard problems [35] possess such energy barriers. This process will only stop after a desired low energy state is acquired.

One main difference between SJ and SA is that in SJ, the system is not allowed to reach an equilibrium state in any temperature, which means to neither completely melt nor freeze. Rapid cooling prevents complete melting while rapid heating melts local regions. This perturbation process will have to explore all search areas, local and global to have good results. For that to happen, the system in SJ will never freeze; it will have continuous transitions between temperature states. Of course, similar to SA, this process prevents algorithms from being trapped too long in any local minimum.

3.2 Successful Applications of SJ

SJ has been implemented in two classes of combinatorial optimization problems: the Quadratic Assignment Problem (QAP) and the Asymmetric Travelling Salesman Problems (ATSP). SJ has been implemented successfully [33] in these two NP-hard type combinatorial optimizations. Amin has also worked with Fernandez twice in combining GA with SJ to achieve better outcome than existing GA results' [36]. In the mobile radio network problem, SJ has shown to work well [33] in regards to time taken to obtain solutions and improved quality of solution.

4   Problem Formulation

The process of docking involves the fitting of ligand to its complimentary protein structure. This process is very useful to the drug design industry. Protein ligand docking problem is an NP Hard Problem. Due to the flexibility of ligand structures, it is hard to dock the ligand to the protein structure as it may exist in different configurations and forms. As the search space for finding the right fitting of ligand is vast, it is time consuming and almost impossible to find the optimal solution within a reasonable time frame.

5   Motivation

Although less known in the field of optimization problems, SJ approaches have proven to be effective in NP-hard optimization problems. We believe that this meta-heuristic should be taken into consideration in solving the ligand docking problem. Surely, a meta-heuristic that can efficiently solve NP-hard problems should be of help. Our intuition of using SJ is threefold. First, it is proven that SA method can effectively solve the protein ligand docking problem. We are of the opinion that SJ, a variant of the SA method, can be a better method to solve the docking problem. Secondly, SJ has ability to converge to the optimal solution within a short time frame. This is very helpful to test our method for protein ligand docking problem, which has huge repositories in protein databank. Finally, SJ is not very hard to implement, hence it is very easy for future researcher to further extend on our work to test on the protein ligand docking problem.

6 Proposed Solution

We use Simulated Jumping to solve the protein ligand docking problem. To represent the ligand structure, we use the databank which contains the geometric 3D structure of the molecule. Then, we develop a program to map the 3D coordinate of the files and represent them in 3D model. Then we identify each of the atoms and their binding. We use scoring function to evaluate the binding energy. Binding energy is evaluated based on the following scoring function:

$$\Delta G_{bind} = \Delta G_{vdw} + \Delta G_{H\text{-}bond} + \Delta G_{hydrophobic} + \Delta G_{rotor}$$

Where ΔG_{vdw} is the van der Waal energy, $\Delta G_{H\text{-}bond}$ is the hydrogen bonding, $\Delta G_{hydrophobic}$ is the hydrophobic bonding, and ΔG_{rotor} is the rotatable bonds.

We start the search space by initializing a sample molecule (Fig. 2). Then we perform rapid cooling and heating where we evaluate the energy in each cycle. Rapid cooling and heating is carried out by adjusting the molecular structures. Various adjustments are made to the rotational degree of freedom of the molecules. Cooling and heating are made by perturbation of the molecular structures. Adjustments are made in such a way that the fitting of the ligand to the protein structure is based on best fitting case, which is the corresponding molecular binding is fitted to be as rigid as possible. In other words, the ligand and protein needs to be conformed to each other structurally before the binding energy is evaluated.

For a ligand to be attached to a protein molecular structure, the pairing between atoms and their respective binding structure needs to adhere to certain conformation. For example, hydrogen bonding requires pairing between two H atoms in order for energy to be evaluated. For a molecule to be adjusted structurally, we need to carefully adjust the rotational bond of each atom and molecule. A single slight adjustment in a rotational bond may adversely affect the entire molecular structure. Therefore, we divide them atom into two different categories. The first category involves individual atom attached to only one other atom. The second category involves highly complex atom which is attached to at least two other atoms. It is the second category which can affect the binding energy of the docking process. Whenever adjustment is made to the second category, we pay particular attention to the entire molecular structure. If the new structure affects the binding energy greatly, we choose to adopt the first solution instead.

For each of the cycle, we accept the new molecular structure if the energy generated is lower than the previous cycle. Otherwise, we accept the previous solution on the assumption that the new solution leads to a poorer solution. In some circumstances, we randomly accept the new solution based on some probability on the understanding that this solution will help us out of the local optima. We use entropy functions to evaluate our probability function. The steps are repeated for many cycles until a final solution is achieved. We assumed that this solution is the most optimum solution for our test.

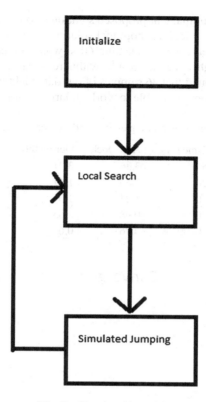

Fig. 2. Simulated jumping

7 Experimental Tests

We benchmark our proposed solution with the state of the art system AutoDock. Auto-Dock is chosen as our benchmark tool because AutoDock uses meta-heuristic techniques as part of its operation and it is freely available. The success rate of docking is measured by calculating the binding energy of the protein ligand docking. We collect a sizeable sample from protein databank (PDB database with 50 random samples comprising simple and complex molecules). In addition to that, the execution time for convergence to the optimal solution for the docking process is calculated by measuring the starting and ending time after N-iterations. We do not aim to yield optimal solution, but instead measuring the success of our algorithm with respect to the running time. Due to the fact that Optimizing Docking Process is a highly computationally intensive operation, we are of the opinion that any solutions that quickly converge to the near optimal solution is the best technique.

As shown in Table 1, our proposed solution is able to outperform AutoDock both in terms of convergence rate and execution time. Our proposed solution is able to obtain better solution within reasonable timeframe. This is due to the fact that our method is able to perform rapid perturbation to the molecular structure, hence making the current solution out of the local optima. Our proposed solution also incorporate a fast heuristic technique, which is able to yield a suboptimal solution within a fast running time. We also observed high convergence rate at time 60 s, which indirectly tells us that our method is able to yield highly accurate results within reasonable time frame (Fig. 3). On the other hand, we also found that 46 samples of our data yield high docking accuracy, which indicates that our method is able to work on large sample size of data.

Table 1. The average binding energy with respect to execution time

Time (s)	AutoDock	Our method
5	3.36	3.21
20	1.56	1.47
60	1.24	1.07
300	0.98	0.85
10000	0.76	0.65

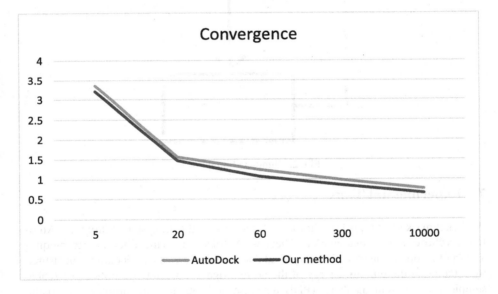

Fig. 3. Convergence graph (Color figure online)

8   Conclusions

In this paper, we have studied various meta-heuristic methods and proposed Simulated Jumping as our method. Meta heuristic methods have proved to be very successful due to their effective searching mechanism whereby they are able to obtain optimal solution within reasonable time frame across a vast search space. Inspired by the huge success Simulated Jumping has obtained in other real world problems, we are of the opinion that Simulated Jumping has great potential to solve molecular drug design problem.

Acknowledgement. This work is carried out within the framework of a research grant funded by Ministry of Higher Education (MOHE) Fundamental Research Grant Scheme (Project Code: FRGS/1/2014/ICT1/TAYLOR/03/1).

References

1. Ewing, T.J.A., Kuntz, I.D.: J. Comput. Chem. **18**, 1175 (1997)
2. Blundell, T.L.: Structure-based drug design. Nature **384**, 23–26 (1996). doi: 10.1038/384023a0
3. Kuntz, I.D.: Structure-based strategies for drug design and discovery. Science **257**, 1078–1082 (1992). doi:10.1126/science.257.5073.1078
4. Jones, G., Willette, P., Glen, R.C., Leach, A.R., Taylor, R.: J. Mol. Biol. **207**, 727 (1997)
5. Tame, J.R.H.: Scoring functions–the first 100 years. J. Comput. Aided Mol. Des. **19**, 445–451 (2005). doi:10.1007/s10822-005-8483-7
6. Gohlke, H., Klebe, G.: Approaches to the description and prediction of the binding affinity of small-molecule ligands to macromolecular receptors. Angew Chemie – Int. Ed. **41**, 2644–2676 (2002)
7. Cosconati, S., Forli, S., Perryman, A.L., et al.: Virtual screening with AutoDock: theory and practice. Expert Opin. Drug Discov. **5**, 597–607 (2010). doi:10.1517/17460441.2010.484460
8. Jones, G., Willett, P., Glen, R.C., et al.: Development and validation of a genetic algorithm for flexible docking. J. Mol. Biol. **267**, 727–748 (1997)
9. Taylor, J.S., Burnett, R.M.: DARWIN: a program for docking flexible molecules **191**, 173–191 (2000)
10. Clark, K.P.: Flexible ligand docking without parameter adjustment across four ligand-receptor complexes. J. Comput. Chem. **16**, 1210–1226 (1995)
11. Schellhammer, I., Rarey, M.: FlexX-Scan: fast, structure-based virtual screening. Proteins Struct. Funct. Genet. **57**, 504–517 (2004). doi:10.1002/prot.20217
12. Mcmartin, C., Bohacek, R.S.: QXP: powerful, rapid computer algorithms for structure-based drug design. J. Comput. Aided Mol. Des. **11**, 333–344 (1997)
13. Welch, W., Ruppert, J., Jain, A.N.: Hammerhead: fast, fully automated docking of flexible ligands to protein binding sites. Chem. Biol. **3**, 449–462 (1996). doi:10.1016/S1074-5521(96)90093-9
14. Chen, G.S., Chern, J.W.: Computer-aided drug design. In: Drug Discovery Research: New Frontiers in the Post-Genomic Era, pp. 89–107 (2006)
15. Song, C.M., Lim, S.J., Tong, J.C.: Recent advances in computer-aided drug design. Brief. Bioinform. **10**, 579–591 (2009)
16. Schneider, G., Fechner, U.: Computer-based de novo design of drug-like molecules. Nat. Rev. Drug Discovery **4**, 649–663 (2005). doi:10.1038/nrd1799
17. Nishibata, Y., Itai, A.: Automatic creation of drug candidate structures based on receptor structure. Starting point for artificial lead generation. Tetrahedron **47**, 8985–8990 (1991). doi: 10.1016/S0040-4020(01)86503-0
18. Lindert, S., Durrant, J.D., Mccammon, J.A.: LigMerge: a fast algorithm to generate models of novel potential ligands from sets of known binders. Chem. Biol. Drug Des. **80**, 358–365 (2012)
19. Gillet, V.J., Newell, W., Mata, P., et al.: SPROUT: recent developments in the de novo design of molecules. J. Chem. Inf. Comput. Sci. **34**, 207–217 (1994). doi:10.1021/ci00017a027

20. Waszkowycz, B., Clark, D.E., Frenkel, D., et al.: PRO_LIGAND: an approach to de novo molecular design. 2. Design of novel molecules from molecular field analysis (MFA) models and pharmacophores. J. Med. Chem. **37**, 3994–4002 (1994). doi:10.1021/jm00049a019

21. Devi, R.V., Sathya, S.S., Coumar, M.S.: Evolutionary algorithms for de novo drug design – A survey. Appl. Soft Comput. **27**, 543–552 (2015)

22. López-Camacho, E., García Godoy, M.J., García-Nieto, J., et al.: Solving molecular flexible docking problems with metaheuristics: a comparative study. Appl. Soft Comput. **28**, 379–393 (2015)

23. Cecchini, M., Kolb, P., Majeux, N., et al.: Automated docking of highly flexible ligands by genetic algorithms: a critical assessment. J. Comput. Chem. **25**, 412–422 (2003)

24. Jones, G., Willett, P., Glen, R.C.: Molecular recognition of receptor sites using a genetic algorithm with a description of desolvation. J. Mol. Biol. **245**, 43–53 (1995)

25. de Magalhães, C.S., Barbosa, H.J.C., Dardenne, L.E.: A genetic algorithm for the ligand-protein docking problem. Genet. Mol. Biol. **27**, 605–610 (2004)

26. Naga, P.L., Murga, R.H., Inza, I., Dizdarevic, S.: Genetic Algorithms for the Travelling Salesman Problem: a review of representations and operators. Artif. Intell. Rev. **13**, 129–170 (1999)

27. Willett, P.: Genetic algorithms in molecular recognition and design. Trends Biotechnol. **13**, 516–521 (1995)

28. Bertsimas, D., Tsitsiklis, J.: Simulated annealing. Stat. Sci. **8**, 10–15 (1993)

29. Rutenbar, R.A.: Simulated annealing algorithms: an overview. IEEE Circuits Devices Mag. **5**, 19–26 (1989). doi:10.1109/101.17235

30. Glover, F., Laguna, M.: Tabu Search (1997). doi:10.1007/978-1-4615-6089-0

31. Dorigo, M., Maniezzo, V., Colorni, A.: Ant system: optimization by a colony of cooperating agents. IEEE Trans. Syst. Man Cybern. Part B Cybern. **26**, 29–41 (1996). doi:10.1109/3477.484436

32. Radcliffe, N., Surry, P.: Formal memetic algorithms. In: Evolutionary Computing, pp. 1–16 (1994)

33. Amin, S.: Simulated jumping. Ann. Oper. Res. **86**, 23–38 (1999)

34. Osman, I.H., Christofides, N.: Capacitated clustering problems by hybrid simulated annealing and tabu search. Int. Trans. Opt. Res. **1**(3), 317–336 (1994). doi:10.1111/1475-3995.d01-43. Print Great Britain

35. Erickson, J.: NP-Hard Problems. Algorithms Course Mater, pp. 1–18 (2009). doi:10.1080/00949658208810560

36. Fernandez-Villacanas, J.L., Amin, S.: Simulated jumping in genetic algorithms for a set of test functions. In: Proceedings of Intelligent Information Systems, IIS 1997 (1997). doi:10.1109/IIS.1997.645223

GLSDock – Drug Design
Using Guided Local Search

Sally Chen Woon Peh[1(✉)] and Jer Lang Hong[2]

[1] School of Biosciences, Taylor's University, Subang Jaya, Malaysia
chenwoonpeh@yahoo.com
[2] School of Computing and IT, Taylor's University, Subang Jaya, Malaysia
jerlang.hong@taylors.edu.my

Abstract. The guided local search method has been successfully applied to a significant number of NP-hard optimization problems, producing results of similar caliber, if not better, compared to those obtained from algorithms specially designed for each singular optimization problem. Ranging from the familiar TSP and QAP to general function optimization problems, GLS sits atop many well-known algorithms such as Genetic Algorithm (GA), Simulated Annealing (SA) and Tabu Search (TS). With lesser parameters to adjust to, GLS is relatively simple to implement and apply in many problems. This paper focuses on the potential applications of GLS in ligand docking problems via drug design. Over the years, computer aided drug design (CADD) has spearheaded the drug design process, whereby much focus has been trained on efficient searching in de novo drug design. Previous and ongoing approaches of meta heuristic methods such as GA, SA & TS have proven feasible, but not without problems. Inspired by the huge success of Guided Local Search (GLS) in solving optimization problems, we incorporated it into the drug design problem in protein ligand docking and have found it to be effective.

Keywords: Bioinformatics · Guided local search · Protein ligand docking · Drug design · Meta-heuristics

1 Introduction

Drugs only bind to receptors with complementary structures. It is why ligand docking is highly dependent on how well a ligand would effectively fit and bind to any binding site. In computer aided drug design, 'docking' is referred to as the prediction of the binding mode between ligand and a known or estimated structure of a receptor (Fig. 1). This process is affected by many variables [1], one of which is the equilibrium state of the ligand-receptor structure. The free energy of binding molecules must be low. Hence, a scoring function or an objective function is required as a representation of the binding energy as an approach to predict ligand docking. Example approaches are Genetic Algorithm, Simulated Annealing and Tabu Search.

Ligand binding can be predicted through 3 aspects: functional site detection, functional site similarity [2] and molecular docking [3]. For the location of binding sites, some approaches focus on geometric matches (i.e. SURFNET; LIGSITE,

© Springer International Publishing Switzerland 2016
O. Gervasi et al. (Eds.): ICCSA 2016, Part III, LNCS 9788, pp. 11–21, 2016.
DOI: 10.1007/978-3-319-42111-7_2

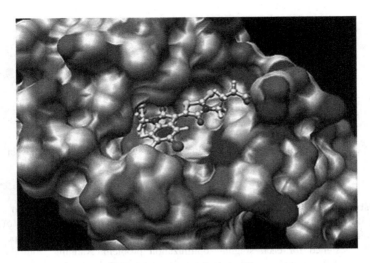

Fig. 1. Protein ligand docking

POCKET, APROPOS, CAST, vdw-fft, Drugsite) [4]. Many are evolutionary tech-niques [5] and are highly effective, because there is an abundance of information on structural and sequence related data. Knowledge on evolutionary of protein and ligand structures can be obtained from chemical databases such as the Protein Data Bank, Cambridge Structural Database, sc-PDB and more.

Prediction is further made complicated with internal flexibility within ligands, receptors and solvent molecules [6]. This is why large amounts of translational and external rotational degrees of freedom are associated with these molecules, contributing to computational complexity. Ligands are non-static in nature with frequent changes in geometry. They may form similar 3D conformation upon binding to receptors, and active conformation is hard to predict. Current computational processing process is not sufficient for thorough combing of vast search spaces for optimal solutions. This issue is becoming increasingly important due to progressive availability of protein structures from high throughput protein purification, X-ray crystallography and Nuclear Magnetic Resonance spectroscopy.

To create efficient docking softwares, both the scoring function and optimization algorithms are key concerns that are crucial. In this paper, we propose a meta-heuristic technique to simulate the molecular docking process. Looking at past successes of meta-heuristic applications in protein ligand docking based approaches, we are positive that GLS, a top optimization metaheuristic solution is able to provide new approaches to the ligand docking problem. The following sections describe various related meta-heuristics, GLS, applications of GLS and the methodology of GLS in ligand docking followed by our conclusion.

2   Related Work

2.1   Meta Heuristic Techniques

Metaheuristic techniques are non-exact optimization techniques that are applied to problems classified as NP-hard or NP Complete (nondeterministic polynomial time complete) as well as combinatorial optimisation problems. They aim to guide underlying heuristics in solving specific problems in terms of improved efficiency. Although great at providing solutions on large scale instances, metaheuristics share a common problem of being computationally expensive and often requires fine-tuning to solve a particular problem. Sections 3 and 4 will cover the capabilities of GLS in overcoming this computational hurdle.

2.1.1   Genetic Algorithm

A widely used heuristic that imitates Darwin's evolution and natural selection, Genetic Algorithms remain the most popular metaheuristic to date. GA was invented by John Holland who works on the principle of evolution whereby human genes that have undergone crossover and mutation subsequently produce future generations with better traits. This technique includes operators such as crossovers to pass on preferred qualities in an individual to offsprings. During rare circumstances, mutation is applied to promote diversity within the new population by changing variables in the parent solutions. After passing the evaluation process, these new individuals rejoin the population to collectively change the composition of the population.

Jones et al. [7] formulated GOLD dock with implementation of GA and have achieved 71 % of a success rate in docking 100 ligands with root mean square of less than 2 Å. It has also been proven that both the traditional GA and Lamarkian GA [8] far exceed the capabilities of the Monte Carlo SA in handling more degrees of freedom. Additionally, it has been used in a wide variety of computational algorithms for molecular recognition such as DIVALI and SSGA. More reviews on GA applications in molecular docking have been compiled by Willette [9].

GA has been used alongside GLS in the hybrid metaheuristic Guided Genetic Algorithm (GGA) [10]. When the current solution can no longer be further improved, GLS modifies the objective function by means of penalties for GA to use in future generations. Needless to say, these penalties greatly affect mutation and crossover operators in GA to introduce in high numbers, specific characteristics to a population. This allows for increased focus in its search for optimum solutions.

2.1.2   Simulated Annealing

Simulated Annealing [11] works are based on the non-linear cooling process of glass where various stages of changes precede the formation of a crystal. It is assumed that different rates of cooling lead to different formations of glass. The system is first 'melted' at a high effective temperature. Annealing is the lowering of the temperature via slow stages to ensure the system 'freezes' so that there are no other changes. This technique performs a check at every interval of cooling where if the energy is lower, the step of cooling is accepted. However, controlled uphill steps are incorporated in the search for generalized iterative improvement in solutions. SA has been incorporated in many other heuristics to optimize problems [12].

In ligand docking, this Metropolis Method (SA) is used for positional search and refinement of the ligand-receptor binding conformation. MCDOCK [13] utilises SA to widely sample the binding site for discovery of the local minimum, and has managed to predict fairly accurately (rms of 0.25–1.84 Å) the binding modes of ligands while taking full flexibility into account. Glide [14] incorporates SA in the optimization of its scoring function and has proven to be superior to Dock, GOLD and FlexX.

2.1.3   Tabu Search

Tabu Search (TS) is a heuristic which works on the principle of searching for the optimal solution without revisiting previous search spaces by checking on neighboring areas with its memory storage. This storage takes the form of a tabu list that contains previously visited solutions. This technique improves on the local search (LS) method by relaxing the few basic rules: a worsening move will only be accepted if there are no other improving moves. In addition to that, Tabu Search encourages exploration of otherwise difficult areas by prohibiting moves that lead to previous search areas.

Unlike other metaheuristics, TS is less common in its application in ligand docking. Baxter et al. pioneered the usage of TS in ligand docking with PRO_LEADS [15]. Found to be on par with SA, GA and Evolutionary Programming (EP), TS was used to sample the conformational space. Although slower in computational time, TS was found to be superior to both FlexX and GOLD in terms of prediction rates and estimation of binding affinities.

Both TS and GLS possess the same function of guiding LS out of the local optima. The penalties in GLS are similar to the restrictions implemented in TS. Like GLS, an initial solution is first created at random to be the current solution, and if is deemed the best after evaluation, it will be added into the Tabu List. Eventually the list will be updated with increasingly better solutions for more intensive search. Because too many penalties would misguide LS, a limited number of penalties are used in later versions of GLS. Old penalties are overwritten and replaced to help the algorithm escape from local minima. A combined approach of TS and GLS was applied to the service network design problem (SNDP) with commendable results [16].

2.1.4   Ant Colony System

The Ant Colony System mimics the behavior of ants in a colony when searching for the shortest route to and back from a feeding source. Every time an ant finds a food source, it will drop pheromones along its path to the food source and back to its nest as a guide for fellow ants. Each ant chooses with high probability to (or not to) follow the original pathway, thus reinforcing the pathway. Upon finding a shorter route, new pheromone trails are laid out. Eventually, the new passageway will be the current preferred route from increased pheromones dropping, due to relatively faster travel. It is assumed that the constant updating of path will eventually lead to an optimal solution. This population based approach is easily applied without major modifications to the Job-shop scheduling problem (SP), Quadratic assignment problems (QAP) and similar versions of the Travelling Salesman Problem (TSP). Tabu lists have also been incorporated in ants as memory storage for improved search results [17].

3 GLS

Many real life problems cannot be realistically solved by a complete search. Similarly, many NP-hard problems are unlikely to be solved via constructive search due to impossibly high computational demand. This led to the development of local search or heuristic methods as an alternative. LS searches in the space of (mostly) randomly generated candidate solutions, then moves on to better 'neighbors' until there are no better solutions than the current option. Solutions of LS can be obtained in a relatively short amount of time, but tend to be trapped in local optima. To continue looking for the global minima, GLS guides LS out of the minima via the implementation of penalties. Certain features are banned so that the algorithm focuses its search in more promising areas. Voudouris [18] surmised the similarity of GLS to the Frequency Based Memory approaches in Tabu Search, in which GLS additionally considers both the structural solution and feedback from local optimization heuristic.

To apply this penalty based metaheuristic algorithm, a candidate solution is first defined as a set of features. Each feature will be associated with a cost and penalty, which are the terms and coefficients from the objective function. When the algorithm settles in local optima, the cost function is augmented by accumulating penalties on selected features. These penalty terms are dynamically manipulated throughout the course of the search to steer the heuristic towards more viable solutions. Naturally, the overall cost will be greatly affected by costly features. This way, GLS is able to focus and distribute its searching efforts into more promising areas besides avoiding the accumulation of unnecessary workforce in any one region of the search space. After iteration of the improvement process, the improved solution is assigned as the current best solution until stopping criterion is met.

Similar to Tabu Search, GLS utilises knowledge gained from the previous searches to guide heuristics out of local optima. It provides more flexibility for exploitation of features of a problem in terms of associated costs. Therefore, GLS is able to converge to a high quality solution much more quickly than other metaheuristics such as Tabu Search or Simulated Annealing.

As mentioned, GLS requires definition of problem features. Voudouris and Tsang (1997) detailed the equations reiterated below. Penalties are first initialized at 0 to be subsequently increased as LS reaches a local optimum. GLS defines a function h that will replace g, the objective function which maps candidate solutions s to numerical values. λ is the parameter to the GLS algorithm; i refers to range of features; p_i is the penalty for feature I and I_i indicates the exhibition of feature i (present or absent of exhibition):

$$h(s) = g(s) + \lambda \times \sum [p_i \times I_i(s)] \tag{1}$$

$$I_i(s) = 1 \; \textit{if s exhibits feature i}; \; 0 \, \textit{otherwise} \tag{2}$$

To take into account the current penalty value of features, $util_i$ (utility of penalizing feature) is defined as follows, with c_i the cost and P_i the current penalty value of feature i.

$$util_i(s*) = I_i(s*) \times \frac{c_i}{1 + P_i} \qquad (3)$$

From the equation above, it is clear that the higher the cost (the greater the c_i), the greater the utility of penalizing it. Conversely, the higher the number of times penalized (the greater the P_i), the lower the utility of penalizing it. The idea is to consider both cost and current penalty in the bid to focus searches in more promising search spaces defined by lower cost features i.e. 'good features' (Fig. 2).

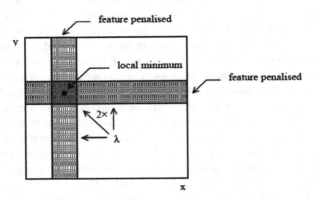

Fig. 2. Penalising features of local minimum to change cost [17]

3.1   Successful Applications of GLS

GLS has been successfully applied in a wide range of well-known NP-hard problems with world class results. In the Radio Link Frequency Assignment Problem, the combination of GGA and GLS hold some of the best results in the CALMA set of benchmark problems, which is the most widely used. GLS also achieved outstanding results in the vehicle routing problem (VRP), another NP-hard problem. In the workforce scheduling problem (WSP), GLS and Fast local search (FLS) achieved the best results in this benchmark problem of minimizing a function involving many variables in the assignment of technicians to jobs [18]. GLS approach in the Optimal Communication Spanning Tree (OCST) problem has also outperformed EA approaches equipped with state-of-the-art search operators [19]. One other highly notable example is the combination of GLS+FLS+2Opt [20] in outperforming Lin-Kernighan algorithm (LK), the specialized algorithm for TSP in average. This GLS combination also achieved superior results compared to SA, TS and GA.

Other than that, general function optimisation problems too benefited from GLS which managed to find consistent solutions in a landscape where local sub-optimals are in abundance. The results show that GLS is capable of defining artificial features for problems without obvious features from the objective function. The Team Orienteering Problem (TOP) found the implementation of GLS to have improved computational time with similar caliber results from other heuristics [21]. Furthermore, GLSSAT (extension of GLS) managed to produce results produced by WalkSAT in the

Max-SAT problem. Once again, a general-purpose algorithm accomplished results comparable to that achieved by a specialized solution to a problem.

In addition to that, GGA obtained results comparable to those produced by GA but with improved robustness in the Generalized Assignment Problem in which agents are assigned to jobs. In 2012, Barbucha introduced agent based GLS [20], which found satisfactory results in the VRP.

4 Motivation

GLS approaches have proven to be effective in a wide variety of optimization problems. It is within our interest to incorporate this meta-heuristic into the ligand docking problem as we believe that it can be a leading tool in the prediction of binding modes and affinities. We are of the opinion that GLS can bring improvements to solving the docking problem. First, it has been proven that this method could yield better results in Travelling Salesman Problem (TSP) than the traditional State of the Art Genetic Algorithm method. Also, it is fast in converging to the optimal solution within a reasonable time frame. Lastly, implementation of GLS for problem solving has proven to be fast and effective besides having a simple and clear approach. As a meta-heuristic, GLS has proven to have wide applications in varied problems with its ability in creating artificial solution features without prior conclusion of present features from the objective function.

5 Proposed Solution

To implement Guided Local Search efficiently, we implement a local search to drive the search to a local optima (Fig. 3). We use the same scoring function to evaluate the binding energy of docking methods. We represent the docking process by constructing the ligand and protein structure using 3D modelling. Once the 3D modelling of the protein and ligand structure is constructed, we then represent the various atoms and bonds between molecule structures.

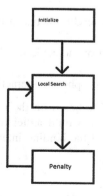

Fig. 3. Flow chart of guided local search

To test our search technique on the docking problem, we divided the docking process into two stages, the first being the orientation of the ligand and the second being the energy generated after binding of ligand and their respective protein structure. Our local search technique alters the orientation of molecules based on two factors, the rotational and degree of freedom. It uses a greedy search that attempts to bind the ligand to the protein structure through best fit. Even the ligand conformation to the protein matched, the respective molecules pairing may eventually lead to non-binding scenario. For such a case, we calculate the binding energy generated after the molecules successfully paired. Every time a ligand is bind to the protein structure, their energy is evaluated. We use scoring function to evaluate the binding energy. Binding energy is evaluated based on the following scoring function:

$$\Delta G_{bind} = \Delta G_{vdw} + \Delta G_{H\text{-bond}} + \Delta G_{hydrophobic} + \Delta G_{rotor}$$

Where ΔG_{vdw} is the van der Waal energy, $\Delta G_{H\text{-bond}}$ is the hydrogen bonding, $\Delta G_{hydrophobic}$ is the hydrophobic bonding, and ΔG_{rotor} is the rotatable bonds.

If the energy reduces a lot, the molecule is altered slightly on the assumption that it has nearly reach local optima. Alteration is made on the molecule rotational degree of freedom by twisting the orientation of some of the atoms. Otherwise, the molecule is altered by a large portion so that the energy is reduced drastically. If the energy increases after binding, the new ligand structure is not accepted on the assumption that this step leads to a poorer solution. Instead, we use the same old ligand structure and perform the same step again.

Under certain circumstances, a molecule is altered randomly even the energy decreases. We perform this step as a penalty function, on the assumption that it could lead the solution out of the local optima. On the other hand, if the solution converges to local optima, we modify the scoring function by adding "extra values" to some of the scoring functions. The penalty we imposed on the scoring function may eventually lead to a better solution. The addition of "extra values" is made based on the following criteria:

Van der Waals Energy
Extra values are added to all the atoms which bind between protein and ligand
Hydrogen Bonding Energy
Extra values are added to the H-H binding between protein and ligand
Rotatable Bonds
Extra values are added to the atom or molecule involved in the rotation
Hydrophobic Interaction Energy
Extra values are added to every hydrophobic interactions

The addition of "extra values" is based on the overall effect the various binding energies generated, for example, we would anticipate that hydrophobic interactions generates the least energy among all the binding interactions, hence a smaller penalty values is assigned.

6   Experimental Evaluation

We evaluate our proposed method with the state of the art system AutoDock. We are of the opinion that AutoDock is the most suitable benchmark for our evaluation since AutoDock uses meta-heuristic techniques as part of its operation and it is freely available. We measure the success rate of docking by calculating the binding energy of the protein ligand docking. We test our method with respect to that of AutoDock on protein databank (PDB database), where we randomly choose 50 random samples comprising simple and complex molecules. In addition to that, we measure the execution time for convergence to the optimal solution in the docking process (Fig. 4).

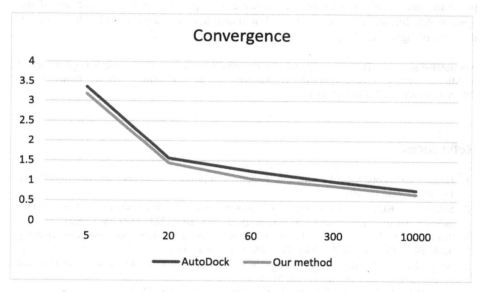

Fig. 4. Convergence graph (Color figure online)

As shown in Table 1, our docking method is more efficient in docking the protein molecular structure. This is due to the fact that we propose a penalty function, where our method impose penalty to the solution randomly, on the assumption that it could jump out from the local optima. On the other hand, we adopt a fast heuristic technique

Table 1. The average binding energy with respect to execution time

Time (s)	AutoDock	Our method
5	3.36	3.18
20	1.56	1.44
60	1.24	1.05
300	0.98	0.87
10000	0.76	0.66

which could converge to the suboptimal solution within reasonable time frame with high accuracy. We believe that our novel proposed method is useful for future protein ligand docking process.

7 Conclusions

It is clear from the results that GLS is a resourceful tool for efficient and effective ligand docking. Similar to past success in application to combinatorial problems, GLS once again proved that it is a potential meta-heuristic for yet another optimization problem. Meta-heuristic techniques are proven to solve many combinatorial optimization problems that involves computationally expensive processing power. We are of the opinion that further refinements on GLS will enable it to stand atop other heuristics in the field of ligand docking.

Acknowledgement. This work is carried out within the framework of a research grant funded by Ministry of Higher Education (MOHE) Fundamental Research Grant Scheme (Project Code: FRGS/1/2014/ICT1/TAYLOR/03/1).

References

1. Huang, S.Y., Zou, X.: Advances and challenges in protein-ligand docking. Int. J. Mol. Sci. **11**, 3016–3034 (2010)
2. Schmitt, S., Kuhn, D., Klebe, G.: A new method to detect related function among proteins independent of sequence and fold homology. J. Mol. Biol. **323**, 387–406 (2002)
3. Taylor, R.D., Jewsbury, P.J., Essex, J.W.: A review of protein small-molecule docking methods. J. Comput. Aided Mol. Des. **16**, 151–166 (2002)
4. An, J., Totrov, M., Abagyan, R.: Comprehensive identification of "druggable" protein ligand binding sites. Genome Inform. **15**, 31–41 (2004)
5. Devi, R.V., Sathya, S.S., Coumar, M.S.: Appl. Soft Comput. **27**, 543–552 (2015)
6. Cecchini, M., Kolb, P., Majeux, N., et al.: Automated docking of highly flexible ligands by genetic algorithms: a critical assessment. J. Comput. Chem. **25**, 412–422 (2003)
7. Jones, G., Willett, P., Glen, R.C., et al.: Development and validation of a genetic algorithm for flexible docking. J. Mol. Biol. **267**, 727–748 (1997)
8. Ross, B.J.: A Lamarckian evolution strategy for genetic algorithms. In: The Practical Handbook of Genetic Algorithms, p. 16 (1999)
9. Willett, P.: Genetic algorithms in molecular recognition and design. Trends Biotechnol. **13**, 516–521 (1995)
10. López-Camacho, E., García Godoy, M.J., García-Nieto, J., et al.: Solving molecular flexible docking problems with metaheuristics: a comparative study. Appl. Soft Comput. **28**, 379–393 (2015). doi:10.1016/j.asoc.2014.10.049
11. Bertsimas, D., Tsitsiklis, J.: Simulated annealing. Stat. Sci. **8**, 10–15 (1993)
12. Suman, B., Kumar, P.: A survey of simulated annealing as a tool for single and multiobjective optimization. J. Oper. Res. Soc. **10**, 1143 (2006)
13. Liu, M., Wang, S.: MCDOCK: a Monte Carlo simulation approach to the molecular docking problem. J. Comput. Aided Mol. Des. **13**, 435–451 (1999)

14. Friesner, R.A., Banks, J.L., Murphy, R.B., et al.: Glide: a new approach for rapid, accurate docking and scoring. 1. Method and assessment of docking accuracy. J. Med. Chem. **47**, 1739–1749 (2004). doi:10.1021/jm0306430
15. Baxter, C.A., Murray, C.W., Clark, D.E., et al.: Flexible docking using tabu search and an empirical estimate of binding affinity. Proteins **15**, 367–382 (1998)
16. Bai, R., Kendall, G., Qu, R., Atkin, J.A.D.: Tabu assisted guided local search approaches for freight service network design. Inf. Sci. (NY) **189**, 266–281 (2012). doi:10.1016/j.ins.2011.11.028
17. Dorigo, M., Maniezzo, V., Colorni, A.: The ant system: optimization by a colony of cooperating agents. IEEE Trans. Syst. Man Cybern. Part B Cybern. **26**(1), 1–13 (1996)
18. Voudouris, C.: Guided local search: an illustrative example in function optimisation. BT Technol. J. **16**, 46–50 (1998)
19. Tsang, E., Voudouris, C.: Fast local search and guided local search and their application to British Telecom's workforce scheduling problem. Oper. Res. Lett. **20**, 119–127 (1997). doi:10.1016/S0167-6377(96)00042-9
20. Barbucha, D.: Agent-based guided local search. Expert Syst. Appl. **39**, 12032–12045 (2012). doi:10.1016/j.eswa.2012.03.074
21. Vansteenwegen, P., Souffriau, W., Berghe, G.V., Van Oudheusdena, D.: A guided local search metaheuristic for the team orienteering problem. Eur. J. Oper. Res. **196**, 118–127 (2009). doi:10.1016/j.ejor.2008.02.037

The Appraisal of Buildable Land
for Property Taxation in the Adopted
General Municipal Plan

Fabrizio Battisti[(⊠)] and Orazio Campo

Department of Architecture and Design (DIAP), Faculty of Architecture,
"Sapienza" University of Rome, Via Flaminia 359, Rome (RM) 00196, Italy
{fabrizio.battisti,orazio.campo}@uniroma1.it

Abstract. In Italy, tax base for "Imposta Municipale Unica" related to the building area - made such by General Plan or its General Variation adopted but not approved - is the value (of the same building area) depending on the building potential of prediction even if not immediately exercisable. However, the building rights can be exercised only after: (i) the final approval of the General Plan/General Variation; (ii) the approval of the Implementation Plan required by Law; (iii) the issuance of certificates of permission building. This has produced in recent years several disputes between owners and local governments; the law did not give univocal solutions: today (2015) there is a conflict of case law relating to consider this areas absolutely as building areas, as well as it isn't defined what estimating procedures should be used. In this paper, through the application of a model of financial mathematics, an approach that overcomes the conflict law related to the appraisal of the building areas included in General Plans/General Variation adopted but not yet approved, is proposed: the appraisal will be performed in relation to the time and variables between the time of the appraisal and the time (alleged) for the completion of the administrative procedure for obtaining authorizations to build.

Keywords: Value of transformation · Land taxes · General municipal plan · Appraisal · Financial mathematics

1 Introduction

Among the issues related to real estate taxes, taxation holds particular relevance for "Imposta Municipale Unica" (IMU, formerly ICI) for the building sites considered as such on the basis of only the General Municipal Plan (GP) or, more frequently, its General Variation (GV)[1], adopted by the City Council (CC) but not approved,

---

The contribution is the result of the joint work of the two authors. Scientific responsibility is equally attributable to all two authors.

---

[1] VG/GP means any modification/revision of a "general" and not "specific" nature, of the GP in force. Since nearly all the municipalities have approved GP and therefore in force, the overall planning practiced by the CC in Italy, as of 2015, only applies to GV; under a formal and substantive standpoint, GP and GV have similar and coincident tools.

© Springer International Publishing Switzerland 2016
O. Gervasi et al. (Eds.): ICCSA 2016, Part III, LNCS 9788, pp. 22–32, 2016.
DOI: 10.1007/978-3-319-42111-7_3

following ADNAP areas (with provision of "adopted but unapproved" building rights), and therefore without building rights, as the actual building rights may only be exercised[2], after the approval of both the GV, issued by the Regional Administration (RA) or provincial (PA) and an Implementation Urban Plan (by CC), as well as qualifying title[3]. The tendency, almost unanimous, of CC in Italy, on the taxation of ADNAP areas, has always been to consider how a tax base for the application of tax rates depending on the value of the estimated building potential, although not immediately exercisable due to the rules of protection that allow for the implementation of the measures with the dual compliance: the plan in force and the adopted plan.

This created several disputes in recent years for which the law did not provide single solutions. Two orientations, contradictory in fact, have emerged in the course of time:

- A first (prevalent) orientation in which it is believed that the purpose IMU is in the simple insertion of the land in the GV adopted with intended use that admits the building rights, because taxed as such. This approach is based on the assumption that the inclusion of the land in the GV as building area determines an obvious and inevitable increase in its value and therefore the IMU withdrawal should be compared to that value, while the actual building potential of the same seems to be completely irrelevant (art. 5 Decree. N. 504/1992; Cassation sentence no. 16751/2004; Cassation sentence no. 19750/2004; art. 11 quaterdecies paragraph 16 Legal Decree no. 203/200 converted by Law no. 248/2005, art. 36 paragraph 2 of Decree no. 223/2006 converted by Law no. 248/2006; Supreme Court with sentence no. 25506/2006; Constitutional Court Ruling no. 266/2008);
- A second orientation in which the circumstance of including the land in the GV to tax it as building land is considered not sufficient, if in fact that land remains without building rights In this case the land must be considered and therefore taxed for IMU, as agricultural land (Cassation sentence no. 21644/2004; Cassation sentence no. 17035/2013).

To date (2016), in this situation of lack of clarity, the main orientation continues to consider the value of ADNAP areas for IMU purposes according to their building rights, even if not exercised. This approach does not consider, among other things, that the ADNAP areas, if held as collateral for credit, are now valued as a function of only their agricultural value. The European Central Bank (ECB) in fact, as part of "Asset Quality Review" (2014), review of the quality of the assets on the balance sheets of the 130 largest European banks to check their state of health, has established criteria and methodology for estimating so-called "Collateral" or of property used to guarantee the loans granted by the banks themselves. With regard to building areas, the ECB has established that they could be considered as such and therefore their value estimated in terms of their building potential, only if the authorization process is completed and

---

[2] As a rule and unless it is an extension of zone B in accordance with Ministerial Decree no. 1444/68.

[3] Building Permit by the relevant technical departments of the CC or alternatively Activity Start Report (DIA) where permitted (where the implementation urban plan provides for building types is still thorough in the aspects as well as urban planning, even building).

therefore any permits issued for construction; in all other cases the valuation principle to consider is the agricultural value (AA.VV. 2014a).

The often burdensome taxation on ADNAP areas, given the uncertainties (upon approval by RA/PA, changes may be officially made, pursuant to Art. 3 of Law 765/1967, which writes off areas for building purposes from the adopted GV) and long bureaucratic process for the completion of the approval process of the various tools required to provide these areas with building rights, together with the "sterility" of the same for the disbursement of credit, are the cause of several negative effects, such as: (i) frequent non-payment of IMU taxes by a significant number of taxpayers owners of ADNAP areas; (ii) cases of avoidance and evasion; (iii) progressive weakening of the vitality of the real estate market on ADNAP areas, which results in contraction (actual and forecast) of real estate development initiatives (AA.VV. 2010b), or in the proposed initiatives in specific urban variant submitted also by private entities, thanks to regulations (e.g. art. 2 of Law no. 179/1992, Art. 8 of Presidential Decree no. 160/2010, art. 14 of Presidential Decree no. 380/2001) which enable the approval of the same specific variant in less time than to those needed for a GV (AA.VV. 2010, AGN International 2014).

A fair taxation on areas ADNAP may be a first and significant step to counter the negative effects mentioned above, as well as to return full and broad programming and planning power for the development of their territory in place a transformation for specific periods of time to the CC (Stanghellini 2012, Van Ommeren and Van Leuvensteijn 2005); to this end, the first step to be performed is the exact determination of their Market Value, by estimating the value of their transformation through the proper application of the principles, criteria and methods of traditional estimation discipline.

2 Context, Methodological Approach and Aims of the Work

An analysis of direct property taxation in 12 European countries (AGN International 2014, European Commission 2013 and 2014, Garnier et al. 2014, Kneller et al. 1999) has shown the existence of two problems concerning the real estate taxation and that both of them imply appraisal issues: (1) definition of the applicable rate in the direct taxation of real estate; (2) appraisal of a specific value which has to be considered such as a taxable income (on which to apply the above mentioned tax rates).

Issues related to the first problem, concerning the definition of tax rates, are responsibility of Legislative and Executive/Administrative Power (Mirrlees and Adam 2011, Piketty and Saez 2012, Profeta et al. 2014): decisions regarding this issues are political; these are mainly determined paying attention to balancing the public accounts (Andrews 2010, Arrondel and Masson 2013, Atkinson et al. 2011, Bertocchi 2011).

Issues related to the second question, concerning the appraisal of asset in the real estate, is a technical matter (in Italy, for example, the Revenue Agency holds the competence to appraisal - for tax purposes - some categories of assets, such as buildings; the Municipalities, instead, usually define the land-values) and may be solved through the use of the tools that the estimation discipline makes available; nowadays, the use of methods and standards recognized in the international arena is growing up.

In this paper, a methodological approach aimed to appraisal the market value (taxable income for tax rate application) of a specific real estate asset - ADNAP area - is defined.

This approach, structured on an Italian specific case, under a methodological point of view may find wider and international use: while the question related to definition of tax rates concerns fiscal policy of every State, appraisal of asset that must be taxed concerns technical issues, of which this paper deals (specifically ADNAP areas).

Referring to the above, the specific aim of this article is to provide elements for estimating the Market Value of an area ADNAP through the method of Transformation Value (also named Hope Value Approach in the international evaluation standard) in accordance with the time distance and variables that exist between the time of the estimate and the time (estimated) of completion of the administrative process for obtaining building authorizations. In particular in the paper are identified: (i) the variables that affect the convertibility of urban area; (ii) time for the conclusion of the planning processes for changing the intended use of the areas[4].

Thereby a solution to "Italian conflict case law" in line with the contents of the Supreme Court of Cassation United Section sentence no. 25506/2006 (to date, in 2015, still not passed) - which notes that the start of a process that will provide an area with building rights, does not correspond to immediate building rights, and that therefore, for tax purposes, it must be regarded as the greater or lesser relevance and potential of its building rights - is possible.

Below: par. 3 will analyze the procedural process for the approval of a GV, and in particular will identify: stages and schedules, main variables; par. 4, based on the results of the analysis, will describe procedures for the proper application of transformation value for estimating an area ADNAP; par. 5 will draw the conclusions of this work.

3   Procedural Analysis of the Process for the Approval of a GP: Stages and Schedules, Main Variables

The following analysis has been implemented in order to identify:

– The stages that make up the procedural process of approving a GV and, consequently, the timing for the conclusion (in favour) of the procedures for its approval;
– The main variables that have implications for the outcome of the process of approving a GV.

The study of phase/time and relevant variables is a prerequisite to the identification of the factors that have an incidence on the market value of an ADNAP area and, in particular, with reference to the combination of interest in the method of calculating the Transformation Value (Campo, 2015):

---

[4] During the estimate of the value of ADNAP area, these identified elements (variables, time), may be transformed into parameters and coefficients to be applied operationally by the assessor; simultaneously a time calculation model "really based" may be defined to reach time for the conclusion of the GV authorization.

– Involve a change in the exponent relative to the time in the so-called combination of interest (phases/schedules);
– Affect the rate of industrial profitability (main variables).

The analysis was conducted by referring to the procedure of a sample of 59 GV concluded in the Lazio region (Provinces of Rome, Rieti and Viterbo) 2000–2015.

3.1 Phases and Schedules

The phases that make up the procedural process for final approval of a GP or a GV are:

1. Adoption phase

   • Adoption of GV by the Town Administration with Town Council Resolution;
   • Publication of the proceedings;
   • Presentation of observations;
   • Investigation on observations;
   • Counter-arguments concerning the observations and the relevant Town Council Resolution.

2. Gathering opinions from the relevant authorities phase

   • Acquisition of all the pertinent opinions by the relevant local authorities[5];
   • Strategic Environmental Assessment Procedure.

3. Approval phase

   • Investigation by the Regional Offices (requests for clarification and additional opinions);
   • Transmission and related opinion of the Regional Technical Committee for the Territory (hereinafter the RTC);
   • If approved (0 % of the sample analyzed);
   • Submission of the opinion to the Regional Council for its approval resolution to end the proceedings;
   • If approved with modifications (100 % of the sample analyzed);
   • Re-submission to the Town Administration;
   • Acceptance of the amendments by the Town Administration with Council Resolution and referral to the Region for Regional Council approval resolution;

---

[5] Such opinions are routinely: (1) opinion art. 89 of Presidential Decree 380/01 (former art. 13 L.64/74), Regional Council Resolution no. 2694/99 and 545/10; (2) prior ASL opinion art. 20-f L.833/78 and art. 1 Regional Law 52/80; (3) regional opinion art. 2 Regional Law 1/86 Residential Uses; (4) regional landscape opinion; (5) opinion of the Basin Authority responsible in case the affected areas fall within the ideological and hydrogeological risk perimeter; (6) Impact Assessment under Directive 2009/147/ EC and 1992/42/ EEC "Habitat", as governed by Law 157/1992 and Presidential Decree 357/1997, as amended by Presidential Decree 120/2003; (7) Parks or Nature Reserves management opinion; (8) BB.AA.CC. Ministry opinion local Superintendent; (9) opinions from government departments and public bodies concerned if the planning instrument changes areas and state-owned property (roads, railways, navigation, etc.).

- In case of non-acceptance or partial acceptance of the changes, Town Council rebuttal resolution to the changes required and return to the Region;
- Investigation of the regional offices of the rebuttal resolution and transmission to the Regional Technical Committee for the Territory;
- Final opinion of the RTC and transmission to the Regional Council;
- The Regional Council final approval resolution that concludes the administrative procedure;
- Publication of the Regional Council Resolution on the Regional Official Bulletin which marks the final validity of the GV;

The study of 59 GV (in Lazio Region) procedures taken as the survey sample, has been further developed in order to determine the time required for the completion of the various stages of the approval process, thus being able to determine the average length of the process. The complexity of the GP and/or GV approval procedure is due to the long time required to reach final approval; an average total length of the process from adopt to approval is about seven years, as reported in Fig. 2.

It should also be noted that for each of the above mentioned phases involving a specific Town Council resolution, it is necessary (especially in medium and large towns) to discuss the measure and their contents through the appropriate board committees, resulting longer time frames for the conclusion of the procedure.

3.2 Main Variables

The main variables that may affect the approval of a GV and/or result in changes with the write-off of ADNAP areas or regulations limiting the urban convertibility are associated with the presence of the following on the ADNAP area:

- Landscape constraints (legal constraints, declarative constraints, landscape plan constraints);
- Archaeological constraints;
- Geological protection requirements (seismic hazard, mechanical properties of the soil);
- Hydro geological protection requirements (danger of flooding, landslide, joint flood and landslide);
- Elements of environmental vulnerability (specific environmental protection provisions in the higher environmental instrumentation order);
- Residential uses;
- Natural parks and reserves;
- Sites of Community Importance (SCI) and Special Protection Areas (SPA).

Although the overall planning on the municipal level should consider all higher environmental and landscape instrumentation order indications/restrictions/requirements as territorial constants (Guarini and Battisti 2014a), the analysis of 59 case studies in the Lazio region showed that, during the preparation and adoption of a GV, choices and subsequent provisions based on "local" reasons prevail (i.e. meeting residential and non settlement demand; completion of areas already partially urbanized regardless of a more complex analysis of the nature of the landscape and environment; research of easy political consensus; will to decentralize residential functions and services to strip city centres), even if not totally consistent with the higher instrumentation order. Consequently, while foreseeing an urban destination for an ADNAP area, in the course of the complex GV approval process, this forecast could be revised if not cancelled; in this case the owner subject to IMU taxation, according to the current orientation prevailing in Italy, until the final conclusion of the approval process, the taxpayer is required to pay the tax in relation to the original building rights conferred by the planning instrument adopted.

4 The Estimate of an ADNAP Area Through the Transformation Value

Through the so-called Transformation Value method (hereinafter Vt), an estimate can be obtained to determine the most accurate and comprehensive evaluation of building land, referred to today (Tajani and Morano 2015). Even today this traditional procedure is more effective than other assessment procedures (Guarini and Battisti, 2014b). The Vt is estimated, analytically, through the following formula:

$$Vt = \frac{Vm(pt) - \sum Kp}{(1 + r')^n}$$

where:

- Vm (pt) = is the Market Value of the property built on the area;
- Σkp = is the sum of all production costs (cost of construction, cost of utilities, technical expenses, general and administrative expenses, concession fees, finance charges, promoter profits, other expenses necessary to build the building);
- r' = is the specific return rate for the work;
- n = is the number of years required to complete work.
- For the implementation of the equation that allows for the Vt estimate of a building area, the Vm (pt) and Σkp estimate can be resolved through the collection, from information sources, of known prices of similar goods.
- If the estimate of an ADNAP area, particular attention should be paid to the estimation of r' and n.

In order to estimate the rate of industrial profitability r', it is to consider that this is the sum of several components, as shown in the following formula (AA.VV. 2010):

$$r' = rrf + \Delta nl + \Delta ra + \Delta ur$$

where:

- rrf = "risk free" profitability rate on a "guaranteed" and "risk-free" investment. This rate is normally equal to the rate on government bonds with consistent maturities with respect to the time horizon of the investment;

– Δnl = change in the rate due to "non liquidity" or the difficulty to quickly convert the value of an investment property into cash; this risk is essentially linked to the time of transaction of goods once made according to the market conditions at the time of marketing of the same goods produced;

	Ascending influence	Descending influence	Relevance
Landscape constraincts			
By Law	if present	if absent	very high
Declarative	if present	if absent	high
From landscape plan	if present	if absent	medium
Archaeological constraints			
All constraints	if present	if absent	high
Geological limitations			
Seismic hazard	if high	if low	very high
Mechanical properties of the soil	if low	if high	medium
Hydro-geological limitations			
Flood danger	if applicable and frequent	if not applicable or infrequent	high
Landslide danger	if applicable and frequent	if not applicable or infrequent	high
Joint flood and landslide danger	if applicable and frequent	if not applicable or infrequent	very high
Elements of environmental vulnerability			
Park Authority established	if present	if absent	medium
SIC and ZPS	if present	if absent	very high
Soil, subsoil, topsoil	if there are any elements of environmental vulnerability	if there are not any elements of environmental vulnerability	medium
Air	if area transformation creates contamination risk	if area transformation does not create contamination risk	low
Water	if area transformation creates contamination risk	if area transformation does not create contamination risk	low
Flora	if area transformation creates alteration risk	if area transformation does not create alteration risk	low
Fauna	if area transformation creates alteration risk	if area transformation does not create alteration risk	low
Population and human health	if area transformation creates pollution risk	if area transformation does not create pollution risk	low

Fig. 1. Ascending and descending influences of the rate of profitability of an ADNAP area

– Δra = change in the rate due to the "risk area", closely linked to the specific characteristics of the investment property. This component thus reflects, if applicable, the initiative promoter's part of profits (extra-profit) (Morano et al. 2015), meaning the individual so starts and manages as well as sells the property; the higher the change an expected result does not materialise, the higher the risk of that investment. It is a type of risk that depends on the characteristics of the housing market and the competitiveness of the object in the same market.

– Δur = change in the rate due to the "urban risk"; it reflects the difficulty and/or temporal uncertainty of obtaining all necessary permits to carry out the task in question; the risks associated to the favourable conclusion of the approval of a GV for an ADNAP area are included in this industrial profitability rate component.

If the components rrf, Δnl and Δra can be defined as a result of the financial and real estate market, the estimation of Δur is particularly significant: in this financial mathematic estimate method that can be answered through the study of ascending and

Months	Phases	
0	Adoption of VG / PRG by the Town Administration with Town Council Resolution	
1	Publication of the proceedings	
1	Presentation of observations	**Adoption**
4	Investigation on observations	
1	Counter-arguments concerning the observations and the relevant Town Council Resolution	
12	Acquisition of all the pertinent opinions by the relevant local authorities	**Opinions from**
12	Strategic Environmental Assessment Procedure	**relevant authorities**
12	Investigation by the Regional Offices (requests for clarification and additional opinions)	
12	Transmission and related opinion of the Regional Technical Committee for Land	
	if approved with modifications	
0,5	Re-submission to the Town Administration	
1	Acceptance of the amendments by the Town Administration with Council Resolution and referral to the Region for Regional Council approval resolution	
2	In case of non-acceptance or partial acceptance of the changes, Town Council rebuttal resolution to the changes required and return to the Region	
6	Investigation of the regional offices of the rebuttal resolution and transmission to the Regional Technical Committee for the Territory	
12	Final opinion of the CTR and transmission to the Regional Council	**Approval**
2	The Regional Council final approval resolution that concludes the administrative procedure	
0,5	Publication of the Regional Council Resolution on the Regional Official Bulletin which marks the final validity of the VG / PRG	
	or if approved	
0,5	Submission of the opinion to the Regional Council for its approval resolution to end the proceedings	
0,5	Publication of the Regional Council Resolution on the Regional Official Bulletin which marks the final validity of the VG / PRG	

Fig. 2. GV approval process time frame

descending influences in relation to the specific ADNAP area and in particular in relation to the presence of limitations/requirements that have been identified in par. "Main variables" on the ADNAP area.

For the purposes of a estimate of the component rru, Fig. 1 shows the analysis of the main ascending and descending influences of the rate of profitability of an ADNAP area. The table shows when and how an element (constraints, limitation, etc.) assumes the character of ascending or descending influence, as well as the relevance[6], which translates into greater or lesser weight of influence[7] for rate estimation (the relevance depends on the capability of the factors considered - constraints, limitation, etc. - to slow down or disallow favorable conclusion of the process by which it attributes the building rights to a land) (Fig. 1).

With reference to the estimate of n, we must assess the time needed for the conclusion of the proceedings. Following is an indication of the time axis derived from the analysis set out in par. 3.1 (Fig. 2).

5   Conclusion

The analysis of the process of approving a GV has highlighted the many variables that make the same procedure uncertain and determine that long lead times for its conclusion.

The proposed evaluation method reflects the results of an analysis conducted on the last 59 GV procedures concluded in the Lazio Region in the period 2000/2015, in valuation parameters that allow for the estimate of such ADNAP areas in relation to their real market value. This can contribute to a fairer taxation and also can have a positive effect on the segment of the real estate market for development.

References

AA.VV.: Asset Quality Review - Phase 2 Manual, European Central Bank (2014a). ISBN:978-92-899-1254-9

AA.VV.: Quaderni dell'Osservatorio. Appunti di economia immobiliare, Agenzia delle Entrate - Osservatorio del Mercato Immobiliare, year 3, no. 1. (2014b)

AA.VV.: Proposed New International Valuation Standards, IVSC, London (2010)

AGN International: Gift and Inheritance Tax – A European Comparison. AGN Website, London (2014)

Andrews, D.: Real House Prices in OECD Countries: The Role of Demand Shocks and Structural and Policy Factors, OECD Economics Department Working Papers, no. 831. OECD Publishing (2010)

---

[6] The relevance depends on the capability of the factors considered - constraints, limitation, etc. - to affect the process by which it attributes the building rights to a land.

[7] In this paper the quantification, in percentage terms, of the ascending and descending influences identified has not addressed; this can be done by the assessor during the estimate. A study by authors on the quantification of this variables even so is ongoing.

Arrondel, L., Masson, A.: Taxing more (large) family bequests: why, when, where? Paris School of Economics WP 2013-17, 14, June 2013

Atkinson, A.B., Piketty, T., Saez, E.: Top incomes in the long run of history. J. Econ. Lit. **49**(1), 3–71 (2011)

Bertocchi, G.: The vanishing bequest tax: the comparative evolution of bequest taxation in historical perspective. Econ. Politics **23**, 107–131 (2011)

Campo, O.: Appraisal of the extraordinary contribution in general regulatory plan of Rome. Int. J. Math. Mod. Methods Appl. Sci. **9**, 404–409 (2015). ISSN: 1998-0140

European Commission: Annual Growth Survey, European Commission Website (2013)

European Commission: Tax reforms in EU member states 2014, tax policy challenges for economic growth and fiscal sustainability. Eur. Econ., no. 6, DGEFA (2014)

Garnier, G., György, E., Heineken, K., Mathé, M., Puglisi, L., Ruà, S., Van Mierlo, A.: A wind of change? reforms of tax systems since the launch of Europe 2020, vol. 53, no. 2, pp. 75–111. De Boeck Supérieur (2014)

Guarini, M.R., Battisti, F.: Benchmarking multi–criteria evaluation methodology's application for the definition of benchmarks in a negotiation–type public–private partnership, a case of study: the integrated action programmes of the Lazio Region. Int. J. Bus. Intell. Data Min. **9** (4), 271–317 (2014a)

Guarini, M.R., Battisti, F.: Evaluation and management of land-development processes based on the public-private partnership. In: Advanced Materials Research, vol. 869, pp. 154–161 (2014b)

Kneller, R., Bleaney, M.F., Gemmell, N.: Fiscal policy and growth: evidence from OECD countries. J. Pub. Econ. **74**(2), 171–190 (1999)

Mirrlees, J., Adam, S.: Tax by Design: The Mirrlees Review, vol. 2. Oxford University Press, Oxford (2011)

Morano, P., Tajani, F., Locurcio, M.: Land use, economic welfare and property values: an analysis of the interdependencies of the real estate market with zonal and macro-economic variables in the municipalities of Apulia Region (Italy). Int. J. Agric. Environ. Inf. Syst. **6**(4), 16–39 (2015). ISSN: 1947-3192

Piketty, T., Saez, E.: Top incomes and the great recession: recent evolutions and policy implications. Paper Presented at the 13th Jacques Polak Annual Research Conference Hosted by the International Monetary Fund, Washington, 8–9 November 2012

Profeta, P., Scabrosetti, S., Winer, S.L.: Wealth transfer taxation: an empirical investigation. Int. Tax Pub. Financ. **21**(4), 720–767 (2014)

Stanghellini, S.: Il negoziato pubblico privato nei progetti urbani. Principi, metodi e tecniche di valutazione. Dei, Rome. (2012)

Tajani, F., Morano, P.: An evaluation model of the financial feasibility of social housing in urban redevelopment. Property Manag. **33**(2), 133–151 (2015). ISSN: 0263-7472

Van Ommeren, J., Van Leuvensteijn, M.: New evidence of the effect of transaction costs on residential mobility*. J. Reg. Sci. **45**(4), 681–702 (2005)

Modeling and Simulating Nutrient Management Practices for the Mobile River Watershed

Vladimir J. Alarcon[1(✉)] and Gretchen F. Sassenrath[2]

[1] Civil Engineering School, Universidad Diego Portales, 441 Ejercito Ave., Santiago, Chile
vladimir.alarcon@udp.cl
[2] Southeast Agricultural Research Center, Kansas State University, Parsons, KS, USA
gsassenrath@ksu.edu

Abstract. In this research, an existing hydrological model of the Mobile River watershed is expanded to include water quality modeling of Nitrate (NO_3) and Total Ammonia (TAM). The Hydrological Simulation Program Fortran is used for modeling the hydrological and the water quality processes. The resulting water quality model is used to implement nutrient management practices scenarios and, via simulation, explore the effects of those management scenarios at the most downstream river (Mobile River). Results show that the implementation of reported Best Management Practices (BMPs) at sub-watershed level (filter strips, stream bank stabilization and fencing) do not work as efficiently as when applied to the entire Mobile River watershed. Removal efficiencies reported for those BMPs at the sub-watershed scale ranged between 10.6 % and 54.0 %. When Filter Strips were applied to agricultural lands throughout the watershed, reductions of NO_3 concentrations ranged from 1.48 % to 12.24 % and TAM concentrations were reduced between 0.84 % and 6.97 %. Applying Stream Bank Stabilization and Fencing to the whole watershed produced removals of NO_3 of up to 14.06 %, and maximum TAM reductions of 8.01 %. The reasons for the discrepancy may be due to the site-specificity of the BMP techniques. This may preclude extrapolating those BMPs to sites where the characteristics are different (topography, soils, stream regime, etc.) reducing the effect of the applied BMPs. Watershed-wide aspects such as sub-basin-to-sub-basin or stream-to-stream interactions, may also reduce the effect of the management practices.

Keywords: Mobile River watershed · Water-quality modeling · HSPF · Nitrate · Total Ammonia · BMP

1 Introduction

Several coastal estuaries in the US Gulf Coast have experienced seasonal hypoxia events in recent years. While the origin of this problem may be related in part to an increase of urban development along the Gulf Coast, the dissolved oxygen depletion occurs mainly due to nutrient contributions from upland watersheds where the main economic activity is agriculture. Nutrients washed-off by rain events from the agricultural fields end up into streams that carry the nutrient-rich water to coastal water bodies. Aquatic vegetation

© Springer International Publishing Switzerland 2016
O. Gervasi et al. (Eds.): ICCSA 2016, Part III, LNCS 9788, pp. 33–43, 2016.
DOI: 10.1007/978-3-319-42111-7_4

growth is enhanced by the availability of nutrients and the dissolved oxygen budget is unbalanced leading to eutrophication and subsequent hypoxia.

Mobile Bay, Alabama, recipient of the sixth-largest freshwater discharge in North America [1] experiences regular hypoxic events during the summer [2]. One of the upland watersheds that drains into Mobile Bay (via secondary streams to Mobile River) has been identified [3] as one of the sources of nutrient input due to its intensive agricultural activity: the Tombigbee River watershed. The increase of agricultural lands could potentially worsen the current situation. It has been estimated [4] that in some portions of the Tombigbee watershed the following land-use/land-cover changes have occurred in a span of 17 years: 34 % increase of agricultural lands, 263 % increase of lands used for grazing or hunting animals (rangeland), and a 16 % decrease of natural forest lands. It is reasonable to infer that nutrient inputs from this watershed may have an impact on the downstream rivers and water bodies. In fact, a recent study [5] has estimated (at a small geographical scale) several management scenarios that may be used for reducing the effects of nutrient and sediment contamination to some of the Tombigbee streams. The Best Management Practices postulated by Kleinschmidt [5] included the following.

- For cropland: filter strips, reduced tillage, stream bank stabilization and fencing, and terraces
- For pasture land: stream bank stabilization and fencing, and terraces

Estimating the effects of nutrient wash-off from agricultural fields to streams and water bodies is not an easy task. Since the sources of those nutrients are distributed in space (non-point sources), complex physical and biochemical processes occur during the transport of those contaminants through soil and water. Watershed models are able to quantify those processes occurring either under natural conditions or due to anthropogenic activities. These models use mechanistic and empirical algorithms to calculate the hydrological processes within a watershed and also the migration of pollutants from point and non-point sources to water bodies. Previous research has shown the use of these types of models for quantification of processes in agricultural watersheds [6]). For the Mobile River watershed, there exists an initial hydrological model [7] that covers all major streams in the watershed, including the Tombigbee River watershed. A water quality model for the entire watershed, however, is yet to be developed.

In this research, the existing hydrological model of the Mobile River watershed [7] is further hydrologically calibrated and validated, and built-up to include water quality modeling of nutrients. The Hydrological Simulation Program Fortran (HSPF, [8]) is used for modeling both: the hydrological and the water quality processes. The Better Assessment Science Integrating Point & Nonpoint Sources (BASINS [9]) GIS system is used to perform most of the geospatial operations. Once the nutrient modeling portion was developed, the water quality model is used to implement the nutrient management practices scenarios [5], and via simulation, explore the effects of those management scenarios at the most downstream river (Mobile River).

2   Methods

2.1   Study Area

The Mobile River is the main contributor of freshwater to Mobile Bay. In its approximately 72 km of length it carries waters coming from the confluence of the Tombigbee and Alabama rivers. The Mobile River has an average annual stream flow of 1840 m^3/s. The Mobile River watershed (Fig. 1) covers approximately 115,000 km^2, extending into

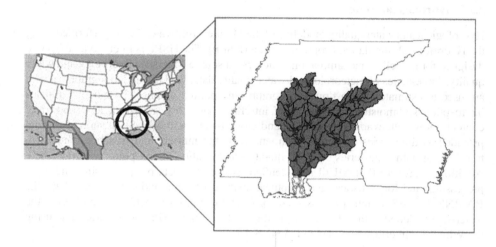

Fig. 1. Mobile River watershed. The drainage basin (fourth-largest in the United States) drains streams located in Alabama, Mississippi, Georgia, and Tennessee.

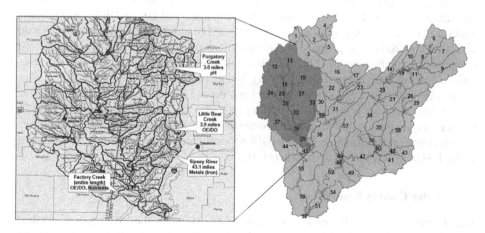

Fig. 2. Tombigbee watershed. The extent of the Tombigbee watershed included in the Tombigbee River Basin Management Plan [5] is shown in relation to the total extent of the Mobile River watershed.

four southeastern states: Alabama, Mississippi, Georgia, and Tennessee. This drainage basin is the fourth-largest in the United States. The river and its contributors have historically provided the principal navigational access for Alabama.

The Tombigbee River watershed is located in northwest Alabama and northeast Mississippi. It drains waters from a watershed (Fig. 2) that cover an area of approximately 35,650 km^2 between both states. Fifty six percent of the watershed is in Mississippi and the rest (44 %) lies in Alabama.

2.2 Hydrological Code

Hydrological and water quality modeling of the Mobile River watershed is performed using the Hydrological Simulation Program Fortran (HSPF). The HSPF is freely-available code designed for modeling and simulation of non-point source watershed hydrology and water quality. Time-series of meteorological/water-quality data, land use and topographical data are used to estimate stream flow hydrographs and polluto-graphs. The model simulates interception, soil moisture, surface runoff, interflow, base flow, snowpack depth and water content, snowmelt, evapo-transpiration, and ground-water recharge. Simulation results are provided as time-series of runoff, sediment load, and nutrient and pesticide concentrations, along with time-series of water quantity and quality, at any point in a watershed. Additional software (WDMUtil and GenScn) is used for data pre-processing and post-processing, and for statistical and graphical analysis of input and output data [10]. The BASINS 4.1 GIS system [9] was used for downloading basic data and delineating the watershed included in this study. The creation of the initial HSPF models was done using the WinHSPF interface included in BASINS 4.1.

2.3 Topographical, Landuse, and Meteorological Datasets

The topographical characterization of the study area was performed using National Elevation Data (NED) (30 m horizontal, 1 m vertical), and USGS DEM (300 m horizontal 1 m vertical) topographical datasets. The NED dataset provides a seamless mosaic elevation data having as primary initial data source the 7.5 min elevation data for the conterminous United States [11]. NED has a consistent projection (geographic), resolution (1 arc second, approximately 30 m), and metric elevation units [12].

The USGS GIRAS is a set of maps of land use and land cover for the conterminous U.S. delineated with a minimum mapping unit of 4 hectares and a maximum of 16 hectares (equivalent to 400 m spatial resolution), generated using the Geographic Information Retrieval and Analysis System (GIRAS) software. Today, they are widely known as the USGS GIRAS land use data sets [10].

2.4 Water Quality Simulations

The Tombigbee River Basin Management Plan [5] established (through watershed modeling) a series of Best Management Practices for several sub-watersheds of the Tombigbee watershed (Fig. 2).

The study applied nine BMPs for each watershed that was modeled, including five different levels of BMP implementation in the subwatershed being modeled: 0 % (no BMP), 25 %, 50 %, 75 %, and 100 %. These percentages represent the proportion of the total acreage of a particular land use (cropland, pasture land) in the sub-watershed in which the BMP would be implemented [5]. The BMPs modeled by the study were: filter strips, reduced tillage, stream bank stabilization and fencing, and terraces. The best BMPs resulting from that exploration (in terms of removal of nutrients) are summarized in Table 1.

Table 1. Average nutrient removal efficiency for Filter Strips, and Streambank Stabilization and Fencing (identified in [5]).

Acreage covered by the BMP	% Nutrient Removal Filter Strip	% Nutrient Removal Stream bank Stabilization and Fencing
25 %	10.6 %	16.0 %
−50 %	29.0 %	36.2 %
75 %	43.4 %	54.0 %
100 %	62.8 %	75.0 %

Table 1 shows that BMP Bank Stabilization and Fencing seem to provide good removal efficiency. The average nutrient removals shown in the table were calculated from results for all BMPs implemented in [5]. It is important to remark that those BMPs were implemented at a localized level (i.e., per sub-watershed). Hence, the results were also obtained at a sub-watershed scale.

In this research, we wanted to test the efficiency of two types of BMPs in the reduction of Nitrate (NO_3) and Total Ammonia (TAM) if these management practices were implemented throughout the entire Mobile River watershed. In order to undertake this computational exploration, a selection of sub-watersheds was first performed to reduce computer processing time.

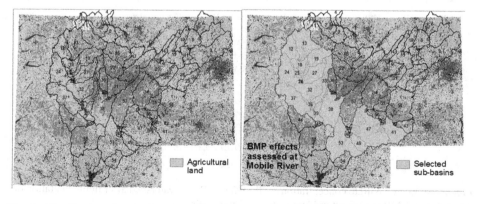

Fig. 3. Distribution of agricultural lands in the Mobile River watershed (left). Sub-basins selected for nutrient simulation are also shown (right). BMP effects were measured at the Mobile River (watershed outlet). (Color figure online)

Figure 3 shows agricultural lands in the Mobile River watershed area (left). It is evident that most of the agricultural activity in the area is concentrated in an agricultural belt that originates in northeastern Mississippi, crosses mid-southern Alabama, and ends in southwestern Georgia. Therefore, the sub-watersheds to which these agricultural lands belong were selected for implementing the BMPs and their corresponding removal efficiency (summarized in Table 1). Three levels of implementation (in terms of percent of agricultural acreage) were simulated: 25 %, 50 %, and 75 %. Those percentages of BMP were selected for ease in comparison with the results reported in [5]. The application of 100 % BMP was not included in this research because it is unrealistic to expect to achieve that level of BMP implementation in such a large area.

3 Results

3.1 Filter Strip BMP

In this section, simulated concentrations for NO_3 and TAM are shown after a Filter Strip BMP was applied to agricultural lands shown in Fig. 3.

Figure 4 shows summarized simulated results for NO_3 concentrations at the Mobile River after implementation of a Filter Strip BMP to agricultural lands. Since the effect of the BMP implementations was not well visualized at daily temporal resolution, the daily results were averaged at monthly and yearly scales. Simulations were generated for the period from January 1st, 1970 to December 31st 1995.

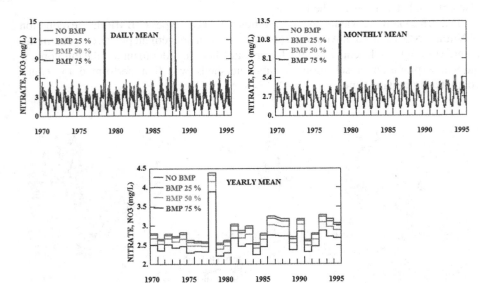

Fig. 4. Simulated Nitrate concentrations at the Mobile River watershed after implementation of a Filter Strip BMP to agricultural lands. (Color figure online)

The model correctly simulates the common seasonal trend for NO_3 concentrations in agricultural lands. Concentration values range between 1 and 6 mg/L, with peaks during summer and minimums in winter. The model simulates four events in which concentrations are higher than 13 mg/L. These uncommon peaks are correlated with four periods of very low flows that occurred in 1978, 1987, 1988, and 1990.

Figure 5 shows simulated results for TAM concentrations occurring at the watershed outlet for four scenarios for the time period 1970/01/01 to 1995/12/31. Since TAM concentrations at a daily level did not show noticeable changes after the implementation of BMPs, Fig. 5 only shows average concentrations at monthly and yearly temporal scales where effects are more significant. TAM concentrations simulated by the model range between and 0.005 and 0.04 mg/L. The model correctly simulates higher TAM concentrations for the crop growing season (March–July). Simulation results show that TAM concentrations through the year are around 0.016 mg/L with the exception of the concentrations during low flow events already identified in the NO_3 simulation.

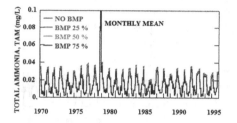 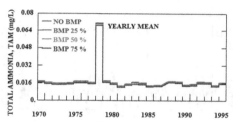

Fig. 5. Simulated TAM concentrations in the Mobile River watershed after implementation of a Filter Strip BMP to agricultural lands. Due to the slight effect of this BMP in TAM concentrations at daily level, only monthly and annual averages are shown. (Color figure online)

In order to assess the effects of the implementation the Filter Strip BMP to agricultural lands, the NO_3 and TAM concentrations produced by the model for each of the Filter Strip BMP percentages (25 %, 50 % and 75 %) were averaged and removal efficiencies were calculated with respect to the scenario where no BMP was implemented. Table 2 summarizes the results.

Table 2. Average nutrient concentration reductions after implementation of a Filter Strip BMP to agricultural lands.

Filter Strip	NO_3 mg/L	TAM mg/L	% NO_3 Reduction	% TAM Reduction
NO BMP	2.86	0.0240	0.00	0.00
BMP 25 %	2.82	0.0238	1.48	0.84
BMP 50 %	2.70	0.0230	5.56	3.16
BMP 75 %	2.52	0.0214	12.24	6.97

The results summarized in Table 2 show that the removal efficiency of the BMP implemented at 25 %, 50 % and 75 %, does not correlate with the results achieved at sub-watershed scale (reported by [5]). Reported removal efficiencies for the BMP Streambank Stabilization range from 16 % to 54 % (see Table 1). The simulation results from our model indicate that the maximum removal efficiency achieved for the Mobile River, if Filter Strips were to apply to areas with the most intensive agricultural activity, is 12 % for NO_3 and 6 % for TAM (at a 75 % BMP level).

3.2   Stream Bank Stabilization and Fencing BMP

Figure 6 shows summarized simulated results for NO_3 concentrations at the Mobile River watershed after implementation of a Stream Bank Stabilization and Fencing BMP to agricultural lands. The effect of the BMP implementations is somewhat visualized at daily temporal resolution; however for better appreciation of the effect of the BMP application the daily results were also averaged at monthly and yearly scales. As in the Filter Strip BMP the period covered by the simulation was 1970/01/01 to 1995/12/31. The charts show a slight increase in the effect of the BMP. At the annual temporal scale, the decrease in NO_3 concentrations (after application of the 75 %-BMP) is around 1 mg/L.

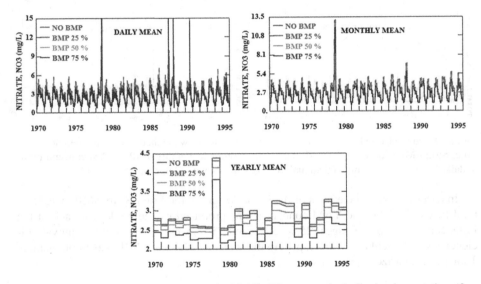

Fig. 6. Simulated NO_3 concentrations at the Mobile River watershed after implementation of a Stream Bank Stabilization and Fencing BMP to agricultural lands. (Color figure online)

Figure 7 shows simulated results for TAM concentrations occurring at the Mobile River estimated by the HSPF model. As in the Filter Strip case, TAM concentrations at a daily level did not show noticeable changes after a Stream Bank Stabilization and Fencing BMP implementation to agricultural lands. For that reason, Fig. 7 only shows mean TAM concentrations at monthly and annual temporal scales.

The application of a Stream Bank Stabilization and Fencing BMP to agricultural lands seem to have a stronger effect on NO_3 and TAM concentrations than the implementation of a Filter Strip BMP. To quantify these effects, Table 3 summarizes average NO_3 and TAM concentrations produced by the model for each of the BMP percentages

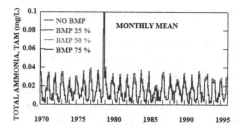
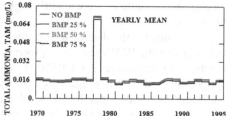

Fig. 7. Simulated TAM concentrations at the Mobile River watershed after implementation of a Stream Bank Stabilization and Fencing BMP to agricultural lands. As in the Filter Strip case, due to the negligible effect in TAM concentrations of the BMP at daily level, only monthly and annual averages are shown. (Color figure online)

(25, 50 and 75 %). Removal efficiencies were calculated with respect to the scenario where no BMP was implemented. The percent reductions achieved for Nitrate range from 1.82 % to 14.06 %, which were slightly higher than the reductions achieved with the Filter Strip BMP. As in the Filter Strip case, percent reductions in TAM concentrations range between 1 % and 8 %, showing that TAM concentrations are not affected significantly by the application of a Stream Bank Stabilization and Fencing BMP to agricultural lands.

Table 3. Nutrient concentration reduction after implementation of a Stream Bank Stabilization and Fencing BMP to agricultural lands.

Action	NO_3 mg/L	TAM mg/L	% NO_3 Reduction	% TAM Reduction
NO BMP	2.86	0.0240	0.00	0.00
BMP 25 %	2.81	0.0237	1.82	1.04
BMP 50 %	2.68	0.0229	6.25	3.37
BMP 75 %	2.46	0.0212	14.06	8.01

4   Conclusions

The computational exploration presented in this paper shows that the implementation of the most efficient management practices identified, at a sub-watershed level, by the Tombigbee River Basin Management Plan [5], do not work as well when applied to the entire Mobile River watershed. While the application of Filter Strip BMPs at sub-watershed scale was reported to show nutrient removals of 10.6 %, 29.0 %, and 43.4 % (for 25 %, 50 % and 75 % BMP application respectively), when Filter Strips were applied to agricultural lands throughout the entire Mobile River watershed, reduction of NO_3 concentrations were only 1.48 %, 5.56 %, and 12.24 % correspondingly. Reductions in TAM concentrations were even more modest: 0.84 %, 3.16 %, and 6.97 %, respectively. Removal efficiencies reported for Stream Bank Stabilization and Fencing (applied at the sub-watershed scale) ranged between 16.0 %, 36.2 %, and 54.0 % for BMP levels of 25 %, 50 % and 75 % BMP, correspondingly. However, applying this BMP to the whole watershed, NO_3 reductions were only 1.82 %, 6.25 %, and 14.06 %, respectively. TAM concentration reductions were slightly higher than in the previous case: 1.04 %, 3.37 %, and 8.01 %.

The reasons for the discrepancy in the outcomes of BMP applications may be due to the fact that the BMPs used in [5] were specific to the sub-watersheds of that study. Therefore, extrapolating those BMPs to sites where the characteristics are different (topography, soils, stream regime, etc.) may change the impact of the applied BMPs. Also, watershed-wide aspects such as sub-basin-to-sub-basin or stream-to-stream interactions, are not accounted for at the sub-watershed scale. These results have important implications for design and implementation of water quality plans.

Acknowledgment. This research was funded by a grant from CONICYT (REDES 140045).

References

1. Park, K., Kim, C., Schroeder, W.W.: Temporal variability in summertime bottom hypoxia in shallow areas of Mobile Bay, Alabama. Estuaries Coasts **30**(1), 54–65 (2007). http://link.springer.com/article/10.1007%2FBF02782967
2. EPA: National Coastal Condition report IV, September 2012. http://www.epa.gov/sites/production/files/201410/documents/0_nccr_4_report_508_bookmarks.pdf
3. EPA, Alabama & Mobile Bay Basin Integrated Assessment of Watershed Health. EPA841-R-14-002 (2014). http://www.mobilebaynep.com/images/uploads/library
4. Alarcon, V.J., McAnally, W.H.: A strategy for estimating nutrient concentrations using remote sensing datasets and hydrological modeling. Int. J. Agric. Environ. Inf. Syst. **3**(1), 1–13 (2012). doi:10.4018/jaeis.2012010101
5. Kleinschmidt Co.: Tombigbee River Basin Management Plan. Alabama Department of Environmental Management (2005). http://www.adem.state.al.us/programs/water/nps/files/TombigbeeBMP.pdf
6. Alarcon, V.J., Sassenrath, G.F.: Sensitivity of nutrient estimations to sediment wash-off using a hydrological model of Cherry Creek Watershed, Kansas, USA. In: Gervasi, O., Murgante, B., Misra, S., Gavrilova, M.L., Rocha, A.M.A.C., Torre, C., Taniar, D., Apduhan, B.O. (eds.) ICCSA 2015. LNCS, vol. 9157, pp. 457–467. Springer, Heidelberg (2015)
7. Alarcon, V.J., McAnally, W., Diaz-Ramirez, J., Martin, J., Cartwright, J.: A hydrological model of the Mobile River Watershed, Southeastern USA. In: Maroulis, G., Simos, T.E. (eds.) Computational Methods in Science and Engineering: Advances in Computational Science, vol. 1148, pp. 641–645 (2009). doi:10.1063/1.3225392
8. Bicknell, B.R., Brian, R., Imhoff, J.C., Kittle Jr., J.L., Jobes, T.H., Donigian Jr., A.S.: HSPF Version 12 User's Manual. National Exposure Research Laboratory. Office of Research and Development U.S. Environmental Protection Agency (2001)
9. Environmental Protection Agency. BASINS: Better Assessment Science Integrating Point & Nonpoint Sources: A Powerful Tool for Managing Watersheds (2008). http://www.epa.gov/waterscience/BASINS/
10. Alarcon, V.J., Hara, C.G.: Scale-dependency and sensitivity of hydrological estimations to land use and topography for a coastal watershed in Mississippi. In: Taniar, D., Gervasi, O., Murgante, B., Pardede, E., Apduhan, B.O. (eds.) ICCSA 2010, Part I. LNCS, vol. 6016, pp. 491–500. Springer, Heidelberg (2010)

11. Lent, M., McKee, L.: Guadalupe River Watershed Loading HSPF Model: Year 3 final progress report. San Francisco Estuary Institute. Richmond, Califormia (2011). http://www.sfei.org/sites/default/files/Guad_HSPF_Model__forSPLRev_17Feb2012.pdf
12. Deliman, P.N., Pack, W.J., Nelson, E.J.: Integration of the Hydrology Simulation Program—FORTRAN (HSPF) WatershedWater Quality Model into the Watershed Modeling System (WMS). Technical report W-99-2, September 1999, US Army Corps of Engineers (1999)

A Geospatial Service Oriented Framework for Disaster Risk Zone Identification

Omprakash Chakraborty[(✉)], Jaydeep Das, Arindam Dasgupta, Pabitra Mitra, and Soumya K. Ghosh

Indian Institute of Technology (IIT), Kharagpur, West Bengal, India
omchakrabarty@gmail.com, jaydeep89das@gmail.com, adgkgp@gmail.com,
pabitra@cse.iitkgp.ernet.in, skgkgp@iitkgp.ac.in

Abstract. Geographical mapping of disaster risk is an important task in disaster planning and preparedness. Heterogeneous data from various sources are integrated to identify regions having high probability of disasters. The nature of data and work-flow for risk assessment are however varying in nature in each scenario. In this work we propose a service oriented architecture to automate the process of disaster mapping. Open Geospatial Consortium Standards are implemented for this purpose. The framework can aid automated risk assessment under complex multi-modal disasters over large scale geospatial locations by integrating heterogeneous data sources. A Case study is presented for flood risk assessment for a coastal region in West Bengal, in Eastern India.

Keywords: Service oriented architecture · Disaster · Risk assessment

1 Introduction

Disasters cause a tremendous adverse effect to both life and property. The gradual evolution of civilizations has led to acquisition of new grounds thus leading to greater disaster prone areas. Thus the identification of the regions with greater risk leading to increased disaster adversaries has gained a crucial importance. As efforts have been incorporated for this purpose, there has been a need for development of a framework to aid the risk identification process and enable its use by citizens to raise the awareness regarding the safety of inhabitants. Risk mappings refers to the depiction of regions having a high probability of being affected by a certain disaster.

As the analysis of the risk areas require processing of data from varied sources of independent providers, Service Oriented Architecture (SOA) [15] serves as a basic platform for the purpose. SOA lays the technical foundation for sharing and utilizing the heterogeneous data sources in the Enterprise Geographical Information Systems (EGIS) [20] framework. As for the handling of varied providers of heterogeneous data, Open Geospatial Consortium (OGC) [18] has issued several standards for sharing the geospatial data repositories in a loosely coupled and interoperable way. The OGC standard Web Feature Service (WFS) [19]

© Springer International Publishing Switzerland 2016
O. Gervasi et al. (Eds.): ICCSA 2016, Part III, LNCS 9788, pp. 44–56, 2016.
DOI: 10.1007/978-3-319-42111-7_5

provide interfaces for spatial vector data retrieval. The Web Processing Services (WPS) [16] provides the standards for geospatial function deployments aimed for processing the retrieved geospatial data.

In recent years a number of efforts have been involved in risk identifications for specific disaster related techniques along with vulnerability concepts [13]. This includes situations like susceptibility analysis for forest fires [1,6], floods [3,4,11] drought [5], earthquakes [12] and cyclones [7]. Researches have also been carried out for risk-based evacuation sub-zones [8] and risk reductions in different fields e.g. agriculture [9] in the risk mapped regions. All these risk depictions reqire a number of geographical constraints [14] to be taken under consideration.

The proposed framework aims at the risk zone identifications for disaster scenarios. The work primarily focuses on the fundamental phases of the risk mapping task relative to disaster and augmented by the necessities of the citizens and their surroundings. The framework also serves as a platform for various algorithms to be implemented in the risk mapping process. Finally it aids in the work-flow executions of the existing models for disaster management [8].

2 Overview of the Framework

The framework utilizes the aspects of Web Feature Services (WFS) and Web Processing Services (WPS) to render the required functionalities by accessing the different databases relative to the input factors. It further processes it to reveal the respective risk zones of a region. The formal description of service compositions are represented in the later section. The Service Oriented Architecture (SOA) framework, as in Fig. 1, is proposed as a means of composing the web services and utilizing them for retrieving the necessary spatial information followed by its further processing. The framework flow model is further illustrated in Fig. 4 for the relevant case study.

3 Automatic Integration of Geospatial Web Services

The framework firstly facilitates automated feature collection through various web services and carry out feature based processes. Besides automation, it also integrates various resources necessary for risk-level analysis. It aids in abstraction towards risk zone identification process, enabling its utilization by citizens. These advantages are brought about in the system by the Service Oriented Architecture (SOA), aimed to enhance and optimize the susceptibility detection for various disasters. The architecture promotes the selection of relevant data services for retrieving relevant information in accordance to the risk analysis objective. The data sets being heterogeneous and being collected from distributed sources required to be integrated into a common Geographical Information System (GIS) platform.

The proposed framework involves the coordination of geo services relevant in the process by exploiting the Open Geospatial Consortium (OGC) based catalog

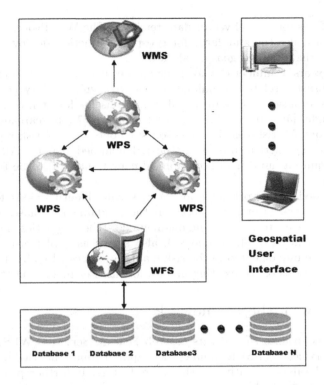

Fig. 1. Overview of risk identification framework

services (CSW). In the process of orchestration the web feature services (WFS) are utilized to retrieve spatial feature data. In order to process of the retrieved geospatial data, OGC also defines web processing services (WPS) for producing required information. This can be conceptualized through a very simple example considering only one feature data. Suppose we need the space proximity of a certain road network, WFS can be firstly utilized to get the road network feature set and then the WPS can be used to process the feature set to get the space proximity. WPS provides many services that can be known through the *getCapabilities()* method (see Sect. 4.3). In respect to finding the space proximity the *Buffer* (see Sect. 4.4) operation can be utilized.

4 Service Oriented (SO) Framework

The proposed SO framework aims at the utilization of the web services to collect the features of the disaster. It further processes it to map out the relative risk zones. The working of the framework can be classified into the following phases:

4.1 Disaster Relative Factor Collection and Corresponding Constraint Depiction

The user interface of the platform provides a means to input various factors related to disaster. It then triggers the feature collection process and imposes effective constraint parameters. The user gives input of the respective factors $(f_1, f_2, f_3, ..., f_n)$ along with their respective constraints $(x_1, x_2, x_3, ..., x_n)$. Once the input is obtained, the selected factors are passed over to trigger the feature collection phase and promote the constraint parameters.

The examples of factors $(f_1, f_2, f_3, ..., f_n)$ are like *elevation, census, road network, forest cover*. The respective constraints $(x_1, x_2, x_3, ..., x_n)$ can be like *height above* 250 m, *population density over 1000, a proximity of* 5 km *from nearest roads*.

4.2 Feature Collection Based on the Disaster Factors

This phase undertakes the utilization of the WFS to obtain the feature set relative to the individual factors in form of Geographic MarkUp Language (GML) which in turn is visualized through the utilization of Web Map Services (WMS). The factors are taken one at a time and the WFS are called for the relative factor. The process is continued iteratively for all the given factors. Once the features are obtained, they are sent for further processing request based on the constraints. The feature are relevant to factors like *waterbodies* feature sets, non-spatial features like *census* and other disaster relevant feature samples like *Land Use Land Cover (LULC) or Standardized Precipitation Index (SPI)* map. The features set can be visualized in the interface and is well enriched with the spatial data sets as well. The resulting map layers thus obtained can be represented as layers $L_i = WMS(f_i)$ $\forall i = 1, 2, ..., n$ The layers thus obtained are $L_1, L_2, ..., L_n$.

4.3 Processing of Individual Feature Sets

Once the factor related features are obtained, the relative constraints are imposed on them utilizing the various aspects of the WPS including execution of various processes like *buffer* creations and *filtering* of the features in accordance to the constraint parameters for the individual layers. These WPS are subject to the various provisional facilities that can be viewed through the *getCapabilities()* request. The *getCapabilities()* operation is supported by the OGC defined WPS and its operations provides access to the live WPS implementations. It also lists the operations and access methods supported by the implementations. The processing can be further implemented using the various measuring algorithms acted upon the inherent feature datasets. Imposing the constraints on the respective layer $L_1, L_2, ..., L_n$ the WPS is called for the *getCapabilities()* request using the following syntax:

Let Op represent a process of WPS e.g. *buffer, filter*. Thus the results can be represented as

$$L_i' = Op_{<x_i>}(L_i) \quad \forall i = 1, 2, ..., n \tag{1}$$

For example, Suppose we need to create a proximity of 10 km around a given road feature (say L_r), the *buffer* process is to be used. Thus the representation of (1) becomes

$$L'_r = Buffer_{<10>}(L_r)$$

This results in the new layer L'_r that contains a buffer region of 10 km around all the roads of the road network obtained from the road feature L_r.

4.4   Integration of the Processed Feature Sets

The framework integrates the processed features to create a new feature set for detection of the risk zones. The WPS results is converted in GML and tranferred to WMS service along with the related WFS feature sets. The processed features as a result of the WPS orchestrations respective to the individual factors are now integrated. Two constraint imposed feature sets are taken at a time and integrated to for a new feature set, which is further integrated with corresponding features. This process is carried on until all the feature sets are integrated to the finalized integrated form. This is more vividly depicted in the case study (see Sect. 5) although it can be conceptualized by the examples of integration of processed results like that of *buffers* and *filters*. *Buffers* and *Filters* both are the part of the OGC based WPS. The *buffer* operation mainly deals with the proximity depiction of features obatined through WFS. It functions by taking the spatial geometry information of the target feature and processes it to create a spatial buffer region at the given distance of the feature thereby revealing the region enclosed within the certain proximity. The *filter* operation, on the other hand, is used to refine the feature set based on a threshold value. Its function is to process the feature based on their attribute data to refine and select only those feature elements that satisfy the given conditions over a threshold parameter. Thus, the integration of two processed feature sets (say L'_i and L'_j) to form a new feature set L'_{ij} can be represented as:

$$L'_{ij} = L'_i \cap L'_j. \tag{2}$$

4.5   Identification of Probable Risk Spots

The final processing is made upon the newly formed integrated feature set consisting all the necessary constraint impositions. The WPS orchestration is further carried out to chalk out the intersections points and identify the suspected high risk areas. In order to compute the mapping regions various other measures can be computed of the integrated layers e.g. the *Drought Vulnerability Index (DVI)* [5], *The Normalized Difference Vegetation Index (NDVI)* [10], *Livelihood Vulnerability Index (LVI)* [17]. The finalized integrated feature set becomes:

$$L'_{1n} = (L'_1 \cap L'_2 \cap .... \cap L'_n). \tag{3}$$

5   Case Study of Execution of the Proposed Framework

The case study aims at the relevant risk zone identifications respective to flood as a disaster. Flood is one of the most common disaster in India. The case study area is a part of West Bengal (WB), in eastern India. The study is mainly oriented to the identify the regions having higher probability of getting affected by flood within the study area. The study area is visualized in the Fig. 2.

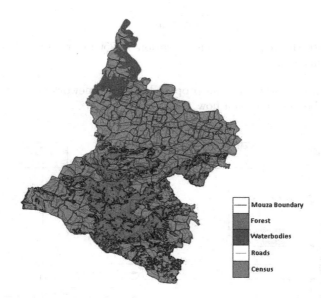

Fig. 2. Geographical study area in West Bengal, India. The area covered is 428.4 sq. km (Color figure online)

The framework provides registering web services in its platform. These web services utilize WFS and WPS to obtain the data related to the factors linked to flooding in the region. Three primary factors are taken for the purpose of risk mapping relative to the disaster of flood.

1. *Distance from waterbodies (rivers)*:
   The distance from waterbodies is one of the key criteria towards the risk. The effective adversities of flood have highest probabilities in the neighboring regions of the water sources. The distance proximity constraint be taken as 'x' e.g. 500 m.
2. *Proximity to settlements*:
   Areas near habitats/settlements are more risk-prone to flood because the damage to property and resources are more adverse rather than flooding over barren lands. The risk lies more in regions having population over a certain threshold 'th' e.g. areas having $census \geq 1000$ as it involves greater

management strategies and more complex plannings in both pre- and post-disaster phases.

3. *Elevation*:

Elevation also plays an important role as the height of a certain region is inversely relative to the water accumulation at the region thereby altering the level of risk. This can also in turn be utilized to extract the feature of water depth of the region. Thus,

$$\frac{1}{Elevation} \propto Risk$$

For this case the regions having elevation $\leq 250\,\mathrm{m}$ are filtered out for possessing higher risk.

The sequence of steps for the proposed SOA framework is shown in Fig. 3. This can also be depicted as a flow model as in Fig. 4.

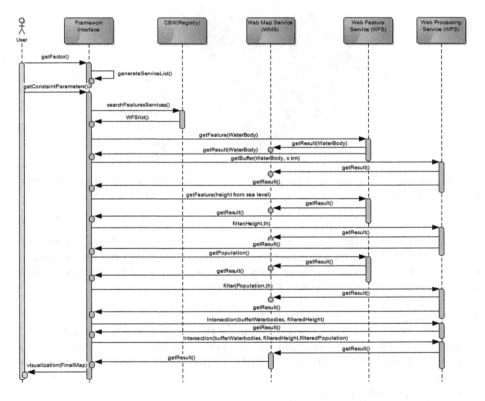

Fig. 3. Sequence of steps for risk assessment for floods

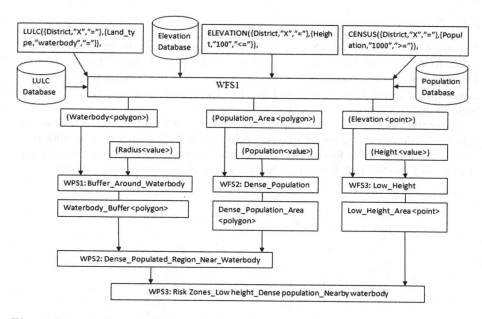

Fig. 4. Example flow model for the proposed framework for the case study of flood risk zone identification

Table 1. The WFS encorporated in the framework for the case study

WFS	Database input	Output
WFS 1	Waterbodies(CS1), Census(CS2), Elevation(CS3)	Waterbody, Population Area, Elevation points
WFS 2: Dense Population	Population Area, Minimum Population Value	Dense Population Area
WFS 3: Low Height	Elevation Points, Maximum Height Value	Low Height Area

Table 2. WPS services utilized for processing of respective feature sets

WPS	Input	Output
WPS 1: Buffer Around Waterbody	Waterbody, Radius	Waterbody Buffer
WPS 2: Dense Populated Region Near Waterbodies	Waterbodies Buffer, Dense Population Area	Intersection of Waterbody Buffer and Dense Population Area
WPS 3: Risk Zones Low height Dense Population Nearby waterbody	Low Height Area, Intersection of Waterbody Buffer and Dense Population Area	Risk Zone Area

– Web Feature Collection: Based on the depicted factors affecting flood damages, the proposed framework firstly carries out the feature collection processes in Table 1.
– Web Processing Services: For further processing of the feature collections to impose the constraint parameters the Web Processing Services are requested as in Table 2.

5.1 Results

The results are presented for various steps of execution of the proposed framework. At first the results of deriving the respective feature sets of the study area are presented. It includes the features of *waterbodies* of the region (Fig. 5), the *population density* data represented by the *census* feature map (Fig. 7) and the regional elevation feature (Fig. 9).

After the feature retrieval, the processed feature set results are presented along side the basic feature maps. These comprise of the *buffer* operation results carried on the waterbodies feature set (Fig. 6), the *filtering.* of the population density (Fig. 8) and the level of elevation (Fig. 10), above the respective thresholds.

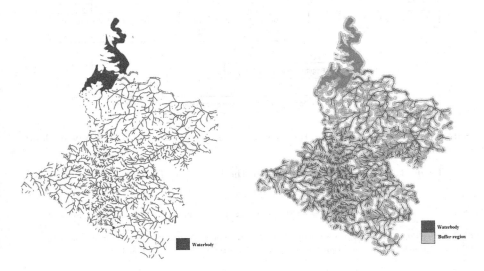

Fig. 5. Distribution of waterbodies within the study area

Fig. 6. Waterbodies feature layer with their respective processed buffer regions around

The processed features are then integrated to detect the intersection points of the constraint impositions and the results are thus presented. First the intersections between the population density and waterbody features are carried out as in Fig. 11 to reveal the densely population areas above the threshold located near the waterbodies. The layer of the water depth intersections is computed next. Figure 12 refers to the new feature layer containing the intersection regions including the filtered features of water depths. Thus, the final risk zones are revealed as in Fig. 13 along with the susceptible zones that are on the verge of risk.

Figures 5, 6, 7, 8, 9, 10, 11, 12 and 13 helps in the visualization of the sequential execution of the proposed SOA framework. The initial features derivations are aided by the WFS whereas the processing operations are carried out by the various WPS operations. The three causal parameters of flood as chosen in

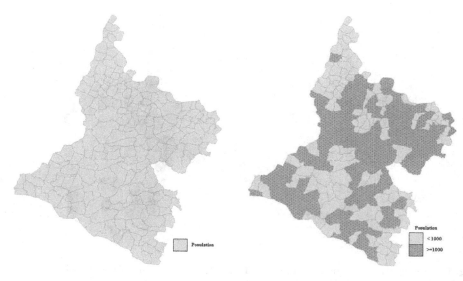

Fig. 7. Census map of the study area enclosing the region-wise population data as attribute

Fig. 8. Regions to depict areas having population ≥ 1000

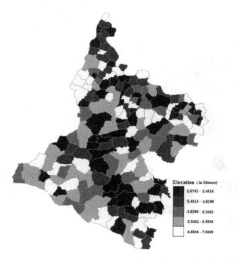

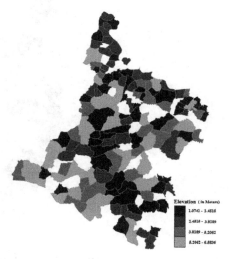

Fig. 9. Map of elevation above sea level within the study area

Fig. 10. Filtering regions to depict areas with water depth ≤ 4.0 m

the case study along with their constraints are thus taken to execution by the framework. The regional proximity from the waterbodies within the study area are depicted utilizing the *buffer* operations. The constraints involving limiting threshold values are taken up by the means of the *filter* operations.

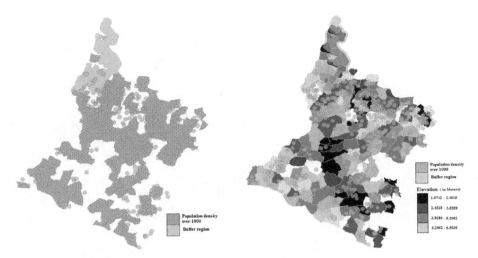

Fig. 11. Intregation for dense populated regions near waterbody

Fig. 12. Intersections of processed feature layers. (Color figure online)

Fig. 13. Depiction of flood in the study area

5.2 Discussion

The results depict and visualize the gradual progress of the risk analysis framework. It revolves around the processing at the individual layer level as well as the integrated feature set level. Once the respective features are processed they contribute towards the integrated results. From the results, it is clear that the large waterbody (as seen in Fig. 5) posseses a great effect towards water level risk in terms of flood assessment. It may also be noted the region under case study also has a varied elevation thus dividing the final risk zones (Fig. 13) into 2 grades as seen.

6 Conclusion

The paper presents a service oriented framework aimed at the depiction of risk zones in a disaster prone region. It utilizes the information of various web services to identify and visualize the susceptible spots. It also brings about the coordination between the different web services to retrieve the features and impose the thresholds. A case study of a region in eastern India is presented to demonstrate the execution of the proposed framework.

References

1. Pourtaghi, Z.S., Pourghasemi, H.R., Rossi, M.: Forest fire susceptibility mapping in the Minudasht forests, Golestan province, Iran. Environ. Earth Sci. **73**, 1515–1533 (2015)
2. Jaiswal, R.K., Mukherjee, S., Raju, K.D., Saxena, R.: Forest fire risk zone mapping from satellite imagery and gis. Int. J. Appl. Earth Obs. Geoinf. **4**, 1–10 (2002)
3. Youssef, A.M., Pradhan, B., Sefry, S.A.: Flash flood susceptibility assessment in Jeddah city (kingdom of Saudi Arabia) using bivariate and multivariate statistical models. Environ. Earth Sci. **75**, 1–16 (2016)
4. Nishiyama, K., Moriyama, T., Morishita, K., Izumi, S., Yokota, I., Tsukahara, K.: Risk analysis of heavy rainfall and associated disaster occurrence based on pattern recognition of meteorological fields using self-organizing map. In: E-proceedings of the 36th IAHR World Congress, The Hague, the Netherlands, 28 June – 3 July, 2015 (2015)
5. Shahid, S., Behrawan, H.: Drought risk assessment in the western part of bangladesh. Nat. Hazards **46**, 391–413 (2008)
6. Pourvali, M., Liang, K., Feng, G., Bai, H., Shaban, K., Khan, S., Ghani, N.: Progressive recovery for network virtualization after large-scale disasters. In: International Conference on Computing, Networking and Communications (ICNC), Kauai, HI, pp. 1–5. IEEE (2016)
7. Kavitha, R., Anuvelavan, S.: Weather master: mobile application of cyclone disaster refinement forecast system in location based on gis using geo-algorithm. Int. J. Sci. Eng. Res. **6**, 88–93 (2015)
8. Hsu, Y.-T., Peeta, S.: Risk-based spatial zone determination problem for stage-based evacuation operations. Transp. Res. Part C: Emerg. Technol. **41**, 73–89 (2014)
9. Nitu, C., Dumitrascu, A., Krapivin, V.F., Mkrtchyan, F.A.: Reducing risks in agriculture. In: 20th International Conference on Control Systems and Computer Science (CSCS), pp. 941–945. IEEE (2015)
10. Shao, Y., Lunetta, R.S., Wheeler, B., Iiames, J.S., Campbell, J.B.: An evaluation of time-series smoothing algorithms for land-cover classifications using modis-ndvi multi-temporal data. Remote Sens. Environ. **174**, 258–265 (2016)

11. Costa, J.R.S, De Almedia Filgueria, H.J., Moreira Da Silva, F.: Application of the analytic hierarchy process as tool to flood risk management in Carnaubais, Rio Grande Do Norte State, Brazil. In: E-Proceedings of the 36th IAHR World Congress, The Hague, the Netherlands, 28 June–3 July 2015 (2015)
12. Salgado-Gálvez, M.A., Romero, D.Z., Velásquez, C.A., Carreño, M.L., Cardona, O.-D., Barbat, A.H.: Urban seismic risk index for Medellín, Colombia, based on probabilistic loss and casualties estimations. Nat. Hazards 80(3), 1995–2021 (2016)
13. Cardona, O.D.: The need for rethinking the concepts of vulnerability and risk from a holistic perspective: a necessary review and criticism for effective risk management. In: Mapping vulnerability: Disasters, Development and People, vol. 17. Earthscan Publishers, London (2004)
14. Tao, K., Wang, L., Qian, X.: Multi-factor constrained analysis method for geological hazard risk. Int. J. Eng. Technol. 8(3), 198 (2016)
15. Erl, T.: Soa: Principles of Service Design, vol. 1. Prentice Hall, Upper Saddle River (2008)
16. Foerster, T., Stoter, J.: Establishing an OGC web processing service for generalization processes. In: ICA workshop on Generalization and Multiple Representation (2006)
17. Hahn, M.B., Riederer, A.M., Foster, S.O.: The livelihood vulnerability index: a pragmatic approach to assessing risks from climate variability and change? a case study in Mozambique. Glob. Environ. Change 19(1), 74–88 (2009)
18. de Vreede, E., Seib, J., Voidrot-Martinez, M.-F.: Open Geospatial Consortium (2014)
19. Vretanos, P.A.: Web feature service implementation specification. Open Geospatial Consortium Specification, pp. 4–94 (2005)
20. Longley, P.A., Goodchild, M.F., Maguire, D.J., Rhind, D.W.: Geographic Information System and Science. Wiley, New York (2001)

Evaluation of Current State of Agricultural Land Using Problem-Oriented Fuzzy Indicators in GIS Environment

Vladimir Badenko[1(✉)], Dmitry Kurtener[2], Victor Yakushev[3],
Allen Torbert[4], and Galina Badenko[5]

[1] St. Petersburg University, St. Petersburg, Russia
vbadenko@gmail.com
[2] European Agrophysical Institute, Amriswil, Switzerland
dmitrykrtnr@gmail.com
[3] Agrophysical Research Institute, St. Petersburg, Russia
VYakushev@agrophys.ru
[4] USDA-ARS National Soil Dynamics Laboratory, Auburn, AL, USA
Allen.Torbert@ars.usda.gov
[5] Peter the Great Saint-Petersburg Polytechnic University, St. Petersburg, Russia
gbadenko@gmail.com

Abstract. Current state of agricultural lands is defined under influence of processes in soil, plants and atmosphere and is described by observation data, complicated models and subjective opinion of experts. Problem-oriented indicators summarize this information in useful form for decision of the same specific problems. In this paper, three groups of problem-oriented indicators are described. The first group is devoted to evaluate agricultural lands with winter crops. Second group of indicators oriented for evaluation of soil disturbance. The third group of indicators oriented for evaluation of the effectiveness of soil amendments. For illustration of the methodology, a small computation was made and outputs are integrated in Geographic Information System.

Keywords: Problem-oriented indicators · Soil disturbance · Damage of winter crops · Fuzzy and crisp models · Spatial extension of fuzzy set theory

1 Introduction

In a general sense, indicators are a subset of the many possible attributes that could be used to quantify the condition of a particular landscape or ecosystem. They can be derived from biophysical, economic, social, management and institutional attributes, and from a range of measurement types [1]. Traditionally, results from physical and statistical analysis of agricultural field have been used as indicators [2, 3].

In many cases, indicators are developed for decision of any problem. For example, recently several indicators have been developed to address a variety of questions and problems related to land evaluation [4, 5], for managing precision agriculture [6, 7], for evaluation of yield maps [8, 9], evaluation of agricultural land suitability [10, 11],

© Springer International Publishing Switzerland 2016
O. Gervasi et al. (Eds.): ICCSA 2016, Part III, LNCS 9788, pp. 57–69, 2016.
DOI: 10.1007/978-3-319-42111-7_6

assessment of soil quality [12, 13], evaluation of resources of agricultural lands [14, 15], zoning of an agricultural fields [16, 17] and other land use planning [18, 19].

Such indicators can be referred to as problem-oriented indicators. Problem-oriented indicators are increasingly being used for diagnostic of current state of agricultural lands and for improvement of decision support processes. Indicators can be valuable tools for evaluation and decision making because they synthesize information and thus can help in the understanding of complex systems [20].

Problem-oriented indicators are the core of the problem-oriented approach [21]. In this approach, current state of agricultural lands is defined by the specific conditions of soil processes, plants, and atmosphere. The conditions are described by observational data, complicated models, and subjective opinion of experts [22]. The problem-oriented indicators summarize this information into a useful form that can be used for decisions regarding these specific problems.

Among the many potential indicators we will focus on three groups.

The first group is devoted to the evaluation of agricultural lands with winter crops. It is well known that damage of winter crops is primarily related to three main factors: (a) influence of adverse agrometeorological conditions, (b) soil temperature in the root area, and (c) winter crop response from negative environmental conditions. The interaction of these factors is very complicated. Therefore, the use of indicators which described damage of winter crops is useful for management of agricultural lands in winter.

The second group of indicators to be considered is those used for evaluation of soil disturbance. Assessment of soil disturbance is very important for making decisions on agricultural and ecological management.

The third group of indicators to be considered is those used for evaluation of the effectiveness of soil amendments. Soil amendments have been shown to be useful for improving soil condition, but it is often difficult to make management decisions as to their usefulness. In Europe as a whole, nearly 60 % of agricultural land has a tendency towards acidification. However, most crops prefer near neutral soil conditions. Traditionally, lime has been added in order to improve acid soils conditions. But precipitation is continual in these regions and after some years the soil conditions becomes acidic again. Therefore, assessment of the effects of additions of liming compounds on soil structure and strength using indicators has a practical interest.

In this paper, three groups of problem-oriented indicators are described. For illustration of the methodology, a small computation was made and outputs are integrated into a Geographic Information System (GIS).

2   Evaluation of Damage of Winter Crops by Frost

The monitoring of frost injury is the most important aspect of the monitoring of agricultural fields for winter crops. Evaluations of frost damage of winter crops has been discussed in many publications [23, 24].

Damage of winter crops is related to three main factors: (a) influence of the adverse agrometeorological conditions, (b) soil temperature in the root area, and (c) winter crops response to the adverse agrometeorological conditions.

Many studies of heat transfer in soil have been performed. Several researchers have observed soil temperature during the winter period and have described two processes which characterize this period: soil freezing and soil thawing [25]. Soil temperature of the root area (at 2 cm depth) is dependent on the depth of the boundary between the frozen and melted top soil layers and, the thickness of the snow cover. Model of soil temperature in the root area has been developed in framework of the general theory of heat transfer in soil [24].

2.1 Modelling Winter Crops Response on the Adverse Agrometeorological Conditions

It is well known [23] that the winter crop resistance to frost is dependent on the kind of crop, and planting density (the number of stems) determines the amount of thermal damage sustained to the winter crop. Also, frost injury is related to the crop's growth stage. For example, if at the end of autumn the winter crop's development is normal, the resistance to frost will be strong. If the winter crop is underdeveloped or outgrowing, then the resistance to frost will be decreased.

The base postulates for developing Indicator of Frost Damage (IFD) are formulated as follows:

(1) IFD is defined as a number in the range from 0 to 1, and modeled by an appropriate membership function.
(2) The choice of a membership function is somewhat arbitrary and should mirror an objective of expert opinion.

The critical temperatures (T_{root}) can be defined as the soil temperature at 2 cm depth at which the winter crop will be destroyed. The Indicator of Frost Damage (IFD) can be described as follows (parameters n_1 and n_2 are defined by Table 1):

$$\text{IDF} = \begin{cases} 0, & T_{root} > n_1, \\ f(T_{root}), & n_2 \leq T_{root} \leq n_1, \\ 1, & T_{root} < n_2 \end{cases} \tag{1}$$

It can be observed if IFD is equal to 1, then there is a maximum impact of frost. If IFD is equal to 0, then the damage to winter crops is negligible. The $f(T_{root})$ in most cases is a linear function.

2.2 Example 1

In this example we used agricultural area located near Saint-Petersburg, Russia, which contains several homogeneous plots. Spatial distribution of winter crops are shown in

Fig. 1. In this example to illustrate this approach, the IFD is modeled by an increasing piecewise-linear membership function. Results of computations are shown in Figs. 2 and 3 for two variants ($T_{root} = -15$ °C and $T_{root} = -20$ °C).

Table 1. Values of parameters n_1 and n_2

Winter crops	n_1, °C	n_2, °C
Barley:		
• Underdeveloped crop	−7	−12
• Normal crop	−14.8	−19.2
• Outgrowing crop	−7	−13
Rye:		
• Underdeveloped crop	−11	−22
• Normal crop	−14	−25
• Outgrowing crop	−11	−22
Wheat with mean frost resistance:		
• Underdeveloped crop	−11	−17
• Normal crop	−14	−19.4
• Outgrowing crop	−10	−15

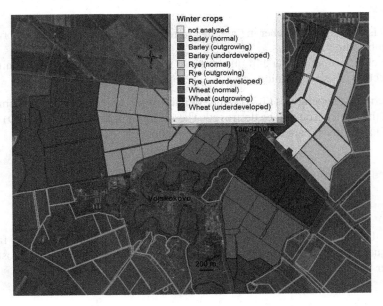

Fig. 1. Allocation of winter crops (Color figure online)

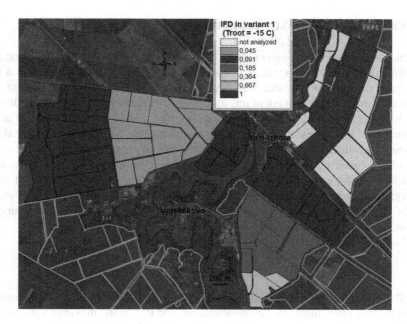

Fig. 2.   IFD in variant 1 ($T_{root} = -15$ °C) (Color figure online)

Fig. 3.   IFD in variant 2 ($T_{root} = -20$ °C) (Color figure online)

3 Evaluation of Soil Amendments

Soil amendments have been shown to be useful for improving soil condition, but it is often difficult to make management decisions as to their usefulness. Recently a tool based on fuzzy indicator model was developed [26, 27]. The effectiveness of soil amendments in the frame of work of this tool can be evaluated by two indicators: Impact Factor Simple (IFS) and Impact Factor Complex (IFC). Using IFS, an estimate of the effectiveness of soil amendments can be determined when only one experimental soil parameter is available. Using IFC, an estimation of soil amendment effectiveness can be calculated based on information about several soil parameters. This methodology was utilized in two case-studies. In the first, evaluation of the effectiveness of using polyacrylamide application as an amendment to reduce subsoil compaction was evaluated [28]. In the second, the evaluation of the organic material "Fluff" as a soil amendment for establishing native prairie grasses was evaluated [29].

IFS is defined as follows [28]:

$$\text{IFS} = (P_{max} - P)/(P_{max} - P_{min}), \quad P_{min} \leq P \leq P_{max}, \tag{2}$$

where P is the soil parameter under consideration (current value), P_{min} is the minimal value of the soil amendment under consideration, and P_{max} is the maximal value for the soil parameter under consideration. In this article the tool [26] is applied for evaluation of the residual effects of additions of calcium compounds on soil structure and strength [30, 31].

3.1 Example 2

In Europe as a whole, nearly 60 % of the agricultural land has a tendency towards acidification [29]. Acidification is a natural process for regions with humid climatic conditions where the precipitation exceeds the evapotranspiration. However, most crops prefer near neutral soil conditions. Traditionally, lime has been added in order to improve soil acidity. But, in these regions the continuous precipitation will result in the return to acidic soil conditions after some years the calcium is washed out.

For a better understanding for agriculture, the effects of the additions of calcium compounds, such as lime and gypsum, on soil strength and structure have been of direct interest [30], [32]. Tensile strength is a particularly useful measure of soil strength because it is sensitive to the micro-structural condition of the soil [29]. The goal of the present example is to illustrate the application of IFS for assessment of the effects of the additions of calcium compounds on soil structure and strength. As the starting point, known models are utilized [29]. The linear regression equations describing the relationships are:

$$S = n_1 + n_2 \, Ca^{2+} + n_3 \, w, \tag{3}$$

$$R = m_1 + m_2 \, Ca^{2+} + m_3 \, w, \tag{4}$$

where S is tensile strength (kPa), R is fracture surface roughness of soil clods (mm), n_i and m_i, i = 1, 2, 3 are empirical coefficients (Table 2), Ca^{2+} is the amount of calcium applied (kg kg^{-1}), and w is the soil water content (kg kg^{-1}). The equations have been obtained in the 0–4 * 10^{-3} kg kg^{-1} range for Ca^{2+} and 0,1–0,22 kg kg^{-1} range for w.

Table 2. Empirical coefficients n_i and m_i, in Eqs. (3) and (4)

In formula for tensile strength		In formula for fracture surface roughness of soil clods	
n_1	9.5	m_1	0.21
n_2	−324	m_2	−8.4
n_3	−18	m_3	0.26

Table 3. Predicted values of tensile strength (S) and IFS on S

Amounts of calcium, kg kg^{-1}	S, kPa	IFS on S
0	6.8	0
0.001	6.476	0.249231
0.002	6.152	0.498462
0.004	5.504	1

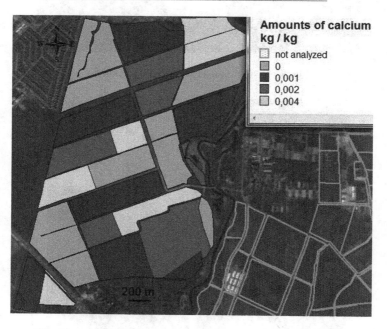

Fig. 4. Distribution of amounts of calcium (Color figure online)

The model describes situations when no tillage occurs (e.g., under long-term pastures) and under natural rainfall conditions. Also, the linear regression equations above give the residual effects calcium compound additions on structurally-degraded soil 6–7 years after the application of the calcium compounds [33]. Taking into account that range for Ca^{2+} is 0–4 * 10^{-3} kg kg^{-1} we calculated predicted values of tensile strength S and IFS on S (Table 3).

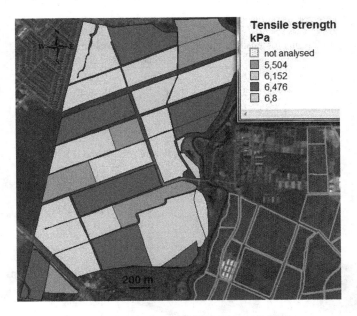

Fig. 5. Distribution of S (Color figure online)

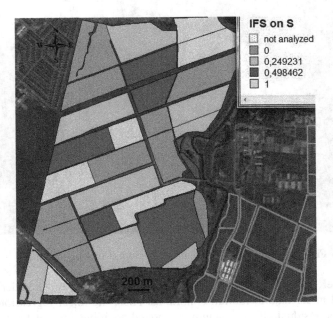

Fig. 6. Distribution of IFS on S (Color figure online)

For this example, we utilized the agricultural territory located near the Saint-Petersburg, Russia, which we used early. It was assumed that the amount of calcium compounds were distributed as shows in Table 3. Distribution of amounts of calcium and S are shown in Figs. 4 and 5. Distribution of IFS on S is given in Fig. 6.

4 Evaluation of Soil Disturbance

Soil disturbance is a great problem and the evaluation of soil disturbance is very important for making decisions on agricultural and ecological management. Small levels of soil disturbance can result in soil surface erosion and soil mass movement. This commonly leads to loss of surface organic matter and a reduction in soil quality. Recently, a new method for potentially evaluating soil disturbance is described [34].

With this method, the usage of two indicators called "Disturbance Factor Simple (DFS)" and "Disturbance Factor Complex (DFC)" were suggested. The DFS is defined as a number in the range from 0 to 1, and modeled by an appropriate membership function. It reflects measured soil parameters which are affected by soil disturbance. In [35], DFS is modeled by an increasing piecewise-linear membership function and can be presented as follows:

$$\text{DFS} = \begin{cases} 0, \text{ if } Y = 0 \\ Y, \text{ if } 0 < Y < 1 \\ 1, \text{ if } Y = 1 \end{cases} \tag{5}$$

$$y = \left(P_n - P_d\right) / \left(P_{max} - P_{min}\right), \quad Y = |y| \tag{6}$$

$$P_{max} = \max\left(P_{max,\,n},\, P_{max,\,d}\right) \tag{7}$$

$$P_{min} = \min\left(P_{min,\,n},\, P_{min,\,d}\right) \tag{8}$$

where P is a soil parameter, P_n is the soil parameter on site in natural conditions (no disturbance) ($P_{min,n} \leq P_n \leq P_{max,n}$), P_d is the soil parameter on site with disturbance ($P_{min,d} \leq P_d \leq P_{max,d}$), P_{min} is the minimal value, P_{max} is the maximal value, and Y is an absolute value (modulus) of y.

The DFC is calculated by combining individual DFS components using fuzzy aggregation algorithms. Using DFC it is possible to assess the combined effect of several DFS to improve the sensitivity of measuring potential soil disturbance impacts.

4.1 Example 3

On the agricultural territory located near Saint-Petersburg, Russia, which we utilized in example 1, there are several fields where processes of soil disturbance are currently occurring. For spatial planning, it is necessary to estimate the level of soil disturbance which has occurred in comparison with the current state of soil on neighboring fields. In this example, weighted coefficients of the significance of soil parameters were

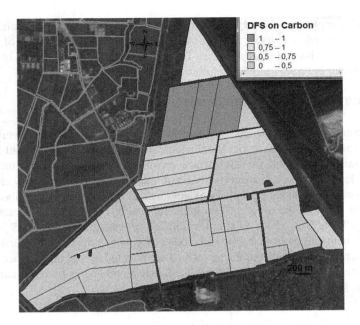

Fig. 7. Allocation of DFS on Carbon (Color figure online)

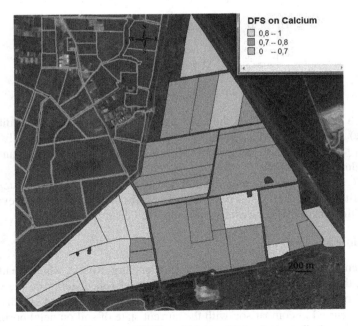

Fig. 8. Allocation of DFS on Calcium (Color figure online)

assigned as equally important. Results of the computation of DFS and DFC in topsoil are shown in Figs. 7, 8 and 9. Here only fields with soil disturbance were analyzed.

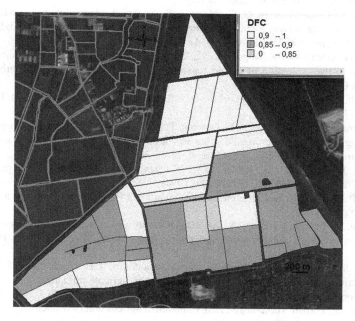

Fig. 9. Allocation of DFC (Color figure online)

5   Conclusions

Current state of agricultural lands was defined under influence of soil processes, plants, and atmosphere and was described by observational data, complicated models and subjective expert opinions. The problem-oriented indicators summarized this information into a useful form for use in decisions regarding these same specific problems.

In this paper, three groups of problem-oriented indicators are described. The first group was devoted to evaluation of agricultural lands growing winter crops. The second group of indicators evaluated soil disturbance. The third group of indicators evaluated of the effectiveness of soil amendments. For illustration of the methodology, a small computation was made and outputs were integrated into GIS.

References

1. Walker, J.: Environmental indicators and sustainable agriculture. In: McVicar, T.R., Li, R., Walker, J., Fitzpatrick, R.W., Changming, L. (eds.) Regional Water and Soil Assessment for Managing Sustainable Agriculture in China and Australia, pp. 323–332. ACT, Canberra (2002)

2. Listopad, C., Masters, R., Drake, J., Weishampel, J., Branquinho, C.: Structural diversity indices based on airborne LiDAR as ecological indicators for managing highly dynamic landscapes. Ecol. Ind. **57**(10), 268–279 (2015)
3. Gaudino, S., Goia, I., Grignani, C., Monaco, S., Sacco, D.: Assessing agro-environmental performance of dairy farms in northwest Italy based on aggregated results from indicators. J. Environ. Manage. **140**(7), 120–134 (2014)
4. Arefiev, N., Garmanov, V., Bogdanov, V., Ryabov, Yu., Terleev, V., Badenko, V.: A market approach to the evaluation of the ecological-economic damage dealt to the urban lands. Procedia Eng. **117**, 26–31 (2015)
5. Burrough, P.A.: Fuzzy mathematical methods for soil survey and land evaluation. J. Soil Sci. **40**(3), 477–492 (1989)
6. Papageorgiou, E.I., Markinos, A.T., Gemtos, T.A.: Fuzzy cognitive map based approach for predicting yield in cotton crop production as a basis for decision support system in precision agriculture application. Appl. Soft Comput. **11**(4), 3643–3657 (2011)
7. Ambuel, J.R., Colvin, T.S., Karlen, D.L.: A fuzzy logic yield simulator for prescription farming. Trans. ASAE **37**(6), 1999–2009 (1994)
8. Hodza, P.: Fuzzy logic and differences between interpretive soil maps. Geoderma **156**(3–4), 189–199 (2010)
9. Badenko, V., Terleev, V., Topaj, A.: AGROTOOL software as an intellectual core of decision support systems in computer aided agriculture. Appl. Mech. Mater. **635–637**, 1688–1691 (2014)
10. Tabeni, S., Yannelli, F.A., Vezzani, N., Mastrantonio, L.E.: Indicators of landscape organization and functionality in semi-arid former agricultural lands under a passive restoration management over two periods of abandonment. Ecol. Ind. **66**(7), 488–496 (2016)
11. Kurtener, D., Torbert, H., Krueger, E.: Evaluation of agricultural land suitability: application of fuzzy indicators. In: Gervasi, O., Murgante, B., Laganà, A., Taniar, D., Mun, Y., Gavrilova, M.L. (eds.) ICCSA 2008, Part I. LNCS, vol. 5072, pp. 475–490. Springer, Heidelberg (2008)
12. Torbert, H.A., Busby, R.R., Gebhart, D.L.: Carbon and nitrogen mineralization of non-composted and composted municipal solid waste in sandy soils. Soil Biol. Biochem. **39**(6), 1277–1283 (2007)
13. Arefiev, N., Terleev, V., Badenko, V.: GIS-based fuzzy method for urban planning. Procedia Eng. **117**, 39–44 (2015)
14. Xu, S., Liu, Y., Qiang, P.: River functional evaluation and regionalization of the Songhua River in Harbin, China. Environ. Earth Sci. **71**(8), 3571–3580 (2014)
15. Medvedev, S., Topaj, A., Badenko, V., Terleev, V.: Medium-term analysis of agroecosystem sustainability under different land use practices by means of dynamic crop simulation. In: Denzer, R., Argent, R.M., Schimak, G., Hřebíček, J. (eds.) ISESS 2015. IFIP AICT, vol. 448, pp. 252–261. Springer, Heidelberg (2015)
16. Klassen, S.P., Villa, J., Adamchuk, V., Serraj, R.: Soil mapping for improved phenotyping of drought resistance in lowland rice fields. Field Crops Res. **167**(10), 112–118 (2014)
17. Syrbe, R.-U., Bastian, O., Röder, M., James, P.: A framework for monitoring landscape functions: The Saxon Academy Landscape Monitoring Approach (SALMA), exemplified by soil investigations in the Kleine Spree floodplain (Saxony, Germany). Landsc. Urban Plann. **79**(2), 190–199 (2007)
18. Laes, E., Meskens, G., Ruan, D., Lu, J., Zhang, G., Wu, F., D'haeseleer, W., Weiler, R.: Fuzzy-set decision support for a Belgian long-term sustainable energy strategy. In: Ruan, D., Hardeman, F., van der Meer, K. (eds.) Intelligent Decision and Policy Making Support Systems. SCI, vol. 117, pp. 271–296. Springer, Heidelberg (2008)

19. Mitchell, J.G., Pearson, L., Bonazinga, A., Dillon, S., Khouri, H., Paxinos, R.: Long lag times and high velocities in the motility of natural assemblages of marine bacteria. Appl. Environ. Microbiol. **61**, 877–882 (1995)

20. Sanò, M., Medina, R.: A systems approach to identify sets of indicators: applications to coastal management. Ecol. Ind. **23**(12), 588–596 (2012)

21. Van der Grift, B., Van Dael, J.G.F.: Problem-oriented approach and the use of indicators. RIZA, Institute for Inland Water Management and Waste Water Treatment, ECE Task Force Project-Secretariat (1999)

22. Watts, D.B., Arriaga, F.J., Torbert, H.A., Gebhart, D.L., Busby, R.R.: Ecosystem Biomass, Carbon, and Nitrogen five years after restoration with municipal solid waste. Agron. J. **104**, 1305–1311 (2012)

23. Moyseychik, V.A.: Agrometeorological conditions and wintering of winter crops. Gidrometeoizdat, Leningrad (1975) 296 p. (In Russian)

24. Yakushev, V., Kurtener, D., Badenko, V.: Monitoring frost injury to winter crops: an intelligent geo-information system approach. In: Blahovec, J., Kutelek, M. (eds.) Physical Methods in Agriculture: Approach to Precision and Quality, pp. 119–137. Kluwer Academic Publishers, Dordrecht (2002)

25. Wu, D., Lai, Y., Zhang, M.: Heat and mass transfer effects of ice growth mechanisms in a fully saturated soil. Int. J. Heat Mass Transf. **86**(7), 699–709 (2015)

26. Krueger, E., Kurtener, D., Torbert, H.A.: Fuzzy indicator approach: development of impact factor of soil amendments. Eur. Agrophys. J. **2**(4), 93–105 (2015)

27. Kurtener, D., Badenko, V.: A GIS methodological framework based on fuzzy sets theory for land use management. J. Braz. Comput. Soc. **6**(3), 26–35 (2000)

28. Busscher, W., Krueger, E., Novak, J., Kurtener, D.: Comparison of soil amendments to decrease high strength in SE USA Coastal Plain soils using fuzzy decision-making analyses. Int. Agrophys. **21**, 225–231 (2007)

29. Grant, C.D., Dexter, A.R., Oades, J.M.: Residual effects of additions of calcium compounds on soil structure and strength. Soil Tillage Res. **22**(3), 283–297 (1992)

30. Chartres, C.J., Green, R.S., Ford, G.W., Rengasamy, P.: The effects of gypsum on macroporosity and crusting of two red duplex soils. Aust. J. Soil Res. **23**, 467–479 (1985)

31. Kurtener, D., Badenko, V.: GIS fuzzy algorithm for evaluation of attribute data quality. Geomat. Info Mag. **15**, 76–79 (2001)

32. Loveday, J.: Relative significance of electrolyte of subterranean clover to dissolved gypsum in relation to soil properties and evaporative conditions. Aust. J. Soil Res. **4**, 55–68 (1976)

33. Khodorkovskii, M.A., Murashov, S.V., Artamonova, T.O., Rakcheeva, L.P., Lyubchik, S., Chusov, A.N.: Investigation of carbon graphite-like structures by laser mass spectrometry. Tech. Phys. **57**(6), 861–864 (2012)

34. Torbert, H.A., Krueger, E., Kurtener, D.: Soil quality assessment using fuzzy modeling. Int. Agrophys. **22**, 1–7 (2008)

35. Torbert, H.A., Kurtener, D., Krueger, E.: Evaluation of soil disturbance using fuzzy indicator approach. Eur. Agrophys. J. **2**(4), 93–102 (2015)

New Prospects of Network-Based Urban Cellular Automata

Yichun Xie[1(✉)] and Hai Lan[2]

[1] Institute for Geospatial Research and Education,
Eastern Michigan University, Ypsilanti, MI 48197, USA
yxie@emich.edu
[2] Department of Geographical Sciences,
University of Maryland, College Park, MD 20740, USA
hlan@umd.edu

Abstract. Cellular automata (CA) build on complexity theory, focus on bottom-up processes that lead to global forms, highlight non-equilibrium or disruptive events, and complement to GIS and remote sensing techniques. Therefore, urban cellular automata (i.e., the applications of CA in urban modeling and simulation) became popular since the early 1990s. However, the CA momentum gradually subsided among urban modelers a decade later due to the criticism on the simplicity and rigidity of urban CA. Michael Batty in his seminal work, The New Science of Cities, interprets cities not simply as places but as systems of networks and flows, which can be examined in better ways by using big data and emerging computational techniques. This book sets a new benchmark for urban CA modeling, which is shedding new lights to dynamic urban modeling. Inspired by the recommendations in Batty's book, a new network-based global urban CA framework is developed in this paper. According to the science of design and planning, this paper describes the global urban CA framework at the micro-, meso-, macro- and global-scales and examines the dynamic flows (i.e. interactions) over the urban networks at the four scales by using terminologies that are often seen in the urban CA modeling literature. In addition, the paper will analyze how big data analytics affect computational implementations of the networked global urban cellular automata, including (1) multi-scales of networked urban CA spaces that lead to modifiable areal unit problems; (2) boundary discords of CA spaces that cause areal unit inconsistency problem; and (3) incomplete data that prevent from a full implementation of the global urban CA framework.

Keywords: Urban cellular automata · Hierarchical urban networks · Big data analytics · Complexity theory

1 Introduction

S. Ulan and J. von Neumann were the first pioneers who developed cellular automata (CA) based turning machines in the late 1940s [1]. Wolfram [2]

© Springer International Publishing Switzerland 2016
O. Gervasi et al. (Eds.): ICCSA 2016, Part III, LNCS 9788, pp. 70–84, 2016.
DOI: 10.1007/978-3-319-42111-7_7

demonstrated that CA can be applied to model and simulate complex natural phenomena. Wolfram [3] later elevated CA as a new science that provides a foundation by which a new look could be taken at a broad range of scientific inquiries. Cities have long been seen as a natural platform on which CA and other complexity theories can be applied or tested [4–10]. The experiment from cells to cities by Batty and Xie [11] was widely regarded as the first operational urban CA model for simulating urban expansion and its subsequent morphological form. The dynamic urban evolutionary modeling (DUEM) was one of the first urban CA simulation machines, which coded various CA components as robust GIS functions and enabled urban modelers examine urban growth under a wide range of development scenarios [11–13].

Alongside the increasing exploration of complexity theories in the urban arena, the concepts of big data and complex networks have developed rapidly in the last decade as new approaches in complexity science [9,14–17]. Batty has been up front of this new development and integrated many of these new concepts into the studies of cities, which becomes the foundation of the book, The New Science of Cities [18]. In Batty's vision, cities as complex systems means that cities are developing like biological organisms evolving in a Darwinian fitness landscape [19]. Cities are networked spatial configurations, over which different sets of actions, interactions and transactions are taking place. This type of complexity is better interpreted through urban models built on complexity theories and supported with big data that are being increasingly collected and available in both spatial and temporal dimensions [19]. Batty's new science of cities sets the benchmark that all urban modelers and geo-complexity researchers will have to follow [18].

The major objective of this paper is to reassess urban CA modeling as a response to Battys new science of cities. In other word, the paper will address how urban CA models could be enhanced and restructured with the two new developments, urban flows and networks, and big data analytics, which were thoroughly examined in the new science of cities [18]. The paper will analyze how the big data science could affect and contribute to computational executions of the networked global urban cellular automata framework. The analyses will cover several important components of urban CA modeling: (1) multi-scales of networked CA spaces that lead to modifiable areal unit problems; (2) boundary discords of CA spaces that cause areal unit inconsistency problem; and (3) incomplete data that prevent from a full implementation of the global urban cellular automata framework. Finally some discussions will be provided concerning realistic integration of big data analytics with the networked global urban cellular automata framework.

2 A Networked Global Urban Cellular Automata Framework

Flows and networks are two irreducible and interdependent elements that compose an urban system and define the ways by which the city functions [18]. From the perspective of network analysis used in transportation geography, an urban network consists of nodes (cities, urban centers or places) and edges (links or transport lines) that enable flows between the nodes. The network can be measured with five groups of interrelated indicators on the basis of the Graph Theory [20]: centrality, clustering, connectivity, accessibility and hierarchy [15,18].

The index of network centrality, which was originated as a geographical concept in the Central Place Theory by Christaller [21], measures which nodes in a network are most central. The network clustering quantifies a degree to which nodes in a network tend to cluster together. A comprehensive review of various measures of network centrality and clustering was provided by Lin [22]. The connectivity of a network is an important measure of its robustness as a network and thorough discussions were provided by Chen et al. [23] and Wang et al. [24]. The network accessibility is defined as interaction opportunities among various nodes in a network [25] and a good survey was conducted by Reggiani et al. [26].

It is worth pointing out that two important quantities can be incorporated in computing these indices: edge weight and node strength. Many different measurements can be used as edge weight or node strength [22]. For instance, traffic flow, transport volume, travel distance and time can be used as edge weights while population, gross domestic product and other socioeconomic indicators as node strengths. Furthermore, weights could be treated as attributes that describe edges (links) and strengths as characteristics that depict nodes. What types of quantities can be used as the weights and strengths depend on the type of network we are studying. For instance, in a commuting network, the nodes are spatial centers of employment and the links could be defined by the volume of commuting. In a public transportation network, the nodes are urban centers and the links could be defined by the frequencies and volumes of public transportation between them. In an information network, the nodes could be Internet hubs or residential centers and the links could be defined by volumes of emails, phone calls, or digital file uploading and downloading [27]. The types of nodes and links in urban networks have significant importance as they represent many aspects of urban life (e.g. cultural, economic, political, transport aspects).

Hierarchy is another important property of urban networks, which is perceptible in studies of cities because they grow from villages into small towns and then into larger urban forms such as a metropolis, megalopolis, or gigalopolis (world city) [18]. A measure of urban network hierarchy can be derived from other four network properties (i.e., centrality, clustering, connectivity and accessibility). Power law scaling based on node centrality or strength is a common technique to discern urban network hierarchy [18]. The notion of hierarchy is the most important factor for renovating urban cellular automata because an urban hierarchy involves all properties of urban networks and flows and also affects every element of urban complex systems from modeling point of view. Furthermore, the notion of hierarchy indicates that urban networks evolve and interact at multiple scales [28]. There are four levels of urban networks from perspectives of urban cellular automata as well as urban planning. The first three levels are analogous to the nested urban networks in The Connected City

How Networks are Shaping the Modern Metropolis [15]: micro-urban network, meso-urban network and macro-urban network. The highest (top) level is the global-urban network.

Micro-urban network could be understood as an extended CA neighborhood, or a community or subdivision within a city. The nodes are objects in study, which could be land-use cells or land parcels (including landscape patches) according to CA terminology. The nodes can also be treated as individual agents in the formation of multiple agent based modeling. The flows in a micro-urban network can be examined as relationships between different types of land uses or land parcels within the geographic extent of a CA neighborhood. The flows determine transition rules of urban cellular automata, which can be formalized by spatial arrangements of existing land-use cells and parcels within the CA neighborhood like the game of life [29], or by a spatial decaying function that follows a statistical distribution [13], or by a modeler-defined rule such as the accessibility to a major road within the neighborhood or to a local employment center that may be beyond the neighborhood [30]. The flows in micro-urban networks are presumably simple but the sum is not trivial, giving rise to the global complexity.

Meso-urban network indicates networks consisting of a city. In other words, a city is treated as a collection of intersecting, interacting, and overlapping (micro-urban) networks that bind the many different parts of the city into an organic whole like a living organism, giving it both form and function [15]. The flows in city-based meso-urban networks contain many different ones, including commuting, education, health, manufacturing, public transit, recreation, retail, telecommunication, etc. The flows over meso-urban networks involve people, economies, infrastructures, institutions, policies and spaces. Meso-urban networks have been the primary subjects of most current urban CA models. A comprehensive review of single-city-based urban cellular automata was provided by Sant and his colleagues [1]. They compared main urban CA models published up to that point of time from the perspectives of model objective, cell space, state, neighborhood, transition rule, constraint, other methods, calibration, and validation. They identified three CA components, neighborhood, transition rule and constraint as the key factors influencing urban CA relaxations. They also grouped the reviewed urban CA models into two categories: the strict transition rules and the relaxed transition rules based on transition potential or probability.

Macro-urban network is a much larger scale of network that links multiple cities together into regional or national urban networks. It is a new filed of research for urban cellular automata. Several preliminary efforts have been reported in the development of urban CA models on the basis of complexity theory and with an attempt to modelling or simulating geographic phenomena in multiple cities and regions. For example, SimPop is an encouraging urban simulation machine for studying the emergence and evolution of the system of cities [31]. The SimPop system supports the perception that the dynamics of a town or a city depends on its ability to interact with other towns and cities, which in turn depends on its relative position in the settlement system (in terms of hierarchical level, specialization, accessibility). The multi-city sustainable regional urban growth simulation (MSRUGS) is another CA-based agent model, which simultaneously considers the spatial variations of a system of cities in terms of population size, development history, water resource endowment and sustainable development potential to determine urban growth potentials of seven cities in West China [32].

There are multiple types of flows over macro-urban networks. Based on the applications of the gravity theory in urban modeling [18, 31–33], the flows across regions that affect interactions among cities over a macro-urban network could be better examined from the viewpoint of location advantage. Location advantage is a country-level phenomenon that can also be evaluated at a regional or sub-country level [34, 35]. The data items pertaining to location advantages can be examined or collected from different perspectives, such as business economics, economic geography and political economy.

Global-urban network is the largest network that links major world cities such as Tokyo, New York, London, Sao Paulo, Seoul, Shanghai, etc. These global urban networks interact and overlap with macro-urban networks, determining the progress of global economy and signifying a frontier for invigorating urban CA models. Cities are typical cases of complex and self-organizing systems [7, 9, 36]. Cities in the twenty-first century are interacting globally and connecting with their hinterlands to form a networked society [27, 37]. It is inevitable that urban modelers have to take into consideration of the globalization of cities or global cities in the formalization of new generation of urban CA models.

Transnational interactions are a unique feature of global-urban networks in comparison with other scales of urban networks. Flows or links crossing country boundaries involve people, capital, information, services and goods [38]. The factors of affecting location advantages discussed at the macro-urban network scale are in general suitable to global-urban networks. Furthermore, there is a uniqueness to the global-urban networks, which is the collection of international firms in these world cities [39, 40]. Based on distribution patterns of international firms, global-urban networks could be formalized into three models. The first one is called the city-by-firm model, which constructs a global-urban network by looking at the ownership linkages from firm headquarters to other branches of the firms [41]. The second one is called the interlocking network model, which defines the degrees of linkage between two cities by counting the number of offices of the same firms in both cities that have sizable functions or capabilities [39, 42]. The third one is called the two-mode city-by-firm model. According to this model, cities are linked by hosting branches of the same firm, whereby firms are connected by co-locating in the same city [43].

Up to this point, four nested urban networks and their conforming flows were discussed, which has established a conceptual as well as an applicable foundation on which a global urban cellular automata framework can be constructed (Fig. 1). The hierarchical networks-based global urban cellular automata consist of four interacting and overlapping urban networks. Over micro-urban networks, classic CA modeling traditions are remained to generate bottom-level dynamics

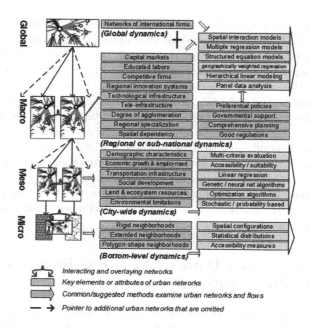

Fig. 1. Hierarchical networks based global urban cellular automata framework (Color figure online)

that lead to a complex global urban system. CA rules could be followed strictly or loosely to some extent. In other words, CA over micro-urban networks is equivalent to an implementation of urban CA at the neighborhood scale or over the bottom-level area, focusing on state and state transition of individual cells over the smallest geographic extent.

For the meso-urban networks, interactions and feedbacks between various parts (i.e., micro-urban networks) of a meso-urban network and overall distribution or equilibrium over the meso-urban network are the primary focuses. A meso-urban CA model is affected by a large group of elements or factors (Fig. 1), including, population change, employment opportunity, economic health, transportation infrastructure, social well-being, available land, natural resources, and environmental constraints. Since many attributes are involved to determine cell transitions, extended (relaxed) CA models are usually employed to simulate or predict urban changes at the meso-scale. Transition rules in these CA models are often modeled with one or several of the following six common groups of techniques (Fig. 1), such as genetic algorithms, land accessibility assessment or suitability analysis, linear regression, multi-criteria evaluation, machine-learning (such as neural network) algorithms, optimization algorithms, and stochastic based processes.

Complex interactions and flows are constantly evolving over macro- and global-urban networks. Beyond a consideration of the elements and flows happening at meso-urban networks, a large set of additional factors have to be examined in order to evaluate flow dynamics between the meso-urban networks nested within a global- or macro-urban network in study (Fig. 1). As explained above, these factors shall be analyzed for the purpose of quantifying location advantages for the cities or meso-urban networks in the global or macro network. It is a challenging undertaking to search for adequate analytical methods or modeling approaches to assess the location advantages. First, the development of urban cellular automata at the global, national or large-regional scale is at an infancy stage. Second, quantitative analysis of location advantages between a large set of nested networks is a new frontier to many other disciplines. Few successful experiences or lessons can be referred as references. Third, a massive (big) dataset is involved to compare location advantages over global- or macro-urban networks. Many data consistency issues arise and computational complexity is astonishingly increased.

3   Challenges of Big Data to Networks-Based Global Urban Cellular Automata

The design, development, and implementation of hierarchical networks-based global urban cellular automata inevitably involve big data sets. We are facing many challenges to preprocess big data in order to feed them into networks-based global CA models at multi-scales. Among them, modifiable areal unit problem (MAUP), (enumeration) areal unit inconsistency problem (AUIP) and missing data problem (MDP) are most critical.

MAUP has long been recognized as an important challenge to geographical information science and spatial analysis and modeling [44,45]. Research findings from a wide array of disciplines confirm the existence of MAUP and its impacts on outcomes of spatial models [46–50]. However, no common solutions to MAUP have been found although a variety of techniques has been proposed. Most of these solutions were developed for special cases and not transferable to other application domains or study areas. These methods in general fall into four categories (Fig. 2): (1) exploratory data mining [46,51] to find critical data items that determine right scale(s) of spatial units; (2) exploratory spatial modeling [52,53]; (3) cell-based spatial reconfiguration to form new analytical areal units [50,54]; and (4) mathematical algorithms to construct new analytical areal units [28,55,56].

The technique assimilating data between different areal units or solving AUIP is called areal interpolation (AI). AI is a geographical data preprocessing, which transfers available attribute data from one spatial unit system (source layer in the terminology of GIS) to a different spatial unit system (target layer) for which attribute data are not existing but are required to feed spatial analysis or modeling with the aid of ancillary information (transfer layer) [57–59]. The critical process of AI is how to extract ancillary information, design a transfer procedure, and assess the transfer accuracy. In general, there are three groups of modern AI algorithms although each group includes many variations.

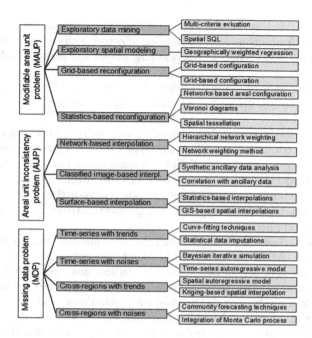

Fig. 2. Challenges of big data to global urban cellular automata and suggested solutions (Color figure online)

The first group is called the network-based interpolation algorithms originated by Xie [57]. This approach is to distribute an attribute (data item) to road segments lying within the source zone boundaries and then intersect these linear features with a target zone. An estimated attribute value for the target zone is derived by summing the attribute values contained by all road segments within the target zones boundary. The second group is called image classification based redistribution algorithm, which employs land covers-uses derived from classified satellite imagery as the ancillary data input [60–62]. The third group is called the surface-based interpolation algorithm [63]. This method uses a set of ancillary points that are believed to provide a reasonable proxy for centers of population density over the source zone. Then a surface is constructed with these points as inputs by applying a GIS spatial interpolation method, and the transfer to the target zone can be achieved by computing volumes with target zone boundaries as barriers.

Global networks-based urban CA models involve a large set of big data that reflects both spatial interactions and temporal changes. Unfortunately, our under-standing of urban systems complexity is an incremental learning process and our traditional research methodology is confined within the scopes of narrow disciplinary boundaries. As a result, gaps inevitably exist in time-series and trans-regional data collected so far. However, complete or comparable datasets in both temporal and spatial dimensions are critical in calibrating and validating macro-urban or global-urban CA models. Thus, it is an important research contest that urban CA modelers have to develop effective techniques to fill or impute missing data that are needed for developing global networks-based urban CA models.

Missing data situations and possible techniques for filling the gaps in the context of constructing urban CA models could be categorized into four types (Fig. 2).

Time-series dataset containing trends: A well written and well documented text by Little and Rubin [64] provides extensive methods, examples, and exercises dealing with substitutions of missing data. Among them, the least square and the maximum likelihood based imputation methods are effective to estimate missing data with linear trends. For non-linear trends, curve-fitting based techniques, such as exponential, modified-exponential, geometric, Gompertz, and parabolic functions can be applied [65].

Time-series dataset containing significant random fluctuations: Auto-regressive model (ARM), which specifies that the outcome of the dependent variable relies linearly on its own previous values, has been often adopted to characterize temporal changes. A Monte Carlo process could be integrated in ARM to provide a mechanism for dealing with situations where dependent samples in time-series contain noises [66]. In other words, Monte Carlo ARM could be used to handle this type of missing data cases. Furthermore, Bayesian iterative simulation based data imputations have been applied as effective techniques of dealing with random fluctuations [67–71].

Cross-regional dataset containing spatial autocorrelation and cross-sectional dependence: Most spatial interpolation methods assume that the spatial variation in the data being modeled is homogeneous across a study area. Kriging, a unique spatial interpolation technique, is based on regionalized variable theory, which could use various semivariograms to characterize different regional trends [72,73]. Moreover, autoregressive models, besides being able to examine time-series dependency of the dependent variable, can also be structured to capture possible spatial interactions across spatial units [74,75].

Cross-regional dataset containing significant local noises: The integration of Monte Carlo process in spatial autoregressive model can be adopted to fill gaps in cross-regional datasets with significant local noises [76]. Bayesian iterative simulation techniques can also be applied in this missing data case [77]. In addition, a worth noting point is that the selection or development of cross-regional imputation methods relies on the research question, i.e., the domain knowledge of a specific application. For example, the economic forecasting and regional planning methods that are built on the top of the economic base theory [78] can be adopted to fill the missing socioeconomic data cross regions. In specifics, the location quotient technique determines the level of basic sector employment by comparing the local economy to the economy of a larger geographic unit like a State or a province, or a country [79]. The assumption of constant share (assuming that each sub-region continues remaining a constant share of the reference region into the future) could be modified by the shift-share analysis.

The shift-share projection technique modifies the constant-share projection formula by adding a shift term to account for differences between local and reference region growth rates that cause an industry's employment to shift into or out of a region [80].

4 Discussion and Conclusion

The new science of cities provides a comprehensive review of urban modeling efforts from the perspective of complexity theories, presents new paradigms of exploring urban growth as systems of networks and flows, and calls for new approaches to design and simulate complex urban systems. This paper expands Battys vision of urban networks and flows in the context of urban cellular automata modeling. According to the science of design and planning [18] (Batty 2013), this paper describes urban networks at the micro-, meso-, macro- and global-scales, which also corresponds to Neals networks of modern metropolis [15] (Neal 2013). Furthermore, the paper examines the dynamic flows (i.e. interactions) over urban networks at the four scales by using terminologies that are often seen in the urban CA modeling arena. In specifics, the paper assesses that the micro-urban networks resemble extended CA neighborhoods; the meso-urban networks are consistent with typical one-city based urban CA models in current literature; the macro-urban networks represent recent efforts in developing regional or large-scale urban CA models although they are at early stages; and the global-urban networks are truly what Dr. Batty sees as complex urban systems, which sets up benchmarks for new generations of urban CA models.

The paper suggests that the four scales of urban networks should be treated as nested complex urban systems, which are intersecting, interacting and overlapping with each other. For this purpose, the paper proposes a hierarchical networks-based global urban cellular automata framework in order to support consistent and transferable urban CA modeling efforts. Furthermore, the paper reveals that the network flows at different scales carry different information, support distinct interactions and generate diverse dynamics and feedbacks. From the implementation point of view, the paper attempts to capture interactions over different urban networks by enlisting key data elements or attributes associated with urban networks at different scales on the basis of literature review and the modeling efforts under the guidance from Batty (Fig. 1).

The paper also recommends analytical and technical approaches to examine complex flows and interactions that are determined by these attributes at four different scales although their overlapping effects are inevitable. The key conceptual and technical foundation for investigating the complex and dynamic flows in the hierarchical networks-based global urban cellular automata framework is how location advantages among urban centers (nodes) can be assessed through the flows over the urban networks. The paper recommends evaluating the location advantages from different perspectives, including business economics, economic geography and political economy. The paper also provides a discussion of analytical methods that are recently developed and also pertinent to urban CA modeling.

Furthermore, the paper elaborates possible challenges that urban modelers expect to face when integrating the concept of cities as systems of networks and flows into urban CA modeling and simulation in the context of big data and its computational complexity. The good side of the adoption of this new concept is that many classic criticisms on urban CA modeling are no longer relevant under the new paradigm of urban cellular automata based on the systems of networks and flows. The contesting part is that the new science of cities places urban modeling into the main stream of geographical information science (GISc). As a result many challenges that have been haunting GISc are applicable to the complexity-theory-based urban modeling, including urban cellular automata. More crucially, these challenges are further confounded by the increasingly growing availability of big data and subsequent computational complexity. The paper tackles three central big data issues from the viewpoint of GISc, MAUP, AUIP and MDP and recommends a good set of possible solutions on the basis of current research findings.

However, the paper suffers several visible limitations. Firstly, in the proposed hierarchical networks based global urban cellular automata framework, a set of key elements (attributes) that quantify flows or interactions over urban networks were advised. It is worth pointing out that this selection was just a small collection of all possible factors that need be considered. The selection of relevant data items should be guided by theoretical and disciplinary considerations. The current set was by no means exhaustive. Secondly, this limitation applies to the recommendation of analytical methods for quantifying the flows, interactions or competitions as well. Thirdly, the survey of currently available methods and techniques for resolving the challenges of MAUP, AUIP and MDP was constrained within GISc and the authors own experience.

Finally the goal of this paper is to propose a new paradigm of urban CA modeling inspired by Battys New Science of Cities. No pilot illustrations or experiments have been completed to prove the concept although preparations of a preliminary demonstration are under way. The paper aims to invite colleagues to reassess urban cellular automata in the era of bid data science and analytics that were well discussed by Batty in the New Science of Cities.

References

1. Santé, I., García, A.M., Miranda, D., Crecente, R.: Cellular automata models for the simulation of real-world urban processes: a review and analysis. Landscape Urban Plan. **96**(2), 108–122 (2010)
2. Wolfram, S.: Cellular automata as models of complexity. Nature **311**(5985), 419–424 (1984)
3. Wolfram, S.: A New Kind of Science, vol. 5. Wolfram Media, Champaign (2002)
4. Holland, J.H.: Adaptation in Natural and Artificial Systems: An Introductory Analysis with Applications to Biology, Control, and Artificial Intelligence. University of Michigan Press, Ann Arbor (1975)
5. Couclelis, H.: Cellular worlds: a framework for modeling micromacro dynamics. Environ. Plan. A **17**(5), 585–596 (1985)

6. Couclelis, H.: From cellular automata to urban models: new principles for model development and implementation. Environ. Plan. B: Plan. Des. **24**(2), 165–174 (1997)
7. Batty, M., Longley, P.A.: Fractal Cities: A Geometry of Form and Function. Academic Press, London, San Diego (1994)
8. Allen, P.: Cities and Regions as Self-organizing Systems: Models of Complexity. Taylor and Francis, London (1997)
9. Portugali, J.: Complexity, Cognition and the City. Springer, Heidelberg (2011)
10. Weidlich, W.: Sociodynamics: A Systematic Approach to Mathematical Modelling in the Social Sciences. Harwood Academic, Reading (2000)
11. Batty, M., Xie, Y.: From cells to cities. Environ. Plan. B: Plan. Des. **21**(7), S31–S48 (1994)
12. Xie, Y.: A generalized model for cellular urban dynamics. Geogr. Anal. **28**(4), 350–373 (1996)
13. Batty, M., Xie, Y., Sun, Z.: Modeling urban dynamics through GIS-based cellular automata. Comput. Environ. Urban Syst. **23**(3), 205–233 (1999)
14. Newman, M., Barabasi, A.L., Watts, D.J.: The Structure and Dynamics of Networks. Princeton University Press, Princeton (2006)
15. Neal, Z.P.: The Connected City: How Networks are Shaping the Modern Metropolis. Routledge, New York (2012)
16. Pijanowski, B.C., Tayyebi, A., Doucette, J., Pekin, B.K., Braun, D., Plourde, J.: A big data urban growth simulation at a national scale: configuring the GIS and neural network based land transformation model to run in a high performance computing (HPC) environment. Environ. Model. Softw. **51**, 250–268 (2014)
17. Tao, S., Corcoran, J., Mateo-Babiano, I., Rohde, D.: Exploring bus rapid transit passenger travel behaviour using big data. Appl. Geogr. **53**, 90–104 (2014)
18. Batty, M.: The New Science of Cities. The MIT Press, Cambridge (2013)
19. Szell, M.: Connecting paradigms. Science **343**, 970–971 (2014)
20. Diestel, R.: Graph Theory, vol. 173. Springer-Verlag, Heidelberg, New York (2005)
21. Christaller, W., Baskin, C.W.: Central Places in Southern Germany. Prentice-Hall, Englewood Cliffs (1966). Translated by Carlisle W. Baskin
22. Lin, J.: Network analysis of Chinas aviation system, statistical and spatial structure. J. Transp. Geogr. **22**, 109–117 (2012)
23. Chen, S., Claramunt, C., Ray, C.: A spatio-temporal modelling approach for the study of the connectivity and accessibility of the Guangzhou metropolitan network. J. Transp. Geogr. **36**, 12–23 (2014)
24. Wang, J., Mo, H., Wang, F.: Evolution of air transport network of China 1930–2012. J. Transp. Geogr. **40**, 145–158 (2014)
25. Hansen, W.G.: How accessibility shapes land use. J. Am. Inst. Planners **25**(2), 73–76 (1959)
26. Reggiani, A., Bucci, P., Russo, G., Haas, A., Nijkamp, P.: Regional labour markets and job accessibility in city network systems in Germany. J. Transp. Geogr. **19**(4), 528–536 (2011)
27. Castells, M.: Rise of the Network Society. Blackwell, Maiden (1996)
28. Xie, Y., Ma, T.: A method for delineating a hierarchically networked structure of urban landscape. Urban Geogr. **36**(6), 947–963 (2015)
29. Gardner, M.: Mathematical games: the fantastic combinations of John Conways new solitaire game life. Sci. Am. **223**(4), 120–123 (1970)
30. Xie, Y., Batty, M., Zhao, K.: Simulating emergent urban form using agent-based modeling: Desakota in the Suzhou-Wuxian region in China. Ann. Assoc. Am. Geogr. **97**(3), 477–495 (2007)

31. Pumain, D.: Multi-agent system modelling for urban systems: the series of SIM-POP models. In: Heppenstall, A.J., Crooks, A.T., See, L.M., Batty, M. (eds.) Agent-Based Models of Geographical Systems, pp. 721–738. Springer, Rotterdam (2012)

32. Xie, Y., Fan, S.: Multi-city sustainable regional urban growth simulationmsrugs: a case study along the mid-section of silk road of China. Stoch. Env. Res. Risk Assess. **28**(4), 829–841 (2014)

33. Chen, Y.: Urban gravity model based on cross-correlation function and fourier analyses of spatio-temporal process. Chaos, Solitons Fractals **41**(2), 603–614 (2009)

34. Porter, M.E.: The competitive advantage of notions. Harvard Bus. Rev. **68**(2), 73–93 (1990)

35. Cuervo-Cazurra, Á., de Holan, P.M., Sanz, L.: Location advantage: emergent and guided co-evolutions. J. Bus. Res. **67**(4), 508–515 (2014)

36. Batty, M.: Cities and Complexity: Understanding Cities through Cellular Automata, Agent-based Models and Fractals. The MIT Press, Cambridge, Massachusetts (2005)

37. Healey, P.: Urban Complexity and Spatial Strategies: Towards a Relational Planning for Our Times. Routledge, London (2006)

38. Holton, R.J.: Global Networks. Palgrave Macmillan, New York (2007)

39. Taylor, P.J.: Specification of the world city network. Geogr. Anal. **33**(2), 181–194 (2001)

40. Tonts, M., Taylor, M.: Corporate location, concentration and performance: large company headquarters in the Australian urban system. Urban Studies **47**(12), 2641–2664 (2010)

41. Rozenblat, C., Pumain, D.: Firm linkages, Innovation and the Evolution of Urban Systems. Routledge, London (2007)

42. Neal, Z.: Structural determinism in the interlocking world city network. Geogr. Anal. **44**(2), 162–170 (2012)

43. Liu, X., Derudder, B.: Analyzing urban networks through the lens of corporate networks: a critical review. Cities **31**, 430–437 (2013)

44. Fotheringham, A.S., Wong, D.W.: The modifiable areal unit problem in multivariate statistical analysis. Environ. Plan. A **23**(7), 1025–1044 (1991)

45. Jelinski, D.E., Wu, J.: The modifiable areal unit problem and implications for landscape ecology. Landscape Ecol. **11**(3), 129–140 (1996)

46. Lembo, A.J., Lew, M.Y., Laba, M., Baveye, P.: Use of spatial SQL to assess the practical significance of the modifiable areal unit problem. Comput. Geosci. **32**(2), 270–274 (2006)

47. Goodchild, M.F.: Scale in GIS: an overview. Geomorphology **130**(1), 5–9 (2011)

48. Lechner, A.M., Langford, W.T., Jones, S.D., Bekessy, S.A., Gordon, A.: Investigating species-environment relationships at multiple scales: differentiating between intrinsic scale and the modifiable areal unit problem. Ecol. Complex. **11**, 91–102 (2012)

49. Mitra, R., Buliung, R.N.: Built environment correlates of active school transportation: neighborhood and the modifiable areal unit problem. J. Transp. Geogr. **20**(1), 51–61 (2012)

50. Houston, D.: Implications of the modifiable areal unit problem for assessing built environment correlates of moderate and vigorous physical activity. Appl. Geogr. **50**, 40–47 (2014)

51. Arsenault, J., Michel, P., Berke, O., Ravel, A., Gosselin, P.: How to choose geographical units in ecological studies: proposal and application to campylobacteriosis. Spat. Spatio-Temporal Epidemiol. **7**, 11–24 (2013)

52. Brunsdon, C., Fotheringham, A.S., Charlton, M.E.: Geographically weighted regression: a method for exploring spatial nonstationarity. Geogr. Anal. **28**(4), 281–298 (1996)
53. Fortheringham, A.S., Brunsdon, C., Charlton, M.: Geographically Weighted Regression: The Analysis of Spatially Varying Relationships. Wiley, Chichester (2002)
54. Götzinger, J., Bárdossy, A.: Comparison of four regionalisation methods for a distributed hydrological model. J. Hydrol. **333**(2), 374–384 (2007)
55. Sabel, C., Kihal, W., Bard, D., Weber, C.: Creation of synthetic homogeneous neighbourhoods using zone design algorithms to explore relationships between asthma and deprivation in Strasbourg, France. Soc. Sci. Med. **91**, 110–121 (2013)
56. Wang, J., Kwan, M.P., Ma, L.: Delimiting service area using adaptive crystal-growth Voronoi diagrams based on weighted planes: a case study in Haizhu district of Guangzhou in China. Appl. Geogr. **50**, 108–119 (2014)
57. Xie, Y.: The overlaid network algorithms for areal interpolation problem. Comput. Environ. Urban Syst. **19**(4), 287–306 (1995)
58. Sridharan, H., Qiu, F.: A spatially disaggregated areal interpolation model using light detection and ranging-derived building volumes. Geogr. Anal. **45**(3), 238–258 (2013)
59. Xie, Y., Crary, D., Bai, Y., Cui, X., Zhang, A.: Modeling grassland ecosystem responses to coupled climate and socioeconomic influences from multi-spatial-and-temporal scales. J. Environ. Inform. (2016, in press)
60. Yuan, Y., Smith, R.M., Limp, W.F.: Remodeling census population with spatial information from LandSat TM imagery. Comput. Environ. Urban Syst. **21**(3), 245–258 (1997)
61. Harris, R.J., Longley, P.A.: New data and approaches for urban analysis: modelling residential densities. Trans. GIS **4**(3), 217–234 (2000)
62. Langford, M.: An evaluation of small area population estimation techniques using open access ancillary data. Geogr. Anal. **45**(3), 324–344 (2013)
63. Zhang, C., Qiu, F.: A point-based intelligent approach to areal interpolation. Prof. Geogr. **63**(2), 262–276 (2011)
64. Little, R.J., Rubin, D.B.: Statistical Analysis with Missing Data. Wiley, Hoboken (2002)
65. Klosterman, R.E.: Community Analysis and Planning Techniques. Rowman & Littlefield Publishers, Savage (1990)
66. Hamilton, J.D.: Time Series Analysis, vol. 2. Princeton University Press, Princeton (1994)
67. Jackman, S.: Estimation and inference via Bayesian simulation: an introduction to Markov Chain Monte Carlo. Am. J. Polit. Sci. **44**(2), 375–404 (2000)
68. Zellner, A., Chen, B.: Bayesian modeling of economies and data requirements. Macroecon. Dyn. **5**(05), 673–700 (2001)
69. Geweke, J.: Contemporary Bayesian Econometrics and Statistics, vol. 537. Wiley, New York (2005)
70. Van Ginkel, J.R., Van der Ark, L.A., Sijtsma, K., Vermunt, J.K.: Two-way imputation: a Bayesian method for estimating missing scores in tests and questionnaires, and an accurate approximation. Comput. Stat. Data Anal. **51**(8), 4013–4027 (2007)
71. Zhang, Z., Dong, F.: Fault detection and diagnosis for missing data systems with a three time-slice dynamic Bayesian network approach. Chemom. Intell. Lab. Syst. **138**, 30–40 (2014)

72. Goovaerts, P.: AUTO-IK: a 2d indicator kriging program for the automated non-parametric modeling of local uncertainty in earth sciences. Comput. Geosci. **35**(6), 1255–1270 (2009)

73. Rivest, M., Marcotte, D., Pasquier, P.: Sparse data integration for the interpolation of concentration measurements using kriging in natural coordinates. J. Hydrol. **416**, 72–82 (2012)

74. Yokoi, T., Ando, A.: One-directional adjacency matrices in spatial autoregressive model: a land price example and Monte Carlo results. Econ. Model. **29**(1), 79–85 (2012)

75. Kyung, M., Ghosh, S.K.: Maximum likelihood estimation for generalized conditionally autoregressive models of spatial data. J. Korean Stat. Soc. **43**(3), 339–353 (2014)

76. Doğan, O., Taşpınar, S.: Spatial autoregressive models with unknown heteroskedasticity: a comparison of Bayesian and robust GMM approach. Reg. Sci. Urban Econ. **45**, 1–21 (2014)

77. Pan, C., Cai, B., Wang, L., Lin, X.: Bayesian semiparametric model for spatially correlated interval-censored survival data. Comput. Stat. Data Anal. **74**, 198–208 (2014)

78. Flegg, A.T., Webber, C., Elliott, M.: On the appropriate use of location quotients in generating regional input-output tables. Reg. Stud. **29**(6), 547–561 (1995)

79. Klosterman, R., Xie, Y.: ECONBASE: Economic Base Analysis, in R. Center for Urban Policy Research, Rutgers University, New Brunswick (1993)

80. Klosterman, R., Xie, Y.: SHFT-SHR: Local Employment Projection, in R. Center for Urban Policy Research, Rutgers University, New Brunswick (1993)

GIS Applications to Support Entry-Exit Inspection and Quarantine Activities

Nicola Ferrè[1(✉)], Qiu Songyin[2], Matteo Mazzucato[1], Andrea Ponzoni[1],
Paolo Mulatti[1], Matteo Morini[1], Ji Fan[3], Liu Xiaofei[2], Dou Shulong[1,2,3],
Lin Xiangmei[2], and Stefano Marangon[1]

[1] Istituto Zooprofilattico Sperimentale delle Venezie, Legnaro (PD), Italy
nferre@izsvenezie.it
[2] Chinese Academy of Inspection and Quarantine, Beijing, China
qiusy@caiq.gov.cn
[3] Shenzhen Entry-Exit Inspection and Quarantine Bureau, Shenzhen, China

Abstract. This paper presents the outcome of a project aimed at developing a system for the management of the information flow regarding the approval and supervision of pre and post-arrival animal quarantine station. In particular, the paper outlines the project strategies and issues identified during the implementation process of a GIS based application in an organisation characterised by a low level of GIS maturity. The tools used for the Business Case evaluation, the assessment of the organisational and technological capacity, the architecture of the webGIS application used for the pilot study and the protocol developed for the quarantine station performance evaluation are presented.

Keywords: Quarantine station · Veterinary · Visual exploratory spatial analysis

1 Introduction

The importation of animals and animal products involves a degree of risk to the importing country represented by the introduction of animals affected by diseases already present or not in the country. However, the threat of disease introduction has been used in some instances to enforce stringent national measures (e.g. import bans) designed primarily to impair trade and protect local livestock industries [1]. In order to achieve a balance between free trade and the protection of human and animal health, the World Organisation for Animal Health (OIE) has defined a set of standards for international trade in animals and animal products. The OIE is an organisation recognised by the World Trade Organization that operates within the framework of the Agreement on the Application of Sanitary and Phytosanitary (SPS).

Measures [2] for commodities derived from livestock [3]. One of the key aims of the OIE standards is to reduce the information gap between importers and exporters by seeking a formal and consistent approach to the assessment of disease risks associated with trade. Therefore, the OIE has developed a set of guidelines to assess risks associated

© Springer International Publishing Switzerland 2016
O. Gervasi et al. (Eds.): ICCSA 2016, Part III, LNCS 9788, pp. 85–97, 2016.
DOI: 10.1007/978-3-319-42111-7_8

with trade in animals and their products, known as import risk analysis[1] (IRA). IRA seeks to reduce a complex underlying process for instance, the introduction of a pathogen to a susceptible population leading to a significant impact on the population, to an understandable and accurate abstract representation [4]. One of the key component of IRA is the release assessment[2]. The release assessment is represented by the description of the biological pathway(s) necessary for an imported commodity to 'release' (that is, introduce) pathogenic agents into a particular environment. Among the factors[3] that should be taken into account in the release assessment, there is the evaluation of the quarantine station performance. A quarantine station is an establishment under the control of the veterinary authority where animals are maintained in isolation with no direct or indirect contact with other animals. The role of a quarantine station is to ensure that there is no transmission of specified pathogen(s) outside the establishment while the animals are undergoing observation [5]. In China, the supervision and administration of the quarantine stations is in charge of the General Administration of Quality Supervision, Inspection and Quarantine (AQSIQ). AQSIQ is responsible for the control of animals and their products which enter or leave China. This includes checks on imported animals, quarantine and certification of animals to be exported. AQSIQ is a central governmental body, with provincial (CIQ) and local branches. Two types of quarantine station are present in China: the quarantine station set up by the AQSIQ and the quarantine station designated by CIQ. According to the AQSIQ decree 122/2009[4], the location of the quarantine station shall comply with standards and requirements issued by AQSIQ. These standards and requirements have the scope to ensure that the station functional boundary provides an adequate separation of housed animals from the surrounding environment. With a view to improve the framework for the evaluation of the quarantine station performance and the related administrative framework, the Chinese Academy of Inspection and Quarantine (CAIQ[5]) has recently launched a project to develop a framework for the management of spatial and no-spatial data of quarantine station. The project called "MoEWebGIS" (MoE) will be developed under the activities of the OIE cooperation project entitled "Capacity development for implementing a Geographic Information System (GIS) applied to surveillance, control and zoning of avian influenza and other emerging avian diseases in China" (http://gis.izsvenezie.it/cooperation/oie/izsve-caiq/index.php). The MoE project has been envisaged during the

---

[1] Risk analysis is an approach to assess both the likelihood and consequences of undesirable events, known as hazards, and used to support decision-making in the face of uncertainty.

[2] The risk assessment process consists of four interrelated steps: (i) Release assessment, (ii) Exposure assessment, (iii) Consequence assessment, and (iv) Risk estimation.

[3] The factors considered in the release assessment analysis are: biological (i.e. species, quarantine), country (i.e. incidence, prevalence), and commodities (i.e. quantity of commodities, effect of processing) [5].

[4] Decree of General Administration of Quality Supervision, Inspection and Quarantine of the People's Republic of China No. 122 "Administrative Measures for the Supervision on the Use of the Quarantine Station of Entry animals" of October 22, 2009.

[5] CAIQ is a national public institute that provides technical support to the policy-making related to inspection and quarantine for China's central government, and provides technical assistance to the law enforcement duties of the AQSIQ.

workshop entitled "GIS applications to support entry-exit inspection and quarantine activities" held in GuangZhou (CHN) on the 14–15th May 2015 and organised by CAIQ. The present paper aims at providing the preliminary results of the MoE project.

2   Methods

For the development of the MoE project the classical waterfall schema has been used [6, 7]. In this paper, the results obtained in the Planning, Analysis, and (partially) Design steps are presented.

For the planning step a Business Case reduced schema derived from the PRINCE2[6] method has been implemented. The schema used consisted of the following elements:

- Executive Summary: highlights the objectives and the key points of the project including also the main benefits expected from the system.
- Reasons: defines the reasons for undertaking the project and explains how the project will enable the achievement of objectives.
- Business Options: analysis and reasoned recommendation for the possible options of do nothing, do the minimal or do a complete system.
- Expected Benefits: the benefits that the project will deliver with respect to the situation as it exists prior to the project.
- Expected Dis-benefits: outcomes perceived as negative by one or more stakeholders.
- Timescale: the period over which the project will run.
- Costs: A summary of the project costs (including maintenance costs) and their funding arrangements.
- Major Risks: a summary of the key risks associated with the project.

For the Analysis step the Use case analysis [8] combined with the As-Is analysis, the Requirement analysis, and the Gap analysis have been adopted.

For the assessment of the CAIQ GIS organisation and technological capacity included in the As-Is analysis, a simplified version of COBIT[7] domain PO (Plan and Organize) integrated with the typical GIS elements, has been used. In particular, the following elements have been evaluated:

- PO2 - Define the Information Architecture (with particular reference to the metadata and data dictionary).
- PO3 - Technological Direction (with particular reference to the assessments of the current vs. planned information systems).
- PO4 - Define the Information Technology Processes, Organisation and Relationships (with particular reference to skills and experience in GIS technologies).
- PO7 - Manage IT Human Resources (with particular reference to personnel expectation and training programme identification).

---

[6] PRINCE2: acronym for PRojects IN Controlled Environments - https://www.prince2.com.

[7] Control Practices: Guidance to Achieve Control Objectives for Successful IT Governance (2nd Edition).

To illustrate the outcome diagrams the software Enterprise architect (Enterprise Architect Ultimate Edition version 10.0 Sparx System) has been used.

For the Design step a pilot project has been designed. The pilot has been structured in the following three elements:

(a) Protocol for data collection.
(b) WebGIS for data editing and presentation.
(c) Protocol for data analysis.

For the definition of the protocol for data collection, a series of interviews have been conducted during the capacity building activities of the already mentioned OIE cooperation project[8].

For the webGIS, a webGIS platform already developed by the Italian team in a past project, has been adapted and deployed for this specific pilot. The architecture of the platform is illustrated in Fig. 1.

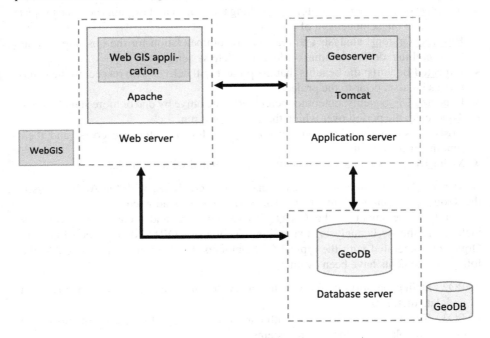

Fig. 1. Web GIS architecture

For the data analysis protocol, the sequence of steps is presented in Fig. 2. The protocol is based on the Simple Additive Weighting (SAW) process [9]. The SAW method is based on the concept of a weighted average in which continuous criteria are

---

[8] Respectively Phase 1: Preparatory activities - Activity 1.2 Baseline assessment of CAIQ GIS organisation and capacity Phase 2: Capacity building - Activity 2.1. Training of two CAIQ scientists on GIS planning at the IZSVe Parent Institute.

standardized to a common numeric range, and then combined by means of a weighted average in order to obtain a score. In our context:

- The criteria are represented by a set of features called Point of Interest (POI). The POIs are premises that directly or indirectly can be exposed to a disease problem derived from the presence of animals in the quarantine station. The list of POIs includes: (i) flocks or herds, (ii) animal hospitals, (iii) animal markets, (iv) slaughterhouses, (v) veterinary research institutes, (vi) rendering plants, (vii) fur and leather plants, (viii) artificial insemination stations, (ix) embryo transfer stations, (x) biosafety disposal stations, (xi) human hospitals, (xii) border posts, (xiii) animal storage depots, and (xiv) airports and ports with an animal transfer area.
- The score, called "Total Impact", is obtained by multiplying the importance weight assigned to each POI class with the classified value of the observed POI and then summing the products over the POIs that fall inside the study area (area of 3 km radius surrounding the candidate quarantine station).

$$Total\ Impact = \sum w_i x_i \times \prod c_j$$

Where:
$Total\ Impact$ = is the impact associated to the POI distribution in the area that surround the candidate quarantine station.
w = is the weight for the i type of POI
x = is the intensity of the observed i POI. The intensity is a quantifiable measure and represents the dimension of the POI. For example, the intensity for the "farm" class of POI is its size (number of animals), for a hospital the number of beds, etc. The values are than re-classified according to a defined range values (i.e.: from 1 to 7)
c = Boolean constraint. Take into account the "critical POIs". The critical POIs are the class of POI that shall not be near a quarantine station. The presence of a critical POI in the area that surrounds the candidate quarantine station leads directly to reject the request to become a quarantine station.

- The weights of relative importance for each class of POIs, at the moment of the paper submission have not been calculated jet. The weights will be calculatedby means of a pair-wise comparison method performed by a panel of CAIQ experts during an expert opinion exercise.

The Total Impact will be compared with a defined threshold value defined by means of a exploratory spatial data analysis. If the impact is above the threshold, the request to become a new quarantine station will be rejected. The threshold value will be defined by means of a scenario analysis. A series of possible POIs distribution (scenario) will be presented to a panel of CAIQ experts during an expert opinion exercise. For each scenario the experts will formulate a judgement that could be (i) approved, (ii) rejected, (iii) more analysis are required. The expert judgements will be then combined to derive the threshold value.

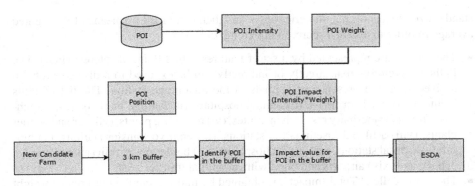

Fig. 2. Protocol for the data analysis

3 Results

The MoE project has been initiated without a formal request for proposals. Therefore, the first step taken by the Chinese and Italian teams that are involved in the OIE cooperation project has been the preparation of an unsolicited proposal. The proposal, written in form of a Business Case document in both English and Chinese languages, used a

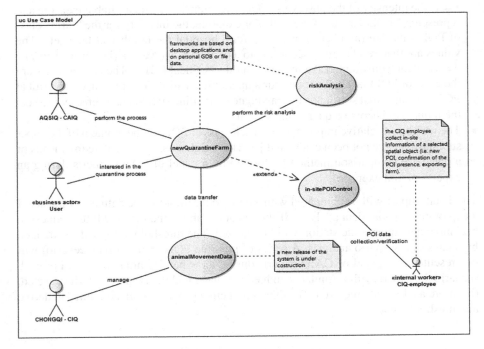

Fig. 3. As-Is analysis - Quarantine Farm approval use case diagram

"simple" language and avoided Information Technology (IT) or GIS specific terminologies. The document has been submitted as an internal document to inform decision-makers of what activities lie ahead and what funding will be necessary. The result of this step has been a commitment from the senior management to proceed with the Analysis and Design phases.

The first activity for the Analysis step has been the assessment of the GIS activities performed by CAIQ for the management of the information flow regarding the approval and supervision of pre and post-arrival quarantine stations. The assessment has been performed by the Italian team composed by a GIS manager, a GIS technician, and a Veterinarian expert in spatial analysis. The team focused its assessment on the most critical resources needed to implement the requested actions, and in particular, on the: (i) GIS office organizational structure, (ii) available spatial data, (iii) GIS applications already in use, (iv) IT infrastructure, and (v) Familiarity on GIS techniques and spatial statistical analysis. The result of the assessment[9] highlighted the necessity to implement the MoE project in order to improve the information flow and quality.

The next step has been the development of the As-Is analysis. From the assessment exercise, the As-Is analysis has been derived. The analysis identified and described (i) layers, (ii) GIS functions, (iii) impact, and (iv) quality of information used by the CAIQ operator for the management of the information flow regarding the approval and supervision of pre and post-arrival quarantine stations. In the As-Is analysis, a principal use case called "Quarantine Farm approval" (Figs. 3 and 4 present, respectively, use case diagram and activity diagram) has been identified. In the use case, the key role of the POI and the on-site action aroused. In particular, the on-site action is of particular importance for the evaluation process because it influences the level of data quality (i.e. completeness, spatial accuracy) and therefore, the reliability of the analysis outcomes (Fig. 5).

The next step has been the development of the First-cut functional analysis. The following classes of functionalities have been identified:

- The new system should be developed on a web-based architecture in order to allow the local CIQs distributed all over China, to access the system functionalities.
- The system should be developed using a service-oriented architecture in order to allow the interoperability with the new coming CAIQ systems.
- The risk analysis component shall be performed in a desktop GIS session and based on a well-defined framework that string together sequences of geoprocessing tools.
- The POI update activity should be performed on a specific survey plan, and based on a well defined framework that shall include a tailored training session for the local CIQ.

Based on the outcome of the As-Is and the first-cut functional analysis, a gap analysis has been developed. The gap analysis has allowed the identification of the function requirements that could not be met by the system previously in place. The following classes of gaps have been identified:

---

[9] The results of the assessment exercise is a proprietary and confidential document, therefore it is not possible to disclosed it in any part.

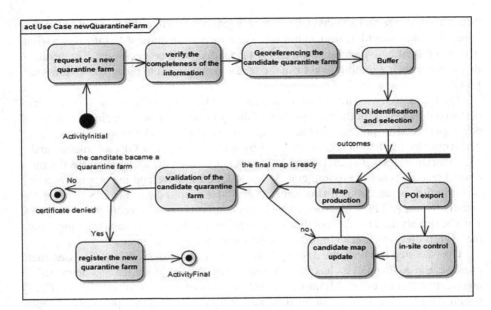

Fig. 4. Activity diagram of the Quarantine Farm approval use case

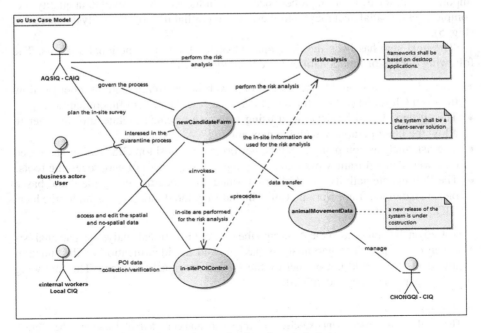

Fig. 5. Quarantine station location analysis use case

1. To shift from a no-structured framework toward a well-know process based on a structured data collection and flow.
   - The database schema shall be documented and tested.
   - The system shall control as much data flow as possible. This means that the data editing shall be done through a web system interface or uploaded through a web service or a well-known file schema.
2. To transpose the primary actor of the process from CAIQ to the local CIQs.
   - The local CIQs shall became the primary actor because he/she becomes responsible for the data collection, validation and use.
   - The CAIQ operator shall supervise the system and perform the risk analysis.
3. To organise the computing system toward a distributed data accessibility web-based system in order to allow the local CIQs to retrieve information useful for their activities.
   - A WebGIS computing technology allows the local authorities to interact with the system in order to retrieve data which are needed for the quarantine and export process.
   - Local authorities will be able to extract data from the designed web-based system in order to perform spatial analysis and reporting.
4. To adopt a well-defined framework for the risk analysis based on the spatial analysis technique in order to improve the results reliability.
5. To classify the POIs.
   - The classification of the POIs in a well-defined structure is essential for the appropriate management of the entire system.
   - A framework for the POI classification implies that the acquired dataset of POIs should be transformed in a schema that assigns a unique identifier to the feature.
   - The POI collection is a process that is activated when a candidate farm is under evaluation. Collected POIs shall be checked and validated by the local CIQs, in order to increase the reliability of data and therefore of the risk assessment. This process shall include the completion of the data set.
6. The feature/attribute shall be transposed in an enterprise environment.
7. The data security shall be introduced in the new framework.

3.1 Pilot Project

To verify the feasibility of the proposed functionalities a pilot test has been designed. The component of the pilot are (i) a WebGIS tool for the collection of the POI spatial/no-spatial information, (ii) a protocol for POI detection and the collection of relate data for the local CIQ operator, and (iii) a protocol for the spatial data analysis. At the moment of the paper submission, the pilot project has been partially developed in some quarantine station surrounding the city of Beijing.

For the pilot study a WebGIS, based on a previous Italian version of the software, has been installed in a server located in Beijing (in Fig. 6 a print screen of the client interface is presented).

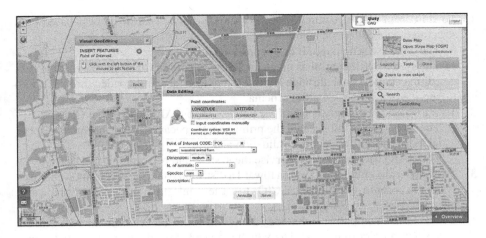

Fig. 6. Print screen of the webGIS tool – client interface

The architectural environment is based on three main components: Web server, Application server and Database server. All components has been installed in the single virtual machine developed in VMware. A suite of Open Source software has been used, in particular:

- The Operating System is Linux 12.04 LTS 64 bit version.
- Apache HTTP Server 2.2 has been installed as Web server and some specific libraries has been enabled, in order to run PHP applications.
- Apache Tomcat 7 has been installed as Application server in order to deploy and run Java applications.
- PostGreSQL version 9.1 has been installed as DBMS (database management system) with the spatial extension POSTGIS 2.0 that add, to the core software, all the functionalities that allow the management of spatial data.

Geo-spatial Server. The java-based application GeoServer 2.4.4 (based on the application server) has been installed in order to allow users to view and edit spatial data. A specific store, directly linked with the geodatabase, has been created and spatial data are published by OGC (Open Geospatial Consortium) standards Web Services (WS) that allow great flexibility in map creation and data sharing. In particular, Web Features Services (WFS) and Web Map Services (WMS) have been used.

WebGIS Application. The development process of the WebGIS application has been followed the same principles of standard web applications. Server side modules allow the database communication and page creation. This part has been developed using PHP 5.3. The client side has been developed using Javascripts (JS) code. Commons functionalities and components are ensured by JS Library (i.e. JQuery) and asynchronous communication with server side is allowed by Asynchronous JavaScript and XML (AJAX) technology. In order to manage spatial data derived by Spatial Services the OpenLayers 2.12 Spatial JS framework has been adopted.

Base Map. In a WebGIS specific services, offered by third-part companies, could be integrated in the application to create the base map layer. Base map layers are necessary to contextualise the position on the map. Currently, the only confirmed base map is OpenStreetMap (www.openstreetmap.org), showed through the portal using the RESTful API (Application programming interface) version 0.6 deployed 17–21 April 2009. Other base maps will be evaluated after the pilot exercise.

Functionalities. The typical GIS functionalities (i.e.: show/hide layers, Scale, Overview, Zoom management, layers interrogation, measurement) together with tailored functionalities (search position on the map, editing point features) are offered in this the application.

The protocol for POI identification and data collection is aimed: (i) to define the sequence of steps that a local CIQ should follow in order to identify POIs that surround the candidate quarantine station, and (ii) to collect the relevant data for each detected POI. Using the functionalities provided by the webGIS, the local CIQ will detect the POI by using both the visual target identification and the surface survey methods. The visual target identification method consist of the identification of any possible POIs in the buffer area by means of a geo-referenced image provided by the webGIS. Once the local CIQ will identify a facility that could represent a POI, he/she will plan the route to reach the identified location. The design of the route shall try to ride as much road that surrounds the candidate quarantine station as possible (unsystematic survey). Along the path, every time he/she will recognise a POI (POI that was not identified from the geo-referenced image) or if he/she will reach one of the targeted POI, he/she will stop the vehicle and will perform the identification of the POI and the related data collection. The data will be recorded in the system by means of the webGIS functionalities.

4   Conclusion

The present paper presents the preliminary results of a project dedicated to the development of a system for the management of spatial and non-spatial information related to the approval of quarantine stations. Once fully operative, the project, that can be considered the first example of a documented GIS application to support import-export veterinary inspection and quarantine activities, will allow:

- Better decision-making - This is associated either with carrying out the data analysis exercise (i.e. better accuracy in quarantine facility identification, risk analysis based on repeatable methodology).
- Improved communication - WebGIS-based maps and visualizations greatly assist in understanding the required analysis on quarantine station location. They can be seen as a new language that improves communication between CAIQ and local CIQ, but most of all, among trade partners.
- Promote the reuse of data - Many organizations that collect and manage data are not able to realize that data can be reused to produce new information. The idea to reuse the "transactional data" (data that support the daily operations of an organization, i.e. describes business events) as "analytical data" (data that support decision-making,

reporting, query, and analysis, i.e. describes business performance) can improve the ERP (Enterprise Resource Planning) capabilities of an organisation.

For CAIQ the proposed GIS development process is a novelty, going beyond the technical dimension of an easier and more efficient management of information. The introduction of a structured GIS-based framework will highlight exploitation areas beyond the narrow context of the existent division of responsibilities between different parts of organizations, as well as cooperation problems, data exchange and utilization. In this sense, it should be expected that an effective GIS introduction in the CAIQ activities would be a breakthrough in terms of functionality and decision-making, will redeploy functional priorities and will lead to a revision of conventional approaches for the character and the efficiency of provided services. To achieve these results, a substantial background and skills, not only in "pure GIS discipline" but also in IT, spatial data analysis and project management are required.

The MoE project, developed within the activities of the OIE cooperation project "Capacity development for implementing a Geographic Information System (GIS) applied to surveillance, control and zoning of avian influenza and other emerging avian diseases in China" is a striking example of how the OIE aims to build the awareness of the scientific communities and improve compliance with its own standards. In fact one of the aims of the OIE it is to strengthen global disease surveillance networks and projects like Moe WebGIS (which promotes capacity building of an Institution) are crucial to foster technological innovation in the management of veterinary surveillance and controls. Moreover, this project highlights the benefits of development cooperation, in particular how the exchange of know - how between two Institutions could provide more opportunities for organizational learning and capacity building development, extending, at the same time, the OIE network of expertise to provide better global geographical coverage for priority diseases and improving the acquaintance of the assessment of disease risks associated with trade.

5   Credits

This project has been developed in the framework the OIE Cooperation Project "Capacity development for implementing a Geographic Information System (GIS) applied to surveillance, control and zoning of avian influenza and other emerging avian diseases in China". Part of research activities were also co-funded by the Italian Ministry of Health.

References

1. Roberts, D., Orden, D., Josling, T.: WTO disciplines on technical barriers to agricultural trade: progress, prospects and implications for developing countries. In: Conference on Agricultural and the New Trade Agenda in the WTO negotiations 2000 WTO, Geneva (1999)
2. WTO: Agreement on the Application of Sanitary and Phytosanitary Measures. World Trade Organisation, Geneva (1995)

3. Ratananakorn, L., Wilson, D.: Zoning and compartmentalisation as risk mitigation measures: an example from poultry production. Revue Scientifique et Technique-Office International des Epizooties, 30, pp. 297–307 (2011)
4. de Vos, C.J., Paisley, L.G., Conraths, F.J., Adkin, A., Hallgren, G.S.: Comparison of veterinary import risk analysis studies. Int. J. Risk Assess. Manag. **15**, 330–348 (2011)
5. OIE: Chapter 2.1. Risk Analysis. Terrestrial Animal Health Code, 12 edn. World Animal Health Organisation, Paris (2012)
6. Huxhold, W.E., Levinsohn, A.G.: Managing Geographic Information System Projects. Oxford University, New York (1995)
7. Harmon, J.E., Anderson, S.J.: The Design and Implementation of Geographic Information Systems. Wiley, New York (2003)
8. Cockburn, A.: Writing Effective Use Cases. Addison Wesley, Boston (1999)
9. Hwang, C.L., Yoon, K.: Multiple Attribute Decision Making-Methods and Applications, A State of the Art Survey. Springer, Heidelberg (1981)

Civil Security in Urban Spaces: Adaptation of the KDE-Method for Optimized Hotspot-Analysis for Emergency and Rescue Services

Julia Gonschorek[✉], Caroline Räbiger, Benjamin Bernhardt,
and Hartmut Asche

Department of Geography, University of Potsdam,
Karl-Liebknecht-Str. 24/25, 14476 Potsdam, Germany
{julia.gonschorek, craebige,
bebernha, gislab}@uni-potsdam.de

Abstract. This article gives insight into a running PhD-project. The focus lies on the adaption of kernel density estimation to optimize hotspot analysis for big geodata in the context of civil security research in urban areas. The parameters of kernel density estimation are commonly set in an exclusively mathematical way or using rules of thumb. The element's spatial component is then left disregarded. This causes an enormous risk for geovisualization in the form of hotspot maps. Urban areas may be declared as police or emergency hotspots despite there only being a small or even not significant random sample of event data. Furthermore there is the risk of over- or under-smoothing of real, existing hotspots in the visual output because of a too small or large cell size for the map grid. That may lead to incorrect tactical, strategic, and operational planning for agencies and organizations with civil security tasks. The method chain presented here offers a relative simple and semi-automatized solution of this problem.

Keywords: Civil security · Kernel density estimation (KDE) · Hotspot analysis · Big data

1 Civil Security – A Public Good

The role of civil security in all its forms is becoming increasingly important in the currently tense refugee crisis. The wide array of topics ranges from disaster protection through the reduction and prevention of crimes to extinguishing fires. Civil security is a public good to be protected and has been the focus of diverse funding projects for the last decade, including amongst others BMBF grant Forschung für die zivile Sicherheit (Research for Civil Security) [3, 4]. As recently as August 2015, a new call was opened, which allows authorities to become more easily involved in an R&D consortium. These developments, amongst others, clearly demonstrate the large need for research and development. Specifically, civil security is described as "the basis of free life and [as] an important factor of economic prosperity in Germany" [4, p. 2]. Federal funding is focused on solutions for the protection of public goods, infrastructure, the

© Springer International Publishing Switzerland 2016
O. Gervasi et al. (Eds.): ICCSA 2016, Part III, LNCS 9788, pp. 98–106, 2016.
DOI: 10.1007/978-3-319-42111-7_9

economy and the civilian population, for example applications and planning tools for the protection and rescue of people, that fit the practice of the responsible emergency services.

As part of the PhD project there exists a solid exchange with the professional fire departments in Cologne and Berlin. Both organizational units have analytics teams that deal exclusively with financial, material and personnel requirements planning. Currently there are analytical tools and software products that support the deployment planning of the police and fire department (e.g. GeoVISTA, RIGEL Analyst and CompStat). In contrast, the adaptation of analysis methods for user-specific requirements, which give mathematically supported insights into past and ongoing operations, falls somewhat short. Therefore, emergency services still consider spatiotemporal analysis to be digitized pinmaps and choropleth maps of the massively available operation data. Usually, little is discernable on these poorly modeled and often misclassified maps, except that the operations are so numerous that, due to the single-point representation (pinmap), not even the city area or infrastructure is visible on the map. The scientifically sound information content of such visual products thus approaches zero.

To enable the testing and adaptation of known scientific analysis methods, the Cologne fire department has provided access to archive databases from all operations conducted between 01.01.2007 and 20.04.2015. Expressed in numbers, these encompass more than 1.04 million emergency calls which resulted in a deployment of fire and rescue service forces in the city of Cologne. Each deployment involves, on average, the use of two to four operation resources, including ambulances, fire engines, turntable ladders, vehicles for emergency physicians and transport of the sick, etc.

The primary data include, inter alia, information on the type of operation (about 140 types are differentiated, for example, accident, fire, injured people), the time at which the operation began and geographic coordinates. The recording of emergencies in the database is performed in the control center for fire and rescue services in accordance with the law on fire safety, providing aid service and disaster protection [2, Sect. 28] of the government of North Rhine-Westphalia. The control center must always be manned, and all fire brigade deployments must be reported here, whether of voluntary, compulsory, company or full-time city fire brigades. When the command center receives an emergency call, the emergency services are alerted. To do so, a dispatcher takes the call in the command center and enters information on the type of emergency, location, persons involved, et cetera into the operations control system. The address information is geocoded in the system. The type of emergency determines which operational forces and resources need to be alarmed. According to the predetermined response order, the responsible station, together with the stand-by team, is alerted. The team moves out and activates the emergency services panel/reporting system in the vehicle (e.g. "Assignment accepted", "Operation location reached", "Transport of sick/injured persons"). The reporting chain, including time information, is automatically added to the database.

"Arguing with numbers guarantees that facts measurable by data can be mathematically confirmed at any time and as often as needed [...] [and they reflect] general rescue service structures and performances as well as their long-term developments in a secured manner" [1, p. 9]. Amongst others, supported by statistical and graphical analysis are:

- Operation scheduling and development planning,
- Requisition,
- Threat assessment and
- Identification of potential risks.

This procedure is established with a view to maintain the desired protection permanently and over the entire operation range. This is understood to be the arrival time or aid deadline, which is usually meant to be the period between the emergency call and the arrival at the emergency site, located at a street. In North-Rhine Westphalia this is five to eight minutes, in Berlin "as needed" [1, p. 25f].

As part of the dissertation project, selected methods applied to the task in question and the user group emergency services are visually processed together with their results. "Objectives of data visualization are the creation of:

1. A deeper understanding of the information contained in the data,
2. A basis for statistical analysis for hotspot identification (clustering methods and density estimation),
3. The identification of spatial and temporal movement patterns and the establishment of contexts" [8, p. 552].

2   KDE Methods for Hotspot Analysis

The geoinformatics research field crime analysis is a pioneer in the use of the method kernel density estimation (KDE) in a non-mathematical environment. For the data analysis of crime, hotspots are somewhat imprecisely defined as "[a]reas of concentrated crime" [19, p. 2]. This approach is founded on the assumption that the accumulation of individual events in close proximity is not random and a connection can be presumed, so it is likely that events of this nature will recur in this environment. The analyses are based on predefined confidence limits (significance levels). Thus, random and non-random distributions can be distinguished. Visually, this can be processed using grid maps or so-called hotspot maps. For this purpose, a more or less fine grid is placed over the area map. Density values are then entered into the resulting cells. The result is an overall image consisting of "valleys" and "hills" according to the chosen type number, limits and interval boundaries.

The events underlying the KDE have a spatial and also temporal reference. If the definition of hotspots mentioned above is extended to cover the latter component, it can be concluded that hotspots show spatially and temporally significant aggregations of events, here operations. Evidence of this non-trivial connection can also be found using KDE by combining different time intervals with the associated spatial information. In this manner, statements can be made about the behavior of hotspots in terms of

- permanence or periodicity,
- mobility or movement patterns and
- aggregation or dissimilation.

The KDE can be described by a function [13, p. 21]:

Let $x_1, x_2, \ldots, x_n$ be real elements of a sample of size n.

A kernel density estimator f with a bandwidth $\lambda > 0$ and the kernel K is then defined as

$$f_\lambda(x) = \frac{1}{n\lambda} \sum_{i=1}^{n} K\left(\frac{x - x_i}{\lambda}\right).$$

Consequently, the key parameters of the method are:

1. *The sample elements.* Usually, all the elements of the sample are used in the method.
2. *The cell size.* The density values depend on the choice of the cell size. Comparing different cell sizes using a consistent data base, the cartographic results differ, sometimes very strongly. Smaller cell sizes result in a continuous surface. Larger cell sizes produce a grainy image in which hotspots can disappear. For comparability between hotspots in different time windows, uniform cell sizes should be chosen. For choosing the *right* or *optimal* cell size, Hengl [11] suggests appropriate methods and claims that an edge length of 20 m to 200 m is representative in most cases.
3. *The bandwidth.* In literature, the bandwidth is also often referred to as *smoothing parameter* or *search radius*. It influences the appearance of the density estimation and is the limiting component of the kernel function to be selected. Known kernel functions include uniform, quartic, triangular, Gaussian, and Epanechnikov, for example.
4. *The kernel function* can be described as a circular window that runs across the study area. This has a fixed radius: the bandwidth. All events within this window are weighted according to their distance from the center of the cell. Events closer to the center point are weighted more strongly than distant events. The kernel function then mathematically describes the weights of the individual events in the window. Finally, the cell is assigned its density value. The choice of bandwidth is essential. The array of methods for its definition ranges from strictly mathematical methods (e.g. adaptive, smoothed cross-validation, biased cross-validation, least squares cross-validation, plug-in and standard deviation) to arbitrary value selection and trial-and-error experiments until the most aesthetically pleasing results or those which best show one's own basic assumptions have been found. Common to all is that the bandwidth is used as a number without a real spatial reference.

The effect on the cartographic result of the choice of cell size and bandwidth is illustrated in Fig. 1. The maps show the same situation and the same base data using the same kernel function (Epanechnikov). Still, each of these geovisualizations leads to different conclusions: Some hotspots appear to include entire city areas of Cologne, others can be determined almost exactly to an intersection. An informed decision support and planning basis for emergency services must be established in a different way because this range of interpretations is not scientifically sound.

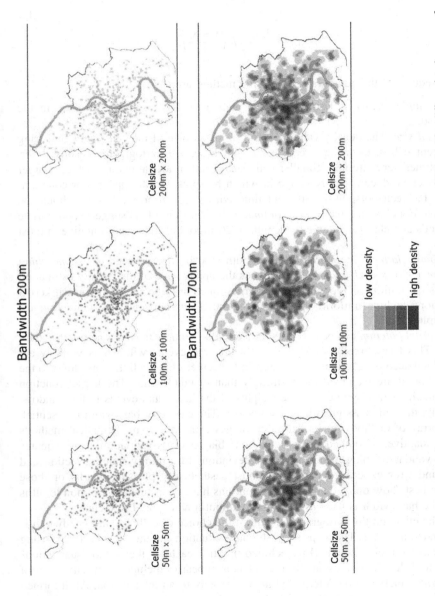

Fig. 1. Uniform data basis, but different parameter settings for hotspot maps enable highly differentiated scopes for interpretation. (Color figure online) (Source: Private draft)

3 Optimization of the Analysis Chain: Adaptation of KDE for Area-Related Issues

One may ask how an estimation method which consists exclusively of numbers without any spatial reference can estimate and thus represent events in space. Or differently: how can a geospatial KDE be usefully described?

If one begins with the preprocessing of data, a logically consistent and mathematically sound path emerges. First, the data of a sample (e.g. an operation type) are partitioned by temporal categories. These elements, so every single fire brigade operation, lie in a defined area section: in the city of Cologne in general, more precisely in a municipality, district or a deployment area with a responsible station. The kernel function works with weighted elements. So it is common to count all events that took place at the same location and set this value as the weight. Normally, the resulting data matrix is directly used in the KDE analysis. If, however, one adds local statistics (also: LISA) such as Local Moran's I, Getis Ord Gi* or others at this point, hotpoints can be identified. Hotpoints are those events that are significant in terms compared to their neighbors. Significance levels in mathematics are, as is known,

1. Significant with a confidence level of 95 % (5 % chance of error)
2. Highly significant with a confidence level of 99 % (1 % chance of error)
3. Very highly significant with a confidence level of 99.9 % (0.1 % chance of error)

Then, the matrix of the weighted data is cleaned by removing all events that are not significant. In the next step, the distances between the hotpoints in each section are determined. The averaged values are used as bandwidth in the KDE and the calculation is made. This process chain is summarized in Fig. 2.

If different temporal sections (years, months, days of the week or times of day) are considered together, the resulting map shows information on permanence or periodicity, mobility, respectively movement patterns, and aggregation or dissimilation. The entire process chain is initially only dependent on the input data and maintains the spatial reference until the data are visualized.

4 Discussion of Results and Outlook

Density estimation methods are used often in geoinformatics. Thus, hotspot analyses are no innovations – particularly in the US and UK, these methods of crime analysis are commonplace and therefore in continuous development [5, 15, 18–20].

Nevertheless, no approach could be found to date in the literature that takes the factor space into account during the analysis of particularly event-dense areas or hotspots. Metric data (e.g. distances or areas) are considered in some methods, but not the location within space itself.

The considerations presented here including the resulting process chain are new and currently in exploration. For this, operation types are selected, partitioned to time windows, and run through the process described in Sect. 3. In addition to the desired result of a higher, mathematically founded density distribution and well-defined

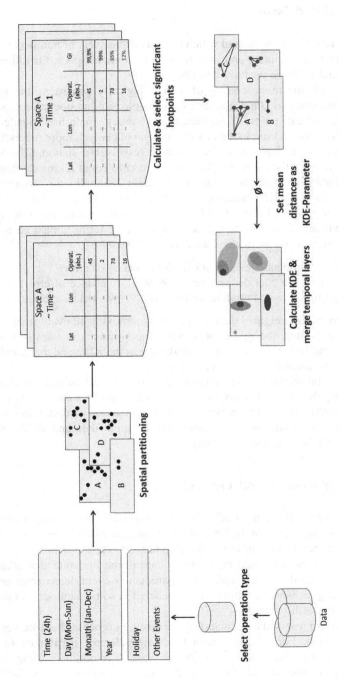

Fig. 2. Modified KDE: Spatial partitioning of the sample and preprocessing of at least significant events permit a choice of parameters with spatial contexts and thus an optimized KDE. (Source: Private draft)

probability spaces, the process also consumes considerably less computing time and memory space. With regard to real-time analysis, this is a major advantage. However, the KDE implementation requires a more comprehensive preprocessing of data. Since these steps can be partially automated, thereby reducing human error, the additional workload for the analysts can be kept comparatively low. Slightly different results are possible depending on the choice of the LISA statistic for determining hotpoints. However, in the tests conducted to date, these are not so strong as to lead to a distortion of the overall result.

When is a hotspot no longer hot? The question of the demarcation between areas with a high operation density and probability and those with significantly lower ones is not insignificant. At this point it should be summarized that areas are estimated from event points. A line should then separate these two areas from each other. An approach from classical cartography can be used: In maps, for example, mountains and valleys are constructed graphically by their color and the definition of a limit, in which the sea level plays the main role. The derivation of such a value is also a subject of the doctoral project. For this purpose, the use of classic cluster methods that work with significance methods, among others, is conceivable. Their known rigid elliptical shapes or convex hulls, however, are not ideal, as they are less suitable for focused hotspot demarcation.

This journal article does not discuss which cartography methods, such as target group oriented coloring and classification, are to be used. These are also study topics in the dissertation project.

Acknowledgements. At this point the main author Julia Gonschorek would like to thank my supervisor Prof. Dr. H. Ashe (Potsdam University) for his patience and constant willingness to talk. Special thanks go to the Cologne Fire Brigade (here Mr. H.-D. Richmann) for their long-term cooperation and provision of data and problems, without which this research would not be possible. The Berlin Fire Brigade (here Mr. S. Karas) should be thanked for the constructive exchange of ideas, which lead to new approaches and considerations. A big thank you goes to Mr. B. Bernhardt and Ms. C. Räbiger for the professional editorial legwork.

References

1. Behrendt, H.: Zahlenspiegel Rettungsdienst – Eine Übersicht über die wichtigsten Kennzahlen im Rettungsdienst. Mendel Verlag, Witten (2008)
2. BHKG: GESETZ ZUR NEUREGELUNG DES BRANDSCHUTZES, DER HILFEL EISTUNG UND DES KATASTROPHESCHUTZES (GV. NRW. 2015/Nr. 48/pp. 886–918) (2015). https://recht.nrw.de/lmi/owa/br_vbl_detail_text?anw_nr=6&vd_id=15416&vd_back=N886&sg=0&menu=1. Accessed 30 Jan 2016
3. BMBF, BUNDESMINISTERIUM FÜR BILDUNG UND FORSCHUNG: Forschung für die zivile Sicherheit in Deutschland – Ergebnisse und Perspektiven (2014). https://www.bmbf.de/pub/erfolgsbroschuere_zivile_sicherheit_2014.pdf. Accessed 30 Jan 2016
4. BMBF, BUNDESMINISTERIUM FÜR BILDUNG UND FORSCHUNG: Forschung für die zivile Sicherheit 2012–2017: Rahmenprogramm der Bundesregierung (2012). http://www.bmbf.de/pub/Rahmenprogramm_Sicherheitsforschung_2012.pdf. Accessed 30 Jan 2016
5. Boba, R.: Crime Analysis and Crime Mapping. Sage Publications Inc., Thousand Oaks (2005)

6. Gonschorek, J., Asche, H., Langer, A., Bernhardt, B., Räbiger, C.: Big data in the field of civil security research: approaches for the visual preprocessing of fire brigade operations. In: Papajorgji, P., Pinet, F. (eds.) International Journal of Agricultural and Environmental Information Systems (IJAEIS), vol. 7. IGI Global (2016, in press)

7. Gonschorek, J., Asche, H., Schernthanner, H., Bernhardt, B., Langer, A., Humpert, M., Räbiger, C.: Big data in civil security research: methods to visualize data for the geovisual analysis of fire brigade operations. In: Gervasi, O., Murgante, B., Misra, S., Gavrilova, M.L., Rocha, A.M.A.C., Torre, C., Taniar, D., Apduhan, B.O. (eds.) ICCSA 2015. LNCS, vol. 9157, pp. 415–425. Springer, Heidelberg (2015)

8. Gonschorek, J., Asche, H., Schernthanner, H., Langer, A., Räbiger, C., Bernhardt, B., Humpert, M.: Big Data in der zivilen Sicherheitsforschung – Methoden zur Datenvisualisierung für die explorative Analyse von Feuerwehreinsatz-daten. In: Strobel, J., Zagel, B., Griesebner, G., Blaschke, T. (eds.) Journal für Angewandte Geoinformatik, AGIT-Symposium, Salzburg, 8–10 Jun 2015, vol. 27. Wichmann Verlag, VDE Verlag GmbH, Berlin/Offenbach (2015)

9. Gonschorek, J., Tyrallová, L.: Geovisualization and geostatistics: a concept for the numerical and visual analysis of geographic mass data. In: Murgante, B., Gervasi, O., Misra, S., Nedjah, N., Rocha, A.M.A., Taniar, D., Apduhan, B.O. (eds.) ICCSA 2012, Part II. LNCS, vol. 7334, pp. 208–219. Springer, Heidelberg (2012)

10. Gonschorek, J., Asche, H.: Geovisuelle Analysen in Raum und Zeit: Methoden zur Unterstützung präventiver Maßnahmen und Einsatzplanung der Kölner Feuerwehr. GIS. BUSINESS Das Magazin für Geoinformation, Heidelberg (4) (2011)

11. Hengl, T.: Finding the right pixel size (2005). http://www.researchgate.net/profile/Tomislav_Hengl/publication/222014409_Finding_the_right_pixel_size/links/53fb3ebf0cf20a4549705c55.pdf. Accessed 30 Jan 2016

12. Keim, D.A., Andrienko, G., Fekete, J.-D., Görg, C., Kohlhammer, J., Melançon, G.: Visual analytics: definition, process, and challenges. In: Kerren, A., Stasko, J.T., Fekete, J.-D., North, C. (eds.) Information Visualization. LNCS, vol. 4950, pp. 154–175. Springer, Heidelberg (2008)

13. Lang, S.: Skript zur Vorlesung Computerintensive Verfahren in der Statistik. Universität Innsbruck (2004). http://www.uibk.ac.at/statistics/personal/lang/publications/compstat_aktuell.pdf. Accessed 31 Jan 2016

14. Lersch, K.M., Hart, T.C.: Space, Time and Crime, 3rd edn. Carolina Academic Press, Durham (2011)

15. McCullagh, M.J.: Detecting Hotspots in Time and Space. In: ISG 2006 (2006). http://www.spatial.cs.umn.edu/Courses/Fall07/8715/papers/mccullagh.pdf. Accessed 30 Jan 2016

16. Roberts, J.C.: Coordinated multiple views for exploratory geovisualization. In: Dodge, M., McDerby, M., Turner, M. (eds.) Geographic Visualization: Concepts. Tools and Applications. Wiley, Chichester (2008)

17. Schmiedel, R., Behrendt, B., Beltzler, B.: Regelwerk zur Bedarfsplanung Rettungsdienst. Mendel Verlag, Witten (2012)

18. Silverman, B.W.: Density Estimation for Statistics and Data Analysis. Chapman & Hall, New York (1986)

19. Smith, S.C., Bruce, C.W.: CrimeStat III – User Workbook. National Institute of Justice, Washington, D.C. (2008). https://www.icpsr.umich.edu/CrimeStat/workbook/CrimeStat_Workbook.pdf. Accessed 31 Jan 2016

20. Williamson, D., McLafferty, S., McGuire, P., Ross, T., Mollenkopf, J., Goldsmith, V., Quinn, S.: Tools in the spatial analysis of crime. In: Hirschfield, A., Bowers, S. (eds.) Mapping and Analysing Crime Data. Taylor & Francis, London/New York (2001)

A Basic Geothematic Map for Land Planning and Modeling (Daunian Subapennine - Apulia Region, Italy)

Giuseppe Spilotro[1], Salvatore Gallicchio[2],
Roberta Pellicani[1(✉)], and Giuseppe Diprizio[2]

[1] Department of European and Mediterranean Cultures,
University of Basilicata, Via San Rocco 3, 75100 Matera, Italy
{giuseppe.spilotro,roberta.pellicani}@unibas.it
[2] Department of Earth and Geoenvironmental Sciences,
University of Bari, Via Orabona 4, 70100 Bari, Italy
{salvatore.gallicchio,giuseppe.diprizio}@uniba.it

Abstract. In this paper a geothematic map at 1:180,000 scale of the Daunian Subapennine, a sector of the Southern Apennines Chain, located in the Apulia Region (Southern Italy) is presented. The map was produced by analyzing, redrawing and integrating the official and historical geological sheets, at different scale, edited by the Italian Geological Service, ISPRA (ex APAT), between the 50s and 60s. The analyzed geological sheets were compiled by different authors, due to the extension of the area to be mapped. For this reason, the geological sheets do not provide an univocal geo-lithological framework of the whole area. The uncertain interpretations of the geological terms contained in the different geological sheets have been checked and updated by using recent geological bibliography. The result is a geolithological map, at regional scale, realized through a homogeneous method, which provides information on the lithologies outcropping in the Daunian Subapennine, useful mainly in the preliminary phase of designing in the engineering and land planning fields.

Keywords: Geothematic map · Geolithological unit · Daunian Subapennine · Southern Apennines · Italy

1 Introduction

The geothematic maps at small scale are a basic tool for the preliminary phase of land planning in regional contexts. It is known that the geological features of a territory are among the most important elements for determining the typology of the "suistanable" operations on the territory, aimed to improve the growth conditions of a region and to preserve the natural dynamic equilibrium of the environment.

The geolithological map represents a geothematic cartography, which highlights the portions of territory characterized by outcropping geological units with same lithological composition and age. In addition, this kind of geothematical map constitutes the basis cartography necessary for producing further maps (for example, landslide hazard map, aquifer vulnerability map, landuse map, etc.), for planning a sustainable territorial development and designing engineering works [1–7].

© Springer International Publishing Switzerland 2016
O. Gervasi et al. (Eds.): ICCSA 2016, Part III, LNCS 9788, pp. 107–119, 2016.
DOI: 10.1007/978-3-319-42111-7_10

Furthermore, in recent times, due to the availability of a wide range of remote sensing data together with data from other sources in digital form and their analysis using GIS, it has become possible to prepare different thematic layers corresponding to the causative factors that are responsible for the occurrence of landslides [8]. The integration of these thematic layers with weights assigned according to their relative influence on the instability phenomenon leads to the generation of a landslide susceptibility and hazard zonation map [9]. In the most of landslide susceptibility predictive models, the geothematic cartography has a important role, because it may be used as basis for producing thematic layers corresponding to mechanical and hydrogeological properties of soils, as plasticity index, water table factor, etc. [7, 10–12].

In this paper, a geothematic map at small scale (1:180,000) of an area located in the western part of Apulia Region, in Southern Italy, was prepared. In particular, a geolithological map was realized. It derived from a project, carried out by Regional Authorities and two Departments of Bari University and of Basilicata University and focused to produce cartography at regional scale for Apulia and Basilicata regions. This project aimed at providing thematic layers for a number of planning activities. Main products of these activities are the landslide hazard assessment and mapping and the infiltration/runoff rate evaluation for hydrological and hydrogeological purposes.

In reason of the costs, coming from the width of the area to be covered, of the subjectivity of the field surveyors and of the not univocal stratigraphic interpretation of the local sequences, it was decided to use the sheets of the geological maps produced, between 1950 and 1960, by the Italian Geological Survey in 1:100,000 scale. Recently, a new edition of the Italian Geological map at 1:50,000 scale is available. In this map it is possible to observe the migration in the legend from units based on the lithology towards units defined according stratigraphic criteria, known as "synthems" and, with their aggregation and disaggregation, respectively, supersynthems and subsynthems. The new geological cartography is produced according to this new scientific insight of the stratigraphic features of the territory, but its use for producing thematic layers needs a further translation made by very specialized geologists.

The use at a regional scale of the old geological sheets could suffer on the other hand from big problems, especially about the homogenization of the lithological units and of their conceptual significance, especially along the border of adjacent sheets.

Therefore, the new geolithological map herein presented was produced in order to overcome the above cited limits of old and new official geological cartography. Furthermore, the chioce of the study area comes from the presence of large extension of multi-lithological and highly tectonized flyschs over and between less disturbed sedimentary units.

2   The Study Area: Geological Setting and Existing Cartography

The study area is located in the western part of Apulia Region, in Southern Italy, which comprises twenty-five municipalities of Foggia province, is presented. This area, known as Daunian Mountains or Daunian Subapennine, is bordered, on the North, by Fortore river, on the East, by "Apulian Tavoliere" Plain, on the West, by Apulian regional boundaries and, finally, on the South, by upper drainage basin of Ofanto river (Fig. 1).

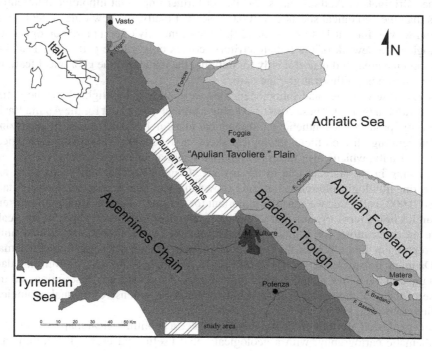

Fig. 1. Regional geological setting of Southern Apennine area with the localization of the study area

The geological framework of Daunian Mountains is closely related to the geological history of Southern Italian Apennines. Indeed, this region represents the Apulian sector of the Apenninic Chain and is made up by Cretaceous to Miocene allochthonous complex overlaying the Plio-Pleistocene terrigenous succession of the Bradanic Trough, which rests in turn on the Mesozoic to Neogene Apulian Foreland units [13, 14].

Therefore, the area of the Daunian Apennines is characterized by a wide variety of lithostratigraphic units interacting with each other, but with different geolithological features and mechanical properties, i.e. rocky successions vs. clays [15–20].

In the study area, the Italian official geological cartography, i.e. the Gelogical Map of Italy at 1:100,000 scale, is composed by four geological sheets. The analysis of these sheets highlighted an inhomogeneous geological framework of the Daunian Mountain territory. Indeed, the sheets of the Gelogical Map of Italy at 1:100,000 scale, which cover the area, were compiled, between the end of the 50s and the 60s, by different working groups by synthetizing original geological surveys, realized on IGM (Military Geographical Institute) topographic basis at 1:25,000 scale, known as "Originals of Author". The different working groups have depicted, for each sheet, stratigraphic and structural features, which are often hard to be correlated, passing from one sheet to another (the same stratigraphical units were often called with several names, described with different features and designed with mismatched areal extension). Moreover, some of the "Originals of Author" maps [21] depict further important lithologic differentiations than the geological sheets at 1:100,000 scale. Finally, the new official geological cartography of Italy at 1:50,000 scale [22], which generally offer a more homogeneous geological framework of the Italian territory, covers only a little part of the Daunian Subapennine area, and is useful only for defining the name and the geological features of the outcropping lithostratigraphic units.

These discrepancies and inhomogeneities were already highlighted for the entire Apulia Region by [23]. These Authors, during a project carried out for the Apulia Basin Authority, produced a synthetic geolithological map, at 1:250,000 scale, by analyzing and integrating all the official geological sheets at 1:100,000 scale of the Geological Survey of Italy, which cover the entire Apulian regional territory.

Tacking in the account that in the several "Originals of Author" maps there are interesting lithologic differentiations (not shown in the official geological sheets of Italy at 1:100,000 scale), that we consider important for a correct understanding of the current dynamics of the territory, and that in the new geological map of Italy at 1:50,000 scale there are indications for a modern reinterpretation of the lithostratigraphical units outcropping in the study area, the Authors of this paper elaborated a geolithological map of Daunian Subapennine, at 1:180,000 scale, by integrating and uniforming data contained in the above mentioned geological maps. Moreover, the description and the name of the geolithological units took into account the extensive regional geological bibliography (e.g. [14]).

In conclusion, the aim of the paper is to produce a geolithological map to be used as geothematic cartography, with a geological detail useful and usable for preliminary studies in several fields, such as geomorphological criticalities, seismic and landslide hazard zonation, engineering geology, etc.

The new geolithological map at 1:180,000 scale of the Daunian area was realized by using a GIS software.

3   Materials and Methodology for Producing the Geothematic Map

The geo-lithological map at 1:180,000 scale was produced by digitizing, elaborating and synthesizing raster images of different sheets of the official Geological Map of Italy published by the Geological Service of Italy [21].

In particular, the following geological sheets at 1:100,000 scale: F° 155 "San Severo" [24], F° 163 "Lucera" [25], F° 174 "Ariano Irpino" [26] and F° 175 "Cerignola" [27] (Fig. 2), and the following "Originals of Author" geological maps at 1:25,000 scale: Tav. 155 III SE "Castello di Dragonara", Tav.163 III NE "Alberona", Tav. 163 III NO "Volturara Appula", Tav. 163 III SE "Biccari", Tav. 163 III SO "San Bartolomeo in Galdo", Tav. 163 III NE "Casalnuovo Monterotaro", Tav. 163 III NO "Colletorto", Tav. 163 III SE "Pietra Montecorvino", Tav. 163 III SO "Celenza Valfortore" [21], were analyzed.

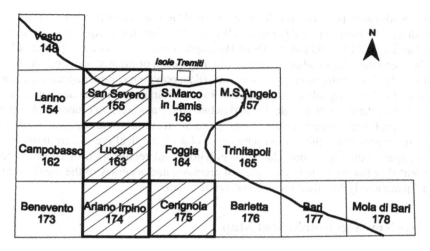

Fig. 2. Sheets of the Geological Map of Italy at 1:100,000 scale covering the study area: F° 155 "San Severo", F° 163 "Lucera", F° 174 "Ariano Irpino" and F° 175 "Cerignola"

The study area is located among the edges of different geological domains, characterized by lithostratigraphic units with different and complex granulometric, diagenetic, and structural features, which depend on the age of the deposits. For this reason, a chronostratigraphic criterion for differentiating these units should be considered.

The comparison and the correlation among the great number of the lithostratigraphic units defined in the above mentioned four geological sheets allowed to uniform their denominations and to reduce the apparent great number of the lithostratigraphical units described in the Daunian Apennines.

Moreover, the analysis of the Originals of Author geological sheets, at 1:25,000 scale, allowed to individuate and differentiate, within the miocenic turbiditic units, the arenaceous portions from the pelitic ones; unfortunately, only a portion of the Daunian Apennines is covered by the Originals of Author geological maps, at 1:25,000 scale, so the above mentioned distinction is partial.

In this way, we defined informal geo-lithologic units taking in the account mainly the lithologic, physical and chemical properties; nevertheless a careful analysis pointed out that these properties are closely related to the age and the geodynamic domains of each unit. These last features have been mainly borrowed from the new geological map of Italy at 1:50,000 scale, in particular from F° 407 "San Bartolomeo in Galdo" and F° 421 "Ascoli Satriano" [17, 28]. Although these are the unique printed sheets of the new map of Italy that cover areas belonging to the Daunian Apennines, they give important indications for a new interpretation and mapping of the lithostratigraphical units outcropping in this region.

Overall, the processing of the Official Geological Maps of Italy and of the Originals of Author led to the identification of 13 geo-lithological units; this allowed us to have a very simplified geolithological map at 1:180,000 scale, useful for preliminary land planning purposes.

It should also be point out that the digitization of Geological Map of Italy at 1:100,000 scale, due to the use of printed materials (JPG raster images), subsequently scanned and georeferenced (UTM WGS84 Northern Hemisphere 33N), shows an imperfect adherence between the edge of adjacent sheets which were not originally designed for the use in GIS. Indeed, sometimes, empty portions (no data polygons) or overlapped areas were produced by digitizing adjacent sheets, due to cartographical mismatches between the boundaries of the geological units located in two adjacent sheets. In order to solve these graphical problems, some simplifications in the digitization of the edges of individual polygons (representing the outcropping areas of a lithological type) were introduced, for example, rounding of boundaries of polygons and rectification of curves. These graphical simplifications led to a degree of precision and accuracy of the digitized map that is qualitatively less than the original printed map.

4   The New Geolithological Map

In order to synthesize and organize at regional scale the geolithological map at 1:180,000 scale (Fig. 3), the different geolithological units were ascribed into the following four main group, according of their chronology and geodynamic domains:

- CRETACEOUS-MIOCENE APENNINE GROUP

  This group comprises geolithological units which outcrops along the Chain and are relative to paleogeographic domains more ancient than Pliocene. These units are represented by: *clayey chaotic complex* and *calcareous marl and shales* relative to Cretaceous-Paleogene pelagic environment belonging to different paleogeographic domains; *quartz-rich arenites, calcareous turbidites* and *siliciclastic turbidites*, referable to deep sea environment relative to Miocene Apennine Foredeep.

- PLIOCENE-PLEISTOCENE FOREDEEP GROUP

  This group includes both Pliocene geolithological units localized on the allochthonous terrains of the external area of the Chain, belonging to wedge top basin domains, than the regressive succession of the Pliocene–Pleistocene foredeep s.s. succession outcropping in the "Apulian Tavoliere" Plain (Bradanic Trough, [29–31]; these units are represented by Pliocene *conglomerate and sandstone*, of transitional environment, by Pliocene – Pleistocenee *silty and marly shales* referable to offshore environment and finally by Pleistocene marine terraced deposits of transitional environment (sand, gravel and calcarenite).

- QUATERNARY CONTINENTAL GROUP

  This group comprises continental Quaternary units which rest unconformably on the substratum; these units are represented by *terraced alluvial deposits, debris deposits, colluvium and eluvial deposits* and, finally, *landslide deposits*.

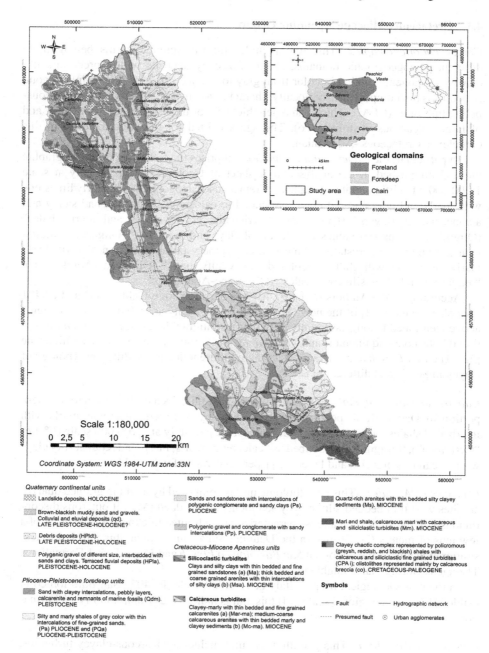

Fig. 3. Geolithological map of Daunian Subapennine at 1:180,000 scale. (Color figure online)

4.1 Cretaceous-Miocene Apennine Group

Clayey chaotic complex. This unit includes lithostratigraphic terms belonging to different paleogeographic domains, characterized by deposits with a predominantly clayey component, ranging in color from gray to red and green, with a mainly chaotic structure. This complex can be locally characterized by the presence of arenaceous-pelitic turbidites (CPAi), which can reach thickness of the order of tens of meters, and of large calcareous olistolites (co). The age specified for the units included in this complex is Cretaceous -Aquitanian.

In particular, the *clayey chaotic complex* comprises the following lithostratigraphic units, distinguished in the analysed four sheet of Geological Map of Italy at scale 1:100,000: "Undifferentiated complex" (Cretaceous-Paleogene); "light marly limestone with marls and gray and red clays associated to calcareous Brecciole and sandy and arenaceous interlayers" (Paleogene); "variegated clayey-shales and marly-shales" (Oligocene), "pink limestones with levels of clays and red marls" (Oligocene-Aquitanian-); "white microcrystalline limestones, cryptocrystalline limestones, varicolored marly limestones with flint, varicolored clayey-silty marls" (Oligocene-Aquitanian-); "red jasper and black siliceous shales" (Oligocene-Aquitanian).

According to the Authors of the F° 407 "S. Bartolomeo in Galdo" [17] and F° 421 "Ascoli Satriano" [28] of the new Geological Map of Italy at 1:50,000 scale, the units above mentioned belong essentially to two lithostratigraphic units: *Red Flysch* Formation (Cretaceous-Aquitanian) and *Variegate Clays* Group, Late Cretaceous-Oligocene [22]. The *Red Flysch* and *Variegate Clays* are different due to the different typology of clayey minerals and different plasticity.

Calcareous marl and shales. This geolithological unit, located in a limited southern portion of Daunian Subapennine, is represented by stratified calcareous marls with interbeded shales; according to the Geological Map of Italy at 1:100,000 scale (F° 175 "Cerignola"), this unit was described as "calcareous marls, marls and reddish silty clays with calcareous breccias and levels of jasper" and was referred to Miocene.

Quartz-Rich Arenites. This geolithologic unit is composed by medium grained quarzo-arenites with thin bedded silty clayey sediments interbedded (Ms). It outcrops discontinuously in the study area and is characterized by a limited thickness, of the order of a few tens of meters (20–30 m). In the Daunian Subapennine area, this unit is located, stratigraphically, at the bottom of *San Bartolomeo Flysch* and *Faeto Flysch* Formations. The age is Burdigalian-Langhian.

In the new Geological Map of Italy at 1:50,000 scale, F° 407 "San Bartolomeo in Galdo" [17], and at regional scale [14] this unit is indicated as *Numidian Flysch* Formation.

Calcareous Turbidites. This geolithologic unit includes calcareous-clayey turbidites, which in the geological map at 1:100,000 scale are attributed to Daunia Formation (Miocene). The unit, considering its lithological features highlighted in the historical sheets at 1:25,000 scale, was divided in two terms: (a) clayey-marly with thin bedded and fine grained calcarenites (Mar-Ma) and (b) medium-coarse calcareous arenites with thin bedded marly and clayey sediments (Mc-ma). In the southern sector of the area, this subdivision is not present, there aren't original of Author. In the sheets F° 407 "San Bartolomeo in Galdo" and F° 421 "Ascoli Satriano" (ISPRA 2011a, 2011c), of the new Geological Map of Italy at 1:50,000 scale, this unit has been referred to the *Faeto Flysch* Formation sensu [32].

Silicoclastic Turbidites. This unit includes deposits represented by arenaceous-pelitic alternations deriving from deep-sea turbiditic processes. The unit, considering its lithological features highlighted in the historical sheets at 1:25,000 scale, was divided in two terms: (a) clays and silty clays with thin bedded and fine grained sandstone (Ma) and (b) thick bedded and coarse grained arenites with thin intercalations of silty clays (Msa).

According to the Geologic map of Italy at 1:100,000 scale, this unit was described as: "miocenic poligenetic conglomerate, silty clays, molasses and sandstones with levels of pudding (Miocene) and molasses, clayey sands and silty clays of Upper Miocene, sandy-arenaceous and clayey-marly formation with lenses of calcarenites of Messinian-Tortonian age".

In the new geological map of Italy at 1:50,000 scale (sheet F° 407 "San Bartolomeo in Galdo"), this unit has been referred to the *San Bartolomeo Flysch* Formation [17].

4.2 Pliocene Wedge-Top Group

Conglomerate. This deposit composed by polygenic gravel and conglomerate with sandy intercalations (Pp) of continental and transitional environments outcrop unconformably on the allochthonous apenninic units along the external area of the chain. According to [10] these deposits belong to the Ariano Unit; In the new Geological Map of Italy they have been referred to the *Supersintema di Ariano Irpino* [28].

Sandstone. This unit consists of sands and sandstones with intercalations of polygenic conglomerate and sandy clays (Ps) of transition environment. This unit rest unconformably on the allochthonous apenninic units and show different type of stratigraphical contacts with the conglomerate. According to [15] these deposits belong to the Ariano Unit. In the new Geological Map of Italy they have been referred to the *Supersintema di Ariano Irpino* [28].

Silty and marly shales of grey color with thin intercalations of fine grained sands (PQa).

This unit outcrops in the external area of the chain and in the Pliocene-Pleistocene foredeep, that in this sector of Italy is known as Tavoliere di Puglia. The unit consists essentially of stiff fissured grey clays and silts deposited in offshore and scarp environments.

The terms autcropping along the external area of the chain have been referred to the Ariano Unit by [10] and to the *Supersintema di Ariano Irpino* by [28].

The terms outcropping in the Tavoliere di Puglia have been called with different terms: *Argille di Montesecco* [24], *Argille e argille marnose grigio-azzurrognole* [21], *Argille ed argille sabbiose grigio-giallastre* [27], *Argille subappennine* [33] In the new Geological Map of Italy at 1:50,000 scale [17, 28] this unit has been called *Argille Subappennine* (*Subapennine Clays*).

Sand, gravel and calcarenite with clayey intercalations and remnants of marine fossils (Qdm).

This unit outcrops in the "Apulian Tavoliere" Plain; it rests unconformably on the *Subapennine Clays* and is represented by terraced deposits of shallow marine and or transitional environment. In the new Geological Map of Italy at 1:50,000 scale, this unit have been grouped in the *Supersintema del Tavoliere di Puglia* [34, 35].

4.3 Quaternary Continental Deposits

Terraced Alluvial Deposits. These deposits are represented by polygenic gravels of different size, interbedded with sands and clays. The pebbles derive primarily from the erosion and rehash of the *Red Flysch*, *Faeto Flysch* and *San Bartolomeo Flysch*. Caliche are locally present in this outcropping. In the new geological map of Italy these deposits have been grouped in the *Supersintema del Tavoliere di Puglia*. The age ranges between the lower Pleistocene and the upper Pleistocene [34, 35].

Debris Deposits. They are generally located at the bottom of hillslopes and are represented by heterometric and disorganized deposits, consisting of medium to coarse gravel and boulders of different composition.

Colluvium and Eluvial Deposits. This deposits are represented by brown-blackish muddy sand and gravels.

Landslides Deposits. They represent ancient and actual slide bodies, affecting mainly the pelitic facies belonging to different geolithologic units. The type of landslide movements has a close correlation with the lithological features of outcropping soils. Indeed, clayey soils with high values of plasticity, such as *clayey chaotic complex*, are affected by flows and/or shallow widespread instability; while, flyschoid formations, with intercalation of lithoid strata in the pelitic facies, are mainly affected by slow complex and composite mass movements.

Landslide deposits represented in this map are only those mapped in the Italian geological sheets at 1:100,000 scale; so, landslides result to be located mainly in the northern sector of the geolithological map. Actually, landslides are spread throughout the Daunian Apennines. Anyway for an exact localization of landslides in the Daunian Subapennine, it would be better to consult specific landslide inventory maps [36].

5   Conclusions

The geolithological map of Daunian Subapennine (Southern Apennine, Apulia Region, Italy) at 1:180,000 scale is herein presented. The aim is to produce a geolithological map to be used as geothematical cartography, with a geological detail useful and usable for preliminary studies in several fields, such as geomorphological criticalities, seismic and landslide hazard zonation, engineering geology, etc. The map was produced by analyzing, redrawing and integrating the official and historical geological sheets, at different scales (1:100,000 and 1:25,000 respectively), edited by the Italian Geological

Service, ISPRA, between the 50s and 60s, by using a procedure for unificating and homogenizating the lithostratigraphic units. In particular, the criteria used for individuating the geolithological units in the new map allowed us to unify different lithostratigraphic units having the same lithology and age and, at the same time, to differentiate a given lithostratigraphic unit if characterized by mappable heterogeneous lithological features. Although the map represents a news in the geological cartography context of the study area, it should be considered as a historical map as it derives mainly from the reinterpretation of geological maps, achieved between the end of 50s and 60s and, secondarily, from the integration of data obtained from literature.

References

1. Doutch, H.F., Feeken, E.H.J.: The geological map as a public utility. Cartography **11**(1), 25–31 (1979). http://www.tandfonline.com/doi/abs/10.1080/00690805.1979.10438057#abstract
2. I.S.P.R.A.: F.° 407 San Bartolomeo in Galdo, Note illustrative e Carta Geologica d'Italia alla scala 1:50.000. Servizio Geologico d'Italia. Litografia Artistica e Cartografica srl, Firenze (2011)
3. Huang, Q., Liu, Y., Li, M., Mao, K., Li, F., Chen, Z., Chen, C., Hu, W.: Thematic maps for county-level land use planning in Contemporary China. J. Maps **8**(2), 185–188 (2012). http://www.tandfonline.com/doi/full/10.1080/17445647.2012.694272#abstract
4. Andriani, G.F., Diprizio, G., Pellegrini, V.: Landslide susceptibility of the La Catola Torrent catchment area (Daunia Apennines, southern Italy): a new complex multi-step approach. In: Lollino, G., et al. (eds.) Engineering Geology for Society and Territory, vol. 5, pp. 387–392. Springer International Publishing, Switzerland (2014). doi:10.1007/978-3-319-09048-1_74
5. Gallicchio, S., Moretti, M., Spalluto, L., Angelini, S.: Geology of the middle and upper Pleistocene marine and continental terraces of the northern Tavoliere di Puglia plain (Apulia, southern Italy). J. Maps **10**(4), 569–575 (2014). doi:10.1080/1744564702014.895436
6. Kong, X., Liu, Y., Liu, X., Chen, Y., Liu, D.: Thematic maps for land consolidation planning in Hubei Province. China. Journal of Maps **10**(1), 26–34 (2014). http://www.tandfonline.com/doi/full/10.1080/17445647.2013.847388#abstract
7. Pellicani, R., Frattini, P., Spilotro, G.: Landslide susceptibility assessment in Apulian Southern Apennine: heuristic vs statistical methods. Environ. Earth Sci. **72**(4), 1097–1108 (2014). doi:10.1007/s12665-013-3026-3
8. Van Westen, C.J.: Training package for geographic informationsystems in slope instability zonation, Part 1, p. 245. ITC publication 15, Enchede (1993)
9. Van Westen, C.J., Van Asch, T.W.J., Soeters, R.: Landslide hazard and risk zonation – why is it still so difficult? Bull. Eng. Geol. Environ. **65**, 167–184 (2006)
10. Stevenson, P.C.: An empirical method for the evaluation of relative landslide risk. Bull. Int. Assoc. Eng. Geol. **16**, 69–72 (1977)
11. Pellicani, R., Van Westen, C.J., Spilotro, G.: Assessing landslide exposure in areas with limited landslide information. Landslides **11**(3), 463–480 (2014). doi:10.1007/s10346-013-0386-4
12. Pellicani, R., Spilotro, G., Van Westen C.J.: Rockfall trajectory modeling combined with heuristic analysis for assessing the rockfall hazard along the Maratea SS18 coastal road (Basilicata, Southern Italy). Landslides (2015). doi:10.1007/s10346-015-0665-3
13. Mostardini, F., Merlini, S.: Appennino centro-meridionale. Sezioni geologiche e proposta di modello strutturale. Mem. Soc. Geol. It. **35**, 177–202 (1986)

14. Patacca, E., Scandone, P.: Geology of the southern Apennines. Boll. Soc. Geol. It. **7** (Special Issue), 75–119 (2007)
15. Dazzaro, L., Di Nocera, S., Pescatore, T., Rapisardi, L., Romeo, M., Russo, B., Senatore, M.R., Torre, M.: Geologia del Margine della Catena appenninica tra il Fiume Fortore e il Torrente Calaggio (Monti della Daunia – Appennino meridionale). Mem. Soc. Geol. It. **41**(411), 422 (1988)
16. Dazzaro, L., Rapisardi, L.: Schema Geologico del margine appenninico tra il Fiume Fortore e il Fiume Ofanto. Mem. Soc. Geol. It. **51**(143), 147 (1996)
17. I.S.P.R.A. (2011) F.° 407 San Bartolomeo in Galdo, Note illustrative e Carta della Pericolosità per Franosità alla scala 1:50.000. Servizio Geologico d'Italia. Litografia Artistica e Cartografica srl, Firenze
18. Vitone, C., Cotecchia, F., Desrues, J., Viggiani, G.: An approach to the interpretation of the mechanical behavior of intensely fissured clays. Soils Found. **49**(3), 355–368 (2009)
19. Vitone, C., Cotecchia, F.: The influence of intense fissuring on the mechanical behavior of clays. Geotechnique **61**(12), 1003–1018 (2011)
20. Cotecchia, F., Vitone, C., Santaloia, F., Pedone, G., Bottiglieri, O.: Slope instability processes in intensely fissured clays: case histories in the Southern Apennines. Landslides (2014). doi: 10.1007/s10346-014-0516-7
21. A.P.A.T.: The complete archive of Italian Geological Cartography. 2 DVD-ROM. Servizio Geologico d'Italia, Roma (2004)
22. A.P.A.T.: Carta Geologica d'Italia alla scala 1:50.000. Catalogo delle formazioni – Unità tradizionali. Quaderni Serie III, 7, Fascicolo 7. S.EL.CA., Firenze (2007)
23. Tropeano M., Pieri, P., Spilotro, G., Delle Rose, M., Diprizio, G., Gallicchio, S., Moretti, M., Sabato, L., Spalluto, L.: Carta Geo-Litologica della Puglia basata sulla elaborazione e sintesi della Carta Geologica d'Italia in scala 1:100,000. Relazione Finale. Dip. Geologia e Geofisica, Università di Bari, pp. 58 (2009)
24. S.G.I.: F° 169 S. Severo, Note Illustrative e Carta Geologica d'Italia alla scala 1:100.000. Poligrafica e Cartevalori, Ercolano (Napoli) (1969)
25. S.G.I.: F° 163 Lucera, Note Illustrative e Carta Geologica d'Italia alla scala 1:100.000. EIRA, Firenze (1964)
26. S.G.I.: F° 174 Ariano Irpino, Note illustrative e Carta Geologica d'Italia alla scala 1:100.000. Lito A. Colitti, Roma (1963).
27. S.G.I.: F°175 Cerignola, Note Illustrative e Carta Geologica d'Italia alla scala 1:100.000. EIRA, Firenze (1963)
28. I.S.P.R.A.: F° 421 Ascoli Satriano, Note illustrative e Carta Geologica alla scala 1:50.000. Servizio Geologico d'Italia. Litografia Artistica e Cartografica srl, Firenze (2011)
29. Migliorini C.: Cenno sullo studio e sulla prospezione petrolifera di una zona dell'Italia meridionale: II Congresso Mondiale Petrolio, Parigi, pp. 1–11 (1937)
30. Casnedi, R., Crescenti, U., Tonna, M.: Evoluzione dell'Avanfossa Adriatica meridionale nel plio-Pleistocene, sulla base di dati di sottosuolo. Mem. Soc. Geol. It. **24**, 243–260 (1982)
31. Tropeano, M., Sabato, L., Pieri, P.: Filling and cannibalization of a foredeep: the bradanic trough (Southern Italy). In: Jones, S.J., Frostick, L.E. (eds.) Sediment flux to basins: causes controls and consequences, vol. 191, pp. 55–79. Geological Society, Special Publications, London (2002)
32. Crostella, A., Vezzani, L.: La Geologia dell'Appennino Foggiano. Boll. Soc. Geol. It. **83**(1), 121–141 (1964)
33. Bonardi, G., Amore, F.O., Ciampo, G., De Capoa, P., Miconnet, P., Perrone, V.: Il complesso liguride autoctono: stato delle conoscenze e problemi aperti sulla sua evoluzione pre-appenninica e i suoi rapporti con l'Arco calabro. Mem. Soc. Geol. It. **41**, 17–35 (1988)

34. I.S.P.R.A.: F.° 408 Foggia, Note illustrative e Carta Geologica alla scala 1:50.000. Servizio Geologico d'Italia. Litografia Artistica e Cartografica srl, Firenze (2011d)
35. I.S.P.R.A.: F.° 396 S. Severo, Note illustrative e Carta Geologica alla scala 1:50.000. Servizio Geologico d'Italia. Litografia Artistica e Cartografica srl, Firenze (2011e)
36. Pellicani, R., Spilotro, G.: Evaluating the quality of landslide inventory maps: comparison between archive and surveyed inventories for the Daunia region (Apulia, Southern Italy). Bull. Eng. Geol. Environ. **74**, 357–367 (2015). doi:10.1007/s10064-014-0639-z

Spatial Modeling and Geovisualization of Rental Prices for Real Estate Portals

Harald Schernthanner[✉], Hartmut Asche, Julia Gonschorek, and Lasse Scheele

Department of Geography, University of Potsdam, Karl-Liebknecht-Strasse 24/25,
14476 Potsdam, Germany
{hschernt,gislab,julia.gonschorek,lasse.scheele}@uni-potsdam.de

Abstract. From a geoinformation science perspective real estate portals apply non-spatial methods to analyse and visualise rental price data. Their approach shows considerable shortcomings. Portal operators neglect real estate agents' mantra that exactly three things are important in real estates: location, location and location [16]. Although real estate portals record the spatial reference of their listed apartments, geocoded address data is used insufficiently for analyses and visualisation, and in many cases the data is just used to "pin" map the listings. To date geoinformation science, spatial statistics and geovisualization play a minor role for real estate portals in analysing and visualising their housing data. This contribution discusses the analytical and geovisual status quo of real estate portals and addresses the most serious deficits of the employed non-spatial methods. Alternative analysing approaches from geostatistics, machine learning and geovisualization demonstrate potentials to optimise real estate portals´ analysing and visualisation capacities.

Keywords: Rental prize · Spatial modeling · Geovisualisation

1 Introduction

Numerous real estate portals apply hedonic regressions to model real estate prices in general and rental prices in particular. Regression results are used as a basis to create suboptimal rental price maps without showing the real spatial price distribution. In fact portal operators ignore Tobler's famous first law of geography: *"Everything is related to everything, but near things are more related than distant things"* [17]. For more than a decade portal operators collect rental price data and use it to offer analytical products, e.g. real estate reports, rental price indices and rental price maps. Spatial analysis and geovisualization techniques are hardly used in the development of those products. A frequently used workflow consists of calculating prices by means of descriptive statistics or hedonic regression and mapping the results on not suitable geometric boundaries.

This article starts giving an overview of the analytical and geovisual state of the art of real estate portals. Next suitable alternatives to optimize modeling and visualization of rental prices, coming from the fields of machine learning, geostatistics and grid mapping, are identified through an in depth literature review and by examining the results of conducted expert interviews. The third section refers to the implementation

© Springer International Publishing Switzerland 2016
O. Gervasi et al. (Eds.): ICCSA 2016, Part III, LNCS 9788, pp. 120–133, 2016.
DOI: 10.1007/978-3-319-42111-7_11

and validation of identified methods, focusing on finding the most suitable model parameters, to achieve the best performance for estimating rental prices. A description on how the best performing price estimation results can be applied as basis to create a prototype of a web-based rental price map is given. Finally, a conclusion, summarizing the major results of the presented research and an outlook is presented.

2   State of the Art of Analyzing and Mapping Rental Prices

To determine how real estate portals analyse and map rental prices, 32 real estate portals were examined by 14 criteria by applying statistical and visualization methods. In addition to this examination, and according to a designed standardized questionnaire, five 45 min interviews with renowned German real estate agents and real estate portal experts were conducted. The analysis examined the 32 largest real estate portals by unique web access [1][1]. Real estate portals have replaced newspapers as the go-to medium to search for rental property. Just to demonstrate the dimensions, the Germany portal Immobilienscout 24.de constantly stores 1.5 million rental properties in its database [10]. The fact that all these offers are geocoded makes them a great, but so far insufficient used source for spatial real estate analysis and visualization.

Summarizing the results of the portals's examination, it can be stated that the most real estate portals offer real estate market indices for clients as banks, investors and public authorities. Those indices neglect the location of real estates and are built by means of the non-spatial statistical method of simple descriptive statistics (mean, median) or by applying hedonic regressions. A common procedure is to map descriptive statistics- or hedonic regression results onto non-appropriate reference geometries, e.g. on ZIP code areas or on city district levels. Trulia.com for example maps the median rental price by ZIP code (cf. Fig. 1). The resulting rental price maps show a distorted distribution of rental prices in space, which does not correspond to the real distribution of price in space.

2.1   Hedonic Regression as the Non-spatial Standard Method of Rental Price Modeling

The most striking result of the analysis is the widespread use of hedonic regressions; the portal operators standard method of rental price modeling. Hedonic regression is a form of price estimation used for products with a regular change in their price-performance ratio (e.g. personal computers or houses) [7]. Applied to the real estate sector, rental price estimation is done via the availability and the intrinsic quality of a property; thus, effects of spatial autocorrelation are neglected. The direct influence of objects for rent is not included in the hedonic regression equation.

---

[1] Alexa.com measures the number of website visitors [1]. According to Alexa.com, the two most visited real estate portals in the UK and the US are Zillow.com and Trulia.com; Immobilienscout 24.de and Immowelt.de are the most visited portals in the German speaking countries.

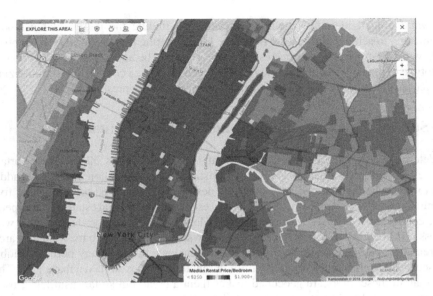

Fig. 1. Example of the US real estate's portal Trulia of mapping a median rental prices on the geometries of New Yorks's city ZIP codea areas [20]

Mostly a semilog variant of the hedonic model is used by real estate portals, where for an object k the price Pk is determined [7]. The resulting price estimations are mapped on non-appropriate reference geometries.

$$lnP_k = \beta_0 + \sum_{j=1}^{m} \beta_j X_{kj} + \in_k \tag{1}$$

2.2   Non-appropriate Reference Geometries to Map Hedonic Regression Results

City districts, ZIP code areas and statistical districts are most commonly used to map hedonic regression results. For example, US portals examined in this research used ZIP code areas, city district limits or core based statistical areas (CBSA) of the US Office of Management and Budget. CBSA for example consist of socioeconomic highly homogeneous areas with at least 10 000 inhabitants centered around at least one urban core [22]. Exemplified on the US portal Zillow.com and on the German portal *Immobilienscout 24,* the incorrect use of reference geometries is specified. Zillow.com publishes their designed polygons called "neighborhood boundaries"[2]. For all bigger U.S. cities, determined by Zillow.com, Zillow builds 7 000 "neighborhood boundaries" [26]. The "neighborhoods boundaries" are based on US zip code regions. However a methodology on how the geometries are formed is neither recognizable nor published by the portal operators. Figure 2 shows an example of three zip code areas (red) in New York's city district of

---

2 Zillow's "neighborhood polygons have been licensed under a Creative Commons license and can be downloaded for free via the portal's website.

Manhattan in comparison with one formed Zillow.com "Neighborhood boundary" named Gramercy (blue) (cf. Fig. 2). For each neighborhood, *Zillow* calculates an hedonic home value index [27]. The German real estate portal Immobilienscout 24.de uses residential quarters named "neighborhoods" that have been developed by the German geomarketing company *Nexiga (former Infas geodata)*. According to *Nexiga*, a neighborhood has a similar socioeconomic structure and in average consists of 400 households; the shape of geometries was based on the shape of German electoral districts [14]. Figure 2 shows an exemplary neighborhood (blue) in comparison with a postal code area used by the German postal service (red) in the city of Potsdam, Germany (cf. Fig. 2).

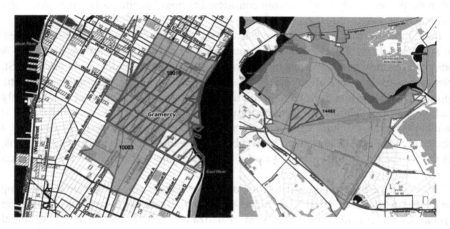

Fig. 2. Left: three zip code areas (red) in New York, USA in comparison to one Zillow neighborhood boundary (blue dashed). Right: postal code area 14482 in the city of Potsdam, Germany (red), compared to a "neighborhood" as used by Immobilienscout 24.de (Color figure online).

3 Selection of Spatial Modeling and Visualization Alternatives

A comprehensive literature survey and the results of the conducted expert interviews serve as a solid scientific base for identifying spatial modeling and geovisualization alternatives. Alternative price modeling approaches were identified in the fields of geostatistics and machine learning, whilst alternative visualization methods have been found in the field of "grid mapping". During the interviews, geostatistical methods from the Kriging family, machine learning methods and a separation from reference geometries have been commented as potential alternatives and starting point to integrate in real estates price modeling and visualization workflow. Interview results have been compared with the results of literature review. Although real estate portals posses a high potential for spatial real estate modeling, specific papers of rental price modeling in the context of real estate portals have not been published yet. Relevant papers, dealing with price modeling under the consideration of spatial autocorrelation have been published by Chica-Olmo [6], Bourassa et al. [3], Tsutsumi et al. [21], Wong et al. [25] Kuntz and

Helbich [11]. The literature review revealed that the most suitable identified alternatives for real estate portals, taking into account spatial-autocorrelation, are methods from the Kriging family: Ordinary Kriging (OK), Kriging with External Drift (KED) and Ordinary co-kriging (OCK) (see [3, 6, 11, 21, 25]). Those methods allow positional exact price estimations based on spatially correlated price data. It should also be noted, that several studies successfully demonstrated experimentations with Geographically Weighted Regression methods (GWR) to predict real estate prizes [3, 13]. Although GWR's prediction accuracies excel hedonic regression results [3], detected minimal lower prediction accuracies of GWR in comparison to Kriging methods [11] caused, that in the present study GWR was not considered as possible alternative to model rental prizes in the context of real estate portals.

Few studies have applied machine learning methods for estimating real estate prices. No study has been found estimating rental prices via machine learning methods. A comparative study of several machine learning algorithms to model house prices in St. Petersburg was done by Antipov and Pokryshevskaya [2]. Gu et al. [8] combine a genetic algorithm with random forest for predicting property prices. Caplin et al. [5] apply machine learning methods for the determination of real estate prices. Park and Bae [15] compare several machine learning algorithms to estimate property prices. As a concluding result of the literature review about applicable machine learning algorithms, a closer look at RF should be done as a possibly suitable method of rental price estimation for real estate portals.

Studies in the field of geovisualisation usually only treat aspects that are worth being considered to develop alternative rental price visualizations. Overall grid mapping approaches are considered as a relevant visualization perspective for mapping rental prices. Furthermore, grid mapping is a possible approach to represent the "real" distribution of rental prices in space. The most relevant literature about grid mapping comes from Trainor [19]. He summarizes several arguments in favor of mapping statistical facts based on grid cells. According to Trainor [19], the most important argument for grid maps are the greater comparability of thematic facts. The U.S. Census Bureau was a pioneer in applying a grid mapping approach and started in 1990's transforming irregular shaped statistical blocks (US Census block) in a national grid.

3.1 Quantitative Data: Rental Property Offers Database

For the evaluation of the identified alternative analysis and visualization methods, a unique multi-temporal and multi-dimensional dataset of rental property offers could be acquired. The database extract made available through the German portals' Immobilienscout 24.de transparency program, is a complete copy of the entire housing rent offers of the German city of Potsdam. The dataset includes all rental property offers of Potsdam, quarterly recorded and ranging from January 2007 to September 2013. Overall, the data set includes 74,098 rental properties. This data is geocoded using the WiGeo EU (Lambert conformal conic projection) projection, and includes 63 different real estate specific variables, 37 of them numeric (area, number of rooms, rent with and without service costs), 16 categorical variables about certain infrastructure (e.g. built-in kitchen, heating type) and ten strings (especial names and internal codes). The dataset structure is similar to other real estate portal databases and therefore, is ideally suited to validate alternatives to the hedonic regression and the mapping on non-appropriate reference geometries. From the 1st quarter of 2013, the rental offers ~3000 data points have been used as in all model runs.

3.2   Kriging as Spatial Alternative to Model Rental Prices

Kriging methods have been identified as the most promising alternative to hedonic regression models. A validation framework was designed in order to determine if Kriging is a useful geospatial price modeling alternative in the operational context of real estate portals. Within this validation framework an exemplary implementation of OK, KED and OCK was realized. Kriging methods model the value of a point by the spatial configuration of its surrounding observation points, by finding a suitable variogram model. Variograms model the semivariance as depended on distance and is defined by the choice of a model function (e.g. gaussian, exponential, spherical, mátern, etc.) and the associated model parameters: nugget, partial sill and range. The spatial structure of a dataset has to be represented by an appropriate choice of the modeling parameters. A detailed description of different Kriging models can be found in Li and Heap [12], Wackernagel [23] or Hengl [9]. All Kriging methods are a variant of the following Kriging formula, estimating the value on position $Z(x_0)$, where n represents the number of data points in the local neighborhood used for the estimation; λ_i is the kriging weight, and $\mu(x_0)$ stands for the mean in the neighborhood within the selected neighborhood and the trend component μ.

$$Z(x_0) = \sum_{i=1}^{n} \lambda_i [Z(x_i) - \mu(x_0)] + \mu \qquad (2)$$

OK uses only primary variable's information. A constant mean of each estimation point in the neighborhood is used as trend component. The residual component results from spatially-weighted deviations from the local average of neighboring observation points. KED models (a synonym to Universal Kriging) are similar to OK models. However, the trend component is calculated using a linear regression of the available secondary variables on the modeling point. This is based on secondary information of the observation points and calculates a Simple Kriging on the corresponding residuals (Li and Heap, 2008). OCK uses autocorrelation between primary variables, and cross-correlation between secondary variables in order to interpolate the primary variables. The main advantage of OCK is that the data of the secondary variables does not have to be present on all observation and modeling points. A detailed description of different Kriging models can be found in Li and Heap [12].

In order to compare the three considered Kriging methods, it was necessary to find a suitable set of secondary variables. To support the selection of secondary variables and to determine the secondary variables importance, RF and a principal component analysis (PCA) were applied. The intersection of RF score values and the PCA analysis was used as co-variable set: Year of construction, number of rooms, floor, space, kitchen (categorical, only for machine learning methods). The selection corresponds to similar studies (see [3, 6, 25]). All model evaluations were done on a historical subset of the data, from the first quarter of 2013.

The most important preprocessing step before the actual model-evaluation is the determination of the presence strength of the spatial autocorrelation. Spatial autocorrelation was tested by applying both, the global Moran's I, Geary's C and the local Getis Ord G tests. By implementing the R packages *spatstat* and *ape*, a weight matrix of inverse distances was built, incorporating 4 points neighboring the observation point. Higher weights were assigned to closer points and lower weights to more distant points. The global autocorrelation was tested applying the Moran's I and the Geary's C hypothesis test, resulting in positive autocorrelation for the 2 global tests. A global Moran's I of 0.81 and Geary's C value of 0.2 could be detected. For a further test of the null hypothesis of Moran's I was applied in addition within binary distance matrices of the observation points within 300 m, 500 m, 1 000 m, 5 000 m and 15 000 m distance from each observation point. By determining the spatial auto correlation to a subset of the observation points (lags) within the defined distance classes, it is possible to examine the behavior of the auto-correlation at different distances. The null hypothesis could be rejected in 4 out of 5 built distance classes. Only the last distance class, of 15 000 m, showed no more spatial autocorrelation. Getis Ord G was applied locally. The results clearly identified areas with negative and strong positive spatial autocorrelation. The test allows further analyses, such as hot and cold spot calculation and their spatial representation.

The performance of the considered methods OK, KED and OCK was be examined taking into account different parameter configurations. A validation framework was designed and implemented within R (*gstat* package), in order to compare OK, KED and OCK methods with each other. The parameters nugget, sill and range of the models have been varied by applying a loop function inside R. Modeling results were validated applying a 5-fold cross validation deriving the mean error (ME); the root mean squared error (RMSE); the mean relative deviation in percentage of data points, with a deviation <10 %; and the models computing time in seconds. These results were compared with hedonic regression results. After 120 KED and OK model runs with 3^3 parameter combinations of KED and after OCK 9 model runs, no more significant changes in the quality of the model could be observed. Spherical, Gaussian and exponential model functions were used. In the OCK model runs only the range value was specified. The remaining parameters (nugget and partial sill) were calculated iteratively by a weighted least square estimation. The reason for the lower number of modeling runs of OCK is the very high computation time from 10 to 12 h. Further model runs have not been realized, as no more significant increases in the model's accuracy could be achieved.

The goal of this comparison was to check the sensitivity of the nugget, sill and range parameters in order to find the most suitable parameter configurations, within the 3 compared Kriging models, applicable on real-world conditions of real estate portals.

An overall result of the model comparison is that all Kriging model outputs showed more precise price estimations than the hedonic regression model. As result of the model comparison, some recommendations can be given on which Kriging model, with which parameter configuration under a real estate portals "real world" condition is best suitable for real estate portals. However, the results retrieved an unexpected fact. The validation showed that OK, although they did not require secondary variables, showed more accurate modeling results than KED and OCK. A tested further variation of co-variable sets didn't show improvements in the models accuracy. KED and OK models show a similar modeling accuracy and performance. The computing time for the entire study area takes up to 60 s. OCK, on the contrary, showed lower modeling accuracy, with up to 100 times higher computing time. The following Table 1 shows the 5 best performing models sorted by an increasing RMS, it's modeling parameters and computational runtime in seconds .

Table 1. Best performing models sorted by an increasing RSME

Method	Nugget	Partial sill	Range	Funct.	RMSE	t (sec)
OK	0.001	0.1	700	Exp	0.788	45.480
OK	0.001	0.1	600	Exp	0.789	50.440
OK	0.001	0.095	600	Exp	0.789	60.030
KED	0.001	0.06	200	Exp	1.069	40.250
OCK			1100	Sph	1.207	34350.4

A graphical comparison of the most accurate Kriging model in terms of the RMSE with a hedonic regression model, in which the regression result was mapped onto a Infas neighborhood polygon (as used by Immobilienscout 24.de), shows the much finer spatial granularity of the Kriging result. This allows recognizing much finer spatial rental price variations (Fig. 3).

Fig. 3. Comparison of OK modelling results in 50 m cell-size, compared with results of a hedonic regression mapped on a "Infas neighborhood" geometry (30 ha.).

3.3   Random Forest as an Alternative to Process Modeling

The machine learning algorithm RF has been identified as a possible alternative to the hedonic regression, although it is not an explicit spatial method. RF is a further development of the conventional simple decision trees that has been developed by Breiman [4]. RF predicts the decision of the final result from a variety of randomized grown "forest" of decision trees [4]. The exemplary dataset includes several categorical variables. As RF can deal with variables of this type and is described to be very robust, it seems to be a very interesting modeling alternative. Its implementation was realized in R using the R data mining GUI "rattle" [24] and the R "random forest" package.

Following Breiman's recommendations to put aside data in order to calculate the out-of-the-bag (OOA) error (a performance measure for assessing the RF models), RF was used exemplarily as a regression model to estimate the base rent per square meter without service cost. The selection of training samples and the decision at which point in the feature space the partitioning takes place. 14 variables of the data set, including categorical variables, which is a strength of RF model, have been used to build the randomized decision trees.

The dependent variable was the base rent per square meter without service costs. 10 model runs have been varied. Following Breiman's [4] recommendations in increments of 50, from 100 to 500 decision trees have been modeled. As measure of the models overall prediction accuracy the mean squared error (MSE) and the root mean squared error (RMSE) was calculated. Furthermore, the logarithmic MSE in% (lnMSE%) for each model run was calculated based on the OOA samples. The lnMSE% calculation is based on the assumption that in every model run a variable is removed from the model and serves to assess the importance of variables in a RF model. A further measure was the nude impurity as a measure of heterogeneity within a node in the model. In the model runs, the error of the prediction rates decreased quickly. This decrease could already be observed from 50 randomized trees. After 50 trees, the decrease lowered and the model reached its best accuracy values with 150 trees.

Overall, it can be said that RF has the potential to serve as an alternative modeling approach. Nevertheless, instead of using RF as a standalone price prediction model, RF has a big potential in predicting punctually "machine trained" observation points as input, when a Kriging model misses input points. Comparisons with hedonic regression results, show that RF models have been more accurate in all model runs.

3.4   Prototype Development of Grid-Maps as Alternative Geovisualisation Approach

Based on the modeling results of the most valid Kriging approach, an exemplary implementation of a grid map approach was realized with the goal to visualize the most approximate distribution of cold rents in space separated from non appropriate geometries. Prior to the implementation of these correctly localized "zoneless" rental price maps, requirements of the rental map, such as the medium and the map reader target group, had to be defined. Map users are apartment seekers, who want to find out the rental price in their preferred residential area; stakeholders from the real estate business, such as real estate agencies or investors; conducting market observations to decide possible investments; and the public sector, that wants to monitor the price structure in their communities but lack appropriate data and evaluation tools. The preferred medium of the target groups is the World Wide Web. Therefore, implementation of a grid map approach has to be done as a web map. As web map, a web-based graphical representation of the data model is understood, where the dataset is made visible by attribution via graphical features. *Leaflet.js*, an open-source JavaScript map library to create mobile interactive maps, was used as visualization framework to implement the grid-map approach. The most important aspect in making a price map is the selection of the appropriate grid-mesh sizes to represent the best approximation of the facts to be displayed on the map on different spatial scales.

It should be noted that per se no ideal mesh size exists and that the mesh size has to be determined in the context of the planned visualisation. GIS software packages, as the proprietary ArcGIS, use a pragmatic, but scientifically not comprehensible approach to determine the mesh size. The extent of a study area is simply divided by the default factor of 250 in north-south and east-west direction. However, this unscientific approach is inadequate for the presented prototype development.

A two-step approach to determine appropriate cell sizes was developed. First, the largest possible scale for the grid-map prototype was determined and second the grid cell sizes for smaller scales was derived from the largest possible scale. Starting point of the approach was the determination of the average number of square meters of the residential buildings in the study area. For that purpose, the Overpass API, a read-only server-side application programming interface to do customized selections of the Open-StreetMap database, was used in order to acquire Potsdam's residential building block. The next step was the calculation of the medium footprint ($247\ m^2$) of the building blocks. This footprint size rooted would mean a grid cell size of 15.72 m. Experimentally interpolations to a raster with such a small cell size however result in extraordinary long interpolation times.

A R script developed by and based on Hengl's [18] research on appropriate grid cell sizes was adapted and extended to determine all in all three resolution levels for the grid map. The medium distance between the present rental property offers (110 m) and the offers' density in the study area ($57.55/km^2$) was calculated. The calculation of the cumulative distance showed that 90 % of the rental property offers have a maximum distance of 20 m. All other offers are farther distant. Optimal mesh sizes for the present application are therefore between 15.75 m, 75 m and 110 m. Finally, based on the calculations made, recommendations regarding the selection of appropriate map scales and zoom levels for 3 grid cell ranges sizes have been made. Zoomlevels have been assigned following OpenStreetMap's zoom levels (Table 2).

Table 2. Recommendations for maps scales and zoom levels

Cellsize ranges (m)	Map-scales	Zoom level
>75–110	1: 35 000–1: 70 000	14 & 13
>50–75	1: 8 000–1: 15 000	16 &15
20–50	1: 4 500, basis for interpolation	

The following figures (cf. Figs. 4, 5, 6 and 7) compare one of the developed prototypes of a real estate grid map to/with the state of the art mapping published by the German portal Immobilienscout 24.de. All in all 19 visualisation prototypes were developed. A much finer spatial distribution of the rental price in all map scales can be observed. Rental prices are only visualized where they actually occur. Topography, industrial and commercial areas are not shown in the prototype, which shows a much more differentiated picture of the "rental landscapes" than the status quo.

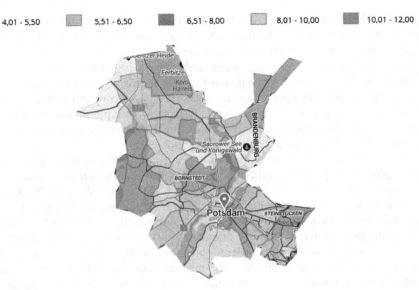

Fig. 4. Rental price map, example for the state of the art at the map scale 1: 36 000. Above: map legend of the Immobilienscout 24.de rental price classes

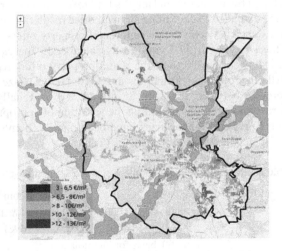

Fig. 5. Developed prototype at the map scale 1.35 000

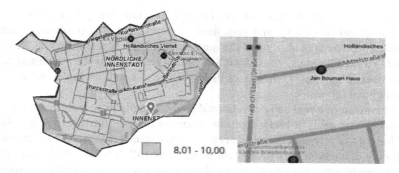

Fig. 6. Rental price maps state of the art at map scales 1: 13 500 (left) and 1: 4 500 (right)

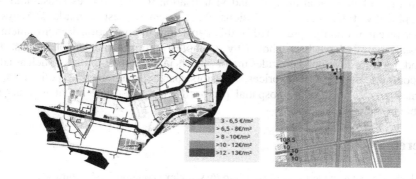

Fig. 7. Developed prototype at the map scales 1. 13 000 (left) and 1: 4 500 (right)

4 Conclusions

This paper presents and discusses novel and innovative approaches for the analysis, visualisation and visual analysis of real-estate rental prices. It demonstrates that any of the new spatial analysis methods presented here proves to be more efficient than the more conventional approaches to rental price determination and visualisation currently in use. The research findings indicate that the analysis results achieved by any of the new methods, ranging from stochastic interpolation to the "random forest" method of machine learning, are more valid than results obtained from traditional statistical methods. By uncovering the "unseen" spatial structures of real-estate markets, they allow for in-depth insight into the largely neglected geographical nature of real-estate rental prices. This is facilitated by a key element implemented in these methods, i.e. precise 2D positional information of real-estate price and their adequate, effective carto-graphic representation.

We expect that the implementation of geospatial and geovisual analysis methods into real-estate web portals will boost both the content, the analytical power and use effectiveness of these portals. To date, real-estate price indices published by such portals often provide the only information on urban real-estate markets. Lacking a well-researched spatial processing and visualisation component, this frequently results in fuzzy and distorted conclusions of the status quo of the housing market. It can be observed that such findings are used without further reflection by the media and politics to present seemingly valid but in fact ill-found insights into an important economic sphere of global relevance.

A comparison of novel and conventional analysis methods reveals that the geostatistical processing, analysis and visualisation approaches presented here help to lay the foundations of a new research area that can be termed spatial real-estate market analysis. The defining element of this R&D field is what we call "geospatial business intelligence". It can be seen that this emerging field of innovative geospatial analysis opens up new and promising research and application perspectives. For the first time, the combination of geospatial analysis and visualisation tools facilitates substantial, i.e. geographically differentiated conclusions about urban real-estate markets. Ways to transfer research findings presented in this paper into the operational environment of real-estate portals have been shown by examples of prototypical applications. The methods toolbox developed provides major players in real-estate-markets, such as advisory committees on property prices, investors, banks, communes or politicians, an informed precise locational geospatial perspective on real-estate rental markets not possible in the past.

References

1. Alexa: Top Sites in Germany (2013). http://www.alexa.com/topsites/countries/
2. Antipov, E.A., Pokryshevskaya, E.B.: Mass appraisal of residential apartments: an application of random forest for valuation and a CART-based approach for model diagnostics. Expert Syst. Appl. **39**(2), 1772–1778 (2012)
3. Bourassa, S.C., Cantoni, E., Hoesli, M.: Predicting house prices with spatial dependence: a comparison of alternative methods. J. Real Estate Res. **32**(2), 139–160 (2010)
4. Breiman, L.: Random forests. Mach. Learn. **45**(1), 5–32 (2001)
5. Caplin, A., Chopra, S., Leahy, J. V., LeCun, Y., Thampy, T.: Machine learning and the spatial structure of house prices and housing returns. SSRN 1316046 (2008)
6. Chica-Olmo, J.: Prediction of housing location price by a multivariate spatial method: cokriging. J. Real Estate Res. **29**(1), 95–114 (2007)
7. Gordon, R.J.: The Measurement of Durable Goods Prices, National Bureau of Economic Research Monograph, 1st edn. University of Chicago Press, Chicago (1990)
8. Gu, J., Zhu, M., Jiang, L.: Housing price forecasting based on genetic algorithm and support vector machine. Expert Syst. Appl. **38**(4), 3383–3386 (2011)
9. Hengl, T.: A Practical Guide to Geostatistical Mapping of Environmental Variables, vol. 140, no. 4, pp. 417–427 (2009)
10. Immobilienportale: Immobilienscout 24 (2012). http://www.immobilienportale.com/uebersicht-immobilienportale/20084-immobilienscout24/
11. Kuntz, M., Helbich, M.: Geostatistical mapping of real estate prices: an empirical comparison of kriging and cokriging. Int. J. Geog. Inf. Sci. **28**(9), 1904–1921 (2014)
12. Li, J., Heap, A. D.: A Review of Spatial Interpolation Methods for Environmental Scientists. Geoscience Australia, Record 2008/23 (2008)

13. Manganelli, B., Pontrandolfi, P., Azzato, A., Murgante, B.: Using geographically weighted regression for housing market segmentation. Int. J. Bus. Intell. Data Min. **9**(2), 161–177 (2014)
14. Nexiga: Datenkatalog Marktinformationen (2014). http://bit.ly/1ySXf4i
15. Park, B., Bae, J.K.: Using machine learning algorithms for housing price prediction: the case of fairfax county, virginia housing data. Expert Syst. Appl. **42**(6), 2928–2934 (2015)
16. Stroisch, J.: Immobilien bewerten leicht gemacht. Haufe-Lexware (2010)
17. Tobler, W.R.: A computer movie simulating urban growth in the Detroit region. Econ. Geogr. **46**(1970), 234–240 (1970)
18. Hengl, T.: Finding the right pixel size. Comput. Geosci. **32**(9), 1283–1298 (2006)
19. Trainor, T.: Common Geographic Boundaries: Small Area Geographies, Administrative, and Grid-based Geographies – One or Many? In: Global Forum on the Integration of Statistical and Geospatial Information, 4-5 August 2014, New York. http://bit.ly/1C3JkHJ. Accessed 3 Mar 2016
20. Trulia: New York Real Estate Market Overview (2016). http://www.trulia.com/real_estate/New_York-New_York/
21. Tsutsumi, M., Shimada, A., Murakami, D.: Land price maps of Tokyo metropolitan area. Procedia Soc. Behav. Sci. **21**, 193–202 (2011)
22. United States Census: Geographic Terms and Concepts-Core Based Statistical Areas and Related Statistical Areas (2010). http://www.census.gov/geo/reference/gtc/gtc_cbsa.html
23. Wackernagel, H.: Multivariate geostatistics: an introduction with applications. In: International Journal of Rock Mechanics and Mining Sciences and Geomechanics Abstracts, vol. 33, no. 8, p. 363A. Elsevier (1996)
24. Williams, G.J.: Rattle: a data mining GUI for R. R J. **1**(2), 45–55 (2009)
25. Wong, S.K., Yiu, C.Y., Chau, K.W.: Trading volume-induced spatial autocorrelation in real estate prices. J. Real Estate Finan. Econ. **46**(4), 596–608 (2013)
26. Zillow: What is a rent Zestimate? (2016). http://www.zillow.com/wikipages/What-is-a-Rent-Zestimate/
27. Zillow: Zillow Home Value Index: methodology (2016). http://www.zillow.com/research/zhvi-methodology-6032/

Constraints-Driven Automatic Geospatial Service Composition: Workflows for the Analysis of Sea-Level Rise Impacts

Samih Al-Areqi[1](✉), Anna-Lena Lamprecht[2], and Tiziana Margaria[2]

[1] Institute for Computer Science, Chair Service and Software Engineering,
Potsdam University, Potsdam, Germany
alareqi@uni-potsdam.de
[2] Lero - The Irish Software Research Centre, University of Limerick, Limerick, Ireland
{anna-lena.lamprecht,tiziana.margaria}@lero.ie

Abstract. Building applications based on the reuse of existing components or services has noticeably increased in the geospatial application domain, but researchers still face a variety of technical challenges designing workflows for their specific objectives and preferences. Hence, means for automatic service composition that provide semantics-based assistance in the workflow design process have become a frequent demand especially of end users who are not IT experts. This paper presents a method for automatic composition of workflows for analyzing the impacts of sea-level rise based on semantic domain modeling. The domain modeling comprises the design of adequate services, the definition of ontologies to provide domain-specific vocabulary for referring to types and services, and the input/output annotation of the services using the terms defined in the ontologies. We use the PROPHETS plugin of the jABC workflow framework to show how users can benefit from such a domain model when they apply its constraints-driven synthesis methods to obtain the workflows that match their intentions.

Keywords: Geospatial services · Service composition · Scientific workflows · Semantic domain modeling · Ontologies · Climate impact analysis

1 Introduction

Like many application domains, the geospatial domain has recently seen a trend towards migrating data analysis software processes from predefined static systems to purpose-specific compositions of existing services, often in the form of workflows. In the geospatial application domain, big geographic data, a lack of interoperability and complex analysis processes constitute particular barriers for a successful and wide reuse of components and services. Service-oriented architecture (SOA) principles and Web Service technologies have been embraced by the geospatial community. Its members have become more aware of the benefits of sharing their data and computational services, and are thus contributing to

© Springer International Publishing Switzerland 2016
O. Gervasi et al. (Eds.): ICCSA 2016, Part III, LNCS 9788, pp. 134–150, 2016.
DOI: 10.1007/978-3-319-42111-7_12

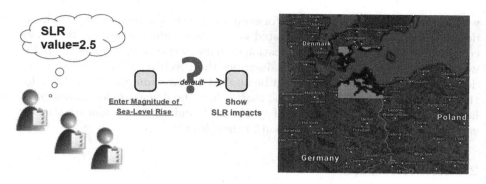

Fig. 1. Example scenario: analyzing the impacts of sea-level rise.

distributed data and services. As a result, scientific data have become increasingly remotely accessible in a distributed fashion through standardized geospatial Web Services [22].

From the perspective of using this distributed infrastructure for developing own software applications tailored to specific needs, however, users still face big challenges with regard to discovering services suitable for the purpose, and with exploring how to (re-)use and compose components correctly. With plenty of generic purpose geospatial services (such as data access services, portrayal services, data transformation services, and location-based services) and specific-purpose services designed to address particular geospatial applications available, it can be very hard and time-consuming for users to identify adequate (combinations of) services manually. As an illustrative example, consider the scenario depicted in Fig. 1, where an arbitrary user wants to analyze the impacts of a sea-level rise of 2.5 m for a particular region. He knows what the initial data in this situation are (magnitude of sea-level rise, the region in question) and what he wants to see in the end (e.g. a map showing the flooded areas), but he does not know which computational steps are needed to carry out the analysis that yields this result. Furthermore, the user might have preferences with regard to the data formats, scale and georeferencing systems used in the process.

A possible approach to overcome this situation is the use of semantics-based automatic workflow composition techniques, which require the available services to be annotated with machine-readable metadata, and are then able to automatically derive possible workflows for a given specification, like the one in the figure. Many such techniques, which are often based on synthesis or planning algorithms, can furthermore take into account the users' specific requirements and preferences, for instance regarding services which should or should not be used, by adding additional constraints to the workflow specification. They are effective means to assist users with different objectives, perspectives, and preferences in their workflow design.

In this paper we show how after an adequate domain modeling of sea-level rise impacts analysis, users can make use of the PROPHETS plugin [31] of the jABC workflow modeling framework [35] for semi-automatic semantics-based workflow design. Section 2 briefly surveys related work on semantic and automatic service composition in the geospatial application domain. Section 3 introduces the jABC workflow modeling framework and describes the PROPHETS plugin. Section 4 demonstrates how PROPHETS can be applied on the example of workflows for the analysis of sea-level rise impacts, that is, how the domain model is designed, how workflows can be specified and synthesized, and especially how constraints can provide further guidance. Section 5 concludes the paper.

2   Related Work

Several works have addressed the construction of domain-specific applications by assembling and reusing geospatial tools and data as services [6,22,30], and many researchers followed the Open Geospatial Consortium's (OGC) Web Service standards [1] in order to increase discoverability and compatibility (e.g. [10,12,33]). However, despite the substantial efforts by the OGC to provide standards for geospatial Web Services and their widespread adoption in the scientific community, they lack a formal semantic description, which would be required to synthesize workflows based on OGC Web services automatically [37,40].

Workflow management systems have been used in the geospatial domain to develop and implement custom processes. For example, jOpera has been applied early in the geospatial domain [4], and also the Kepler scientific workflow system [27] has been used to implement distributed geospatial data processing workflows using Web Services [15] and in particular OGC services [7,32]. Other works used BPEL-based business workflow technology to orchestrate geospatial services [14]. However, they only comprised means to simplify the (manual) workflow composition process syntactically, and learning how to apply these technologies to build a system based on services remained complex for application experts, in particular with the heterogeneity and the interoperability challenges of geospatial data.

Recently, more attempts were undertaken towards semantic and automatic geospatial service composition using AI planning and program synthesis techniques [9,38]. For example, [8] presented an approach that integrates planning methods and semantic annotation to improve the robustness of geospatial Web services composition based on geodata quality requirements. Already earlier, several works used OWL and OWL-S techniques to describe the functional capabilities of geospatial services [13,19,20,39]. For example, OWL was introduced into Kepler to enable automatic structural data transformation in the data flow among services and OWL-S was adopted to automate the composition of geospatial Web services [39].

The successful application of all these techniques for (semi-)automatic workflow composition depends on the provisioning of adequate meta-information about the involved technical entities (services, data types) of the target application. Ontologies or taxonomies are frequently used structures to represent this information. For example, an ontology model to address the semantic discovery and retrieval of geospatial services has already been described by [11,18] almost a decade ago, and also the service taxonomies of [5,25] where designed to handle the semantic discovery of geospatial services. Another example is the service taxonomy model that has been introduced by the International Organization for Standardization (ISO) in ISO19119 [5], which is however restricted to OGC services.

In our work, we combine an intuitive graphical formalism for the manual composition of services into workflows with additional functionality, again embedded in a very intuitively usable plugin, to apply a synthesis algorithm to combine services automatically according to an abstract specification. In this paper we focus on the application of this framework on an example that deals with the semi-automatic composition of workflows for the analysis of the impacts of sea-level rise. Since no application-specific ontological models about the services and types in this domain were available, we designed the required taxonomies ourselves. In contrast to many related approaches, which either support OGC or not, they comprise OCG-compliant as well as non-OGC terms. Thus, the domain model enables the user to apply a greater range of user objectives, perspectives and input/output preferences during the synthesis process.

3   The jABC Framework and PROPHETS

The multi-purpose process modeling and execution framework jABC [35] is the current reference implementation of the eXtreme Model-Driven Design (XMDD) paradigm [29], which advocates the rigorous use of user-level models throughout the software development process and software life cycle. The service concept of jABC is very close to an intuitive understanding of service as something that is required to be ubiquitously accessible (location-agnostic) and mechanically configurable [16]. The term "service" is used to denote functional building blocks (SIBs), which are viewed as independent from their location, the program entity, and hardware-platform which provides them. The jABC provides a comprehensive and intuitive graphical user interface in which users easily develop workflow applications by composing reusable building blocks into hierarchical (flow-)graph structures (called Service Logic Graphs, or SLGs) that are executable models of the application. The workflow development process is furthermore supported by a set of plugins providing additional functionalities, so that the SLGs can be analyzed, verified, executed, and compiled directly in the jABC.

Figure 2 gives an impression of the jABC in action: The SLG on the canvas has been created using SIBs from the library (displayed in the upper left of the window) in a drag & drop fashion, and connecting them with labeled branches representing the flow of control. After the parameters of the SIBs have been configured (in the SIB inspector in the lower left), the workflow is ready for execution. The small window in the upper right corner of the figure is the control panel of the Tracer plugin that steers the execution of the models. The third window in the figure shows the result of the workflow that also has been opened during execution.

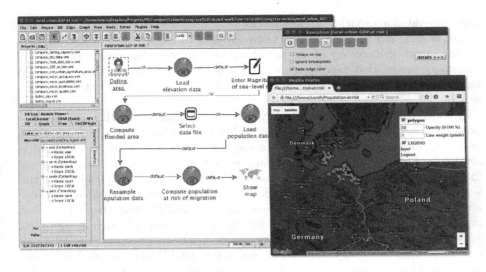

Fig. 2. GUI of the jABC framework in action.

One of the plugins providing additional functionality to the jABC is PROPHETS (Process Realization and Optimization Platform using a Human-readable Expression of Temporal-logic Synthesis) [31], which follows the Loose Programming paradigm of [24] to facilitate semantics-based semi-automatic workflow design in addition to manual workflow construction. With PROPHETS, workflow designers are not required any more to implement the entire workflow manually. Instead, they can just provide a sketch of the intended workflow, together with a set of constraints that further specify the analysis objectives. The plugin then applies a synthesis algorithm [34] to this abstract specification, and returns a set of possible implementations to the user, who can select the one to be inserted into the workflow. As illustrated in Fig. 3, working with PROPHETS consists of basically two phases:

1. In the **domain modeling phase**, domain experts provide resources (services, data) and the corresponding metadata. Concretely, domain modeling for PROPHETS involves the following steps:
   - Integrating the services (in fact, provisioning SIB libraries for jABC),
   - defining service and type taxonomies to provide a controlled vocabulary for referring to entities in the domain model (taxonomies can be seen a special kind of ontologies, namely ontologies with only *is-a* relations),
   - describing the behavior of the services interfaces (inputs and outputs) and
   - possibly also defining constraints that express additional knowledge and requirements about the application domain.
2. In the **workflow design phase**, the workflow designer can then indicate one or more branches between SIBs as loosely specified and apply the synthesis framework provided by PROPHETS to replace them by appropriate

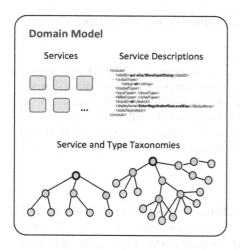
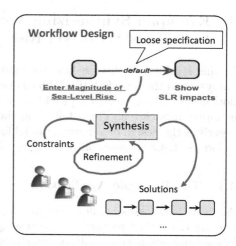

Fig. 3. Phases of automatic workflow design with PROPHETS.

concrete service sequences. He can also define additional constraints to be taken into account by the synthesis. For this purpose, PROPHETS provides a constraint editor with natural-language constraint templates, which users can apply without having knowledge of the underlying logic. Currently, 12 templates are available in PROPHETS (covering frequently applied constraints such as service/type avoidance, service redundancy avoidance and service/type enforcement), but users can also define additional templates or concrete formulas directly in an "advanced" mode of the editor.

The input specification for the synthesis algorithm that PROPHETS applies is then simply a conjunction of all available constraints:

– The *start constraint*, that is, the set of data types that are available at the beginning of the loosely specified branch (determined by using a data-flow analysis method),
– the *goal constraint*, that is, the set of data types that are required by the SIB at the and of the loosely specified branch, and
– any *further workflow constraints*, that is, constraints from the domain model and additional constraints provided by the user.

PROPHETS automatically transforms the domain model and constraints into a Semantic Linear Time Logic (SLTL) formula [34]. Although a number of solutions can result from the synthesis execution, obtaining actually adequate solutions also depends on the provided constraints. In the next section, we show in detail how users can define different types of constraints for the example of sea-level rise impact analysis and how this influences the solutions that are generated.

4   Example: Synthesizing Workflows for Assessing Impacts of Sea-Level Rise

In this section, we discuss how we model the domain of sea-level rise impact analysis and then use PROPHETS to apply techniques for semi-automatic workflow design. First, Sect. 4.1 provides additional background information on the example application. Then, following the two phases described above, Sect. 4.2 describes the design of a domain model for the SLR impact analysis applications, before Sect. 4.3 focusses on performing the actual SLR workflow design.

4.1   The Example Application

Analyzing and assessing potential impacts of climate change are critical and challenging tasks that require the processing of large and heterogeneous datasets. These analyses are particularly demanding because of the multi-scale and multiobjective nature of environmental modeling for climate change impact assessment [26]. For the example of sea-level rise (SLR) that we focus on in this paper [2], climate change is assessed with respect to the potential loss of agricultural production, calories available and effect for food security, but also with respect to properties of rural and urban damage functions. To this end, heterogeneous data (such as, e.g., elevation, land-use, population density or yield data) has to be used, which comes in different formats and at different scales, requiring adequate integration and aggregation.

Several tools and applications have been developed to analyze the risk index of climate impacts, such as data creation, conversion, and visualization tools. The scientific tools that we used for our application address the analysis of the impacts of sea-level rise. These tools are used in the ci:grasp[1] climate information platform [36]. They are based on scripts in the GNU R language that comprises several tools for spatial analysis. The srtmtools package [21] used for the data analysis provides the methods required to produce results as presented on ci:grasp. It combines various tools that are based on different packages, such as the raster package tool[2] for data reading, writing, manipulating, analyzing and modeling of gridded spatial data, the Gdal tool[3] for data conversion, and other packages for data visualization such as Png[4] and plotGoogleMaps [17]. The current SLR workflow scenario comprises six different applications for the computation of potential land loss (ha), population at risk of migration, rural and urban GDP at risk, potential yield loss, potential production affected ($), and potential caloric energy loss [3]. According to different objectives and user preferences to assess SLR impacts, each workflow application has several variations.

---

[1] http://www.cigrasp.org.
[2] http://cran.r-project.org/web/packages/raster/.
[3] http://www.gdal.org.
[4] http://www.rforge.net/png.

4.2   Domain Modeling

In the following we describe the domain modeling for the SLR application example, that is, the provisioning and the description of the services and the design of taxonomies for services and data types, in greater detail. Note that the taxonomies have been designed from scratch, that is, they have not been derived from any existing ontology models.

Services. The first step in setting up the domain model for the SLR impacts analysis workflows was to turn the SLR tools mentioned in the previous section into services adequate for (re-)use as SIBs in the jABC framework. We used the jETI (Java Execution Tool Integration) platform [28], with which this was a straightforward process once the desired functionalities had been identified. After the provisioning of the actual SIBs, the next crucial task for the domain modeler was then to define adequate modules to be used by the synthesis algorithm and annotating them with semantic meta-information.

Each module is linked to a concrete SIB that provides the implementation, but the module's names and the names used to describe input and output data types for the synthesis algorithm are symbolic. This makes it possible to use own domain-specific terminology for the module descriptions that is decoupled from the terminology used for the SIBs implementation, and in particular also allows for polymorphism in the sense that one SIB can be used by several modules. This is in particular useful for SIBs that provide quite generic functionality, and where several modules with specific functionality can be defined based on the same underlying implementation. In fact, modules with specific, unambiguous functionality typically lead to better synthesis results (cf. [23]).

We would define just one module for services with very specific functionality (e.g. computational services), whereas for more general services (e.g. data loading, data resampling, data format converting or data projection transformation services) several modules were defined. For example, only the module Compute population-risk has been defined for the Compute population at risk of migration SIB. As another example, four modules refer to the load raster data SIB, namely load-landuse-data, load-yield-data, load-GDP-data, and load-population-data.

In addition to the modules we have defined based on the standard SIB libraries of jABC (such as the data input modules Enter-magnitude-of-sea-level-rise, Define-area-coordinates and Select-vector-GDPdata), we defined 37 different modules for SIBs that were created specifically for the SLR application example. Table 1 shows the names, descriptions, and input/output information for 30 of them.

Taxonomies. Service and type taxonomies are used to provide abstract classifications for the terms used in the module descriptions, which are in particular useful for the formulation of constraints about groups of data types or services. In addition to particular features of the concrete applications, it is often also

Table 1. Selected modules of the SLR application example.

Name	Description	Inputs	Outputs
Enter-magnitude-of-sea-level-rise	Enter the value of sea level rise	no input	SLR
Define-area-coordinates	Enter coordinates values for the investigated region	no input	Coordinates
Load-SRTMelevation-data	Download the digital elevation model (DEM) for the selected area	Coordinates	SRTM-data
Select elevation data-file	Select a file contain elevation data	no input	raster-elevation-datafile
Load-elevation-data	Load elevation data from a raster data file	raster-elevation-datafile	usrelevationdata
Select raster file	Select a file contain raster object with a certain resolution	no input	raster-object resolution
Select GDP data-file	Select a file contain GDP data	no input	GDP-datafile
Convert-vector-to-raster-GDP	Convert vector data format to raster data format	GDP-datafile	GDPdata
Load-GDP-data	Load GDP data from a raster data file	GDP-datafile	GDPdata
Load-population-data	Load population data from a raster data file	raster-Population-datafile	populationdata.rds
Load-landuse data	Load landuse data from a raster data file	raster-landuse-datafile	landusedata.rds
Compute-flooded-areas	Compute the flooded areas for a region based on its DEM	Elevation-data and SLR	slrlandloss
Create-master-resolution60	Create a master-grid of raster data resolution 60	Elevation-data	RS60m
Create-master-resolution90	Create a master-grid of raster data resolution 90	raster-object resolution	RS90m
Resample-resolution	Resample raster data resolution to a particular scale	slrlandloss and Data-scales	resampled-data
Resample-landuse-data	Resample land use data with resampled data	resampled-data and landusedata.rds	landuse-sample
Resample-GDP-data	Resample GDP data with resampled data	resampled-data and GDPdata.rds	GDP-sample
Resample-population-data	Resample population data with resampled data	resampled-data and population-data.rds	population-sample
Compute-GDP-risk	Estimates potential economic damage in coastal communities	GDPdata and GDP-sample	slrGDP-loss
Compute-population-risk	Estimates the number of people that would be affected	populationdata.rds and population-sample	slrpopulation-risk
Compute-landloss(ha)	Estimates the area that will be potentially inundated	resampled-data	slrlandlossha
Compute landloss-class	Define the type of land affected, from 1–16 different land types	resampled-data, slrlanduse-sample and class-number	slrlandclassloss
Compute yieldloss	Compute actual and potential production value affected in USD	yielddata.rds and yield-sample	slrYieldlossUSD
ReadCRS-data	Read a predefined data of a particular georeferencing system from a file	no input	Georeferencing systems
Transform-to-EPS	Transform and convert data to a certain georeferencing system	Georeferencing systems and outputdata	output-result
Show google map	Generate an interactive map output using the Google Maps API	output-result	google-map
Generate Png outputs	Create static map in Png format	output-result	Png-file
Generate Pdf outputs	Create static map in Pdf format	output-result	Pdf-file
Produce GeoTIFF output	Create a geo-referenced file (GeoTIFF, ASCII) which can be used for further external GIS processing	output-result	Geo-referenced-file
Produce text output	Create a text file containing some summary and statistic information	output-result	Text-file

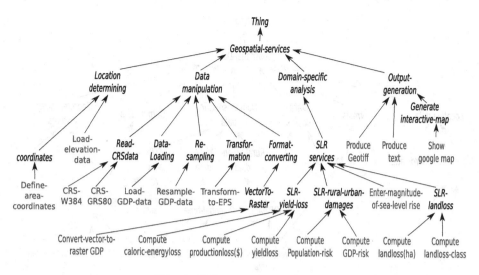

Fig. 4. SLR service taxonomy.

useful to incorporate knowledge from the application domain in general, in this case the geospatial application domain.

Figure 4 illustrates the service taxonomy that we defined for SLR application example. (Due to the limited space in the paper, we only depict the excerpts of the taxonomy that are relevant for the examples discussed in the next section.) Under the general *geospatial services* class, it defines four subclasses: *domain-specific analysis, data manipulation location determining* and *output generation* services. While *data manipulation, location determining* and *output generation* services could be reused in the whole domain of geospatial applications, the domain-specific *SLR services* class comprises services specifically for SLR impacts analysis. Thus, it contains SLR-related specific application services such as *SLR-landloss, SLR-urban-rural-damages* and *SLR-yieldloss*. The leaves of the service taxonomy tree correspond to concrete modules of the domain model, as described above.

Figure 5 illustrates the type taxonomy defined for the SLR application example, again reduced to the parts that are relevant to the discussed example. Unlike some existing models, in our type taxonomy we also treat geospatial features (e.g. format, resolution and georeferencing systems) as types. We classified *geospatial data* into six classes in order to address major characteristics of geospatial data. The data that is only used for SLR impact analysis is included in the class of *SLR data*. Data that might be used in several domains, such as population data, urban-rural data, land use data, yield data and calories data are categorized as *domain specific data*. Based on the *domain-specific data* classes, users can indicate via constraints which data are required for a specific analysis of SLR impacts, or avoid unnecessary data to reduce the time required.

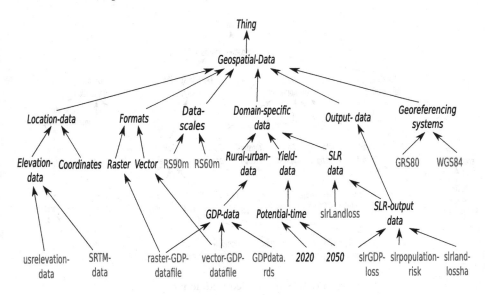

Fig. 5. SLR type taxonomy.

To enable users to consider different formats of geospatial data and handling their constraints regarding to the available data formats, we define the *Formats* class, which comprises various data formats (only Raster and Vector classes are depicted in Fig. 5). On the other hand, to address the respective data resolution, the *Data-scales* class is used to enable users to specify a certain scale of data resolution. Assuring the consistency of data projection is treated by defining a class for available *Georeferencing systems*. The *location data* class, which comprises *Elevation-data* and *coordinates* data is used to enable users with several options to determine location of the region. The *Output-data* class is used to group outputs of computation services, which are again inputs for the *output generation* services (see last four rows in Table 1). Again due to the limited space, only excerpts of the type taxonomy are depicted.

Constraints. During the domain modeling phase, general constraints that should be applied for the whole application domain (in this case SLR impacts analysis) are defined. For instance, we define the following constraints to avoid redundancy of some services and to ensure that an output generation service is used as a last service in the workflow:

- Do not use module Enter magnitude of sea-level rise more than once.
- Do not use module *Location-determining* more than once.
- Do not use module Select-raster file more than once.
- If module *ReadCRS-data* is used, module Transform to EPS has to be used next.
- Use module *output-generation* as last module in the solution.

4.3   Workflow Synthesis

In this section, we show how workflow synthesis can be applied to the example from Fig. 1, and how users benefit from the domain modeling for an adequate workflow design. As shown in Fig. 3, the overall synthesis process of PROPHETS performs the following steps: (1) interpreting the branch between the enter magnitude of sea-level rise SIB at the beginning and show SLR impacts SIB at the end as loose specification, (2) enabling users to edit the constraints, (3) generating the possible solutions, and (4) inserting the selected solution in the loose branch automatically.

Typically there are many different possibilities for workflows implementing this specification, and the adequate solutions depend on user requirements such as, analysis objectives (e.g. land loss, rural/urban damages, or yield loss), and user preferences (e.g. showing specific damages such as GDP loss or using certain type of data formats and resolution). In the following we show how the iterative adding of constraints helps the user to narrow down the set of obtained solutions to those that are of interest to him.

When we start the synthesis with only the domain constraints (which were defined in the domain modeling phase), first solutions are found in search depth 7, where 12 possible implementations of the specifications are detected. When we proceed to greater search depth, we find more and more solutions, for example 72 when searching up to search depth 10, and 768 solutions for depth 15. While all these are technically possible and correct combinations of services that solve the request, most of them are not adequate solutions, that is, they do not represent what the user actually wants to see. By adding constraints that express the user's intentions, uninteresting solutions can be excluded and adequate solutions be enforced. To demonstrate this impact of constraints, we defined three refinement levels based on different user requirements and preferences, reflecting different sorts of preferences. The numbers given in Table 2 illustrate their impact on the obtained solutions:

(A) This refinement is designed to find and compose services related to user objectives. We added constraints that concern the intended goal of SLR analysis by pointing the synthesis algorithm to one of the main classes of SLR analysis services. For instance, we can focus on rural urban damages analysis by adding the following constraint that enforces the use of at least

Table 2. Impact of constraints on the synthesis solutions

Refinement	Number of solutions					
	Depth 10	Depth 11	Depth 12	Depth 13	Depth 14	Depth 15
-	72	192	264	312	504	768
A	0	48	120	168	360	624
B	0	24	24	24	96	96
C	0	0	0	0	1	1

one module from the *SLR-services* class from the service taxonomy in the solution:

– Enforce the use of module *SLR-rural-urban-damages.*

As shown in Table 2, with constraints of refinement A, the synthesis starts to find solutions only at a search depth of 11, and less solutions are returned. At search depth 15, still 624 possible implementations remain.

(B) This refinement is used to tailor the specification closer to the user's specific perspective. Concretely, we add the following constraints to explicitly include and exclude particular (groups of) modules from the solutions:

– Enforce the use of module *compute-GDP-risk.*

– Do not use the module *SLR-landloss.*

– Do not use the module *Load-SRTM-elevation data.*

This reduces the number of obtained solutions further, in fact an already manageable set of 96 possible implementations is obtained (which interestingly does not change until a search depth of 18).

(C) Finally, the constraints of this refinement focus on constraints and preferences regarding data formats, scales and georeferencing systems to address the variety of existing SLR impact data input and data output formats. Exemplarily, we use the following constraints to define some preferences on the desired input and output data:

– Enforce the use of module *VectorToRaster.*

– Enforce the existence of type *RS90m.*

– Enforce the existence of type *WG384.*

– Enforce the use of module *Generate-interactive-map.*

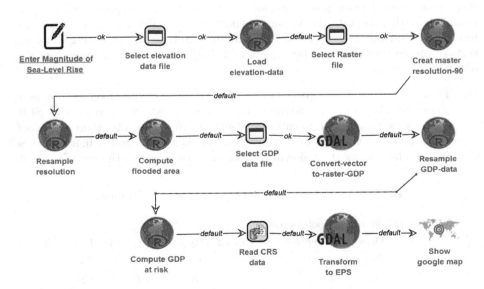

Fig. 6. Synthesized workflow.

Now, only one 14-step solution remains up until a search depth of 15. This solution, depicted in Fig. 6, comprises the computational steps that are needed to carry out the analysis that adequately matches the requirements and preferences expressed by the constraints.

Note that there is neither a guarantee that such refinements will lead to exactly one solution, nor that there will be a solution at all. What is returned depends on what is possible according to the domain model for the current specification together with all additional constraints. Usually many solutions are technically possible, but constraints are needed exclude useless and unintended service combinations, and to guide the synthesis to what is actually desired. However if the constraints demand more than the services in the domain model can provide (so-called over-specification), no solutions will be found. In this case, blocking constraints have to be removed or the domain model revised or extended by additional functionality in order to obtain solutions again.

5   Conclusion

We demonstrated in this paper how an adequate semantic domain modeling, comprising the design of services and the provisioning of semantic meta-information about the services in terms according to domain ontologies, can enable the automatic design of workflows for analyzing the impacts of sea-level rise. We showed how through the PROPHETS synthesis plugin of the jABC workflow modeling framework users can easily benefit from the domain model and easily obtain adequate workflows. In this setup, workflow designers are not required any more to implement the entire workflow manually. Instead, they can just provide a sketch of the intended workflow, together with a set of constraints that further specify the analysis objectives. PROPHETS then applies a synthesis algorithm to this abstract specification, and returns a set of possible implementations to the user. We showed how through the successive application of more and more constraints that the set of solutions that is returned by the synthesis algorithm can be reduced from an unmanageably large set of initial solutions down to a manageable set of actually adequate solutions that match the user's intents.

The technicalities involved in a correct and adequate workflow design and service composition frequently impose great challenges on researchers working in the geospatial application domain, especially if they are not IT experts or trained programmers. Hiding the technical complexity of geospatial service composition from the user, this approach greatly simplifies the design of correct and adequate workflows, and thus addresses one of the great challenges for these researchers. The successful application of such methodologies crucially depends on adequate domain modeling, which is by no means trivial. Although ontologies for supporting semantic service discovery have seen much progress in the last years, there is still a significant need for further development. Accordingly, a significant part of our future work is going to focus on the evaluation of existing models and on continuing the design of new domain-specific ontologies in collaboration with experts from the application domain.

Acknowledgments. We are very grateful to Steffen Kriewald, Dominik Reusser, and Markus Wrobel from the climate change and development group at the Potsdam Institute for Climate Impact Research (PIK), who provided us with the tools and data sets required for the scenario. We also thank Hartmut Asche from the department of Geography at University of Potsdam for his valuable feedback.

This work was supported, in part, by Science Foundation Ireland grant 13/RC/2094 and co-funded under the European Regional Development Fund through the Southern & Eastern Regional Operational Programme to Lero - the Irish Software Research Centre (www.lero.ie).

References

1. OGC Web services standards. http://www.opengeospatial.org/standards. Accessed 30 Feb 2016
2. Al-Areqi, S., Kriewald, S., Lamprecht, A.L., Reusser, D., Wrobel, M., Margaria, T.: Agile workflows for climate impact risk assessment based on the ci:grasp platform and the jABC modeling framework. In: International Environmental Modelling and Software Society (iEMSs). 7th Intl. Congress on Env. Modelling and Software (2014)
3. Al-Areqi, S., Kriewald, S., Lamprecht, A.-L., Reusser, D., Wrobel, M., Margaria, T.: Towards a flexible assessment of climate impacts: the example of agile workflows for the ci:grasp platform. In: Margaria, T., Steffen, B. (eds.) ISoLA 2014, Part II. LNCS, vol. 8803, pp. 420–435. Springer, Heidelberg (2014)
4. Alonso, G., Hagen, C.: Geo-opera: workflow concepts for spatial processes. In: Scholl, M.O., Voisard, A. (eds.) SSD 1997. LNCS, vol. 1262, pp. 238–258. Springer, Heidelberg (1997)
5. Bai, Y., Di, L., Wei, Y.: A taxonomy of geospatial services for global service discovery and interoperability. Comput. Geosci. **35**(4), 783–790 (2009)
6. Bernard, L., Ostländer, N.: Assessing climate change vulnerability in the arctic using geographic information services in spatial data infrastructures. Clim. Change **87**(1–2), 263–281 (2008)
7. Chen, N., Di, L., Yu, G., Gong, J.: Geo-processing workflow driven wildfire hot pixel detection under sensor web environment. Comput. Geosci. **36**(3), 362–372 (2010)
8. Cruz, S.A., Monteiro, A.M., Santos, R.: Automated geospatial web services composition based on geodata quality requirements. Comput. Geosci. **47**, 60–74 (2012)
9. Farnaghi, M., Mansourian, A.: Automatic composition of WSMO based geospatial semantic web services using artificial intelligence planning. J. Spat. Sci. **58**(2), 235–250 (2013)
10. Foerster, T., Schaeffer, B., Brauner, J., Jirka, S.: Integrating OGC web processing services into geospatial mass-market applications. In: International Conference on Advanced Geographic Information Systems and Web Services, 2009, GEOWS 2009, pp. 98–103. IEEE (2009)
11. Gone, M., Schade, S.: Towards semantic composition of geospatial web services-using WSMO instead of BPEL. Int. J. Spat. Data Infrastruct. Res. **3**, 192–214 (2008)

12. Granell, C., Díaz, L., Gould, M.: Service-oriented applications for environmental models: reusable geospatial services. Environ. Model. Softw. **25**(2), 182–198 (2010)
13. Granell, C., Lemmens, R., Gould, M., Wytzisk, A., De By, R., Van Oosterom, P.: Integrating semantic and syntactic descriptions to chain geographic services. IEEE Internet Comput. **10**(5), 42–52 (2006)
14. Hobona, G., Fairbairn, D., Hiden, H., James, P.: Orchestration of grid-enabled geospatial web services in geoscientific workflows. IEEE Trans. Autom. Sci. Eng. **7**(2), 407–411 (2010)
15. Jaeger, E., Altintas, I., Zhang, J., Ludäscher, B., Pennington, D., Michener, W.: A scientific workflow approach to distributed geospatial data processing using web services. In: SSDBM, Citeseer, pp. 87–90 (2005)
16. Jung, G., Margaria, T., Nagel, R., Schubert, W., Steffen, B., Voigt, H.: SCA and jABC: bringing a service-oriented paradigm to web-service construction. In: Margaria, T., Steffen, B. (eds.) ISoLA 2008. CCIS, vol. 17, pp. 139–154. Springer, Heidelberg (2009)
17. Kilibarda, M.: A plotGoogleMaps tutorial (2013)
18. Klien, E., Lutz, M., Kuhn, W.: Ontology-based discovery of geographic information services an application in disaster management. Comput. Environ. Urban Syst. **30**(1), 102–123 (2006)
19. Klusch, M., Gerber, A., Schmidt, M.: Semantic web service composition planning with OWLS-Xplan. In: Proceedings of the AAAI Fall Symposium on Semantic Web and Agents. AAAI Press, Arlington (2005)
20. Kolas, D., Hebeler, J., Dean, M.: Geospatial semantic web: architecture of ontologies. In: Rodríguez, M.A., Cruz, I., Levashkin, S., Egenhofer, M. (eds.) GeoS 2005. LNCS, vol. 3799, pp. 183–194. Springer, Heidelberg (2005)
21. Kriewald, S.: srtmtools: SRTM tools, r package version 2013–00.0.1 (2013)
22. Lake, R., Farley, J.: Infrastructure for the geospatial web. In: Scharl, A., Tochtermann, K. (eds.) The Geospatial Web, pp. 15–26. Springer, Heidelberg (2007)
23. Lamprecht, A.-L. (ed.): User-Level Workflow Design - A Bioinformatics Perspective. LNCS, vol. 8311. Springer, Heidelberg (2013)
24. Lamprecht, A.L., Naujokat, S., Margaria, T., Steffen, B.: Synthesis-based loose programming. In: 2010 Seventh International Conference on the Quality of Information and Communications Technology (QUATIC), pp. 262–267. IEEE (2010)
25. Li, H., Wang, Y., Cheng, P.: Semantic description for the taxonomy of the geospatial services. Boletim de Ciências Geodésicas **21**(3) (2015)
26. Lissner, T.K., Reusser, D.E., Schewe, J., Lakes, T., Kropp, J.P.: Climate impacts on human livelihoods: where uncertainty matters in projections of water availability. Earth Syst. Dyn. Discuss. **5**, 403–442 (2014)
27. Ludäscher, B., Altintas, I., Berkley, C., Higgins, D., Jaeger, E., Jones, M., Lee, E.A., Tao, J., Zhao, Y.: Scientific workflow management and the Kepler system. Concurrency Comput. Pract. Experience **18**(10), 1039–1065 (2006)
28. Margaria, T., Nagel, R., Steffen, B.: jETI: a tool for remote tool integration. In: Halbwachs, N., Zuck, L.D. (eds.) TACAS 2005. LNCS, vol. 3440, pp. 557–562. Springer, Heidelberg (2005). http://www.springerlink.com/content/h9x6m1x21g5lknkx
29. Margaria, T., Steffen, B.: Service-orientation: conquering complexity with XMDD. In: Hinchey, M., Coyle, L. (eds.) Conquering Complexity, pp. 217–236. Springer, London (2012). http://dx.doi.org/10.1007/978-1-4471-2297-5_10
30. Mineter, M.J., Jarvis, C., Dowers, S.: From stand-alone programs towards grid-aware services and components: a case study in agricultural modelling with interpolated climate data. Environ. Model. Softw. **18**(4), 379–391 (2003)

31. Naujokat, S., Lamprecht, A.-L., Steffen, B.: Loose programming with PROPHETS. In: de Lara, J., Zisman, A. (eds.) Fundamental Approaches to Software Engineering. LNCS, vol. 7212, pp. 94–98. Springer, Heidelberg (2012)

32. Pratt, A., Peters, C., Siddeswara, G., Lee, B., Terhorst, A.: Exposing the Kepler scientific workflow system as an OGC web processing service. In: Proceedings of iEMSs (International Environmental Modelling and Software Society) Congress (2010)

33. Rautenbach, V., Coetzee, S., Iwaniak, A.: Orchestrating OGC web services to produce thematic maps in a spatial information infrastructure. Comput. Environ. Urban Syst. **37**, 107–120 (2013)

34. Steffen, B., Margaria, T., Freitag, B.: Module configuration by minimal model construction. Technical report, Fakultr Mathematik und Informatik, Universit Passau (1993)

35. Steffen, B., Margaria, T., Nagel, R., Jörges, S., Kubczak, C.: Model-driven development with the jABC. In: Bin, E., Ziv, A., Ur, S. (eds.) HVC 2006. LNCS, vol. 4383, pp. 92–108. Springer, Heidelberg (2007). http://dx.doi.org/10.1007/978-3-540-70889-6_7

36. Wrobel, M., Bisaro, A., Reusser, D., Kropp, J.P.: Novel approaches for web-based access to climate change adaptation information – MEDIATION adaptation platform and ci:grasp-2. In: Hřebíček, J., Schimak, G., Kubásek, M., Rizzoli, A.E. (eds.) ISESS 2013. IFIP AICT, vol. 413, pp. 489–499. Springer, Heidelberg (2013). http://dx.doi.org/10.1007/978-3-642-41151-9_45

37. Yue, P.: Semantic Web-Based Intelligent Geospatial Web Services. Springer, New York (2013)

38. Yue, P., Baumann, P., Bugbee, K., Jiang, L.: Towards intelligent GIServices. Earth Sci. Inf. **8**(3), 463–481 (2015)

39. Yue, P., Di, L., Yang, W., Yu, G., Zhao, P.: Semantics-based automatic composition of geospatial web service chains. Comput. Geosci. **33**(5), 649–665 (2007)

40. Zhang, C., Zhao, T., Li, W.: Conceptual frameworks of geospatial semantic web. In: Zhang, C., Zhao, T., Li, W. (eds.) Geospatial Semantic Web, pp. 35–56. Springer, Heidelberg (2015)

Urban Growth and Real Estate Income.
A Comparison of Analytical Models

Massimiliano Bencardino[1(✉)], Maria Fiorella Granata[2],
Antonio Nesticò[3], and Luca Salvati[4]

[1] Department of Political, Social and Communication Sciences,
University of Salerno, Via Giovanni Paolo II, 132-84084 Fisciano, SA, Italy
mbencardino@unisa.it
[2] Department of Architecture, University of Palermo, Palermo, Italy
maria.granata@unipa.it
[3] Department of Civil Engineering, University of Salerno, Fisciano, SA, Italy
anestico@unisa.it
[4] Council for Agricultural Research and Economics (CREA-RPS), Rome, Italy
luca.salvati@entecra.it

Abstract. Urban growth processes are notoriously complex, depending on vastly different demographic, socio-cultural and economic factors. The analysis is even more complex in the metropolitan areas, since they are the result of ancient agglomeration processes in a phase of intensive development of settlement and, more recently, of the formation of urban polycentrism. Investigation requires collection, analysis and processing of useful information at homogeneous territorial units, based on already consolidated models or through new validating protocols.

The present paper analyzes urban growth based on a micro-scale approach, identifying homogeneous local districts for localization features that can affect real estate market value for residential use. These features include the location of the housing unit compared with the city center, the level of infrastructure, the presence of community facilities and shops, but also the external environment quality in terms of availability of green public and air pollution degree. Geographic Information System applications are used to process the available dataset to identify, at first, the demographic evolution of Naples as a functional urban region according to the life cycle model proposed by Van den Berg and, then, the real estate dynamics of the metropolitan in the light of income flows which each asset is capable of producing.

Understanding the spatio-temporal evolution of real estate property values can be useful to explain the intimate mechanism of urban growth at the metropolitan scale.

Keywords: Urban growth · Real estate value · GIS · Territorial planning · Italy

---

This paper is to be attributed in equal parts to the four authors.

© Springer International Publishing Switzerland 2016
O. Gervasi et al. (Eds.): ICCSA 2016, Part III, LNCS 9788, pp. 151–166, 2016.
DOI: 10.1007/978-3-319-42111-7_13

1   Introduction

The analysis of city's growth through a restricted number of developmental stages is a well established research strategy in urban studies. The need to give a logical scheme to the "counter-urbanization" and the slowdown of the demographic and industrial growth of large urban areas, since the 1970s in the United States and slightly later in Europe, has given rise to different approaches and analytical models (Norton 1979; Berry 1980; Hall and Hay 1980). The existence of different phases of urban growth, possibly corresponding to different stages of industrialization, namely urbanization (concentration), suburbanization (industrialization) and disurbanization (decentralization), have been well captured in the Van den Berg model, as in Fig. 1 (Van den Berg et al. 1982; Cecchini 1989). The model of the city life cycle culminates in a fourth period, called reurbanization, in which a return to demographic growth of the city center is expected. So, the theory suppose a cyclical organization of urban growth.

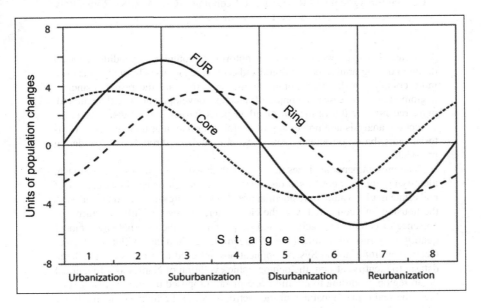

Fig. 1. Urban growth phases according to Van den Berg model (1982; edited)

Since the 1970s, a slowdown in demographic dynamics has been observed in northern Italy. Simultaneously, we have recorded an increase of settlements in the smaller towns and rural areas, with the consequent expansion of urban boundaries and the formation of polycentric metropolitan areas (Cafiero 1988). However, declining population growth rates has not reduced the sprawl of urban areas - showing important variations at the local scale. So, the urban sprawl has shown all its pervasiveness and its capacity to transform the traditional compact structure of the Italian urban network in an «urban anarchy» of spontaneous and not, young and old, whether or not authorized interventions (Baioni 2006). It is a phenomenon that has changed the landscape of non-urban peripheries and territories. We have seen the emergence of «city-regions» as areas of urban decentralization, where it failed the traditional conflict between city and countryside and new urban-rural lifestyles have appeared (Bolocan Goldstein 2008).

Driven by diffusive dynamics of the population and by a reorganization of the productive system and by the transformation of the economic system towards a tertiary economy (Bencardino 2015), the city has taken, at first, the form of «diffused city» and then, especially for large cities, of «metropolitan archipelago». This last is the form of a city that tends to reshape itself through an ordered urban shape in the current unstructured space without borders (Indovina 2009).

Have the cities come to an end? Being the current economic reality quite complex, it is difficult to imagine if there is a return to the urban centrality, as envisioned by Van den Berg, and a rethinking of the same according to the new paradigm of «compact city» (Indovina 2009) or if, by contrast, the cities will dissolve in a formless and multifunctional regional space, as a result of the end of the Fordist cycle in the developed countries (Petrillo 2009).

Today, in many European countries, including Italy, there is a large turmoil on the topic of land use (Bencardino 2015). In addition, several cities have started urban regeneration actions (Nesticò and De Mare 2014), especially related to the paradigm of «smart city» (Bencardino and Greco 2014). So, the policies that have led us to the current situation are bound to change in the near future and that, inevitably, the new centrality will result in a rethinking of urban functions and their spatial distribution.

2   The Housing Dimension in Strategic Planning

Since the early 1980s, the system of European cities has been the subject of comparative studies dealing with changes in the urban hierarchy and spatial modifications of economic functions and productive specialization (Bonavero 2000) and, more recently, on the connection between networks of cities for regional competitiveness (Dematteis 2005; Dansero 2012) or, specifically, on the redefining of urban boundaries for a comparison of the same (Dühr 2005; OECD 2012; Bretagnolle 2013). In all these cases it was necessary to address the problem of delimitation of urban areas.

In Italy, the metropolitan areas, established by Law no. 142/1990, included the main municipalities (metropolitan cities) and "the other municipalities whose the settlements have tight relationships with them as to the economic, social, essential services activities, cultural relationships and territorial characteristics". The debate on what were the boundaries of metropolitan areas does not lay a summary. They were not realized, missing a shared model of governance for large urban areas. With the law n. 56/2014, the need to give a new definition to the metropolitan cities, provinces and associations of municipalities has been reaffirmed. According to this rule, the underground city of Naples, the subject of this work, it is the territorial collectivity of large area which replaced the province of Naples starting from January 1, 2015.

But even this additional regulatory definition solves the question about the functions, the responsibilities (Camagni 2014) and the boundaries af a vast area (Calafati 2014).

In particular, it is not clear what are the boundaries of a «diffuse city», i.e. the space in which the planning instruments are efficient in managing the interdependence of population centers that compose and characterize metropolitan areas or to define enforcement actions to «land take» or – even – to prepare actions of an area that is homogeneous with respect to its functional specializations.

Then, the difficulty of identifying both the size and the optimal analytical scale to examine and represent the urban complexity is in addition to highlighting issues. These are fundamental to define the measures aimed to increase the competitiveness of the urban system.

However, in this study the optimal analytical area of investigation is the «Napoli *de facto*» (Calafati 2014), which includes a smaller area of the metropolitan city, but more homogeneous compared to the socio-economic processes that characterize it (Fig. 2). In fact the considered municipalities (ranging from *Monte di Procida* in *Giugliano, Acerra* up to *Herculaneum*), form a territorial unit which differs from the excluded areas (i.e. *Nola*, the Vesuvius area, the city of *Castellammare di Stabia* and *Sorrento* peninsula) by population density and functional characteristics. These areas gravitate around other urban districts (*Nola, Torre del Greco* or *Castellammare di Stabia*) distinct from Naples or are characterized by a real estate market with dissimilar dynamics, e.g. owing to tourism specialization or second-home speculation (cf. *Sorrento* coast).

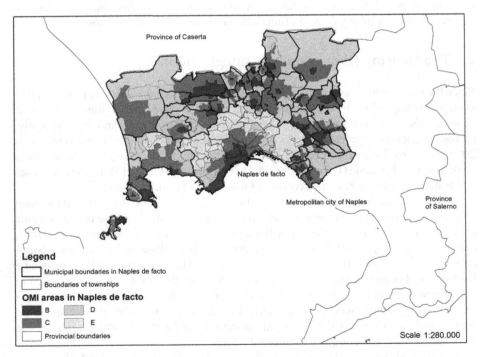

Fig. 2. OMI areas in Naples *de facto* and metropolitan city of Naples (Color figure online)

The Naples *de facto* lends itself to an evolutionary analysis as a real «functional urban region» according to the model of Van den Berg. The development of this analysis is the first purpose of the research. Subsequently, we plan to examine the distribution of value added that urban real estate are able to generate in the considered vast area and to evaluate their measure even in comparison to the diachronic measurements on the resident population.

We emphasize that the use of the databases provided by the Observatory of the Real Estate Market (OMI) Land Agency solves the issue of selecting the optimal investigation analytical scale. Indeed the IMO areas, as well as return a homogenous framework for the characteristics of the housing market, allow the aggregation of data on the resident population, as gathered from official statistics units.

3 Analysing Changes in Resident Population at the Local Scale

The spatial analysis of changes in population between 2001 and 2011 in Naples *de facto* is carried out at the scale of homogeneous territorial areas of the Property Market Observatory (micro-zones OMI), aggregating the values obtained from Census geographic units of Istat as in Fig. 3.

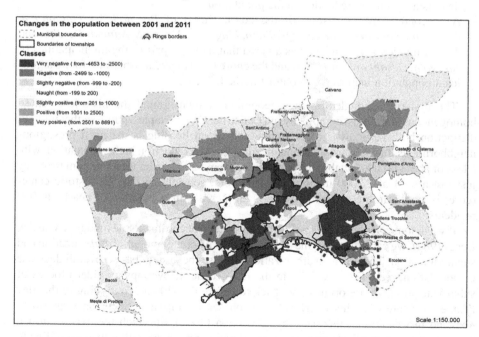

Fig. 3. Change in the population in Naples *de facto* (Our elaboration) (Color figure online)

Given the lack of homogeneity between the OMI territorial bases and the Istat one, we obtained a values of the population at the scale of the OMI zones, deriving a population density per unit area by census fractions. So, we associated with each OMI zone a density as a weighted average of the density of each included Census unit. The accuracy of this transformation is quantified as an average deviation lower then 1 %.

Specifically, the survey area, which includes the City of Naples and surrounding municipalities, consists of 318 micro-zones, each of which expresses uniform levels of the local real estate market on the basis of urban, socioeconomic and service facilities characteristics.

The population changes make the scenario of Fig. 3, where three distinct homogeneous and concentric zones are classified.

- A «core», composed of the *Avvocata-Mercato, Vicaria* and part of the *Stella-S.Carlo all'Arena* municipalities (the historical center), where there are no significant changes of the population. Inside this core there was a slight increase in population in coastal areas and a slight loss for the more internal ones.
- A «first ring», which is formed by the *Fuorigrotta, Soccavo, Capodimonte, Colle Aminei* and *Piscinola-Scampia* neighborhoods, by *S. Ferdinando-Chaia-Posillipo, Vomero-Arenella, Miano-Secondigliano-San Pietro a Patierno* and *Ponticelli-Barra-S. Giovanni a Teduccio* urban districts, from the coast strip composed of the municipalities of *San Giorgio a Cremano, Portici* and *Herculaneum*, as well as by the municipalities of *Arzano, Casoria, Grumo Nevano* and *Frattamaggiore*. Here there is a significant reduction in the population.
- A «second ring», consisting of the coastal areas of *Pozzuoli*, by *Giuliano in Campania, Quarto, Villaricca, Qualiano, Mugnano, Melito, S. Antimo, Cardito*, and *Acerra* municipalities, as well as an area that includes part of the municipalities of *Afragola, Casalnuovo di Napoli* and the entire town of *Volla*, where we found significant population increases, in contrast to the first ring.

The first ring, though drawings geographic crown around the historical center, homogeneous and almost completely internal to the City of Naples, puts together without distinction both large areas of economic and social housing is the most prestigious neighborhoods. In this ring, we found the most marked reduction in the population, with a loss of about 60,000 inhabitants between 2001 and 2011, in an urban system that they just lost a total of 22,000 inhabitants. Therefore, considering that the historic center (core) is stable, in the outer ring there is an increase in population of nearly 40,000 residents.

If we assume Naples *de facto* as an urban system, according to the model of Van den Berg (1982), we can affirm that the area is affected by contemporary disurbanization and suburbanization. There is a disurbanization because the system has an overall decrease of population, recorded especially in the last decade. It overlaps an older process of suburbanization, whose origins date back to before the '70 and continuing to this day (Table 1). Moreover, this overlapping of phases is consistent with the scenario of «beyond urbanization», theorized by Champion (2001), in which the existing urban systems evolve according to a range of overlapping processes, rather than according to a single progressive trajectory (Kabisch and Haase 2009).

Table 1. Evolution of the population in Naples de facto according to Van den Berg model

	Pop. 1971	Pop. 1981	Pop. 1991	Pop. 2001	Pop. 2011
Naples municipality	1.226.604	1.212.387	1.067.365	1.004.500	962.003
Urban ring	812.624	1.006.618	1.146.261	1.237.818	1.258.281
Naples *de facto*	2.039.228	2.219.005	2.213.626	2.242.318	2.220.284

	Var. '71-'81	Var. '81-'91	Var. '91-'01	Var. '01-'11
Naples municipality	−14.217	−145.022	−62.865	−42.497
Urban ring	193.994	139.643	91.557	20.463
Naples *de facto*	179.777	−5.379	28.692	−22.034

In fact, in the 1970s, the whole urban ring was affected by growing population, contrary to what happened in downtown Naples. At that stage, the expansion was targeting to areas of new suburbanization, i.e. towards the Campi Flegrei and the north-western and north-eastern quadrant. However, the demographic reduction of the City of Naples has not resulted in a weakening of the strong functional polarization of the chief town (Amato 2008). Instead, in a chaotic and disorderly way, suburbs that have incorporated the industrial areas, active or abandoned, and the main axis of connection were born.

Then, other municipalities have had growth rates gradually decreasing until losing population in the 1980s (*Frattamaggiore, Cercola e Portici*), which in the 1990s will

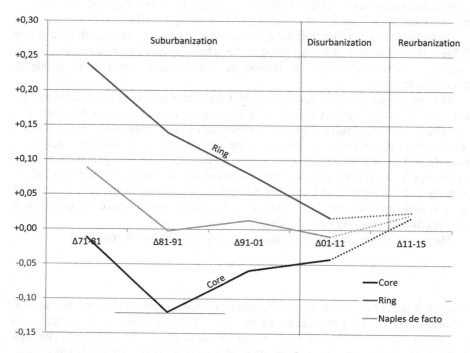

Fig. 4. Population growth rate in Naples. (Color figure online)

be accompanied by *Grumo Nevano, Arzano, Casavatore, Herculaneum* and *Pomigliano d'Arco*, and then also by *Marano di Napoli, Casoria, Cercola, Massa di Somma, San Sebastiano al Vesuvio* and *St. Anastasia* in the 2000s. Therefore, the system gets closer to a stabilization, due to the slowdown in the growth of the suburbs.

By contrast, in the scenario just described, we observed signs of re-urbanization. In fact, analyzing the population growth rates according to the model of Van den Berg (Fig. 4), we find that the rate of decrease of the City of Naples ("core") are in turnaround since the 80 s and could become positive in the next census.

So, analyzing the trend of the population at the scale of the OMI areas, we recorded a change of sign of these rates in the latest statistical surveys in a portion of the municipality, represented by the historic core of the city (Fig. 3).

Ultimately, the analysis based on the Van den Berg criteria show the signals of a possible next change of phase in the life cycle of the city. Furthermore, a reversal in the suburbanization processes towards a new urbanization (obviously different from that of the '50 s) is confirmed in many cities of the Campania Region (Bencardino and Bencardino 2015).

4 Temporal Evolution and Spatial Distribution of Real Estate Values

The topic of urban competitiveness can be investigated according to the territorial capital, distinguished in four major classes: 1. natural and cultural capital, 2. settlement capital, 3. cognitive capital and 4. social capital (Camagni 2009).

As regards the rate of territorial capital included in property stocks, with the aim of defining the economic processes related to the urban income of wide areas and to seek correlations with demographic dynamics, the rental market for housing in Naples *de facto* is analysed.

The logical and functional relationship that exists between urban growth and income levels of a territory is widely known, as has been explained in literature for some time now (Derycke 1972; Orefice 1984). These earning capacities are summarised in terms of urban real estate values (Nesticò and Galante 2015), which are possible to achieve – as explained by the discipline of estimates – through *direct* (or synthetic) estimate procedures, based on the synthetic appreciation of the most likely market value of the property to be estimated, and *indirect* estimation procedures (or the capitalisation of income, sometimes incorrectly called analytical), which consist in discounting the future incomes that the property can provide.

The indirect estimate can be summarised in the well-known formulation:

$$V = \frac{R}{r},$$

in which R is the average, annual and continuous income and r represents the capitalisation rate. In turn, the determination of the income R can be made: (i) synthetically, on the basis of rents detected by the similar assets of reference; and (ii) analytically, through the financial results of production, or rather, by resorting to the corporate balance sheet.

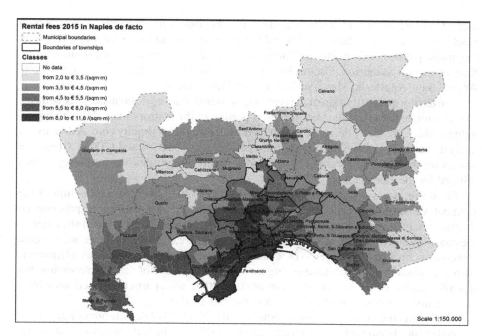

Fig. 5. Average rent per square meter and per month of residential properties in Naples *de facto* in 2015. (Our elaboration) (Color figure online)

In this way, upon agreement of a measure of the rate r, the examination of rents in the study area can also provide information about asset values V, which are heavily dependent on socio-economic factors of the territory (Gabrielli et al. 2015; Morano et al. 2015; Tajani and Morano 2015).

Tables relating to rents and OMI geometries with the perimeters of individual bands and of 318 cadastral micro-zones are used as reference. The first operation on the data relates to the correction of gaps and overlaps in geometry perimeters. Due to the lack of a GIS logic in the OMI information, a digital database in which the elements – consisting of numerical values and corresponding surfaces – result as being in a biunivocal relationship by means of an identification code, is constructed.

The available quotations are the minimum and maximum rent values for ordinary residential housing units. They are rents intervals expressing ordinary conditions, so they do not include quotations referring to properties of particular value or degradation or which, however, have characteristics which are not common for the building typology of the relevant zone. A geo-referenced map of the average rents for housing units to 2013 has been created. Due to the stagnation of prices and rents that has characterised the residential market over in the last two years, the map referring to rents in 2013 shown in Fig. 5 can be considered representative also of the situation in 2015.

A concentration of higher rents emerges within the city centre of Naples, with more micro-zones (Vomero, Arenella, Chiaia, Posillipo and San Ferdinando) where rents assume values included in the range between 8.0 to € 11.6 (m^2 month). As you move away from the city centre there is a gradual reduction of the housing income, which reach the lowest levels between 2.0 and € 3.5 (m² month) in the outer areas, that is in the municipalities of Giugliano in Campania, Qualiano, Villaricca, Calvizzano, Melito, Sant'Antimo, Casandrino, Grumo Nevano, Frattaminore, Crispano, Cardito, Caivano, Acerra, Sant'Anastasia, Massa di Somma and Herculaneum.

The territorial distribution of the rents explains how Naples *de facto* actually behaves as one, unique urban system with a central portion of the metropolitan plot with higher values, which gradually decrease as you go further away from the city core. A diachronic analysis has been carried out for the time periods (2008–2013) and (2003–2013). The results are shown respectively in Figs. 6 and 7. Figure 8 represents the frequency ranges utilized for the calculations.

From Fig. 6 we can see that, for the period 2008-2013, there is a reduction of the nominal values of the rents in almost all the studied areas, with more pronounced declines (between −32 % and −18 %) in downtown Naples, along limited stretches of the coastline and disaggregated portions of the inland areas. There are large areas where *R* records contractions between −18 % and 0 %, interrupted by narrow areas where rents increase, albeit in a limited manner. Figure 6 is representative of the long recession that has affected the major western economies since 2008, with particularly serious effects on the construction sector and for urban real estate markets.

The results regarding changes in rents over the 2003–2013 decade are of interest for understanding latent patterns of urban expansion (Fig. 7). In fact, the area of study has

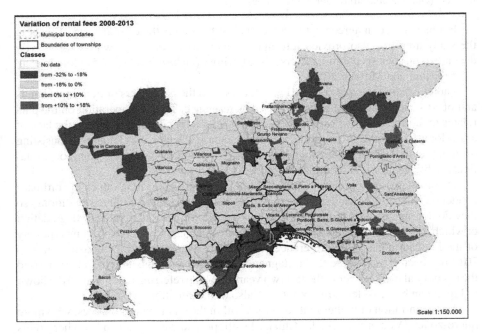

Fig. 6. Differential in the average rental fees of residential property in the period 2008–2013 compared with 2008. (Our elaboration) (Color figure online)

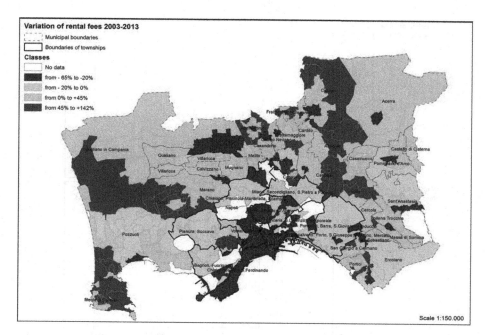

Fig. 7. Differential in the average rental fees of residential property in the period 2003–2013 compared to 2003. (Our elaboration) (Color figure online)

a territorial division which can be assimilated to that of Naples *de facto* as inferred by the Van den Berg model, already implemented for the examination of demographic flows in the Istat survey covering the 2001–2011 decade (Fig. 3): with a *central core*, roughly the historic centre of Naples, where there are no significant changes in population and at the same time there are strong declines in rents (from −65 % to −20 %); a *first band*, where there is a significant reduction of the population and simultaneously a decline in rents in the range of −20 % to zero percentage points; a *second band* which, unlike the first, highlights substantial increases of the population and corresponding prevailing increments of rents, especially in the 0 %–5 % and 5 %–12 % ranges (see Fig. 8).

Our results are consistent with merchant rules governing real estate dynamics. The central core of the city of Naples is given by saturated residential neighbourhoods in terms of building capacity. Here, speculative logics led to particularly high housing market values, with the same number of inhabitants. Consequently, the recent economic crisis has caused significant reductions in rents, from just −65 % to −20 % in nominal terms over the 2003–2013 period.

The situation is different in the first and second bands shown in Fig. 3, where the trend of the resident population is capable to affect the profile of real estate prices, despite the current economic downturn. In fact, in the first band, the general decline in rents can be associated with a reduction of the resident population, while in the second, the rents increase widely, even to a sustained extent, directly related to the positive demographic

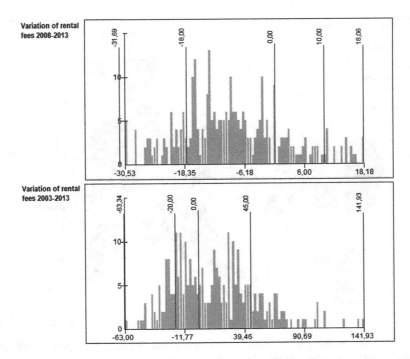

Fig. 8. Statistical distribution of the variables shown in Figs. 6 and 7

balance. Evidently, particular situations occurring in individual urban districts should be considered in relation to requalification projects (Calabrò and Della Spina 2014; Guarini et al. 2014; De Mare et al. 2015; Nesticò et al. 2015) or specific factors concerning the local socioeconomic context.

5 Conclusions

Our study paper deals with the question of the size and optimal analytical scale for the examination and representation of urban complexity, using real estate values as a paradigmatic variable. It is a problem that can be solved only by defining homogenous districts compared to the set of social-economic processes that characterised them. In this regard, useful references come from the areas that are left to the base of comparative investigations of civil buildings market values. Thus, with regard to the Naples *de facto* which includes an area less vast than the entire Neapolitan subway city but certainly more closely inter-related, the study is conducted on the basis of micro-zones of the Property Market Observatory, where similar extrinsic characteristics can be found (urban, infrastructural, availability of services, etc.) able to affect the formation mechanisms of housing prices.

Two studies are carried out. A first analysis of the spatial variations of the resident population between 2001 and 2011, the years relating to the last two Istat censuses. Once the problem of differing territorial bases OMI and Istat is solved, the implementation of the Van den Berg model allows us to distinguish three different homogeneous and concentric zones (Fig. 3): a *central core*, composed of the municipalities of Avvocata-Mercato, Vicaria and part of the municipalities of Stella-S. Carlo to Arena (the old town), where there is no relevant change of the resident population; a *first band*, with a significant reduction of inhabitants; a *second band*, in which there is a considerable increase in the population.

The second study concerns the spatial distribution of urban real estate income and the corresponding time evolution over in the 2003–2013 period, comparable to the decade of the latest Istat surveys and close enough to date.

Thus, it is clear that the analysis of the real estate market is capable of explaining the logic of urban growth. Specifically, the rents spatial distribution shows how the Naples *de facto* behaves effectively as one, unique urban system, with a central portion of the metropolitan fabric having higher values, which are gradually decreasing as you go further away from the city "core".

The diachronic analysis is summarized in Figs. 6 and 7. The first exposes the recessionary economic effects on urban incomes in the years after 2008. Figure 7 provides the variations in the rents between 2003 and 2013, with the aim of obtaining correlations with demographic flows throughout the decade. Important results emerge. The real estate income differentials tend to be distributed throughout the Naples *de facto* according to territorial distribution deriving from the Van den Berg's model, already implemented for the examination of demographic flows. Thus, in the *central core*, no significant changes in the population are associated with sharp falls in rents; in the *first band*, there is a reduction in population and a concurrent reduction in rents; in the *second band*, significant population increases are observed with consequent rent increases. And moreover, it is explained that these results are consistent with market rules governing real estate dynamics: namely with speculative logics that in the core of the city of Naples have created a real estate bubble; and with temporal variations of the real estate profitability conditioned by its demographic balance.

The spatial distribution of asset values and the diachronic analysis of income and population flows provide essential elements for proper urban planning, therefore to delimit future areas of intervention and to identify specific projects, plans or investment programmes for each portion of the urban fabric (Guarini and Battisti 2014; Cilona and Granata 2015; Napoli 2015; Nesticò and Pipolo 2015). In this perspective, the definition in quantitative terms of the functional relationships between the variables should be deeply analyzed to characterise an economic evaluation model supporting urban policies.

References

Amato, F.: La periferia italiana al plurale il caso del Napoletano. In: Sommella, R. (ed.) Le città del Mezzogiorno. Politiche, dinamiche, attori, pp. 219–242. FrancoAngeli, Milano (2008)
Baioni, M.: Diffusione, dispersione, anarchia urbanistica. In: Gibelli, M.C., Salzano, E. (eds.) No sprawl. Alinea, Firenze (2006)

Bencardino, M.: Consumo di suolo e sprawl urbano: drivers ed azioni politiche di contrasto. In: Bollettino della S.G.I., Serie XIII, Vol. VIII, Fasc. 2, pp. 217–237 (2015)

Bencardino, M.: Demographic changes and urban sprawl in two middle-sized cities of campania region (Italy). In: Gervasi, O., Murgante, B., Misra, S., Gavrilova, M.L., Rocha, A.M.A.C., Torre, C., Taniar, D., Apduhan, B.O. (eds.) ICCSA 2015. LNCS, vol. 9158, pp. 3–18. Springer, Heidelberg (2015). doi:10.1007/978-3-319-21410-8_1

Bencardino, M., Greco, I.: Smart communities. social innovation at the service of the smart cities. In: TeMA. Journal of Land Use Mobility and Environment, SI/2014, pp. 39–51, University of Naples "Federico II" Print (2014). doi:10.6092/1970-9870/2533

Berry, B.J.L.: Urbanization and counterurbanization in the United States. In: The Annals of the American Academy of Political and Social Science, vol. 451, pp. 13–20. Changing Cities (1980)

Bolocan Goldstein, M.A.: Città senza confini, territori senza gerarchie. In: AA.VV. Rapporto 2008. L'Italia delle città, tra malessere e trasfigurazione, Società Geografica Italiana, Roma (2008)

Bonavero, P.: Traiettorie della ricerca urbana europea. in Bonavero, P. (eds.) Il Sistema Urbano Europeo fra Gerarchia e Policentrismo, pp. 8–14, Working Paper Eupolis, Torino (2000)

Bretagnolle, A., et al.: Naming U.M.Z.: a database now operational for urban studies, ESPON Database 2013 project, Final Report (2013). Retrieved from: https://www.espon.eu/.../ESPON2013Database/3.5_TR_Naming_UMZ.pdf

Cafiero, S.: Il ruolo delle città per lo sviluppo. In: Rivista economica del Mezzogiorno, a. II, n. 1 (1988)

Calabrò, F., Della Spina, L.: The cultural and environmental resources for sustainable development of rural areas in economically disadvantaged contexts. Economic-appraisals issues of a model of management for the valorisation of public assets. In: ICEESD 2013, Advanced Materials Research, vols. 869–870, pp. 43–48 (2014). doi:10.4028/www.scientific.net/AMR.869-870.43, Trans Tech Publications, Switzerland

Calafati, A.: Nuova perimetrazione e nuove funzioni per le Città metropolitane. Il caso di Napoli, Centro Studi - Unione Industriali di Napoli (2014). Retrieved from: http://www.lavoce.info/archives/17288/citta-metropolitane-delrio-province/

Camagni, R.: Città metropolitane? No, solo province indebolite (2014). Retrieved from: http://www.lavoce.info/archives/17288/citta-metropolitane-delrio-province/

Camagni, R.: Per un concetto di capitale territoriale, in Borri, D., Ferlaino, F. (eds.), Crescita e sviluppo regionale: strumenti, sistemi ed azioni, FrancoAngeli, Milano (2009)

Cecchini, D.: Stadi di sviluppo del sistema urbano italiano. In: Rivista economica del Mezzogiorno, a. III, n. 4 (1989)

Champion, T.: Urbanization, suburbanization, counterurbanization and reurbanization. In: Padison, R. (ed.) Handbook of Urban Studies. SAGE Publication, London (2001)

Cilona, T., Granata, M.F.: Multicriteria prioritization for multistage implementation of complex urban renewal projects. In: Gervasi, O., Murgante, B., Misra, S., Gavrilova, M.L., Rocha, A.M.A.C., Torre, C., Taniar, D., Apduhan, B.O. (eds.) ICCSA 2015. LNCS, vol. 9157, pp. 3–19. Springer, Heidelberg (2015). doi:10.1007/978-3-319-21470-2_1

Dansero, E.: L'Europa Delle Regioni e Delle reti. I nuovi modelli di organizzazione territoriale nello spazio unificato europeo, UTET, Milano (2012)

De Mare, G., Granata, M.F., Nesticò, A.: Weak and Strong Compensation for the Prioritization of Public Investments: Multidimensional Analysis for Pools. Sustainability 7(12), 16022–16038 (2015). doi:10.3390/su71215798. ISSN: 2071-1050, MDPI AG, Basel, Switzerland

Dematteis, G.: Verso un policentrismo europeo: metropoli, città reticolari, reti di città. In: Moccia, D., et al. (eds.) Metropoli In-Transizione, Innovazioni, pianificazioni e governance per lo sviluppo delle grandi aree urbane del Mezzogiorno, Urbanistica Dossier, n. 75, INU Edizioni (2005)

Derycke, P.H.: Economia urbana, Il Mulino, Bologna (1972)

Dühr, S.: Potentials for polycentric development in Europe, The ESPON 1.1.1 project report. Planning Practice and Research 20(2), 235–239 (2005)

Gabrielli, L., Giuffrida, S., Trovato, M.R.: From surface to core: a multi-layer approach for the real estate market analysis of a central area in catania. In: Gervasi, O., Murgante, B., Misra, S., Gavrilova, M.L., Rocha, A.M.A.C., Torre, C., Taniar, D., Apduhan, B.O. (eds.) ICCSA 2015. LNCS, vol. 9157, pp. 284–300. Springer, Heidelberg (2015). doi:10.1007/978-3-319-21470-2_20

Greco, I., Bencardino, M.: The paradigm of the modern city: SMART and SENSEable Cities for smart, inclusive and sustainable growth. In: Murgante, B., Misra, S., Rocha, A.M.A., Torre, C., Rocha, J.G., Falcão, M.I., Taniar, D., Apduhan, B.O., Gervasi, O. (eds.) ICCSA 2014, Part II. LNCS, vol. 8580, pp. 579–597. Springer, Heidelberg (2014). doi:10.1007/978-3-319-09129-7_42

Guarini, M.R., Battisti, F., Buccarini, C.: Rome: re-qualification program for the street markets in public-private partnership. A further proposal for the Flaminio II street market. In: Zhang, X., et al. (eds.) GCCSEE 2013, Advanced Materials Research, vols. 838–841, pp. 2928–2933, (2014). ISSN: 10226680, doi:10.4028/www.scientific.net/AMR.838-841.2928, Trans Tech Publications, Switzerland

Guarini, M.R., Battisti, F.: Social Housing and redevelopment of building complexes on brownfield sites: the financial sustainability of residential projects for vulnerable social groups. In: Xu, Q., et al. (eds.) EESD 2013, Advanced Materials Research, vols. 869–870, pp 3–13 (2014). ISSN: 10226680, doi:10.4028/www.scientific.net/AMR.869-870.03, Trans Tech Publications, Switzerland

Hall, P., Hay, D.: Growth Centers in the European Urban System. Heinemann, London (1980)

Indovina, F.: Dalla Città Diffusa All'arcipelago Metropolitano. FrancoAngeli, Milano (2009)

Kabisch, N., Haase, D.: Diversifying European agglomerations. Evidence of urban population trends for the 21st century. Population Space Place 17(3), 236–253 (2011). doi:10.1002/psp.600

Morano, P., Tajani, F., Locurcio, M.: Land use, economic welfare and property values. An analysis of the interdependencies of the real estate market with zonal and socio-economic variables in the municipalities of the Region of Puglia (Italy). Int. J. Agric. Environ. Inform. Syst. 6, 16–39 (2015). doi:10.4018/IJAEIS.2015100102. ISSN: 1947-3192

Napoli, G.: Financial sustainability and morphogenesis of urban transformation project. In: Gervasi, O., Murgante, B., Misra, S., Gavrilova, M.L., Rocha, A.M.A.C., Torre, C., Taniar, D., Apduhan, B.O. (eds.) ICCSA 2015. LNCS, vol. 9157, pp. 178–193. Springer, Heidelberg (2015). doi:10.1007/978-3-319-21470-2_13

Nesticò, A., De Mare, G.: Government tools for urban regeneration: the cities plan in italy. a critical analysis of the results and the proposed alternative. In: Murgante, B., Misra, S., Rocha, A.M.A., Torre, C., Rocha, J.G., Falcão, M.I., Taniar, D., Apduhan, B.O., Gervasi, O. (eds.) ICCSA 2014, Part II. LNCS, vol. 8580, pp. 547–562. Springer, Heidelberg (2014). doi: 10.1007/978-3-319-09129-7_40

Nesticò, A., Galante, M.: An estimate model for the equalisation of real estate tax: A case study. Int. J. Bus. Intell. Data Min. 10(1), 19–32 (2015). doi:10.1504/IJBIDM.2015.069038. ISSN: 17438187, Inderscience Enterprises Ltd., Genève, Switzerland

Nesticò, A., Macchiaroli, M., Pipolo, O.: Costs and Benefits in the Recovery of Historic Buildings: The Application of an Economic Model. Sustainability 7(11), 14661–14676 (2015). doi: 10.3390/su71114661. ISSN: 2071-1050, MDPI AG, Basel, Switzerland

Nesticò, A., Pipolo, O.: A protocol for sustainable building interventions: financial analysis and environmental effects. Int. J. Bus. Intell. Data Min. **10**(3), 199–212 (2015). doi:10.1504/IJBIDM.2015.071325. ISSN: 17438187, Inderscience Enterprises Ltd., Genève, Switzerland

Norton, R.D.: City Life-Cycles and American Urban Policy. Academic Press, New York (1979)

OECD: Redefining "Urban": A New Way to Measure Metropolitan Areas, OECD Publishing, Paris (2012). doi:10.1787/9789264174108-en, Retrieved from: http://www.oe-cd.org/gov/regional-policy/all.pdf

Orefice, M.: Estimo. UTET, Torino (1984)

Petrillo, A.: Storicizzare lo sprawl? In: AA.VV., La città: bisogni, desideri, diritti. La città diffusa: stili di vita e popolazioni metropolitane, FrancoAngeli, Milano (2009)

Tajani, F., Morano, P.: An evaluation model of the financial feasibility of social housing in urban redevelopment. Property Management **2**, 133–151 (2015). doi:10.1108/PM-02-2014-0007. ISSN: 0263-7472

Van den Berg, L., et al.: Urban Europe, a Study of Growth and Decline. Elsevier Ltd., London (1982)

A Digital Platform to Support Citizen-Government Interactions from Volunteered Geographic Information and Crowdsourcing in Mexico City

Rodrigo Tapia-McClung[✉]

Centro de Investigación en Geografía y Geomática "Ing. Jorge L. Tamayo",
A.C. Contoy #137 esq. Chemax, Col. Lomas de Padierna, Tlalpan,
14240 Mexico, D.F., Mexico
rtapia@centrogeo.org.mx

Abstract. This paper presents two case studies of volunteered geographic information processes in two different neighborhoods in Mexico City. Both cases deal with citizen empowerment and actions directed for the improvement of their local surroundings. They are constructed in a bottom-up fashion: from the citizens towards the local authorities. A digital platform was developed to support user-generated data collection for both cases; the second being an evolution of the first that incorporates several enhancements. The collection of enough citizen data is useful to focus efforts to negotiate with the authorities in detected regions and matters that need attention. Citizen-generated maps are useful communication tools to convey messages to the authorities, as the identification of these locations and situations provide a better picture of what, from the citizens' perspective, is significantly deviated from the government's point of view. The platform incorporates a way to validate official data, a voting strategy as a first approach to assess the credibility of citizen-contributed observations and crowdsourced information on parcel records.

Keywords: Volunteered geographic information · Crowdsourcing · Citizen science · Citizen empowerment · Geospatial web platform · Web mapping

1 Introduction

Geographic data collection has experienced a paradigm shift in the sense that users not only consume, but also generate new data. It has been progressively easier for people to participate in mapping processes, thus effectively becoming a *citizen mapper*. Recent tendencies show that it is not necessary for people to be certified in the field of cartography to actually be able to contribute to local cartographic records and collaborate in the collection and updating of geographic data.

Traditionally, governmental agencies have been in charge of this task, following strict protocols and adhering to quality assurance standards in order to provide the best data available for different purposes: population, cadastral, vehicle registrars, businesses, censuses, natural resources, etc. It has become increasingly common for citizens to collect geographic data that does not necessarily conform to these governmental

O. Gervasi et al. (Eds.): ICCSA 2016, Part III, LNCS 9788, pp. 167–182, 2016.
DOI: 10.1007/978-3-319-42111-7_14

standards. Apart from collecting, with help of all the mapping technologies available on the web [1], it is also common for users to disclose part of this data, in what constitutes one of the pillars of neogeography: "sharing location information with friends and visitors, help shape context, and conveying understanding through knowledge of place" [2].

Volunteered geographic information (VGI) is, as defined by Goodchild, "the widespread engagement of large numbers of private citizens, often with little in the way of formal qualifications, in the creation of geographic information" [3]. It is often heard in contexts related to collective mapping activities or crowdsourcing, which involves "generating a map using informal social networks and web 2.0 technology" [4].

Crowdsourcing and VGI are two common terms. Crowdsourcing can be found in many different topics, not just geographical information, and "implies a coordinated bottom-up grassroots effort to contribute information" [5]. For some, VGI represents an "unprecedented shift in the content, characteristics, and modes of geographic information creation, sharing, dissemination and use" [6]. Others, like Harvey [7], propose that not all crowdsourced data is volunteer data and suggest making a distinction when data is collected with an "opt-in" or an "opt-out" agreement. Nonetheless, both rely on many users contributing data and are strong advocates of the "wisdom of the crowds" and collective intelligence: the idea of whether a product created collectively is better than the best individual product [8, 9].

It is also becoming more common for citizens to be able to compare what the official figures tell with what they observe and experience in their everyday life, without necessarily challenging the existence of official records. These types of citizen participation and public input have found a good niche in the report of social incidents and have given individuals direct access to establish a dialogue with the authorities. Data from different sources can be analyzed to propose different scenarios and possible courses of action that could improve citizens' quality of life without waiting for the government to provide official data, thus fostering the existence of different mechanisms that can be used to study societal issues.

Citizen-government interactions can be broadly separated as *bottom-up* or *top-down*. The former can be understood as "actions conceptualised, incepted, developed and led by members of the local community" [10]. Indeed, the appropriation of space by citizens has gained much momentum and has also proven to be an effective means for them to negotiate with the government for the procurement of benefits.

Conversely, top-down social processes focus on providing the structure of said processes for the building blocks of society. Namely, the government lays the infrastructure so the development in question can take place. In other words, it can be understood that bottom-up processes give rise to the existence of institutions because they are needed by the common practice of citizens. In the opposite case, institutions are determined by laws written by political leaders [11]. Johnson and Sieber [12] call the former 'citizen-to-government' (C2G) and the latter 'government-to-citizen' (G2C) processes.

With these types of action, the net effect from the societal point of view is citizen empowerment. The possibility for collaboration opens up and the citizenry has the opportunity to get involved in the betterment of its community.

VGI and crowdsourced information have the advantage of having low costs of entry, which in turn can represent a problem, as mentioned above, of users with little to no technical knowledge collecting data. This raises issues related to the quality, accuracy, and credibility of VGI [9, 13, 14]. In contrast, other advantages include how fast data can be made available, especially distributing it over the web, and how citizens, while becoming empowered, can challenge and exert some pressure on the government so that certain information gets collected or is made available to them.

In this paper, two C2G processes are presented in which the citizenry engages in data collection of information not necessarily available from official sources. One goal is to promote actions from the citizens towards local authorities to try and implement actions that can help improve their neighborhood. For this, a digital platform was constructed that serves specific purposes for each case, although the second is derived from the first and incorporates several enhancements.

The rest of this paper is organized as follows: the next section gives a short review of other platforms. Each case study is presented along with their main findings and lessons learned. Preliminary results are discussed and lastly, conclusions are laid out together with some ideas for future work.

2   Other Platforms for Participatory Mapping

Citizen observation and reporting platforms have existed for quite some time now. They have a wide range of interests and applications –for both spatial and non-spatial data– but all seem to conform to the idea that many users can share knowledge and, by doing so, it becomes easier to achieve a proposed goal.

iNaturalist allows users to record observations about the natural world and make them available on the internet or through mobile apps [15]. By creating a community of people that shares common interests, and amassing a very large number of observations on different topics, their goal is to be able to provide a living record of life on Earth that can be used to monitor biodiversity. OpenTreeMap is a "collaborative platform for crowdsourced tree inventory, ecosystem service calculations, urban forestry analysis, and community management", also providing users with mobile apps [16]. GeoKey is a platform for participatory mapping that allows the creation of customized projects [17]. With a web-based backend interface, it allows users to collect, share and discuss data reports. GeoKey projects can be configured to use mobile devices to collect data with the EpiCollect+ mobile app [18]. Ushahidi was originally built to crowdsource crisis mapping [19]. Since its beginnings in 2008, it has come a long way to establishing a reputation as a good platform for "human rights activism, crisis response and civilian empowerment" [20]. It is available on the web, offers mobile apps and can be configured to receive data from various different sources. GeoCitizen is a platform for community-based spatial planning that involves governments and citizens [21]. ArgooMap is a tool designed to support structured discussions with spatial reference objects and has been used in different planning exercises [22]. It was re-implemented from a former Java-based incarnation (ArguMap) using Web 2.0 services and Google Maps API. Mapchat is also an open source platform that allows the integration of maps with real-time (and asynchronous) online discussions among several users [23].

In the case of Mexico, the city of Monterrey has a map that captures citizens' reports on diverse incidents in the public space [24]. The platform receives, validates, canalizes, keeps track and publishes citizen's reports, but no further information is provided on how this is actually done. In a sense, data curators and platform seem to act as a liaison between citizen's complaints and governmental attention.

This sample of apps and platforms can aid in exercises of participatory citizen mapping. Most of them are available as open source software and while it is possible to contribute and implement new functionality, it is also a cumbersome process to localize all the frontends to cater for users in different languages (Ushahidi is the exception that already provides some translations). Not all users are bilingual and this is a key factor in terms of how well or fast the platform is adopted. Most of these platforms focus on point data but GeoKey allows other geometries, although one still needs to develop a front end to interact with the data via its web API.

3   Citizen to Government (C2G) or Bottom-up Cases

While all of the platforms presented in the previous section allow users to contribute data, some of them open the door to citizen-government interactions. In the context of this research, bottom-up processes are those that are created from the citizenry, which stem from the needs of the neighborhood and are aimed towards an interaction with local authorities. One goal is to provide information to help in the negotiation of that which is of interest to the citizens and that governments can help ameliorate.

Two bottom-up cases are presented in this section for which a digital platform was developed. The first is an example of the collaboration with citizens of a central neighborhood in Mexico City that were looking for ways to collect spatial data from their surroundings to help them justify actions and support negotiations with local authorities for neighborhood enhancements. This is what they called their 'citizen agenda'. It was decided to try and implement basic functionality in order to be able to capture citizens' perceptions as opposed to using a full blown platform and try to tweak and adapt it for a particular use. A complete suite of participatory mapping would have been too complicated for citizens to manage because this exercise started with the simple requirement of an online map and eventually needed the inclusion of more complex functionality and interactivity. However, the lack of personnel placed citizens in charge of the overall management the platform, so it had to be kept as simple as possible, yet functional and powerful. The second case is a derivative of the first in which neighbors from a different municipality were eager to partake in participatory data collection. The existing platform was adapted to include enhancements suitable for these neighbors' specific requirements.

The societal trend of citizen empowerment motivates the presentation of these two cases. Even though these are based in Mexico City, they are representative of what can happen in other cities. Many neighborhoods have to deal with issues related to business licenses, security, land use, city services, appropriate mobility, real estate and waste management. It is not just the location of the neighborhood that matters, but rather the general behavior of the local activities that can be found someplace else. The recognition that citizens can and want to make a difference in local processes is very well expressed in these examples.

3.1 The Roma Neighborhood

This first case is an exercise carried out in the Roma neighborhood in central Mexico City, from mid-2014 to mid-2015. The Roma is a central neighborhood covering an area of about 3.7 km^2 (1.4 mi^2) with an approximate population of 45,000 inhabitants and 322 city blocks.

The Roma project is an essential one, because even though a digital platform was specifically tailored for this case, it set the basis for a more general model. The motivation for this project is that citizens in this neighborhood had already perceived many problems in their territory, but did not have the tools to properly show their locations and importance to others. They were aware that having such information would be valuable to articulate proposals and find solutions. There had been previous efforts to create paper maps pinpointing these locations and situations, but with little success. The idea of collecting citizen perceptions, making them publicly available, analyzing and using them to support proposals and solutions were the driving forces behind the project.

This was a direct collaboration with a civil group. Geared and spearheaded by the citizens' interests and needs, support was provided to help them understand why spatial conscience was useful to map and how it could be done. A digital platform was built to help collect data, visualize and make it available for the creation of a citizen agenda that would then be taken to the local authorities for negotiation. The goal was for the citizen group to take a 'snapshot' of their neighborhood and evaluate its current conditions. With that as a starting point, they would be able to use the special value of spatial data to identify *what* could be done, and *where*. The idea was for them to be able to analyze the current state of the neighborhood to detect issues and opportunities.

For this, an agreement on what was to be observed was needed. The list of variables to collect in the field was defined together with the citizen group in an iterative process of workshops and meetings that concluded with six main categories (businesses, services, real estate, security, mobility and waste) for a total of 42 variables (see Fig. 2). Some variables required teams of citizens to go out in the field at different times to capture the dynamics of their environment during day and night.

The citizen group was in charge of promoting the utilization and adoption of the platform in the neighborhood. When the project started, a strong social base already existed and this was a key factor in the adoption of the platform and successful completion of data collection, as it made it easier to communicate the purpose and the usefulness of the goal to the rest of the participants.

During data collection, users had access to an online map in which they could report their observations. Names, addresses, and a picture were able to be attached to a location on the map. Neighbors and volunteers were both in charge of data collection and responsible for data quality. Some of them acted as users and managers at the same time. Teams worked in the field using sheets of paper and, during a second stage, information was entered into the platform. Paper backups were useful to keep a record of the work carried out and also in clarifying information in case there were discrepancies with what was entered online.

The neighbors' organization wanted to avoid fake data as much as possible. Mattioli raises the issue that interventions put forward by citizens are often times not appropriately treated by the authorities due to their nature by giving examples of urban agriculture and urban abandonment maps in which users are not able to directly enter data, but rather get in touch with data curators to provide updates [25]. A conclusion is that complete control on data collection can help improve its accuracy, but can also discourage spontaneous participation. Taking this into account, a user authentication process was implemented so that any visitor could navigate, explore and see collected data on the platform, but only authenticated users could add new information. Knowing which user entered which records was deemed a reasonable way to discourage adding false information.

Database points are displayed on the map using a clustering strategy for each category that dynamically changes when zooming in or out. This serves as a visual aid for identifying areas where certain kinds of observations accumulate and is a much more efficient way to display data in terms of the platform's performance. Otherwise, users would experience lags when zooming in or out, or while panning.

The platform also includes some analysis capabilities. Users can create heat maps on the fly with information stored on the live database, overlap two of them and, as a very basic exploratory data analysis tool, compare if there is a spatial relationship between two variables (Fig. 1).

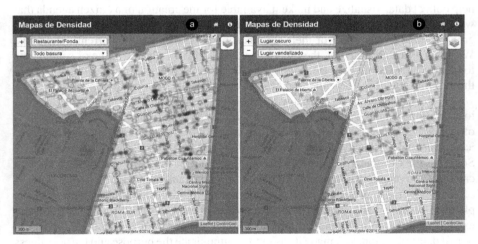

Fig. 1. Heat map intersections for the Roma exercise for (a) restaurants and waste locations and (b) dark places and vandalized places. Selected variables are shaded with warm/cool color ramps.

This was important as it enabled citizens to identify, at a very basic level, if their perceptions were showing some sort of spatial autocorrelation that could help them focus their efforts on specific regions and variables. The purpose was for this to become an essential tool for citizens to consult and review when analyzing the spatial behavior of the variables they were interested in, so it could provide them with initial evidence to support the construction of their citizen agenda.

The digital platform was developed using open-source software and freely available tools such as PostgreSQL, Leaflet JS, PHP, Bootstrap and jQuery. In terms of the user interface, the platform was built as a web app, functional in all operating systems and browsers, on both desktop and mobile environments.

Once the data collection phase was completed, it was time for citizens to analyze all this information. They formed commissions for each of the different categories under study, with the task of finding ways to portray the spatial findings they had obtained, so they could fit their proposals for action with the local authorities with evidence of the neighborhood's perception. Spatial data is an essential part of the citizen's agenda that is being built to negotiate with the authorities. Once their proposals are ready, they will also be publicly available on the project's website for others to consult.

Lessons from the Roma Project. Several things were learned from this study. First, even though only authenticated users were able to add data to the platform, a mechanism to correct information or delete data points was not implemented. As mentioned above, teams of citizens observers and volunteers collected data and were managed by another neighbor. If a mistake was detected, users could inform their fellow managers who in turn could contact the database administrator to make specific changes. There were, however, very few mistakes that needed correction. The two-step process used for data collection and entering was likely helpful in this matter, as records were tacitly double-checked before being added to the database. Second, the existence of a strong social base that pushes forward the adoption of the cause, and at the same time fosters the neighbors' interest, is essential. It would be very difficult to make a project like this work without such group. People in charge of the citizen side of the project were very upbeat about talking to other neighbors, recruiting them, inviting passersby and promoting the exercise. Additionally, this project was surrounded by public presentations, town-hall meetings, media communications and other types of diffusion mechanisms, which are paramount to reach different citizen groups that can and should get involved in participatory processes.

Another lesson is the need to provide users with access to the platform through different gadgets and operating systems. Most users accessed from desktop systems and a certain reluctance to adopt mobile devices was observed. This may have been due to the lack of smartphone and tablet native apps and of users not being eager to access web apps.

3.2   The Lindavista Neighborhood

The experience with the Roma neighborhood triggered the beginning of new collaborations with other sectors of society. In response to media coverage of the Roma study, neighbors from a different municipality called out their interest in having something similar.

The Lindavista neighborhood is located north of Mexico City, covering an area of about 2.1 km^2 (0.8 mi^2) with an approximate population of 15,000 inhabitants and 122 city blocks. It was perceived that for the Roma case there were too many variables to be collected. Citizens' perceptions of a wide variety of topics have a broad range of variability, dependent on factors such as age, gender or personal interests. Fortunately, for this case, neighbors knew better what they wanted to map and it was much easier to come up with a limited set of variables to collect. Four categories with fewer subcategories were available.

While Roma neighbors were interested in gathering a lot of data to be able to construct an agenda, Lindavista citizens were already using paper maps to collect data on the field, had a clearer idea of what issues they wanted to tackle and a better understanding of the type of process in which they could participate. In a sense, Roma neighbors needed a more elaborate explanation on how to study their space and how maps could help them convey their message, as opposed to Lindavista neighbors who were more advanced in their spatial process and were already collecting information on paper maps.

The variables used in both bottom-up cases are shown in Fig. 2. In the center are those that are common to both studies, to the left those that are only for Roma and to the right two that are used only for Lindavista. Nodes on this diagram are symbolized with patterns representing different variables' categories.

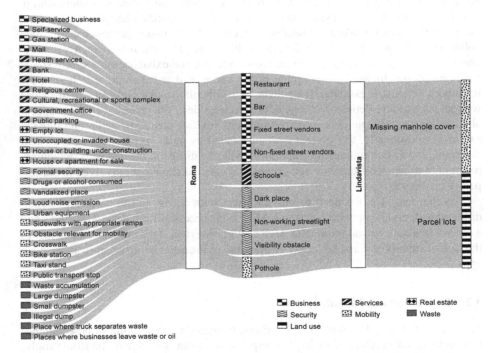

Fig. 2. Categories and corresponding subcategories used in the Roma and Lindavista projects.

Lindavista first considers four categories (businesses, schools, security and mobility) for a total of 14 variables (about one-third of those in the previous exercise) that can be collected in a more consistent way by different observers. This improves the data collection process from the Roma case as, from the citizens' perspective, it allows them to better discern whether a particular situation they observe fits in a given subcategory or not. For Roma, schools were one subcategory of the *Services* category, while for Lindavista the *Schools* category had different subcategories: elementary, middle, high school, university and combined (marked with an * in the diagram). An additional category for land use was considered for Lindavista to collect cadastral data on parcel lots.

This second case is relevant because it gave the opportunity to gain more experience in working with the citizenry and allowed enhancements to the platform. It posed the need to deal with data collection for cadastral data and gave another opportunity to stress the need to use more objective variables. The following features that had not been considered before are incorporated into the platform: the inclusion of official sources of data for citizens to validate; the ability of users to (dis)agree with an observation by voting; the possibility for users to collect cadastral data difficult to obtain from the authorities.

Data Validation. The national government publishes a nationwide survey of economic units. It is useful to locate and identify establishments that fit into the first two categories (businesses and schools) of the Lindavista study. This database is available from the National Institute of Statistics and Geography [26]. In the case of the Roma, this database was not used because citizens wanted to thoroughly collect what was in their neighborhood. They argued that the inclusion of this information would not be very useful since it was most likely outdated, and the dynamics of the neighborhood would not be accurately reflected. For the Lindavista exercise, on the other hand, this was perceived differently and citizens were glad they could be given a lot of records to begin with, so they did not have to start from scratch. Elements from the database of economic units matching the *Business* and *Schools* categories were added to the interactive map, and their associated information was available for display on click.

Again, visitors can navigate and explore data on the web app. Authenticated users have additional functionality when clicking markers. If the official category matches one of the desired ones, the user can validate it. If data is incorrect, the user can edit the information and update the data point. If the marker is not relevant for the categories under scrutiny (for instance, a karate school), the user can delete the observation. Figure 3a shows an example of the validation, edition or deletion of data points.

User Voting. The credibility and accuracy of volunteered geographic information are topics that have been widely recognized and are open for debate. In Fonte et al. [27], some guidelines for assessing data quality and volunteer credibility are proposed. Here, a first attempt to measure the reliability of citizen contributions is introduced with a voting strategy. The idea is to capture users' perceptions to quantify how many people (dis)agree with existing reports and obtain an empirical degree of trustworthiness of the observations and, indirectly, of contributors. With this approach, quality assessment is carried out by citizens themselves. Only authenticated users have access to this functionality, otherwise one runs the risk of letting a single user vote many times in favor or against a particular observation.

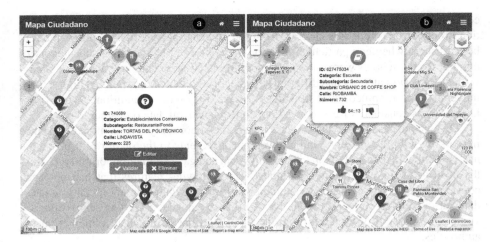

Fig. 3. Enhancements for the Lindavista case study: (a) validation, edition or deletion of data points; (b) the voting strategy.

Once a user casts a vote, it is not possible to up- or down-vote again for the same data point, but users can change their opinion. Again, this is to prevent users from voting several times on a single observation and try to better capture the general feeling of a larger group of users. Figure 3b shows an example of the voting strategy implemented for this exercise.

Parcel Lot Information. Citizens in this neighborhood are also interested in obtaining information with respect to how aligned the land use in this urban area is with respect to the current land use policy for the city. Particularly, there are strong concerns regarding the maximum allowed levels of construction in each parcel lot. In accordance with local law, it is possible to access public records that contain relevant information for this matter, but it is a cumbersome process that can take a long time to achieve and obtained data may not be readily available in open formats.

The parcels' geometry was obtained, overlaid on top of the base map and each polygon was initialized with a *null* value for the number of levels of construction. Users can click on polygons and obtain information while authenticated users can edit the number of levels.

Parcel lots are colored in accordance with the map legend and clicking on a polygon displays a popup window (also shaded with the color corresponding to the number of levels) with the lot id, cadastral account, and current levels. The color schemes help the user get a better idea of the information and the map is a fast and an eye-catching way to get a quick overview of the spatial distribution of levels of construction. Once completed, this information will be very useful in order to compare citizens' observations with official maps. With respect to government data, "it is crucial to distinguish between government transparency and open data" [28]. This remark is framed in the Open Government Working Group's eight principles of government data, which clearly define what characterizes open data [29]. It is often the case that data obtained through the Transparency Office in Mexico is precisely that: transparent but not necessarily open. Unfortunately, initial approaches with local authorities to obtain an official land use map have not been very successful. However, the citizens' drive to obtain and make this information publicly available will likely serve as an example for the authorities to push towards open data. Figure 4 shows an example of the spatial distribution of parcel lots with different levels of construction in the study area.

Fig. 4. Parcel lot data collection interface for Lindavista.

Lessons from the Lindavista Project. In this second case, citizens' demands are more specific than in the previous one, because they want to promptly negotiate with the local authorities in specific matters. In a sense, they already have an 'agenda' and are looking into ways of supporting it with the spatial dimension from around the neighborhood. In contrast, the Roma was using collected information to build their agenda. The quality and accuracy of collected information are very important topics in VGI. With the approach presented here, users' opinions will be useful in clarifying the veracity and usefulness of reports for identification of issues, opportunities, and proposals. The inclusion of official sources of information is an important step as it provides the platform with a start, rather than building its database from scratch. This was well received by citizens, as it is useful for them and accelerates field data collection.

The spatial distribution of levels of constructions will be helpful to identify the type and characteristics of buildings that might be violating the law, as perceived by citizens. It will also serve to identify the type of land use that corresponds to those parcel lots and will help citizens provide clear examples to the authorities about what concerns them.

The number and type of variables used in this case are more concise than in the Roma. Being less subjective helps users better place their observations and provide more useful and meaningful data.

4   Preliminary Results

The Roma project aimed at developing a platform to help citizens put points on a map and allowed them to voice their opinions and support the construction of a citizen agenda. This agenda is intended to help steer negotiations with their local authorities to improve their neighborhood. Around 200 volunteers collected little over 6 thousand points on 41 variables inside the perimeter of the locality in about two months. Once this was completed, the citizens' organization decided to pause data collection to be able to concentrate on analyzing all the information they had collected and find out the current state of the neighborhood. Clusters of observations were useful to identify areas on the map that need attention. With the use of the dynamic generation of heat maps from these observations, citizens can display cross-correlation layers of the observed variables, useful in refining regions where different variables intersect. This supports the identification of locations, stimulates possible courses of action for the citizen agenda and helps in the generation of agreements between citizens, as they can prioritize places that need attention.

It is worth noting that citizens' analyses are not focused on individual reports for different variables, but rather in the collection of observations and perceptions of different users. It is this sense that the knowledge of the masses, the "wisdom of the crowds" and collective intelligence play an important role in the collaborative process.

The implementation of this platform was a successful endeavor in terms of citizen groups adopting it. As mentioned above, the existence of a strong social base that promotes it was essential. It would have been very difficult to bring together all the volunteers that donated their time, effort and knowledge to these projects.

Nonetheless, there were some concerns raised during data collection, as some of the variables were considered troublesome and is best to try and collect objective variables. That being said, the observations reported on the map represent users' perceptions of what happens in their neighborhood. There is always a tendency to judge what goes on in the territory. An example of this is that neighbors in the Roma strongly believed that waste accumulation was directly related with the existence of restaurants. A quick look at the heat map that compares waste accumulation and restaurants reveals that even though there are some places where they intersect, it is not the norm. The spatial intersection of variables available within the platform can help neighbors debunk certain beliefs of what goes on in the territory.

The Lindavista case is an ongoing one at the time of writing. As discussed above, the inclusion of other sources of information proved to be useful for citizens to speed up data collection. The strategy to let users vote and obtain an empirical degree of trust for users and contributions is useful, but it is in the long run when its true value will be seen, once many users have voted on several observations.

With the exception GeoKey, the reviewed platforms work with point data and none allow the possibility to collect data on parcel lots. Once data collection on levels of construction is completed, it is expected that it will either soothe citizens because there are not many deviations from the land use law, it will make evident the local government has done something wrong, or it will be a compromise where some regions are within the law and some are not. Regardless of the outcome of this comparison, this is a process that will definitely set the ground to empower citizens in this neighborhood.

The fact that the platform was not available as a native app for mobile devices may have had an impact on its appropriation by users, as the preferred method of access was on desktop systems. Initial data shows that roughly 10 % of the users accessed the platform from smartphones, only 6 % used tablets and 84 % were on desktop systems.

5   Conclusions

The platform supporting the exercises presented here not only provides the possibility to store complaints about issues, but rather boosts and supports the strong collaboration that can exist between citizens and authorities in a C2G process. It represents a way in which the population, by expressing perceptions, can be and feel more involved in what is happening around it. It also aids in the construction of citizen agendas that contain proposals that can be supported by local governments. This effectively constitutes a way in which individuals are empowered. They no longer are at the expense of the government's good will to act: they become involved. As citizens participated in different stages of the study, from the problem definition through data collection and analysis, this project fits Haklay's proposed typology of an extreme citizen science exercise [30].

These two bottom-up cases have been a substantial endeavor in terms of collaborating with the community that gave opportunities to reflect on several aspects regarding VGI and public participation together with how science and technology can relate to citizen issues and communication. They deal with how it is possible for citizens to collect geospatial information that can serve several purposes. First, it helps in the generation of data that may not be on the government's agenda; or that may in fact be, but for which no formal collection strategies have been implemented. Second, it can pose a challenge to current government structures in terms of the collection and availability of data that should be accessible through them in an open and transparent way. Third, it shows that citizen groups can be empowered by the added value that spatial data can provide them. Specifically, it can help them negotiate citizen agendas for the development of public policies with the government and their local authorities, which are routed for the improvement of the citizens' quality of life.

On the other hand, if local governments decide to follow up on those issues raised by citizens, a fruitful dialogue can be initiated between both parts in which authorities can inform back to the citizenry on the advancements on those issues that need attention. This is what Johnson and Sieber call 'citizen-to-government-to-citizen' (C2G2C) processes [12]. Authorities can also benefit from these practices by recognizing that many eyes on specific situations can be better than just their own. A long-term goal is to have a generic platform that can be adapted and applied to heterogeneous cases that would allow its utilization in top-down social processes as well.

Further research is needed in order to understand the low percentage of access from mobile devices even though the platform is functional on these systems. Field data collection can and should take advantage of mobile devices. To reach a greater audience, it is useful to provide native mobile apps and complement them with websites.

Perhaps it would be advantageous to be able to report on social networks and platforms that users are already familiar and acquainted with. Other platforms reviewed for this work already provide mobile apps and work very well, but are also specifically tailored for each particular case. A notable exception seems to be GeoKey, although a downfall seems to be its mobile counterpart, which has been adapted from a previous study and is now under development, but looks like a promising alternative.

During the final review of this paper, FixMyStreet [31] came across. It is an online service with mobile apps that allows users to report local problems in their neighborhoods. One of its most important features is that it communicates users' observations with the city council and the department in charge of overseeing specific types of reports so they can be taken care of. It is possible to deploy a local instance based on it (like several cities in different countries already have). Being a full blown participatory platform, it works best with a dedicated team of administrators, designers, developers, and data curators to keep track of reports and their evolution. It is an effort that aligns nicely with the spirit of this research and will certainly be interesting to try to merge the proposed platform with what FixMyStreet can do. One disadvantage is that it focuses on point data that somehow needs to be communicated to the authorities. This workflow implies that direct contact with the authorities has already been established since citizen reports are directed to the proper department for attention. For this research, however, such interaction with the government has not yet been established and is one of the goals. Additionally, as mentioned before, citizens are currently more interested in detecting regions that need attention, rather than taking care of specific reports.

It is also worth noting that the way the platform has been developed responds to specific requirements and demands. The social process that drives the needs of technological developments is far more important, even though there may be technological shortcomings that prevent the obtainment of better data. The proposed platform pretends to provide a way to help tackle issues and needs of social groups. This guides the platform development that is mounted on current technological trends but does not respond to an application of software developers' amusements.

Future work includes improving the data model to let users delete or edit their own reports, manage the reported number of levels of construction more efficiently, include visualizations for historical data of votes, optimize the reading of cadastral geometry into the map, and look into the strengths and weaknesses of other platforms to have better functionality.

References

1. Haklay, M., Singleton, A., Parker, C.: Web mapping 2.0: the neogeography of the GeoWeb. Geogr. Compass. **2**, 2011–2039 (2008)
2. Turner, A.J.: Introduction to Neogeography. O'Reilly Media, Sebastopol (2006)
3. Goodchild, M.F.: Citizens as sensors: the world of volunteered geography. GeoJournal **69**, 211–221 (2007)
4. Heipke, C.: Crowdsourcing geospatial data. ISPRS J. Photogram. Remote Sens. **65**, 550–557 (2010)

5. Crooks, A., Pfoser, D., Jenkins, A., Croitoru, A., Stefanidis, A., Smith, D., Karagiorgou, S., Efentakis, A., Lamprianidis, G.: Crowdsourcing urban form and function. Int. J. Geogr. Inf. Sci. **29**, 720–741 (2015)
6. Sui, D., Goodchild, M., Elwood, S.: Volunteered geographic information, the exaflood, and the growing digital divide. In: Crowdsourcing Geographic Knowledge: Volunteered Geographic Information (VGI) in Theory and Practice. pp. 1–12. Springer Netherlands (2013)
7. Harvey, F.: To volunteer or to contribute locational information? Towards truth in labeling for crowdsourced geographic information. In: Sui, D., Elwood, S., Goodchild, M. (eds.) Crowdsourcing Geographic Knowledge: Volunteered Geographic Information (VGI) in Theory and Practice, pp. 31–42. Springer, Heidelberg (2013)
8. Surowiecki, J.: The Wisdom of Crowds: Why the Many are Smarter Than the Few and How Collective Wisdom Shapes Business Economies Societies and Nations. Doubleday, New York (2004)
9. Spielman, S.E.: Spatial collective intelligence? Credibility, accuracy, and volunteered geographic information. Cartogr. Geogr. Inf. Sci. **41**, 115–124 (2014)
10. Jovchelovitch, S., Priego-Hernandez, J.: Bottom-up social development in favelas of Rio de Janeiro: a toolkit. http://eprints.lse.ac.uk/62563/1/ToolkitSocialDevelopmentLSE2015.pdf
11. Kapiga, K.: Bottom-up and top-down approaches to development. http://global_se.scotblogs.wooster.edu/2011/06/26/bottom-up-and-top-down-approaches-to-development/
12. Johnson, P.A., Sieber, R.E.: Situating the adoption of VGI by government. In: Sui, D., Sarah, E., Goodchild, M. (eds.) Crowdsourcing Geographic Knowledge: Volunteered Geographic Information (VGI) in Theory and Practice, pp. 65–81. Springer, Heidelberg (2013)
13. Haklay, M.: How good is volunteered geographical information? A comparative study of OpenStreetMap and Ordnance Survey datasets. Environ. Plan. B Plan. Des. **37**, 682–703 (2010)
14. Flanagin, A.J., Metzger, M.J.: The credibility of volunteered geographic information. GeoJournal **72**, 137–148 (2008)
15. California Academy of Sciences: iNaturalist. http://www.inaturalist.org/
16. OpenTreeMap: OpenTreeMap. https://www.opentreemap.org/
17. University College London: GeoKey. http://geokey.org.uk/
18. Imperial College London: EpiCollect+. http://www.epicollect.net/
19. Usahidi: Usahidi. https://www.ushahidi.com/
20. Brandon Rosage: Ushahidi History. https://www.dropbox.com/s/cm20y1tgteaoot4/History.pdf?dl=0
21. Atzmanstorfer, K., Resl, R., Eitzinger, A., Izurieta, X.: The GeoCitizen-approach: community-based spatial planning – an Ecuadorian case study. Cartogr. Geogr. Inf. Sci. **41**, 1–12 (2014)
22. Rinner, C., Keßler, C., Andrulis, S.: The use of web 2.0 concepts to support deliberation in spatial decision-making. Comput. Environ. Urban Syst. **32**, 386–395 (2008)
23. Hall, B.G., Leahy, M.G., Chipeniuk, R., M., H.: Map Chat http://mapchat.ca
24. CIC (Centro de Integración Ciudadana): Tehuan. http://tehuan.cic.mx
25. Mattioli, C.: Crowd sourced maps: cognitive instruments for urban planning and tools to enhance citizens' Participation. In: Contin, A., Paolini, P., Salerno, R. (eds.) Innovative Technologies in Urban Mapping, pp. 145–156. Springer, Heidelberg (2014)
26. INEGI (Instituto Nacional de Estadística y Geografía): Directorio Estadístico Nacional de Unidades Económicas. http://www.inegi.org.mx/est/contenidos/proyectos/denue/presentacion.aspx

27. Fonte, C.C., Bastin, L., See, L., Foody, G., Estima, J.: Good practice guidelines for assessing VGI data quality. In: The 18th AGILE International Conference on Geographic Information Science (2015)
28. ECLA (Economic Commission for Latin America): Big data and open data as sustainability tools: a working paper prepared by the Economic Commission for Latin America and the Caribbean. Santiago, Chile (2014)
29. Open Government Working Group: The Annotated 8 Principles of Open Government Data. http://opengovdata.org/
30. Haklay, M.: Citizen science and volunteered geographic information: overview and typology of participation. In: Sui, D., Sarah, E., Goodchild, M. (eds.) Crowdsourcing Geographic Knowledge: Volunteered Geographic Information (VGI) in Theory and Practice, pp. 105–122. Springer, Heidelberg (2013)
31. mySociety: Fix My Street. https://www.fixmystreet.com/

Future Cities Urban Transformation and Sustainable Development

Teresa Cilona$^{(\boxtimes)}$

Department of Architecture, University of Palermo, Palermo, Italy
teresa.cilona@unipa.it

Abstract. The complex urban reality, in continuous evolution, are character-ized by buildings, facilities, equipment, human capital, social capital and the ability to create sustainable economic development. Today, urban planning is called to respond to the new needs of the community, for this reason it is necessary to avoid the mistakes made in the past and think of a plan to be adapted to the change. All this is possible through the implementation of par-ticipatory strategic actions which ensure high levels of quality of life as well as responsible management of land resources. This outlines the concept of sus-tainable development and resilient cities, forcefully entered in today's urban paradigm, becoming the key to activate the competitiveness of cities. In this work, particular attention is paid to sustainable mobility in Italy, in the knowledge that the innovations of the mobility and transport system is needed to ensure the livability of future cities.

Keywords: Resilience · Change · Sustainable mobility · Participation

1 Introduction

The ONU program, established in 1992 in Rio de Janeiro, called Agenda 21 on sustainable development, defines the action to be taken, at all levels of government and administration of the territory, from global to local, in every area and in any business, in which the human presence has impacts on the environment [1]. A key element of this action plan is the involvement of stakeholders that operate on a given territory[1] [2]. The Agenda 21 aims to: 1. promote the participation of institutional, economic and social components to the definition of strategies, objectives, sustainable development instruments and actions; 2. Quantify, share and give evidence to the environmental problems and the critical area; 3. Monitor the effects of actions and public policies in the direction of sustainability.

---

[1] They can be subdivided into three categories: information/communication: is an informative approach; the administration informs, communicates to stakeholders the choices and solutions decided by the - consultation/listening: it is an approach that provides both the information phase is the phase of listening to stakeholders. The observations will then be considered by the administration for a possible redefinition of policies - collaboration/involvement: it is an approach that provides a path aimed to take joint decisions between government and stakeholders.

© Springer International Publishing Switzerland 2016
O. Gervasi et al. (Eds.): ICCSA 2016, Part III, LNCS 9788, pp. 183–197, 2016.
DOI: 10.1007/978-3-319-42111-7_15

The issue of sustainability, which is closely linked to scientific research and technological development (monitoring, analysis, evaluation and representation), is intended to ensure the fulfillment of the essential needs of social classes and the poorest countries and to grant to future generations conditions and opportunities for development at least equal to those of the developed countries [3].

Over the years, sustainable development has been defined in several ways. In 1987, the WCED, World Commission on Environment and Development, defines as sustainable development all actions that meet the needs of the present generation without compromising the ability of future generations to meet their own.

In 1991, the IUCN, International Union for Conservation of Nature affirms that the development has to ensure the satisfaction of quality of life, staying within the limits of the carrying capacity of the ecosystems that sustain it.

In 1994, ICLEI, the International Council for Local Environmental Initiatives, defines what is sustainable a sort of development that offers environmental services, the basic social and economic services to all members of a community without threatening the operation of natural systems, buildings and society as which depends on the provision of such services. It is precisely in 1994 that the international Council for local environmental initiatives (ICLEI), organized from May 24 to 27, the first European conference on sustainable cities, also known as conference Aalborg, under the joint sponsorship of the European Commission and the city itself [4].

Here it was signed by 80 European local administrations and 253 representatives of international organizations, national governments, scientific institutes, consultants and individuals, a document known as the *Aalborg Charter* or *Charter of European Cities for a lasting and sustainable development*[2].

We also recall, in 2007, the Leipzig Charter on Sustainable European Cities, signed by Ministers from 27 European countries. In the document the urban development policy strategies are set out in order to ensure high quality of public spaces and urban landscapes; modernizing infrastructure networks and improving energy efficiency; pursuing strategies for upgrading the physical environment; strengthening the local economy and the local labor market; promote the efficient and accessible urban transport [5].

And again, in 2007, the Green Book on Urban Mobility aimed at the creation of a specific European action plan. This document aims to develop a new culture of urban mobility in line with the objectives of sustainable development [6].

---

[2] The Aalborg Charter, consists essentially of three parts that define, respectively: 1. the so-called principle of the Declaration; 2. Campaign of Sustainable European Cities; 3. the local plans of action for a sustainable urban model contained in Agenda 21. With a verification process the Aalborg Charter is activated and all those activities are driving the urban ecosystem towards balance and those that move away it. In addition, it aims to preserve the natural capital, biodiversity, human health, air quality, the water, the soil in order to ensure, in the long life and well-being of humans and animals and plant. The principles of sustainable development are "rethought" in a more flexible set, a local creative process, in which the individuality of each city emerges, the latter called to develop their own strategies, to implement them and to inform each other and experiences.

2 Resilient Cities: New Ways of Planning for Sustainable Development

Economic growth and the development of modern society brought with them a number of problems that menace the possibilities for future development [7].

This happened because very often the true nature of the relationship between man and environment has been ignored and in fact it is necessary to recognize that man depends on the environment in which he lives, and therefore he cannot change it to alter the balance of nature[3] [8].

In order to make cities adapt to changes we need to identify, through the evaluation of risks and costs, the actions needed to be made. Ultimately, we have to find the ways to design resilient city and seeking finance [9].

The resilience is generally conceived as the ability of a system (for example system of cities) to absorb disturbance and reorganize in order to maintain essentially the same *function, structure, identity* and *feedback* [10].

Or, as the cities are complex systems that require the expertise of various disciplines it is necessary, as stated in Table 1, consider the 5 W - who, what, when, where, why - if we hope to create cities that are really resilient [11–13].

Table 1. Urban regeneration capacity for designing resilient cities

		QUESTIONS TO CONSIDER
WHO?		Who determines what is desirable for an urban system?
		Whose resilience is prioritized?
		Who is included (and excluded) from the urban system?
WHAT?	T	What pertubations should the urban system be resilient to?
	R	What networks and sectors are included in the urban system?
	A	Is the focus on generic or specific resilience?
WHEN?	D	Is the focus on rapid-onset disturbances or slow-onset changes?
	E	Is the focus on short-term resilience or long-term resilience?
	O	Is the focus on the resilience of present or future generations?
WHERE?	F	Where are the spatial boundaries of the urban system?
	F	Is the resilience of some areas prioritized over others?
	S	Does building resilience in some areas affect resilience elsewhere?
WHY?	?	What is the goal of building urban resilience?
		What are the underlying motivations for building urban resilience?
		Is the focus on process or outcome?

---

[3] The environment has three main functions in the economy in general and man's sustenance: 1. provides us the natural resources that it uses as an engine of all its activities; 2. receives and assimilates the waste resulting from human activities; 3. ensures the survival of mankind while providing living space and space for recreation.

Speaking of resilient cities it is necessary that local authorities were aware of the problems that climate change poses and the challenges they are facing, putting behind the urban design principles of prevention, not only of the present risks, but also those future, in relation to the vulnerability either natural or induced by human activities on their territory.

You have to find a new way to plan urban areas to obviate the risks and damages of climate change [14–16].

Recent studies talk of a real danger for the cities, where the interventions of physical transformation lead to uncontrollable phenomena, with gentrification effects [16] for the historical centers of segregation of urban neighborhoods or destruction of the architectural heritage. It is necessary, therefore, that the change in the control and in the management strategies matches a conscious renewal in the city values.

This is the real challenge of the twenty-first century: a resilient city combines its historical identity with the change, the old and the new values, rationality and emotions, conservation and development [17].

The new territorial planning, in Italy, a country rich in history and with a built heritage of great artistic value, should therefore be combined with storage and intelligent retrieval of historic centers and the existing building, while as concerns the suburbs and new settlements, we must rise from the current methods of construction of rigid and cities vulnerable to resilient and adaptable methods of city planning.

A city that evolves taking into consideration the starting cultural conditions, accepting the elements imposed by the change, increasing the ability to adapt [18]. In this sense, resilience understood as flexibility, involvement for the collective identity.

3   Action Plans and Areas of Intervention

In 2001, the Commission of the United Nations, said that sustainable development goes from through the improvement of the response capacity of local systems to adverse *shocks* and, at the same time, through the impact of containment measures and development interventions, resulting in strong pressures anthropogenic, on territorial systems, they may affect these capacities. Hence the assumption that a sustainable city is thus a resilient city[4] [19]. The resilient cities, is an urban system that is not limited to adapt to climate change (especially global warming) that in recent decades make it increasingly vulnerable cities with ever more dramatic consequences and very heavy costs [20].

The resilient city is changed by building social responses, economic and environmental new enabling it to resist in the long run to re-stress environment and history. In order to prepare local areas to climate change and build resilient cities [21–24] Plans of Action should be drawn and identify areas of action.

The areas are summarized as follows: 1. land-use planning; 2. hydrogeological structure; 3. urban planning; 4. water cycle in urban areas, conservation of the resource

---

[4] The difference between the two concepts is that sustainability is a goal to be achieved to ensure the welfare of future generations (Bruntland Report on Sustainable Development, UN 2987), resilience, on the contrary covering a way in which we face change, can be a It was more or less desirable.

and its quality; 5. Green multifunctional urban; 6. health and social services and civil protection. The Plan should include concrete actions that lead to a lower risk exposure and, at the same time, the ability to manage urban space through cross-winds of remediation and environmental restoration.

In Italy, the Municipality of Bologna, with the project called LIFE BlueAp[5], as part of the *Mayor Adapt*[6] promoted by the European Commission, has started, first, the path to building a resilient city, which is able to protect its citizens, the land and infrastructure from the risks related to climate change [25].

The process provides involvement for stakeholder in developing the local adaptation plan, Table 2, including advice on concrete actions for the prevention and best practices to be adopted.

Table 2. Guidelines for local adaptation plans, European project LIFE ACT

The project phases and instruments produced		
Phase 1	Start the process - To start	Technical management and organizational commitment political, estimates of the necessary financial resources
Phase 2	Collect Information	Collection of information available to avoid the mistakes made earlier
Phase 3	Vulnerability assessment and risk	Vulnerability assessment and risk, identify priorities for action
Phase 4	Describe the Plan	Setting purpose and targets
Phase 5	Implement the Plan	Identifying financial resources, drivers, constraints define responsibilities
Phase 6	Assess, supervise, and update the Plan	Progress is supervised and updated

In addition, to ensure a future for our cities, we must also focus the attention to areas subject to hydrogeological risk (Fig. 1), through interdisciplinary contributions, which are centered on the maintenance and protection of the environment.

4   Urban Transformation Policies and Sustainable Mobility

In the face of the great changes of the cities, urban planning and environmental policies must change radically.

It is not only to promote new interventions of urban regeneration by widening the spread possible, but to question every part of the city whose operation involves a

---

[5] Bologna local urban environment adaptation plan.

[6] The initiative "Mayors Adapt, The Covenant of Mayors Initiative on Adaptation to Climate Change" was launched 19 March 2014 by the European Commission as part of the Strategy. "*Mayors Adapt*" aims to increase support to local actions, to provide a platform for greater commitment and networking the city by raising public awareness of the adaptation measures to climate changes that are necessary.

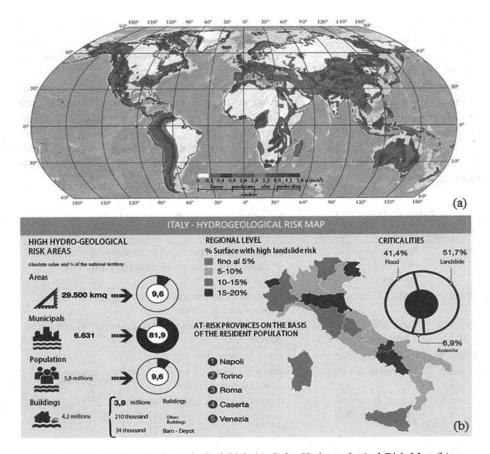

Fig. 1. World Map Hydrogeological Risk (a); Italy, Hydrogeological Risk Map (b)

shortfall between available resources and those needed for sustainable growth, bearing in mind along with the environmental resources also as energy. A strategy that for these reasons can be defined urban regeneration as resilience [26].

Resilience should be based on uncertainty, typical of crisis times to suggest new solutions and face collective challenges for the future, following the strategic directions drawn with Europe 2020. Thanks to the crisis, we are called to face, the opportunities of transformation and innovation and therefore opportunities to rethink the very future sustainable of the city.

Analyzing the urban phenomenon, we must now considering the trend of putting the focus of discussions of territorial policies, the social phenomena related the mobility system [27].

In Italy, the current mobility model, based on the use of private cars is cause to criticality economic, environmental and social. The traffic of our city is an economic problem in terms of resource consumption do not replenished. Is a environmental problem because it generates negative effects in terms of pollutant emissions, impacts on health of the planet and people. Is a social problem because it seriously affects the quality of life and safety of citizens. The mobility system to be sustainable must be given to public transport, where pedestrian traffic and cycling play a central role. Should ensure the improvement of the service in terms of accessibility to places, persons, services. Function without endangering the environmental balance or threaten human health.

It is therefore necessary to rethink a city capable of using the environmental challenges of the near future, to transform and innovate its spatial and functional organization, maximizing its energy efficiency and reducing dependence on non-renewable energy.

For these reasons, the Ministry of Environment through the Fund for Sustainable Mobility has launched programs for the enhancement of local public transport (TPL), with environmentally friendly vehicles such as cycling, more streamlined distribution processes of goods in urban areas, promoting flexible transport and of the car sharing.

The main objective is the "decarbonisation" through: 1. intelligent traffic management, 2. the creation of "grid" infrastructure for electric mobility, 3. better logistics, 4. a reduction in $CO2$ emissions for road vehicles, aviation and maritime sector, 5. use of green cars [28].

Moreover, the experimental and industrial research, has made available to the community two different types of innovative products: a new generation of electric vehicles (partially or totally) and the so-called Intelligent Transportation Systems (ITS). These systems are based on information technology and telecommunications are able to support the management and control of the private mobility and the public transport service (by cell phone at satellite tracking, from broadcasting and communication short-range at Internet, from sensors for traffic detection and image processors to electronic payment devices, to the display technologies and GIS digital mapping). Offer the ability to control traffic, as well as driver assistance, safety of all road users, driving comfort and estimated emissions [29].

Between sustainable urban mobility projects we recall SPARTACUS (System for Planning and Research in Towns and Cities for Urban Sustainability); PROPOLIS (Planning and Research Of Policies for and Use and transport for Increasing Urban Sustainability); the Italian project OSIMOS (Italian Observatory on Sustainable Mobility) and MVMANT (Movement Ant), best described in the next paragraph.

Reading the ranking of the Italian cities on sustainable mobility it is clear how the country it is divided into two parts [30]. There is on the subject, a strong attention and sensitivity among the cities of the north, which does not happen in the southern cities (Fig. 2).

A study on Euromobility[7] has analyzed the habits and tastes of the public on a new mobility management [31]. A job that starts from citizen reactions to the innovations introduced in the system of mobility (car sharing, bike sharing, mobility manager, etc.), And their effectiveness, has verified the implementation of city health in relation to the

---

[7] Non-profit association that promotes the country new forms of mobility and transport, both individually and collectively, systems and more sustainable facilities for the benefit of the quality of life of citizens and greater respect of the environment as possible.

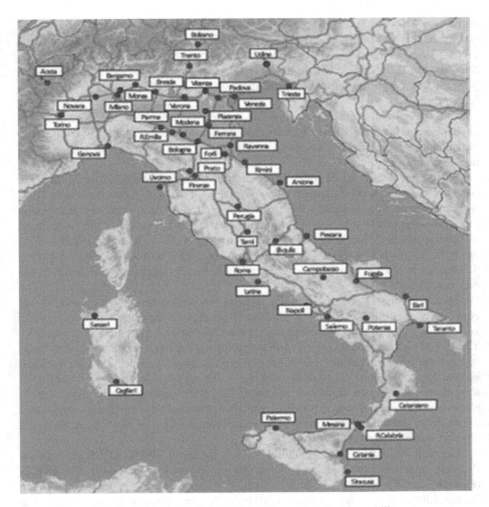

Fig. 2. Italy - Map of the Italian cities on sustainable mobility

presence of car new generation or powered with alternative fuels (LPG, CNG), controlled the supply of public transport, and the area dedicated to bike paths, examined the status of implementation of the management tools and traffic planning, quantified the number of parking lots exchange and paid, the number of accidents and of the mortality rate, the number of cars per square kilometer, analyzed the air quality and the promotion and communication initiatives in favor of sustainable mobility.

The survey found that over 55 % of residents in these cities use the car, although in the last year, perhaps about the economic crisis, increased use of public transport and that of the extra-urban cycle. In the last year there was the spread of bike sharing and the strong increase in car LPG and CNG: over 7 cars out of 100 are gas.

Clear directions for policy-makers, emerging from this our study, are those of paying more attention to the citizens, who often prove forward-looking, for the proposed capacity and sensitivity of their own administrators: 83 % thinks that the spread of bike sharing can be a valuable contribution to reduce congestion and pollution in the city and about 80 % would like a fleet of bicycles even in their own town or, where they already exist, would that they increased the number parking where to find an available bike. It is important, therefore, that the politics develop a new culture of mobility in order to make it more livable the urban environment.

5 Sustainable Mobility in Sicily

There can be no economic development and improving the quality of life in cities if nothing is done radically about our mode of transport and therefore on our habits. It is in this case (only) a question of money: it is a cultural and political question, of foresight, of openness, environmental sensitivity and social equity that we can not help but meet the ever more pressing economic crisis, ecological and energy that characterize our times and are all faces of the same coin.

In theme of sustainable mobility, try to change direction two Sicilian towns, Palermo and Ragusa.

Palermo is the second European city with nearly 800,000 inhabitants, for traffic congestion. Occupies the 33rd place, in 2015, in the national rankings on sustainable mobility in Italy (Table 3). Palermo administration, in reducing the presence of private cars, and polluting emissions from transport, proposes a set of shared services through the strengthening of public transport with low emissions, as well as the adoption of zones limited to traffic, an unprecedented proposal taxi and a new tram line (Fig. 3). Specifically, the Sicilian capital, inaugurated in October 2015, the electric car sharing in partnership with Renault, Enel and Amat (local public transport company).

A new service listed on the National circuit *I Guido Car Sharing* that requires the use of more than 100 cars with a low environmental impact.

The car sharing service has been joined to chat of bike sharing Citizen, with 420 bicycles available, 20 of which with pedal assistance. As well as to report in the past months the realization of more than 100 km of cycle paths.

Table 3. Urban mobility of Italy - ranking of the fifty major cities

1°	Venezia	11	Verona	21	Rimini	31	Ancona	41	Monza
2°	Brescia	12	Genova	22	Vicenza	32	Napoli	42	Catania
3°	Torino	13	Udine	23	Trento	**33**	**Palermo**	43	Messina
4°	Parma	14	R. Emilia	24	Novara	34	Terni	44	Taranto
5°	Milano	15	Piacenza	25	Perugia	35	Foggia	45	Siracusa
6°	Firenze	16	Forlì	26	Ravenna	36	Aosta	46	Sassari
7°	Bologna	17	Roma	27	Bari	37	Latina	47	Catanzaro
8°	Padova	18	Modena	28	Prato	38	Salerno	48	L'Aquila
9°	Bergamo	19	Bolzano	29	Trieste	39	Livorno	49	Potenza
10	Cagliari	20	Ferrara	30	Pescara	40	Campobasso	50	R. Calabria

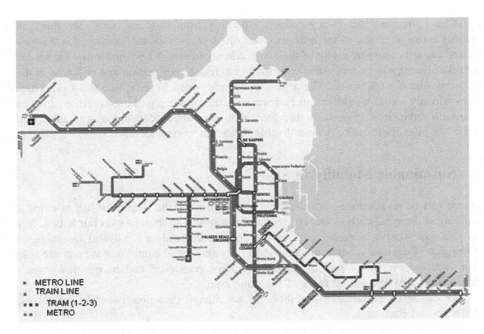

Fig. 3. Palermo - Map of the sustainable mobility

To complete the offer of alternative mobility we nave to remember the opening, on December 30, 2015 of the tramlines. A solution that gives the tram in the Sicilian capital after decades of waiting (the first projects date back to the nineties).

There are four new lines with a total extension of 17 km, integrated with the underground rail system connecting the center to the periphery (17 tram, with 250 seats (56 seats), at a cost of 1.40 euro for 90 min, tris is a limit within which you can make several trips). To make efficient transport tram, bus lanes, the *intelligent* traffic lights that give priority to public transport and the frequency of trips every 7 min during peak hours. And again, two further developments on sustainable mobility of Palermo. The introduction of two limited traffic zones and sharing taxi service, a proposal made in agreement with the trade associations, which allows you to share a taxi to the fixed cost of 2 euro. The service includes five urban routes and one town (8 euro for the extra-urban journey of Cinisi airport).

Also the city of Ragusa (Fig. 4) with 72,976 inhabitants, thanks to the project called MVMANT of Edisonweb, pointes to a new urban mobility to solve the problem of city traffic and improve the quality of life of residents and tourists [32].

The project selected by Fiware/Frontiercities, acceleration program financed by the EU, plans to address the management of urban transport through the use of an algorithm[8].

---

[8] Software House elaborated by Mirabella Imbaccari with a team of physicists and computer technicians of Catania led by Riccardo D'angelo.

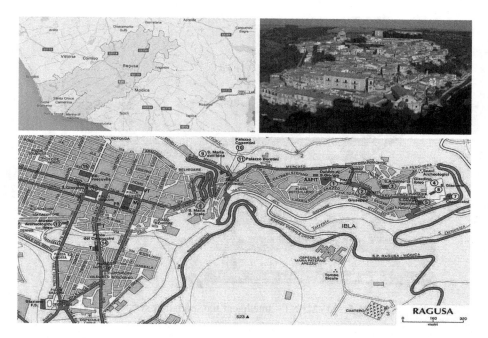

Fig. 4. Ragusa

MVMANT is the acronym derived from Movement (Movement) and ANT (ant) the insect that according ethologists shows more intelligence and organization in the overall management of their displacement distances.

The shared transport service experimented for three weeks, the model is the Particular Cuban taxi, taxis moving along the city's main arteries and allows the transport of several people at once. Through the smartphone, you can book one seat in the vehicle, estimate waiting time and pay for your ticket thanks the mobile device (Fig. 5). The system is also promoting the territory. With monitors that are inside of the cars you can get information on the cultural and commercial centers of the city, book a ticket, a visit or a purchase. The sponsorship is Mercedes-Benz Vans that for experimentation has provided four vehicles representing different mobility solutions.

The Sicilian city has reacted very well to the possibility of having an alternative and sustainable mobility service. A survey shows that about 88 % of a sample of respondents would be interested to leave one of the (media) family cars if he had the chance to take advantage of this service constantly. Also for families the cost savings was estimated at 6,000 euro per year. To date other Italian and European cities have shown interest the model, such as Modena, Dubai and Berlin.

Fig. 5. MVMANT - Intelligent transportation system

6 Conclusions

The city of today, it is subjected to forces of different nature (climate change, wars, earthquakes, technological development, globalization). The resilience may offer new growth scenarios and a new way to look at the processes of change. The city must progress without losing its roots, must be able to absorb adverse disturbances of any kind, including unexpected.

In Italy, the challenge to the resilient city takes on a particular connotation, linked to the discomfort to the attrition of the beauty of the city and the countryside in general [33].

The beauty expresses what survives in the continuous transformations of the urban structure. Although our country has been the manifesto of cultural wealth, today, we are seeing an inability to value our cultural heritage and decline the concept of identity in contemporary world because we have lost the cultural references, now elsewhere.

Today if at the basis of the crisis of the city there is a loss of the sense of identity it is essential to develop processes that develop awareness and sense of belonging. We have to find ways and places to implement this intrinsic capacity for renewal of the social and urban system through material and immaterial spaces of sharing. The practice of resilience is a process that requires sharing and social participation, while institutional flexibility to adapt projects and actions political, economic change [34–36]. The urban recovery path has been a management effort to unexpected scenarios and only in a few cases it was the result of a strategic plan and concerted effort to territorial identity renewal.

In the light of these reflections, cities should be interpreted as social, economic, political laboratories, and areas in which to imagine and experience renewed forms of governance of the complex processes that occur in and out of the [37] increasingly fluid borders.

The housing and urban regeneration must be reconsidered as the actuation value of the particular resources and capabilities of the places. The need to make choices of urban based on the protection of identity management objectives of the places and at the same time the socio-economic development of the city can be met through the search for a new balance between the constraints of protection and transformation pressures [38] in this way, the attention to the recovery of the architectural and urban heritage must be able to interpret the meaning of architecture and return it to the needs of new organizations of the urban space [39].

If we think about the future of our cities, such actions lead back to the enhancement of the low spontaneous actions, guided by policies and processes organized and institutionalized, so to a *governance* that must accept the multiple knowledge and powers, interests and skills.

In conclusion, there is not a panacea for a resilient and sustainable cities, as well as an optimal situation does not appear explicitly in the best of any other [40].

The dynamic interaction objectives, accessibility, orientation towards change, however, can lead to a resilient, an open, modular, flexible city, able to accept the new proposals from time.

Acknowledgements. The author would like to thank the leaders of the Ministry for the Environment, Land and Sea, Assogas liquids, the Ecogas, Bicincittà.

References

1. General Assembly – United Nations – Resolution adopted by the general Assembly on 27 July 2012, A/Res n.66/288. The future we want (2012)
2. Baruzzi, V., Bigi, M., Martini M., Saltarelli E.: Layman's Report, ACT – Adapting climate change in time, LIFE 08 ENV/IT/000436 (2013)
3. Meerow, S., Newell, J.P., Stults, M.: Defining urban resilience: a review. Landsc. Urban Plan. (2015)
4. ICLEI How does the Urban NEXUS Approach and inter-sectoral coordination make cities more resilient? A.3. SESSION DESCRIPTION – 5TH Global Forum on Urban Resilience & Adaptation, Bonn, Germany, 29–31 May 2014
5. European Ministers responsible for Urban Development: Leipzig Charter on Sustainable European Cities. Final Draft (2007). http://ec.europa.eu/regional_policy/archive/the-mes/urban/leipzig_charter.pdf
6. Eur-Lex, Access to European Union Law, Libro verde per una nuova mobilità urbana. http://eur-lex.europa.eu/legal-content/IT/TXT
7. http://www.a21coordinamento.it/GliimpegnidiAgenda21/DaRiodeJaneiroadAalborg
8. Tiezzi, E., Marchettini, N.: Che cos'è lo sviluppo sostenibile? Donzelli Editore, Roma, pp. 86–90 (1999)
9. Desouza, K. C., Flanery, T. H.: Designing, planning, and managing resilient cities: a conceptual framework, Cities, journal homepage (2013). www.elsevier.com/locate/cities

10. Energia Ambiente e Innovazione - Città resilienti: un nuovo modello per il futuro/Anno 2011, n. 3, Maggio-Giugno (2011)
11. Walker, B., Holling, C.S., Carpenter, S.R., Kinzig, A.: Resilience, adaptability and transformability in social – ecological systems (2004). Published here under licence by The Resilience Alliance
12. Meerow, S., Newell, J.P., Stults, M.: Defining urban resilience: a review. Landsc. Urban Plan. **147**, 38–49 (2016). (p. 46)
13. Folke, C., Carpenter, S.R., Walker, B., Sheffer, M., Chapin, T., Rockstrom, J.: Resilience thinking - integrating resilience, adaptability and transformability. Ecol. Soc. **15**, 1–20 (2010)
14. Chelleri, L.: Urban Resilience Trade-offs: Challenges in Applying an Integrated Approach to Urban Resilience, 16 April 2015
15. Fabbricatti, K.: Le sfide della città interculturale. La teoria della resilienza per il governo dei cambiamenti: La teoria della resilienza per il governo dei cambiamenti, FrancoAngeli (2013)
16. Walker, B., Gunderson, L., Kinzig, A., Folke, C., Carpenter, S., Schultz, L.: A handful of heuristic and some propositions for understanding resilience in socio-ecological systems. Ecological and Society (2006)
17. European Sustainable Cities and Towns Conference. http://ec.europa.eu/environment/-urban/aalborg.htm
18. Sustainable Cities International – Canadian International Development Agency: Indicators for Sustainability. How cities are monitoring and evaluating their success. Vancouver November (2012)
19. Chourabi, H., Nam, T., Walker, S., Gil-Garcia, J.R., Mellouli, S., Nahon, K., Pardo, T.A., Scholl, H.J.: Understanding smart cities: an integrative framework. In: 45th Hawaii International Conference on System Sciences (2012)
20. Walker, B., Holling, C.S., Carpenter, S.R., Kinzig, A.: Resilience, adaptability and transformability in social-ecological systems. Ecology and Society (2004)
21. Brundtland Report: Our common future. World Commission on Environment and Development (WCED). Oxford University Press, Oxford (1987)
22. Carpenter, S.R., Whalcher, B.H., Anderies, J.M., Abel, N.: Resilience management in social-ecological systems: a working hypothesis for a participatory approach (2014)
23. http://italiaecosostenibile.it/il-rapporto-brundtland (1987)
24. L'adattamento per la città resiliente, Progetto BlueAp, Bologna e il cambiamento climatico, ECOSCIENZA Numero 5 (2014)
25. Ipcc (Intergovernmental Panel on Climate Change), Climate Change 2007: Synthesis Report. Contribution of Working Groups I, II and III to the Fourth Assessment Report of the Intergovernmental Panel on Climate Change (2007). (Core Writing Team, Pachauri R.K. and Reisinger A., Geneva, Switzerland)
26. Minghetti, A., Africani, P., Paselli, E., Lorenzini, L., Scagliarini, S., Ferrari, E., Poggiali, M., Bizzarri Cristina, C., Cigarini, A.: "Sistema di gestione e pubblicazione degli strumenti di pianificazione urbanistica" in "Atti 14a Conferenza ESRI Italia" (2013)
27. Galderisi, A.: Un modello interpretativo della resilienza urbana. PLANUM J. Urbanism **II** (27) (2013). ISSN 1723-0993
28. Galderisi, A.: Climate change adaptation. Challenges and opportunities for a smart urban growth. TeMa, J. Land Use Mob. Environ. **7**(1) (2014)
29. Chelleri, L., Kunath, A., Minucci, G., Olazabal, M., Waters, J.J., Yumalogava, L. Multidisciplinary perspective on urban resilience. In: Workshop report, BC3, Basque Centre for Climate Change (2012)
30. Quarto rapporto sulla mobilita sostenibile in 50 citta italiane redatto da euromobility (2011). http://www.arpat.toscana.it

31. Misurare la sostenibilità, dati osservatorio. http://www.euromobility.org/dati-osservatorio-2015

32. Frontier Cities, European Cities Driving the future internet, Smart mobility a Ragusa (2016). http://ragusa.mvmant.com

33. Chelleri, L., Kunath, A., Minucci, G., Olazabal, M., Waters, J.J., Yumalogava, L.: Multidisciplinary perspective on urban resilience. In: Workshop report, BC3, Basque Centre for Climate Change (2012)

34. CIAT Report for Cities Impacts & Adaptation Tool. Climate Center, University of Michigan (2015). http://graham-maps.miserver.it.umich.edu/ciat/report.xhtml

35. EEA: Urban adaptation to climate change in Europe - Challenges and opportunities for cities together with supportive national and European policies. EEA Report No 2 (2012). http://www.eea.europa.eu/publications/urban-adaptation-to-climate-change

36. Moir, E., Moonen, T., Clark G.: What are future cities? Origins, meanings and uses (2014). https://www.gov.uk/government/uploads/system/uploads/attachment_data/file/337549/14-820-what-are-future-cities.pdf

37. Settis, S.: Paesaggio, Costituzione Cemento - La battaglia per l'ambiente contro il degrado civile, Passaggi Einaudi (2010)

38. Lazzarini, A.: Dare forma allo spazio del convivere, Architettura e politica: un incrocio di sguardi, Ocula 13 (2012)

39. Cilona, T., Gestione partecipata, integrazione sociale e rigenerazione urbana. Un caso studio. Urbanistica Informazioni, speciale ISSUE, INU, IX giornata di studi, Infrastrutture blu e verdi, reti virtuali, culturali e sociali (2015)

40. Lynch, K.: A Theory of Good City Form. MIT Press, Cambridge (1981)

A Seismic Risk Model for Italy

Alessandro Rasulo[1]([✉]), Maria Antonietta Fortuna[2],
and Barbara Borzi[3]

[1] University of Cassino and Southern Lazio, Cassino, Italy
a.rasulo@unicas.it
[2] Graduate Student at University of Cassino and Southern Lazio, Casino, Italy
[3] European Centre for Training and Research in Earthquake Engineering,
Pavia, Italy
barbara.borzi@eucentre.it

Abstract. All Italian territory, with the only exception of the island of Sardinia, is subject to seismic events, making this threat a major concern for decision makers in charge for the implementation of risk reduction policies. Seismic risk vary greatly across the country and within single administrative regions. Seismic risk maps at national scale are therefore a useful tool for representing the expected adverse outcomes due to the seismic events and for programming the appropriate mitigation measures. The production of those maps is a complex task that involves the combination of data coming from different field of expertise. The aim of the study is to show how the already available information can be combined together in a Geographical Information System (GIS) tool. The results provide a reliable representation of the seismic risk at national scale to be used when planning the mitigation measures to be undertaken in order to improve the level of preparedness in case of an earthquake.

Keywords: Risk analysis · Reliability engineering · Earthquake engineering · Socio-economic modeling · Seismic hazard · Seismic fragility functions

1 Introduction

Seismic risk analysis at territorial scale may be defined as the estimation of the potential damages due to earthquakes expected in a region in a specified period of time. Since urban areas are the places where generally the human activities are concentrated, main attention in those studies is focused on the existing building stock (which represents the physical environment cities are constructed of), mostly composed by structures not compliant with modern seismic design criteria [9, 18–20].

The analysis of the expected loss an earthquake can produce over a territory, must necessarily take into account the uncertainty that are involved into the forecast [27]. It is worth noticing that during the risk evaluation some of the uncertainties are inherent (the randomness of the seismic phenomena is such that no one can say when, where and how intense will be the next earthquake), others uncertainties are epistemic and can theoretically be reduced, but equally practically persistent (the wider is the area object of the study, the looser will necessarily be the inventory of all the goods subject at risk considered in the analysis).

O. Gervasi et al. (Eds.): ICCSA 2016, Part III, LNCS 9788, pp. 198–213, 2016.
DOI: 10.1007/978-3-319-42111-7_16

Usually the developing of a seismic risk analysis is a complex task that involves many disciplines including geophysics and geology (in order to take in account past seismicity, seismo-tectonic framework, wave propagation as well as soil effects), survey (in order to collect data about the building stock), structural analysis (in order to assess the building response under seismic loads) and social and economic sciences (in order to evaluate socio-economic consequences of an earthquake) [12, 16, 17, 25, 26, 28].

The standard definition of seismic risk is the probability or likelihood of a damage, due to an earthquake, and consequent loss to a specified class of elements at risk over a specified period of time. In order to keep the problem of computing the risk tractable, it is tackled initially decomposing the task in specialized (simpler) components, conditionally independent and conventionally referred as hazard (pertaining to the likelihood of the seismic shaking on ground), vulnerability (pertaining to the susceptibility to damage of the built environment) and exposition (containing the socio-economic evaluation of the potential losses) and then recursively applying the total probability theorem in order to aggregate together the separate components. Hence the risk can be expressed by a convolution integral [8].

The objective of the study presented here is the definition of the seismic risk for the whole Italian territory with the latest available data. It represents the natural continuation of the study presented in [28], where just a specific place, the city of Cassino, was examined.

2 Seismic Hazard

The scope of the seismic hazard analysis is the definition of the occurrence of a specified measure of the intensity of the ground motion at a site. When a probabilistic approach is followed, the representation of the seismic hazard can lead to produce either: the probability of exceeding a specified ground motion (in this case the output is generally a set of curves showing the exceedance probabilities of various ground motions at a site) or the ground motion that has a specified probability of being exceeded over a particular time period (in this case the output is a set of maps showing the estimated magnitude distribution of ground motion that has a specific exceedance probability over a specified time period within a region).

Numerous studies have coped with the analysis of Italian seismic hazard both at local and national level. After 2004 those studies assumed more momentum, since the definition of seismic input to be employed in seismic design was compulsory associated with the likelihood of reaching some levels of seismic accelerations at site. Therefore, the probabilistic hazard analysis conducted by the INGV [23] has received officially legal recognition in Italy as the national reference for engineering applications. The results have been mapped on national scale over a 0.05° grid for various annual frequencies of exceedance (the reciprocal of the return period: T_r, varying from 30 to 2500 years) presenting peak ground acceleration (PGA) and spectral ordinates in acceleration for various natural periods ($S_a(T_n)$: T_n: varying from 0.1 to 2.0 s.).

In total 90 maps have been produced. As shown in Fig. 1(a), where one of those maps is presented, Italy is one of the Mediterranean countries with a high seismic hazard. The seismo-tectonic framework sees the peninsula at the convergence of the

African and Eurasian plates. The highest seismicity is concentrated in the central-southern part of the peninsula, along the Apennine ridge, in Calabria and Sicily and in some northern areas, like Friuli, part of Veneto and western Liguria. Only Sardinia is not particularly affected by seismic events.

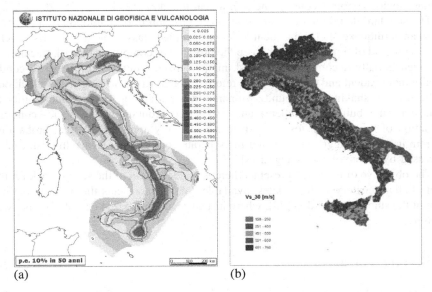

(a)   (b)

Fig. 1. (a) Seismic hazard maps of Italy: PGA at Tr = 475 year, and (b) Shear wave velocity of the upper 30 m soil, V_{s30}. (Color figure online)

All the results are evaluated on bedrock and flat surfaces since the hazard analysis conducted by the INGV does not includes seismic amplification effects due to soil stratigraphy or topography. A simple way to introduce both those effects has been recently discussed in [30] on the basis of data made available by the United States Geological Service (USGS) [31, 32], even if in the present study only the stratigraphic effects have been considered.

The method consists in using the topographic slope as a proxy for shear wave velocity in the top 30 m (V_{s30}): steep topographies (i.e., large altimetry gradient values) have been associated to hard rock sites with high propagation velocity whereas plain areas (i.e., zero or very low gradients) have been associated to thick alluvial deposits whit lower velocities. Those data are presented in Fig. 1(b) and, despite the very simplified approach followed, they are consistent with a more geological-sound approach, such as the information derived from the 1:100,000 geology map published by the Italian Geological Service [24]. Indeed the only main known differences between the two approaches has been identified in Italy in areas where there are flat near-surface rock plateaux that are necessarily undetected by the altimetry-based Wald and Allen [31] approach (e.g., the Karst areas present around the city of Trieste in the North and along the Salento peninsula, in the South).

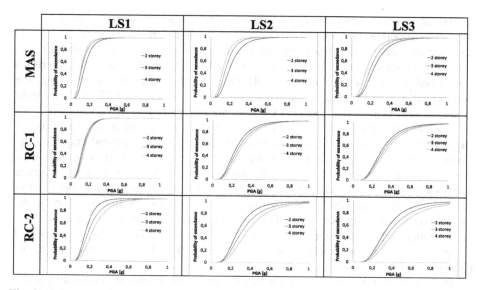

Fig. 2. Seismic fragility functions used in this study. MAS (masonry build.), RC-1 (reinforced concrete build. Non-seismically designed), RC-2 (reinforced concrete build. Seismically designed with minimal provisions). LS1, LS2 and LS3 (slight, significant and collapse limit states). (Color figure online)

The procedure followed can have some ulterior shortcomings: the V_{s30} parameter was recently criticized as not being the best indicator of the actual soil amplification conditions [33, 34] especially in a country, like Italy, with a very complex geology setting, and the employment of low-resolution geological data can induce to neglect locally the presence of averse soil conditions (necessarily un-surveyed in a large-scale geologic map). However at the time of the release of this study, the seismic micro-zonation studies, aimed at the definition of the soil conditions in prospective of seismic hazard exercises, were still not covering the whole national territory and the grade of approximation provided by defining the soil characteristics with the already available data was considered consistent with the scope of the study and the other simplifications necessarily introduced in tackling the other relevant components of the analysis.

3   Seismic Vulnerability

Seismic Vulnerability is intended to represent the damage susceptibility due to seismic excitation of the physical objects under scrutiny. As already explained in the introduction, the decomposition of the problem in specialized components permits to treat the value of the measure of the seismic input at the base of the structure as deterministic (since its occurrence probability at site has been specifically tackled within a separate task, i.e. the seismic hazard component of the study).

The main issue within the seismic vulnerability component stems out from the fact that while structural engineering has refined methods for the analysis of the seismic response of a single item (like a specific building or a structural component) [18–20], when it is under scrutiny a bulk of items, like a building stock, whose characteristics are defined in looser terms (a class of buildings generally is obtained grouping together buildings that have in common the same structural type, number of floors, age, technique of construction …), statistical methods are required.

Therefore the vulnerability of a structures is often expressed in terms of fragility curves that take into account the uncertainties in the seismic demand and capacity. Those functions can be constructed on the basis of the observed damages experienced in past seismic events. In the past, such empirical approach has been largely adopted worldwide: in Italy, for example, the data retrieved from the post-earthquake surveys conducted with 1st and 2nd level assessment forms issued by National Group for the Defense against Earthquakes (GNDT) have been largely utilized [1, 7, 11].

The major shortcoming of the empirical approach is however its accuracy and completeness, since the database of damage observations may not cover the whole national territory.

In the present study (see Fig. 2) the fragility curves have been built according to the SP-BELA approach [2, 3]. According to this methodology the displacement capacity of the buildings at different damage levels (limit states) is produced, relating the displacement capacity to the material and geometrical properties. Three limit state conditions have been taken into account: slight damage (LS1), significant damage (LS2) and collapse (LS3).

The slight damage limit condition refers to the situation where the building can be used after the earthquake without the need for repair and/or strengthening. If a building deforms beyond the significant damage limit state it cannot be used after the earthquake without retrofitting. Furthermore, at this level of damage it might not be economically advantageous to repair the building. If the collapse limit condition is achieved, the building becomes unsafe for its occupants as it is no longer capable of sustaining any further lateral force nor the gravity loads for which it has been designed. The aforementioned limit states can be assumed equivalent to the definitions contained in Eurocode 8, as follows: LS1: Damage Limitation (DL), LS2: Significant Damage (SD) and LS3: Near Collapse (NC).

In order to fit fragility functions to exposure data, in the case of masonry buildings (MAS), four separate building classes have been defined as a function of the number of storeys (from 1 to 4), whilst for reinforced concrete the building classes have been defined considering the number of storeys (from 1 to 4) and the period of construction. The year of seismic classification of each municipality has then been used so that the non-seismically designed (RC-1) and seismically designed buildings (RC-2) could be separated. In this way, the evolution of seismic design in Italy and the ensuing changes to the lateral resistance and the response mechanism of the building stock could be included in the model.

4 Exposure

Essentially the exposure component in a risk study deals with the enumeration of the population of items that are subject to the risk and their relevant aspects in relation to the analysis. Obviously this kind of information has necessarily to interact with both hazard and vulnerability components of the study.

Depending on the extension of the scope of the analysis, exposure may include a single building with its occupants and contents, or the entire constructed environment in a specified area, inclusive of buildings and lifelines (infrastructural systems forming networks and delivering services and goods to a community).

In order to facilitate information collection about the existing facilities in a region, a standardization of the inventory is deemed, providing a systematic classification of the structures according to their type, occupancy and function, as discussed in the conclusions.

In Italy the general characteristics of the building stock are provided by the Census: the housing data are regularly collected since 1861 and with more statistical rigor since 1931 with 'ad hoc' statistics [40, 41] and since 1951 routinely associated with the decennial population census recording. The data utilized in the present study are obtained from the 15th General Census of the Population and Dwellings (ISTAT 2011) [21].

The Census data are collected and aggregated at different levels: the basic unit for data collection is the single household and dwelling, but each dwelling is classified as being located within a building, of a given construction type (RC, Masonry, Other), with a given number of storeys (1, 2, 3, 4 +) and age of construction ($\leq$ 1919, 1919/1945, 1946/1960, 1961/1970, 1971/1980, 1981/1990, 1990/2000, 2000/2005, hx00A0;$\geq$ 2005). In order to protect privacy, the collected data are disclosed only in aggregated format whose minimum territorial extension is the Census tract. A Census tract is a small, relatively permanent statistical subdivision of an administrative municipality, designed to be relatively homogeneous with respect to population characteristics, economic status and living conditions. Additionally, tracts have to be delimited by some clearly identifiable physical boundaries (such as rivers, streets, rail-tracks, ...) in order to assure that statistical surveyors will immediately recognize the census area where the data are recorded. In highly urbanized areas, a census tract generally has the dimensions of a building block, whilst in rural areas it can be significantly larger.

Census include also, for each municipality, one or more fictitious census tracts, used to store data about homeless people and those that were temporarily cast away from their habitual residence due to emergency situations (like the 2009 L'Aquila earthquake's evacuees).

Further details about the elaboration of the exposure data are discussed in the next section. In Fig. 3 the percentage of structural type at municipality level is mapped.

5 Application Results

The case analyzed in this paper is represented by the whole Italian territory, with exclusion of Sardinia which is commonly recognized as not seismically prone.

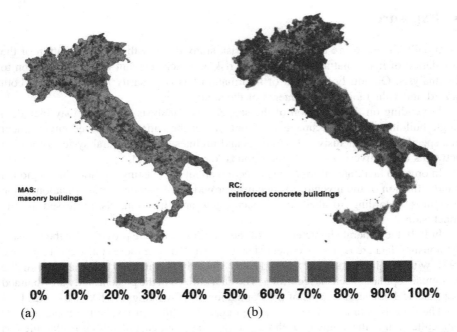

Fig. 3. Exposure: percentage of number of buildings for structural type aggregated at municipality level. (a) MAS (masonry build.) (b) RC (reinforced concrete build.) (Color figure online)

All the components of the seismic risk have been handled within a Geographical Information System (GIS) which has been linked to a series of external resources written in Matlab language [39] for performing the required analysis (essentially: pre-processing of data and risk computation).

In the present study hazard data where constituted by 9 maps, each one composed by 10'751 spatial points (plus the data coming from the hazard special studies that have covered some of the minor islands), whilst exposure data where coming from 341'862 census tracts (relative to 7'724 municipalities). The data, after processing, were displayed in aggregated format at municipality level only.

Figures 4 through 6 report the maps, for the entire national territory and with a municipality resolution, of the probability of exceeding in a 50 year time interval of the three specified limit states (LS1: slight damage, LS2: significant damage and LS3: collapse) for all the nine structural building classes considered in the study (obtained by combination of structural type, period of construction and number of storeys). The plotted results represent the class-specific damage potential (sometime referred as damage function). It has been obtained by convolution solely of fragility and hazard (including the soil conditions), since it does not consider the actual exposition data (local consistence of the considered building classes at the site where hazard has ben considered).

As it is obvious the geographic location plays a crucial role: as a consequence of the framework of seismo-genetic sources along Italy, the specific risk maps reproduces

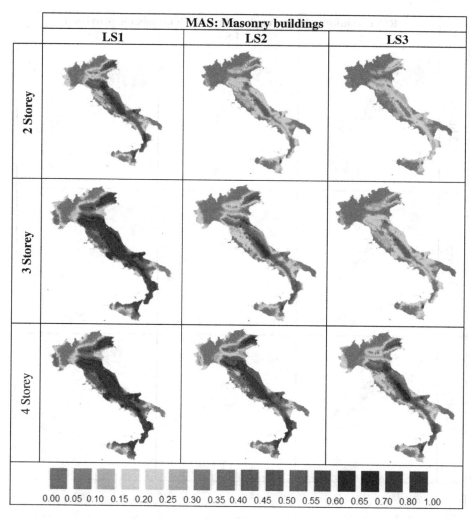

Fig. 4. Class-specific damage potential for masonry buildings. (Color figure online)

closely the geographical outline of the hazard maps. Also the vulnerability of the single building class plays an important role on the final results, with the specific damage potential decreasing from masonry to reinforced concrete structural types.

Then on the basis of the damage potentials, the specific risk has been calculated, including the exposition data. The computation has been performed through a weighed average: for each limit state and municipality, the class-specific damage potentials have been summed up through a normalizing factor which accounts for the percentage of presence in the municipality of that class (in terms of number of buildings).

In order to have a term of comparison, the specific risk which represent the probability of exceedance of the specified limit state of the existing buildings,

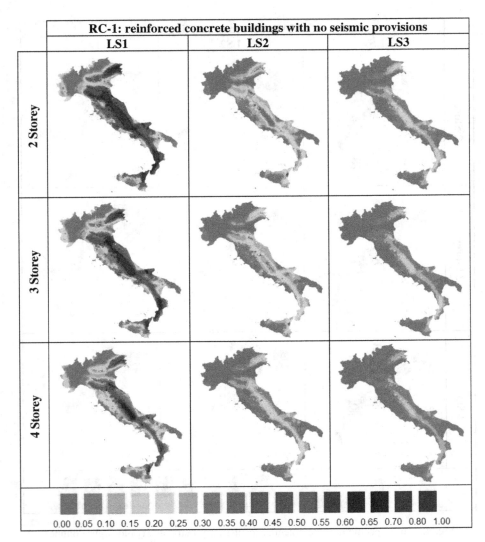

Fig. 5. Class-specific damage potential for reinforced concrete buildings with no seismic provisions. (Color figure online)

$P_{ex,50}(LS_i)$ (i = 1,2,3), has been divided by the probability of occurrence of the seismic action used in the design of the new residential buildings (and the assessment of the existing ones), $P_{new,50}(LS_i)$ (Fig. 7).

According to Italian seismic rules NTC-08 [10] this probability is given for the three limit states as follows:

$P_{new,50}(LS_1) = 0.63$ $P_{new,50}(LS_2) = 0.10$ $P_{new,50}(LS_3) = 0.05$.

Therefore the obtained index, $I = P_{ex,50}(LS_i)/P_{new,50}(LS_i)$, represents a comparative measure between the expected capacity (numerator) and the expected legally-

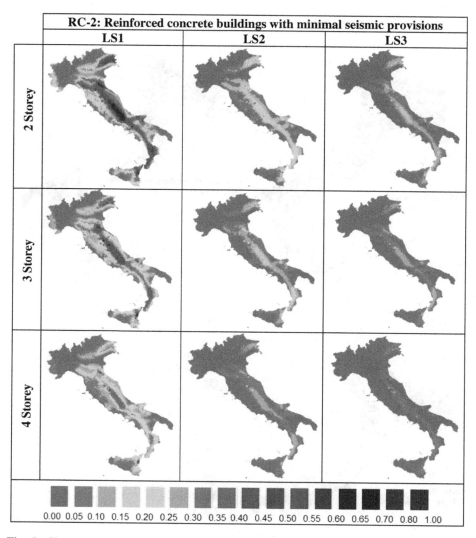

Fig. 6. Class-specific damage potential for reinforced concrete buildings with minimal seismic provisions. (Color figure online)

compulsory demand (denominator) in terms of probability of exceedance (the highest is the index, the less safe is the structure).

Obviously the new structures, which have at least to legally comply with the prescribed demand, are designed with additional conservative measures (contained in the NTC-08 [10] provisions and represented by load and resistance safety factors, over-strength factors, capacity design rules, minimum design requirements), so that the eventual cases where the ratio is less than unity ($I < 1.0$), do not necessarily imply that an existing structure is safer than a new one. It is worth noticing, even if the discussion

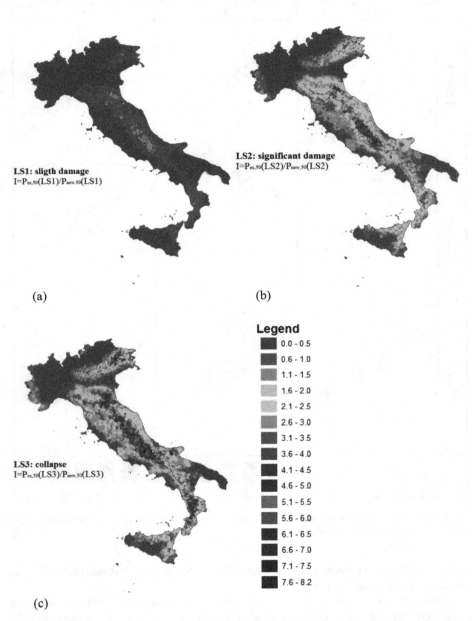

(a)

(b)

(c)

Legend

■	0.0 - 0.5
■	0.6 - 1.0
■	1.1 - 1.5
■	1.6 - 2.0
■	2.1 - 2.5
■	2.6 - 3.0
■	3.1 - 3.5
■	3.6 - 4.0
■	4.1 - 4.5
■	4.6 - 5.0
■	5.1 - 5.5
■	5.6 - 6.0
■	6.1 - 6.5
■	6.6 - 7.0
■	7.1 - 7.5
■	7.6 - 8.2

Fig. 7. Index of exceedance of a specified limit state at municipality level, normalized with the design thresholds imposed by the code for new buildings: $I = P_{ex,50}(LS_i)/P_{new,50}(LS_i)$. (a) LS1: slight damage; (b) LS2: significant damage; (c) LS3: collapse. (Color figure online)

of the issue would fall well beyond the scope of this paper, that recent studies [30] evidenced as the NTC-08 [10] seismic input, assumed here as a reference, could possibly be inappropriate for a safety-consistent representation of the seismic action.

As shown in Fig. 7, while the ratio between capacity and demand is minimal for LS1 (and apparently favorable for existing constructions, since the index I is generally below unity), it deepens as the level of damage increases: peaks of I around 4 for LS2 or around 8 at LS3 in high hazard regions whilst still remains around unity in low-seismicity areas. This kind of result was somehow expected and can be easily explained.

Indeed the slight damage (LS1) is mostly conditioned by the quality of the details of non-structural components (whose design is controlled by architectural or climatic rather than seismic or structural considerations). Furthermore, while the occurrence of significant damage (LS2) and collapse (LS3) in high-seismicity areas is strongly conditioned by the presence in the design of seismic provisions and considerations about the expected mechanism of collapse under seismic actions (usually neglected in conceiving existing structures, designed when most of the sites were not recognized earthquake-prone and eventually enforcing looser seismic provisions), the occurrence of the same limit states (LS2 and LS3) in low-seismicity areas is proportionally less conditioned by the seismic considerations since the structural design can be more influenced by the gravity loads than the horizontal ones.

Another important outcome of the study is represented by the absolute risk (Fig. 8), that can be either expressed in terms of number of buildings or their surface expected to reach or exceed a specified limit state in a definite lapse of time. The riskiest situations, as it was obvious since the beginning of the study, reside in the urban agglomerates, but the chosen risk indicator not necessarily reflect merely the dimensions of the municipalities or their location in a high seismically prone area). Indeed the ten worst cases (in descending order of risk) are: Rome, Palermo, Naples, Bologna, Florence, Catania, Verona, Modena, Forlì, and Messina.

This kind of data can possibly be correlated with the expected economic losses due to earthquake occurrence [4–6, 22]: a rough estimate that can be established on the basis of actual construction/repair costs is of the order of € 2–3 billion per year, in line with the expenses sustained in the past 50 years [13, 14]. The GAR program estimated an average annual total loss for earthquakes in Italy of 9.7 billion of US$.

This issue is becoming an argument of interest in national civil protection debate, since the option of issuing a compulsory or a semi-mandatory insurance as a means of complementary coverage against the damages for properties [35–38].

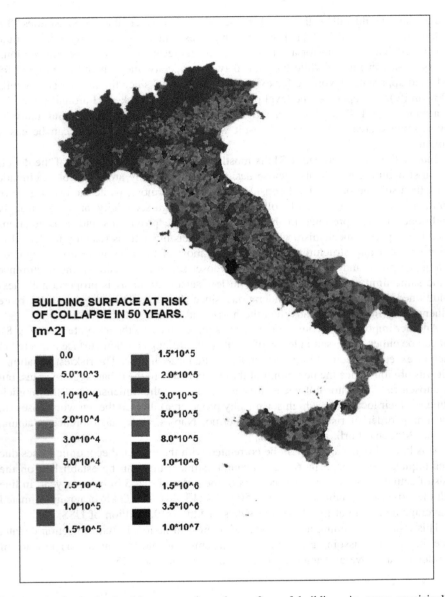

Fig. 8. Absolute seismic risk expressed as the surface of buildings in every municipality expected to reach the limit state of collapse (LS3) in 50 years. (Color figure online)

6 Conclusions

Earthquake risk may be thought of as encompassing three main factors that interact together: the level of the seismic hazard, the vulnerability to this hazard of the built environment (where human activities are concentrated) and the number of people and the value of assets exposed to the building damage due to a seismic action.

Most of Italian peninsula is subject to medium to high seismic hazard: in Mediterranean basin only Greece and western Turkey have higher probabilities of occurrences of the seismic shaking. Existing buildings stock in Italy is particularly vulnerable to the effects of earthquake action as a consequence of several factors such as the late adoption of either modern building codes and seismic classification of the territory and the low rate of renovation of the buildings. Less than 23 % of the residential dwellings present in Italy have been built after 1980, when a major revision of seismic provisions was undergone as a consequence of Irpina earthquake. If one adds to the above aspects also the high urban density and the inestimable value of the Italian cultural heritage, it is highly probable to conclude that the country is one with the highest seismic risk in the world.

The work presented herein consisted in the assessment of the seismic risk map of the whole Italian territory (Sardinia excluded) using a state-of-the-art evaluation procedure and the latest available data. The analytical procedure follows a quantitative rather than a qualitative approach, since it has been based on a physically sound representation of the seismic shaking severity at site that has been measured through a ground motion parameter and on damage functions derived from a mechanical model aimed at predicting the structural response of the classes of buildings under the seismic action.

The key issue in developing reliable risk maps still remain the inclusion in the study of a dependable and fit-for-the-purpose inventory of the building stock. Indeed risk analysis conducted at a so large scale cannot benefit of specific surveys that can be conducted only locally on limited portions of the territory and with a great monetary and human effort, so that nation-wide studies have so far used databases built for other purposes such as census or taxation. By this viewpoint, Eurostat institution has granted that the census exercises conducted by EU national states provide comparable and reliable information all around Europe, including harmonized high-quality data that describe the built environment. However it is evident that an inventory suitable for seismic evaluations requires to carry the sort of information that are of interest for the purpose of the study. As outlined in the paper exposure, data has to be combined with hazard and vulnerability components of the analysis, so that a substantial engineering judgment is required in collecting, interpreting and elaborating the data made available from different sources. Efforts are currently made at international level to develop an uniform inventory of the assets exposed to natural and man-made hazards (see for instance the UNISDR global exposure dataset [29] or the EU European building inventory framework [15]), but Italian experience teaches that the problem is quite arduous. Indeed the solution has to tackle the extreme variety of construction techniques (Italian building stock has a not indifferent percentage of vernacular and heritage architecture built following local traditions) and the different implementation of design rules and construction standards throughout the country.

References

1. Benedetti, D., Petrini, V.: On Seismic Vulnerability of Masonry Buildings: Proposal of an Evaluation Procedure. L'industria delle costruzioni, Milano (1984)
2. Borzi, B., Crowley, H., Pinho, R.: Simplified pushover-based earthquake loss assessment (SP-BELA) for masonry buildings. Int. J. Archit. Heritage 2(4), 353–376 (2008)
3. Borzi, B., Pinho, R., Crowley, H.: Simplified pushover-based vulnerability analysis for large scale assessment of RC buildings. Eng. Struct. 30(3), 804–820 (2008)
4. Braga, F., Gigliotti, R., Monti, G., Morelli, F., Nuti, C., Salvatore, W., Vanzi, I.: Post-seismic assessment of existing constructions: evaluation of the shakemaps for identifying exclusion zones in Emilia. Earthq. Struct. 8(1), 37–56 (2015)
5. Braga, F., Gigliotti, R., Monti, G., Morelli, F., Nuti, C., Salvatore, W., Vanzi, I.: Speedup of post earthquake community recovery: the case of precast industrial buildings after the Emilia 2012 earthquake. Bull. Earthq. Eng. 12(5), 2405–2418 (2014)
6. Carbonara, S., Cerasa, D., Sclocco, T., Spacone, E.: A preliminary estimate of the rebuilding costs for the towns of the Abruzzo region affected by the April 2009 earthquake: an alternate approach to current legislative procedures. In: Gervasi, O., et al. (eds.) ICCSA 2015. LNCS, vol. 9157, pp. 269–283. Springer, Heidelberg (2015)
7. CNR-GNDT: seismic risk for public buildings, part I, Methodological aspects, Gruppo Nazionale per la Difesa dai Terremoti, Roma (1994) (in Italian)
8. Cornell, C.A., Krawinkler, H.: Progress and challenges in seismic performance assessment. PEER Center News 3(2). Berkeley (2000) http://peer.berkeley.edu/news/2000spring/performance.html
9. Decanini, L., De Sortis, A., Goretti, A., Langenbach, R., Mollaioli, F., Rasulo, A.: Performance of masonry buildings during the 2002 Molise, Italy, earthquake. Earthq. Spectra 20(S1), 191–220 (2004)
10. Decree of Ministry of Infrastructure: Nuove norme tecniche per le costruzioni. Istituto Poligrafico dello Stato, Roma (2008) (in Italian)
11. Di Pasquale, G., Goretti, A., Dolce, M., Martinelli, A.: Confronto fra differenti modelli di vulnerabilità degli edifici. X Congresso Nazionale "L'ingegneria Sismica in Italia", Potenza-Matera (2001)
12. Dolce, M., Masi, A., Marino, M., Vona, M.: Earthquake damage scenarios of the building stock of Potenza (Southern Italy) including site effects. Bull. Earthq. Eng. 1(1), 115–140 (2003)
13. Dolce, M.: Mitigation of seismic risk in Italy following the 2002 S. Giuliano Earthq. Earthq. Tsunamis Geotech. Geol. Earthq. Eng. 11, 67–89 (2009)
14. Dolce, M.: The Italian national seismic prevention program. In: Proceedings of 15th World Conference on Earthquake Engineering, Lisbon, Portugal, 24–28 September 2012
15. European Union: European building inventory framework; EUR-27603-EN, Joint Research Centre (2015)
16. Faccioli, E., Pessina, V. (eds): The Catania Project: Earthquake Damage Scenarios for High Risk Areas of the Mediterranean, p. 225. CNR—Gruppo Nazionale per la Difesa dai Terremoti, Rome (2000)
17. Faccioli, E., Pessina, V., Calvi, G.M., Borzi, B.: A study on damage scenarios for residential buildings in Catania city. J. Seismolog. 3(3), 327–343 (1999). ftp://ingv.it/pro/gndt/Pubblicazioni/Faccioli_copertina.htm
18. Grande, E., Rasulo, A.: A simple approach for seismic retrofit of low-rise concentric X-braced steel frames. J. Constr. Steel Res. 107, 162–172 (2015)

19. Grande, E., Rasulo, A.: Seismic assessment of concentric X-braced steel frames. Eng. Struct. **49**, 983–995 (2013)

20. Grande, E., Imbimbo, M., Rasulo, A.: Experimental response of RC beams strengthened in shear by FRP sheets. Open Civil Eng. J. **7**, 127–135 (2013)

21. Istituto Nazionale di Statistica (ISTAT): 15° Censimento della popolazione e delle abitazioni 2011 – Dati definitivi, Roma (2012) (in Italian)

22. Liel, A.B., Ross, B., Corotis, R.B., Camata, G., Sutton, J., Holtzman, R., Spacone, E.: Perceptions of decision-making roles and priorities that affect rebuilding after disaster: the example of L'Aquila, Italy. Earthq. Spectra **29**(3), 843–868 (2013)

23. Meletti C., Montaldo V.: Stime di pericolosità sismica per diverse probabilità di superamento in 50 anni: valori di ag. Progetto DPC-INGV S1, Deliverable D2 (2007). http://esse1.mi.ingv.it/d2.html

24. Michelini, A., Faenza, L., Lauciani, V., Malagnini, L.: ShakeMap implementation in Italy, Seismol. Res. Lett. **79**(5), 688–697 (2008)

25. Nuti, C., Rasulo, A., Vanzi, I.: Seismic safety evaluation of electric power supply at Urban level. Earthq. Eng. Struct. Dyn. **36**(2), 245–263 (2007)

26. Nuti, C., Rasulo, A., Vanzi, I.: Seismic safety of network structures and infrastructures. Struct. Infrastruct. Eng.: Maintenance Manag. Life-Cycle **6**(1–2), 95–110 (2010)

27. Rasulo, A., Goretti, A., Nuti, C.: Performance of lifelines during the 2002 Molise, Italy, earthquake. Earthq. Spectra **20**(S1), 301–314 (2004)

28. Rasulo, A., Testa, C., Borzi, B.: Seismic risk analysis at Urban scale in Italy. In: Gervasi, O., et al. (eds.) ICCSA 2015. LNCS, vol. 9157, pp. 403–414. Springer, Heidelberg (2015)

29. United Nations Office for Disaster Risk Reduction (UNISDR): Global Assessment Report on Disaster Risk Reduction 2015 (GAR15). United Nations General Secretariat (2015)

30. Vanzi, I., Marano, G.C., Monti, G., Nuti, C.: A synthetic formulation for the Italian seismic hazard and code implications for the seismic risk. Soil Dyn. Earthq. Eng. **77**, 111–122 (2015)

31. Wald, D.J., Allen, T.I.: Topographic slope as a proxy for seismic site conditions and amplification. Bull. Seismol. Soc. Am. **97**(5), 1379–1395 (2007)

32. Wald, D.J., Earle, P.S., Quitoriano, V.: Topographic slope as a proxy for seismic site correction and amplification. EOS Trans. AGU **85**(47), F1424 (2004)

33. Gallipoli, M.R., Mucciarelli, M.: Comparison of site classification from Vs30, Vs10, and HVSR in Italy. Bull. Seismol. Soc. Am. **99**(1), 340–351 (2009)

34. Wald, L.A., Mori, J.: Evaluation of methods for estimating linear site-response amplifications in the Los Angeles region. Bull. Seismol. Soc. Am. **90**, S32–S42 (2000)

35. Servizio Studi della Camera dei deputati: Documentazione per l'esame di progetti di legge. Disposizioni urgenti per il riordino della Protezione civile. D.L. 59/2012 - A.C. 5203-A. Dossier D12059C, Roma, 28 May 2012

36. Friedman, D.G.: Natural hazard risk assessment for an insurance program. Geneva Pap. Risk Insur. **9**(30), 57–128 (1984)

37. Kousky, C., Cooke, R.: Explaining the failure to insure catastrophic risks. Geneva Pap. Risk Insur. – Iss. Pract. **37**, 206–227 (2012)

38. Gizzi, F.T., Potenza, M.R., Zotta, C.: The insurance market of natural hazards for residential properties in Italy. Open J. Earthq. Res. **5**, 35–61 (2016)

39. MathWorks, Inc. Matlab reference manual, version 6.5. MathWorks, Inc., Natick, Mass, USA (2002)

40. Istat. Indagine sulle abitazioni al 21 aprile 1931, Firenze (1936)

41. Istat. L'Italia in 150 anni. Sommario di statistiche storiche 1861–2010, Roma (2011)

Depreciation Methods for Firm's Assets

Vincenzo Del Giudice[1], Benedetto Manganelli[2]([⊠]),
and Pierfrancesco De Paola[1]

[1] University of Naples "Federico II",
Piazzale Vincenzo Tecchio, 80125 Naples, Italy
{vincenzo.delgiudice,pierfrancesco.depaola}@unina.it
[2] University of Basilicata, Viale dell'Ateneo Lucano, 85100 Potenza, Italy
benedetto.manganelli@unibas.it

Abstract. This study focuses on the analytical description of depreciation methods applied to firm's equipments for corporate accounting and balance sheet. Depreciation is a systematic allocation of fixed asset cost over its useful life. Several methods are applied to estimate depreciation cost: Straight-Line Method, Sum of Years Digits Method, Declining Balance Method, Declining Balance Method switched to Straight-Line Method, Interest Methods (Sinking Fund Method and Annuity Method), Usage Methods (Machine Hours Method and Production Units Method), Depletion Method. Each of these depreciation methods have been examined in detail, concluding this work with an analytical and critical comparison between them.

Keywords: Depreciation methods · Firm's assets · Corporate accounting · Balance sheet

1 Introduction

The depreciation of fixed assets, tangible and intangible, as well as the possible recovery of value thereof, is a particularly complex issue that requires a careful and prudent analysis in preparation of corporate financial statement.

Concretely, the main difficulties are related to identification of moment when must be detected the write-down, depreciation's amount attributable to income statement, the moment when recognize an eventual recovery value. Other issues are also found from the fiscal point of view due to the misalignment between book values and tax values of corporate assets.

Generally, tangible assets are recorded in the company's balance sheet in terms of purchase prices or production costs. In fact, for a tangible asset the book value is inclusive of actual purchase price (usually indicated in contract or invoice, to net of any discounts), additional purchasing fees and all those other possible charges that firm has to sustain so that asset may be used. Instead, when an asset is produced in economy, its book value is equal to sum of manufacturing costs, including direct costs (material and direct labour, design costs, costs for external supplies, etc.) and a share of manufacturing overheads, as well financial charges for the period of its construction (calculated until the asset is ready to use).

O. Gervasi et al. (Eds.): ICCSA 2016, Part III, LNCS 9788, pp. 214–227, 2016.
DOI: 10.1007/978-3-319-42111-7_17

On the other hand, the cost of an intangible asset or a fixed asset, whose utilization is limited in time, it must be systematically amortized each year in relation to its remaining useful life. Amortization begins when the asset is available for use or, in any case, when produces economic benefits for the firm [17].

The causes that can lead to a loss of asset value may be different [5–7, 20]: (1) normal physical wear and tear, related to intensity of use, preventive maintenance and periodic checks, also considering that some wear depends on time and not from use; (2) customization or usage, with a expected rate of wear and tear for each year or month; (3) abnormal occurrences, as accidents or physical defects; (4) technological changes and industrial development that can increase the obsolescence; (5) changes in the production factors; (6) variations in the purchasing power of money or benchmark interest rates.

For a generic asset the basic parameters to be considered during calculation of depreciation's amounts are the original production cost or purchase price, useful life, salvage value at the end of useful life, depreciable cost or original production cost/purchase price less the expected salvage value, book value for each kth year (also referred as written-down value) equal to original production cost/purchase price less the accumulated depreciation until k^{th} year [14–16].

That being stated, in the international literature is hardly possible to find a complete and comprehensive dissertation on various methods of depreciation used in operational practice. As well as is not simple a comparison between depreciation methods. For these reasons, this paper aims first of all to present the genesis of analytical formulations for each existing depreciation method, in some cases introducing some innovative implementations using the fuzzy-logic, and then to compare different depreciation methods between them. Other aim of this paper is provide a valid support for firms in construction of automated mechanisms useful for evaluation and depreciation of fixed assets, constantly assisting the decision-making in the management of tangible and intangible assets.

2 Depreciation Methods

Most commonly methods applied for determination of depreciation cost are the followings:

- Straight-Line Method;
- Sum of Years Digits Method;
- Declining Balance Method;
- Declining Balance switched to Straight-Line Method;
- Interest Methods: Sinking Fund Method, Annuity Method;
- Usage Methods: Machine Hours Method, Production Units Method;
- Depletion Method.

Below we will then be examined in detail these depreciation methods [1, 8–11, 13, 19].

2.1   Straight-Line Method

The simplest and most used depreciation method is certainly straight-line, where the annual depreciation is constant, and in addition, for determination of depreciation it is assumed that value loss is directly proportional to asset's age:

$$D_k = \frac{(P - S)}{n} \tag{1}$$

$$D_k^* = \pm \frac{k(P - S)}{n} \tag{2}$$

$$BV_k = P - \frac{k(P - S)}{n} \tag{3}$$

In Eqs. (1), (2) and (3): D_k is the annual depreciation for k^{th} year ($k = 1, 2, ..., n$); P is the purchase price of firm's asset; S is the final salvage value in n^{th} year; n is the asset's useful life expressed in years; D_k^* is the cumulative depreciation for k^{th} year ($k = 1, 2, ..., n$); BV_k is the book value for k^{th} year ($k = 1, 2, ..., n$).

Due to the uncertainty of parameters n and S, the literature [12] suggests a fuzzy approach for definition of these parameters. At first, on base of Delphi fuzzy method can consider triangular fuzzy numbers for final salvage value and n term [4]: $\tilde{S} = (S_1, S_2, S_3)$, $\tilde{n} = (n_1, n_2, n_3)$. Then, can rewrite Eq. (1) considering the fuzzy approach [2]:

$$\tilde{D} = \frac{\left(P - \tilde{S}\right)}{\tilde{n}} = \left(D_1, D_2, D_3,\right) \tag{4}$$

At this point, is possible to transform fuzzy triangular numbers into a numerical interval with following assumptions:

$$\tilde{Z} = (Z_1, Z_2, Z_3)$$

$$\mu_{\tilde{z}}(x) = \begin{cases} \frac{(x - Z_1)}{(Z_2 - Z_1)}, Z_1 \leq x \leq Z_2; \frac{(Z_3 - x)}{(Z_3 - Z_2)}, Z_2 \leq x \leq Z_3; 0, otherwise \end{cases} \tag{5}$$

$$Z_\alpha = \{x | \mu_{\tilde{z}}(x) \geq \alpha\} = [Z_L^\alpha, Z_R^\alpha] \tag{6}$$

$$Z_\alpha = [A_1 + \alpha(Z_2 - Z_1), A_3 + \alpha(Z_3 - Z_2)], \alpha \in [0, 1] \tag{7}$$

Therefore, S and n terms can be rewritten using α-cut method and arithmetic intervals:

$$\tilde{S} = (S_1, S_2, S_3) \Rightarrow S_\alpha = [S_L^\alpha, S_R^\alpha] = [S_1 + Z(S_2 - S_1), S_3 + Z(S_3 - S_2)] \tag{8}$$

$$\tilde{n} = (n_1, n_2, n_3) \Rightarrow n_\alpha = \left[n_L^\alpha, n_R^\alpha\right] = [n_1 + \alpha(n_2 - n_1), n_3 + \alpha(n_3 - n_2)] \qquad (9)$$

$$D_\alpha = \left[D_L^\alpha, D_R^\alpha\right] = \left[\frac{P - S_R^\alpha}{n_R^\alpha}, \frac{P - S_L^\alpha}{n_L^\alpha}\right], \alpha \in [0, 1] \qquad (10)$$

2.2 Sum of Years Digits Method

For to obtain depreciation for each year of asset's useful life by this method, the corresponding numbers to each possible year of life are first mentioned in reverse order. After, depreciation for each year is obtained like ratio between the numbers listed in reverse order for that year and the sum of years (SOY) of asset's useful life.

Therefore, in this method the depreciation is maximum in the first year and decreases until the last year (minimum depreciation). For this reason, the method is also defined as accelerated depreciation method.

The d_k depreciation rate for every k^{th} year is given by:

$$d_k = \left[\frac{(n - k + 1)}{SOY}\right] \qquad (11)$$

Where n is the asset's useful life and SOY is the sum of years digits of the useful life [n(n + 1)/2]. Thus, Eq. (11) becomes:

$$d_k = \left[\frac{2(n - k + 1)}{n(n + 1)}\right] \qquad (12)$$

The annual depreciation (D_k) for k^{th} year is calculated multiplying the depreciation rate by the total depreciation amount, this last given by difference between purchase price and salvage value:

$$D_k = d_k \cdot (P - S) = (P - S) \cdot \left[\frac{2(n - k + 1)}{n(n + 1)}\right] \qquad (13)$$

Book value (BV_k) and cumulative depreciation (D_k^*) are obtained by following equations:

$$BV_k = P - \left[\frac{2(P - S)}{n}\right] \cdot k + \left[\frac{(P - S)}{n(n + 1)}\right] \cdot k \cdot (k + 1) \qquad (14)$$

$$D_k^* = P - BV_k \qquad (15)$$

In all equations, the terms have same meaning as Sect. 2.1.

The fuzzy depreciation value can be obtained with following assumptions (see paragraph 2.1):

$$\widetilde{S} = (S_1, S_2, S_3) \Rightarrow S_\alpha = \left[S_L^\alpha, S_R^\alpha\right] = [S_1 + Z(S_2 - S_1), S_3 + Z(S_3 - S_2)] \tag{16}$$

$$\widetilde{n} = (n_1, n_2, n_3) \Rightarrow n_\alpha = \left[n_L^\alpha, n_R^\alpha\right] = [n_1 + \alpha(n_2 - n_1), n_3 + \alpha(n_3 - n_2)] \tag{17}$$

$$D_k^\alpha = \left[D_{kL}^\alpha, D_{kR}^\alpha\right] = \left[2\left(\frac{n_L^\alpha - k + 1}{n_R^\alpha(n_R^\alpha + 1)}\right)(P - S_R^\alpha), 2\left(\frac{n_R^\alpha - k + 1}{n_L^\alpha(n_L^\alpha + 1)}\right)(P - S_L^\alpha)\right], \alpha \in [0, 1] \tag{18}$$

2.3   Declining Balance Method

The Declining Balance method, sometimes called "Constant Percent Method" or "Matheson Formula" or also "Written-Down Value Method", assumes that annual depreciation cost is a fixed percentage of book value at beginning of each year.

The relationship between depreciation and book value is constant over asset's useful life and is designated by R parameter $(0 < R < 1)$.

Also in this method, the highest depreciation occurs in the first year and decreases with a fixed rate every year:

$$D_1 = P \cdot R$$

$$D_2 = [P - (P \cdot R)] \cdot R$$

and iterating the reasoning, depreciation for k^{th} year is given by following formula:

$$D_k = P \cdot (1 - R)^{k-1} \cdot R \tag{19}$$

Follow the formulas for cumulative depreciation (D_k^*), book value (BV_k) and final salvage value (or book value of asset at n^{th} year, BV_n):

$$D_k^* = P \cdot \left[1 - (1 - R)^k\right] \tag{20}$$

$$BV_k = P \cdot (1 - R)^k \tag{21}$$

$$BV_n = P \cdot (1 - n)^n \tag{22}$$

The Declining Balance method is rather easy to apply, but has two weaknesses points: a) the annual depreciation cost is different for each year and, analytically, this is an inconvenience; b) the asset cannot never be depreciated to a zero value because the S term is not used in formulas (19), (20), (21) and (22).

For these reasons, in Declining Balance method the book value can be cut under salvage value considered when S is greater than zero. There are no serious difficulties for switching between declining balance method to every other slower depreciation method (such as the Straight-Line method), because the switch can be done so that we can reach a zero value (or any other) for the book value when it has reached the n year. Consequently, is simple to switch to a zero value when BV_k becomes equal to S.

The depreciation rate depends on type of perishable properties and is indicated in term of straight-line depreciation rate equal to $1/n$ (for a null salvage value). Since depreciation rates for Declining Balance are greater of Straight-Line depreciation rates, the declining balance method is often referred as an accelerated depreciation method. When using a depreciation rate (R) equal to $2/n$, the method is called as Double Declining Balance.

Generally, for all types of depreciable assets except for real estate the R depreciation rate is equal to $2/n$, for all used depreciable assets and new real estate R is equal to $1,5/n$, and for rental property R is equal to $1,25/n$.

Literature suggests the following general formula used for calculating depreciation rate [18]:

$$R = \left[1 - \sqrt[n]{(S/P)}\right] \cdot 100 \tag{23}$$

The issue of depreciation value due to fuzziness of n term, for the case of Double Declining Balance, can be address with following relations:

$$\tilde{n} = (n_1, n_2, n_3) \Rightarrow n_\alpha = \left[n_L^\alpha, n_R^\alpha\right] = \left[n_1 + \alpha(n_2 - n_1), n_3 + \alpha(n_3 - n_2)\right] \tag{24}$$

$$D_k^\alpha = \left[D_{kL}^\alpha, D_{kR}^\alpha\right] = \left[2\left(\frac{2P}{n_R^\alpha}\right)\left(1 - \frac{2}{n_L^\alpha}\right)^{k-1}, 2\left(\frac{2P}{n_L^\alpha}\right)\left(1 - \frac{2}{n_R^\alpha}\right)^{k-1}\right], \alpha \in [0,1] \tag{25}$$

2.4   Declining Balance Switched to Straight-Line Method

Sometimes, especially for tax reasons, it may be advantageous switch from Declining Balance to Straight-Line method.

In fact, as already mentioned, Declining Balance method never reaches a book value equal to zero. For this reason, is possible switch from this approach to Straight-Line method, so that book value is zero or any other considered salvage value. The switching operation can be made only by an accelerated depreciation method, being not possible to switch from a slower to a faster depreciation method.

The switching between two methods is carried out in correspondence of the year where depreciation obtainable by the Straight-Line method becomes higher of depreciation obtained with Declining Balance method (or in correspondence of the year where the deviation obtained with two methods, applied separately, is minimum). Therefore, up to the switch, annual depreciation is calculated by Declining Balance formulas.

Analytically the switch will be made when:

$$1 - \sum_{k=1}^{z} R \cdot (1 - R)^{k-1} \div (n - z + 1) = R \cdot (1 - R)^{z-1} \qquad (26)$$

Alternatively:

$$R \cdot [1 + (1 - R) + (1 - R)^2 + \ldots + (1 - R)^{(z-1)}] = R \cdot \left[\frac{(1 - R)^z - 1}{(1 - R) - 1} \right] = 1 - (1 - R)^z$$

By performing the replacements:
$(1 - R)^z = (n - z + 1) \cdot R \cdot (1 - R)^{z-1}$, which becomes: $z = [(1/R) + 2]$
In last formulas z is the year where switching begins.

2.5   Sinking-Fund Method

The Sinking-Fund method is a technique for asset's depreciation in which the firm sets aside a money amount to invest annually so that money, plus the interest earned in the fund, will be enough for to replace the asset at the end of its useful life.

Thus the annual depreciation, for each year, has two components: first component is the fixed monetary amount that is deposited into the sinking-fund (A), second component is the interest earned on the amount accumulated in sinking-fund until the beginning of that year.

In this method, the lowest depreciation occurs in the first year and gradually increases until final year.

The first component of depreciation (A) at the end of each year is given by:

$$A = (P - S) \cdot \left(\frac{A}{F}, i, n \right) = \frac{i \cdot (P - S)}{(1 + i)^n - 1} \qquad (27)$$

Where i is the interest rate per year.

Depreciation for 1st year is only equal to A term, because this is the monetary amount to be deposited in sinking-fund at the end of 1st year and hence there is no interest accumulated until that moment:

$$D_1 = A \cdot (1 + i)^0 = A$$

Consequently, the book value at the end of 1st year is:

$$BV_1 = P - D_1 = P - A \cdot (1 + i)^0$$

Depreciation for 2nd year is equal to A amount to be deposited in sinking-fund at the end of 2nd year, plus the interest earned on the amount accumulated till beginning of 2nd year:

$$D_2 = A + D_1 \cdot i = A + A \cdot i = A \cdot (1+i)$$

Therefore, the book value, at the end of 2nd year, is equal to:

$$BV_2 = BV_1 - D_2 = P - A \cdot (1+i)^0 - A \cdot (1+i)^1 = P - \left[(1+i)^0 + (1+i)^1 \right]$$

Similarly, depreciation and book value for 3rd year are:

$$D_3 = A + (D_1 + D_2) \cdot i = A \cdot (1+i)^2$$

$$BV_3 = BV_2 - D_3 = P - A \cdot (1+i)^0 - A \cdot (1+i)^1 - A \cdot (1+i)^2$$

$$BV_3 = P - \left[(1+i)^0 + (1+i)^1 + (1+i)^2 \right]$$

Iterating and generalizing the process, the depreciation and the book value for k^{th} year are obtained from following formulas, where i_k is an appropriate interest rate for each year:

$$D_k = A \cdot (1+i)^{k-1} \tag{28}$$

$$D_k = \frac{i_k \cdot (1+i_k)^{k-1} \cdot (P-S)}{(1+i_k)^n - 1}, k = 1, 2, \ldots, n \tag{29}$$

$$BV_k = P - \left[(1+i_1)^0 + (1+i_2)^1 + \ldots + (1+i_{n-2})^{n-2} + (1+i_{n-1})^{n-1} \right] \tag{30}$$

The last formula also can be shown as geometric series (uniform series of compound factor):

$$BV_k = P - A \cdot \left(\frac{F}{A}, i, k \right) \tag{31}$$

$$BV_k = P - (P-S) \cdot \frac{(1+i_k)^k - 1}{(1+i_k)^n - 1}, k = 1, 2, \ldots, n. \tag{32}$$

The method is used with help of a Sinking-Fund Table, but is seldom used because it is somewhat complicated. However, the method is appropriate in any industrial sectors, such as regulated utilities, where the return on investment is fixed and the long-lived assets are expensive.

Finally, assuming that n, S and i_k are fuzzy, the annual depreciation with fuzzy approach is obtained as:

$$\tilde{S} = (S_1, S_2, S_3) \Rightarrow S_\alpha = \left[S_L^\alpha, S_R^\alpha\right] = [S_1 + \alpha(S_2 - S_1), S_3 + \alpha(S_3 - S_2)] \tag{33}$$

$$\tilde{n} = (n_1, n_2, n_3) \Rightarrow n_\alpha = \left[n_L^\alpha, n_R^\alpha\right] = [n_1 + \alpha(n_2 - n_1), n_3 + \alpha(n_3 - n_2)] \tag{34}$$

$$\tilde{i}_k = \left(i_{k_1}, i_{k_2}, i_{k_3}\right) \Rightarrow i_{k_\alpha} = \left[i_{k_L^\alpha}, i_{k_R^\alpha}\right] = \left[i_{k_1} + \alpha\left(i_{k_2} - i_{k_1}\right), i_{k_3} + \alpha\left(i_{k_3} - i_{k_2}\right)\right] \tag{35}$$

$$D_k^\alpha = \left[D_{kL}^\alpha, D_{kR}^\alpha\right] = \left[\frac{i_{k_L^\alpha}\left(1 + i_{k_L^\alpha}\right)^{k-1}(P - S_R^\alpha)}{\left(1 + i_{k_R^\alpha}\right)^{n_R^\alpha} - 1}, \frac{i_{k_R^\alpha}\left(1 + i_{k_R^\alpha}\right)^{k-1}(P - S_L^\alpha)}{\left(1 + i_{k_L^\alpha}\right)^{n_L^\alpha} - 1}\right], \alpha \in [0, 1]. \tag{36}$$

2.6   Annuity Method

The Annuity method equalizes each year's sum of depreciation and an imputed interest charge calculated with a constant rate on the asset's undepreciated book value [3].

The Annuity method, as Sinking-Fund method, computes depreciation using compound interest and produces an increasing annual depreciation charge.

In particular, this method differs from Sinking-Fund for logic and formula because the latter produces an annual depreciation equal to the increase in a hypothetical interest-earning asset-replacement fund. In other words, depreciation is charged every year and refers to interest losing or reduction in the original cost of fixed assets.

Generally depreciation's amount is calculated with help of an Annuity Table, although this method is not frequently used in practice for its complex calculation (as Sinking-Fund method also).

Annuity method considers the following relation for calculating yearly depreciation:

$$D_{year} = \left\{P - \left[\frac{S}{(1 + i)^n}\right]\right\} \cdot [(1 + i)^n \cdot i] / [(1 + i)^n - 1] \tag{37}$$

Where in formula (37): n is the asset's useful life, i is the imputed interest rate, P and S are, respectively, the purchase price and salvage value.

2.7   Machine Hours Method

In some case, the depreciation of an equipment factory can be calculated on the basis of its functional performance over time.

Then, this method requires a prior estimation of the asset's functional efficiency over time, as well as its depreciable life and the entire amount of service hours of equipment.

Main advantage of method is that unit cost of depreciation is constant and gives low depreciation costs during periods of low production.

The hourly depreciation rate is determined dividing the purchase cost of equipment (minus its final salvage value) by estimated total number of hours utilized every year. So depreciation rate is calculated as for Straight-Line method, except that useful life is expressed in terms of use hours:

$$D_k = (Use\ Hours)_k \cdot \frac{(P - S)}{(Total\ Use\ Hours)_n} \tag{38}$$

$$D_k^* = \sum_1^k (Use\ Hours)_k \cdot \frac{(P - S)}{(Total\ Use\ Hours)_n} \tag{39}$$

$$BV_k = P - D_k^* = P - \left[\sum_1^k (Use\ Hours)_k \cdot \frac{(P - S)}{(Total\ Use\ Hours)_n} \right] \tag{40}$$

2.8   Production Units Method

The Production Units method (also called "Productive-Output method") corresponds to Machine Hours method with the difference that equipment's service life is expressed in terms of production units rather than use hours:

$$D_k = (Production\ Units)_k \cdot \frac{(P - S)}{(Total\ Production\ Units)_n} \tag{41}$$

$$D_k^* = \sum_1^k (Production\ Units)_k \cdot \frac{(P - S)}{(Total\ Production\ Units)_n} \tag{42}$$

$$BV_k = P - D_k^* = P - \left[\sum_1^k (Production\ Units)_k \cdot \frac{(P - S)}{(Total\ Production\ Units)_n} \right] \tag{43}$$

2.9   Depletion Method

This method is similar to Machine Hours and Production Units method. Indeed, Depletion Method is a special case of these two depreciation methods because in Depletion Method is estimated the available quantities in advance of a resource, rather than working life of an equipment (estimated in terms of hours or production units).

Depletion Method is mostly used for natural resources such as quarries, mines, gas and oil, etc., where the quantity of resources can be obtained on the basis of minerals availability.

The depreciation formulas, for a generic k^{th} year, are the following:

$$R = \frac{(Cost\ of\ Mines)}{(Estimated\ Minerals\ to\ be\ Extracted)} \tag{44}$$

$$D_k = (Quantity\ of\ Minerals)_k \cdot R \tag{45}$$

3 Comparison and Impact of Depreciation Methods

Main depreciation methods have been so far examined.

The depreciable amount (or the total depreciation amount charged over an asset's useful life) is the same regardless of depreciation method, but the adoption of a particular depreciation method has however relevant effects on the firm's accounting in each year of asset's life.

Figure 1 and Table 1 show, for a same fixed asset, the effect of different depreciation methods on yearly depreciation amount (with exception of Usage methods and Depletion method, inasmuch the first methods give results in function of actual use or production related to the specific asset considered, while depletion method isn't generally referred for equipment but for primary resources only).

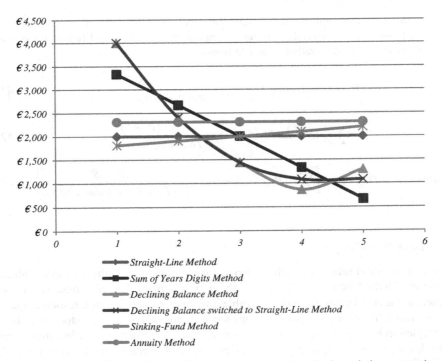

Fig. 1. Graphical comparison of depreciation methods (yearly depreciation amount)

Table 1. Analytical comparison of depreciation methods (yearly depreciation amount)

	Year 1	Year 2	Year 3	Year 4	Year 5
Straight-Line Method	2000	2000	2000	2000	2000
Sum of Years Digits Method	3333	2667	2000	1333	667
Declining Balance Method	4000	2400	1440	864	1296
Declining Balance Switched to Straight-Line Method	4000	2400	1440	1080	1080
Sinking-Fund Method	1810	1900	1995	2095	2200
Annuity Method	2310	2310	2310	2310	2310
Machine Hours Method	–	–	–	–	–
Production Units Method	–	–	–	–	–
Depletion Method	–	–	–	–	–

The comparison of different depreciation methods is based on the following data: fixed asset cost (purchase price) equal to 10,000 Euros, salvage value nil, asset's useful life equal to 5 years and depreciation rate equal to 0.40 (for calculating depreciation using Double Declining Balance method).

The obtained results demonstrate that some depreciation methods are easy to use but may not reflect true economic benefits of asset, others methods are more complex but they may present possible bias in the basic parameters (for example as interest rate, depreciation rate, etc.).

This comparison between the different depreciation methods suggests that amortization does not necessarily require the application of the straight-line method, as is often done in practice.

Although straight-line method is the preferred method for calculation of depreciation in the firms, it is based on assumption that usefulness of a depreciable asset was apportioned to the same extent for each year of asset's useful life. The straight-line is the most common depreciation method because is easy to apply and facilitates the process of interpretation of financial statements, facilitating comparisons between corporates, but it is acceptable only for specific types of assets.

Regarding methods based on declining of depreciation shares (including in this class: Sum of Years Digits, Declining Balance, Declining Balance switched to Straight-Line, Interest Methods), they are based on assumption that the company benefits an increased usefulness in the early years of asset life, because its technical efficiency tends to decrease with age and the maintenance costs tend to increase for the aging process of same asset. More in detail, these methods allow to depreciate about two-thirds of the original value of an asset in the first half of its useful life.

Only in limited and documented circumstances, Usage methods and Depletion method can represent better the distribution of usefulness obtainable from the asset over its economic life, such as when want to assign to each period the depreciation share determined by the ratio between quantities produced during the year and the amount of total production expected during entire useful life of asset.

4 Concluding Remarks

In corporate accounting and balance sheet, depreciation is a systematic allocation of fixed asset cost over its useful life. It is a way of matching the fixed asset cost with the income (or other economic benefits) generated over same useful life.

Without depreciation, the entire fixed asset cost should be recognized in the year of its purchase and this would provide an unreal framework of profitability for firm.

The asset useful life is not related to its "physical life", but to its "economic life", that is the period when is expected that the asset will be useful for firm. This period is normally always less than physical life.

In practice, each depreciation method produces, for firm, a different pattern of expenses over time. In this sense, the paper has examined analytical formulas for main classes of depreciation methods, showing that reduction of asset value frequently tends to be greater in the earlier years (with exception of straight-line method) with costs and maintenance that increase with raising age.

Definitely, from a practical point of view, this paper is able to help construction of automated mechanisms for evaluation and depreciation of fixed assets, assisting decision making in the management of tangible and intangible assets.

References

1. BS ISO 15868-5:2008: Building and constructed assets - Service life planning; Part 5 - Life cycle costing. British Standards Institution, United Kingdom (2008)
2. Buckley, J.J.: Fuzzy Statistic. Springer, Hiedelberg, Germany (2004)
3. Carmichael, D.R., Graham, L.: Accountants' Handbook. Wiley, New York (2009)
4. Chang, P.T., Huang, L.C., Lin, H.J.: The fuzzy Delphi via fuzzy statistics and membership function fitting and an application to human resources. Fuzzy Sets Syst. **112**, 511–520 (2000)
5. Del Giudice, V.: Estimo e Valutazione Economica dei Progetti. Paolo Loffredo Iniziative Editoriali (2015)
6. Del Giudice, V., De Paola, P.: Undivided real estate shares: appraisal and interactions with capital markets. Appl. Mech. Mater. **584-586**, 2522–2527 (2014). Trans Tech Pubblications
7. Del Giudice, V., De Paola, P.: The assessment of damages to scientific building: the case of the "Science Centre" museum in Naples. Adv. Mater. Res. **1030-1032**, 889–895 (2014). Trans Tech Pubblications
8. FAS Manual Depreciation. State of Idaho FAS Fixed Asset System, Rev. 25 November 2013 (2013)
9. International Accounting Standards Board Exposure Draft. Fair Value Measurement, May 2009
10. International Accounting Standards Board. International Financial Reporting Standards (IFRS including International Accounting Standards (IASs) and interpretations as at 1 January 2008. International Accounting Standard No. 16. Property, plant, and equipment. International Accounting Standard No. 2. Inventories. International Accounting Standard No. 17. Leases. International Accounting Standard No. 20. Accounting for Government Grants and Disclosure of Government Assistance. International Accounting Standard No. 36. Impairment of assets. International Accounting Standard No. 40. Investment Property (2008)

11. International Valuation Standards (2007)
12. Khalili, S., Mehrjerdi, Y.Z., Zare, H.K.: Choosing the best method of depreciating assets and after-tax economic analysis under uncertainty using fuzzy approach. Decis. Sci. Lett. **3**(4), 457–466 (2014). Growing Science
13. Liapis, K.J., Kantianis, D.D.: Depreciation methods and life-cycle costing (LCC) methodology. Procedia Econ. Financ. **19**, 314–324 (2015)
14. Manganelli, B., Morano, P., Tajani, F.: Companies in liquidation. a model for the assessment of the value of used machinery. WSEAS Trans. Bus. Econ. **11**, 683–691 (2014)
15. Manganelli, B.: Maintenance, building depreciation and land rent. Appl. Mech. Mater. **357-360**, 2207–2214 (2013). (Architecture, building materials and engineering management)
16. Morano, P., Tajani, F., Locurcio, M.: Land use, economic welfare and property values: an analysis of the interdependencies of the real estate market with zonal and macro-economic variables in the municipalities of Apulia Region (Italy). Int. J.Agric. Environ. Inf. Sys. **6**(4) (2015)
17. Organismo Italiano di Contabilità, Principi Contabili, no. 16
18. Periasamy, P.: A Textbook of Financial, Cost and Management Accounting. Himalaya Publishing House, Mumbai (2010)
19. Peterson, R.H.: Accounting for Fixed Assets, 2nd edn. Wiley, New York (2002)
20. Manganelli, B.: Economic life prediction of concrete structure. Adv. Mater. Res. **919-921**, 1447–1450 (2014). (Advanced Construction Technologies)

A Density Dependent Host-Parasitoid Model with Allee and Refuge Effects

Burcin Kulahcioglu$^{(\boxtimes)}$ and Unal Ufuktepe

Department of Mathematics, Izmir University of Economics, Balcova, Izmir, Turkey
{burcin.kulahcioglu,unal.ufuktepe}@ieu.edu.tr

Abstract. A discrete-time density dependent model is studied by Sophia et al. in [3]. In this paper, we use this model and try to develop it by adding Allee and Refuge effects. With Allee Effect the intraspecific cooperation, with Refuge Effect environment heterogeneity are taken into account. We make the stability analysis of the resulting models together with some numerical simulations.

Keywords: Host-Parasitoid model · Density dependence · Allee effect · Refuge effect

1 Introduction

For discrete time one dimensional population modelling, there are two types: linear and nonlinear. First is linear model in which growth rate can be about birth, death, immigration, emigration rate and it is independent from the density. These models are in the form: $x_{n+1} = rx_n$, where r is growth rate and it is constant.

The second one is nonlinear model in which growth is dependent to density. A population model is said to be density dependent if the per-capita growth rate of the population changes according to the density. The general form is $x_{n+1} = f(x_n)x_n$. The classical approach suggests that f, namely fitness function (per-capita growth rate), should be chosen as a decreasing function because of intraspecific competition and capacity constraint.

When host parasitoid interaction is considered, parasitoid density heavily depends on host population and generally can not live without their host. So, adding density dependence to the host population makes the model more realistic. To be density dependent, in the absence of parasitoid the host population should be in the form of nonlinear one dimensional model.

An insect parasitoid is an organism whose larvae develops in or on its host (insect), damages to it and kills it eventually. They are smaller than host and specialized in their choices. Since they usually target certain groups, in the absence of their target group they can not survive. Using these biological properties, host-parasitoid interaction can be modelled mathematically as the following form:

$$H_{t+1} = bH_t f(H_t, P_t)$$
$$P_{t+1} = cH_t[1 - f(H_t, P_t)] \tag{1}$$

O. Gervasi et al. (Eds.): ICCSA 2016, Part III, LNCS 9788, pp. 228–239, 2016.
DOI: 10.1007/978-3-319-42111-7_18

In this system of equation H_t denotes the number or density of host (prey) species, P_t denotes the density (number) of parasitoid (predator) at time t. Time period can be in terms of hours, days, months or years. $b > 0$ is the parameter of reproduction rate of host, $c > 0$ is the parameter denoting average number of egg (larvae) released by parasitoid on a single host. $f(H_t, P_t)$ stands for probability of not to be parasitized and $1 - f(H_t, P_t)$ is the probability of being parasitized at time t.

In 1980, Wang [1] asked a remarkable question: "Does the ordering of density dependence and parasitism in the host life cycle have a significant effect on the dynamics of the interaction?" May et al. tried to answer this question giving three different model types according to the order and make numerical simulations of these models in [2].

2   The Model

May et al. [2] concluded that the sequence of density dependence and host parasitoid interaction have a marked effect on the population dynamics by examining three different host-parasitoid model. Another conclusion is that the most frequent choice will be between Model2 and Model3.

For the case given as Model3, parasitism acts first, followed by the density dependence but only on the survivors from parasitism (i.e. $H_t f(P_t)$) and the model type is:

$$\begin{aligned} H_{t+1} &= H_t g(H_t, f(P_t))f(P_t) \\ P_{t+1} &= H_t[1 - f(P_t)] \end{aligned} \tag{2}$$

Letting $f(P_t) = e^{-bP_t}$ and $g(H_t, f(P_t)) = \frac{\lambda}{1+kH_t e^{-bP_t}}$ and adding β multiplier to the second equation one can get:

$$\begin{aligned} H_{t+1} &= \frac{\lambda H_t}{1+kH_t e^{-bP_t}} e^{-bP_t} \\ P_{t+1} &= \beta H_t[1 - e^{-bP_t}] \end{aligned} \tag{3}$$

Sophia et al. use this model in [3] and make the stability analysis. In model (3) note that the host population in the absence of the parasitoid is modeled by Beverton-Holt equation $\frac{\lambda H}{1+kH}$. Beverton-Holt Model is obtained if a decreasing rational function is used as fitness (density function). $\beta > 0$ is the parameter denoting average number of egg (larvae) released by parasitoid on a single host. All parameters are positive.

2.1   Stability Analysis of Model (3)

Theorem 1. *For the system (3)*
(i) if $\lambda < 1$ the only fixed point is (0,0) which is locally asymptotically stable.
(ii) if $1 < \lambda < 1 + \frac{k}{\beta b}$ there are two fixed points: (0,0) is unstable and $(\frac{\lambda-1}{k}, 0)$ is locally asymptotically stable.

*(iii) if $\lambda > 1 + \frac{k}{\beta b}$ there are three non-negative fixed points which are $(0,0), (\frac{\lambda-1}{k}, 0)$
and coexistence fixed point (H^*, P^*) which can not be found explicitly.
(iv) if $\lambda = 1$, $(0,0)$ coincides with $(\frac{\lambda-1}{k}, 0)$ and it is unstable.
(v) if $\lambda = 1 + \frac{k}{\beta b}$ then $(\frac{\lambda-1}{k}, 0)$ is unstable.*

Proof. (i), (ii) and (iii) can be written from [3].
(iv) If $\lambda = 1$ the eigenvalues are $\lambda_1 = 1$ and $\lambda_2 = 0$.
We can use Center Manifold Theorem [4]. This theorem can be applied for the
fixed point $(0, 0)$. Since our fixed point is $(0, 0)$ let H=x, P=y.
    First let us write our system in the following way:

$$x_{n+1} = A\ x_n + f(x_n, y_n)$$
$$y_{n+1} = B\ y_n + g(x_n, y_n)$$

    In this case A=1, B=0 and the remaining parts are f and g. Our eigenvectors
are unit vectors so the h function can be in the form:

$$h(x) = c_1 x^2 + c_2 x^3 + O(x^4).$$

The following functional equation will be solved n order to find c_1 and c_2:

$$h[Ax + f(x, h(x))] - Bh(x) - g(x, h(x)) = 0$$

In our case

$$f(x, y) = \frac{x(1 - by + \frac{b^2 y^2}{2})}{(1 + kx(1 - by + \frac{b^2 y^2}{2}))} - x$$
$$g(x, y) = \beta x(by - (\frac{b^2 y^2}{2})) \tag{4}$$

The result is $c_1 = 0$ and $c_2 = 0$.
    Plugging these values, now we are interested in the new equation:

$$P(x) = x - kx^2 + k^2 x^3 - k^3 z^4 + O(z^5)$$

Since $P'(0) = 1$ and $P''(0) = -2k \neq 0$, we conclude that $(0, 0)$ is semistable
(unstable) if $\lambda = 1$ (Fig. 1)
    (v) if $\lambda = 1 + \frac{k}{\beta b}$ then the eigenvalues $\lambda_1 < 1$ and $\lambda_2 = 1$.
To use Center Manifold Theorem let $x = H - (\frac{\lambda-1}{k})$ and $y = P$.
    The corresponding Jacobian Matrix is

$$J* = \begin{pmatrix} \frac{1}{\lambda} & \frac{b-\lambda b}{\lambda k} \\ 0 & \frac{(-1+\lambda)b\beta}{k} \end{pmatrix}$$

We can rewrite our system of equations:

$$x_{t+1} = \frac{1}{\lambda} x_t + \frac{b-\lambda b}{\lambda k} y_t + f(x_t, y_t)$$
$$y_{t+1} = \frac{(-1+\lambda)b\beta}{k} y_t + g(x_t, y_t) \tag{5}$$

Following theorem will be used to find the h function.

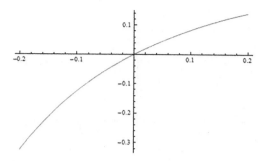

Fig. 1. The map P on the center manifold y = h(x) with k=2

Theorem 2. *(Invariant Manifolds Theorem) [9], see also [10, 11] Suppose that $F \in C^2$. Then there exist C^2 stable W^s and unstable W^u manifolds tangent to E^s and E^u, respectively, at $X = 0$ and C^1 center manifold W^c tangent to E^c at $X = 0$. Moreover, the manifolds W^c, W^s, and W^u are all invariant.*

Since invariant manifold is tangent to the corresponding eigenspace by Theorem 2, let us assume that the map h takes the form

$$h(y) = \frac{-b}{k}y + c_1 y^2 + c_2 y^3 + O(y^4).$$

Solving the functional equation:

$$\frac{1}{\lambda}h(y) + \frac{b - \lambda b}{\lambda k}y + f(h(y), y) = h(y + g(h(y), y))$$

we get

$$c_1 = \frac{-2\lambda b^3 \beta - b^2 k + \lambda b^2 k)}{2(-1+\lambda)k^2}$$
$$c_2 = \frac{-3\lambda^2 b^5 \beta^2 + \lambda^2 b^4 \beta k}{(-1+\lambda)^2 k^3}$$

(6)

The new equation is:

$$P(y) = y - \frac{b^2 \beta y^2}{k} + \left(\frac{b^3 \beta}{2k} - \frac{b\beta \left(2\lambda b^3 \beta - b^2 k + \lambda b^2 k\right)}{2(-1+\lambda)k^2} \right) y^3$$

$$+ \left(\frac{b^2 \beta \left(2\lambda b^3 \beta - b^2 k + \lambda b^2 k\right)}{4(-1+\lambda)k^2} + \frac{\lambda^2 b\beta \left(3b^5 \beta^2 + b^4 \beta k\right)}{(-1+\lambda)^2 k^3} \right) y^4 + O[y]^5$$

$P'(0) = 1$ and $P''(0) = -2k \neq 0$. So, the fixed point $(\frac{\lambda-1}{k}, 0)$ when $\lambda > 1$ and $\frac{(-1+\lambda)b\beta}{k} = 1$ is semistable (unstable) (Fig. 2).

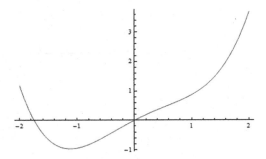

Fig. 2. The map P on the center manifold x = h(y) with $\lambda = 3$, $k = 2$, $b = 0.5$, $\beta = 2$

2.2   Numerical Simulations for the Model 3

In this section numerical values are given to the parameters λ, b, k and β. According to these values coexistence fixed points are obtained approximately. And their stability behavior are investigated for these parameter values.

The coexistence fixed point can not be found explicitly. However it can be written as $\left(\frac{\lambda y - 1}{ky}, \frac{\log y}{-b}\right)$ for $0 < y < 1$ where $y = e^{-bP^*}$. And it exists if $\lambda > 1 + \frac{k}{\beta b}$. So, the following parameters are chosen in order to satisfy this condition.

For the parameters $\lambda = 3$, $b = 0.4$, $k = 0.2$, $\beta = 0.3$ the system will be discussed. $(0, 0)$ fixed point always exists. Since $\lambda > 1$ the other fixed point $\left(\frac{\lambda - 1}{k}, 0\right)$ will be taken into consideration, but it is unstable. Finally, there will be coexistence point (H^*, P^*) when $\lambda = e^{bP} + \frac{kP}{\beta(1 - e^{-bP})}$. Although we can not solve it directly, using numerical approximation we get fixed point $(H^*, P^*) \approx (9.06, 0.43)$ which is locally stable (Fig. 3).

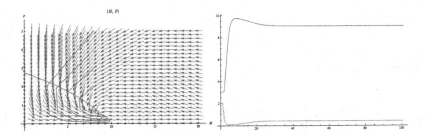

Fig. 3. Plane diagrams and time series diagrams with the parameters $\lambda = 3$, $b = 0.4$, $k = 0.2$, $\beta = 0.3$ (Color figure online)

Next, the system will be discussed for the parameters $\lambda = 5$, $b = 0.7$, $k = 0.2$, $\beta = 0.3$ $(0, 0)$ and $\left(\frac{\lambda - 1}{k}, 0\right)$ are again unstable. For these parameter values, the coexistence fixed point is $(H^*, P^*) \approx (8.22, 1.73)$ which is unstable (Fig. 4).

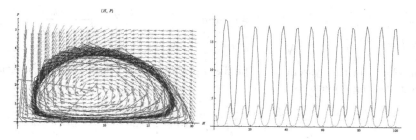

Fig. 4. Plane diagrams and time series diagrams with the parameters $\lambda = 5$, $b = 0.7$, $k = 0.2$, $\beta = 0.3$ (Color figure online)

3   The Model with Allee Effect

Allee Effect is a causal relationship between the number of individuals in a population and their overall individual fitness [5]. Here just a simple idea works: "The more the merrier" [6]. The classical approach focuses on the intra specific competition. Because of the limited capacity and resources, if the population is small then each individual can get much more amount of resources. But the classical idea is lacking the term cooperation and the cost of rarity. If the population is too small, individuals have some difficulties on hunting, protecting themselves, foraging and even finding mates to reproduction. These difficulties are caused by lack of cooperation and can be thought as the cost of rarity [7].

Although there are many functions differ greatly in their ability to describe demographic Allee effects, for the Allee effect due to the mate limitation $I(x) = \frac{sx}{1+sx}$, where s is an individual's searching efficiency, is used. When Mate Limitation Allee Effect is added to the host population in the model we get:

$$
\begin{aligned}
H_{t+1} &= \frac{\lambda H_t}{1+kH_t e^{-bP_t}} e^{-bP_t} \frac{sH_t}{1+sH_t} \\
P_{t+1} &= \beta H_t [1 - e^{-bP_t}]
\end{aligned}
\tag{7}
$$

3.1   Stability Analysis of Model (7)

Theorem 3. *The system (7) has*
(i) (0, 0) as non-negative fixed point for all values of parameters and always locally asymptotically stable.
(ii) two extra non-negative fixed points $(\frac{-k-s+\lambda s-\sqrt{-4ks+(k+s-\lambda s)^2}}{2ks}, 0)$ and $(\frac{-k-s+\lambda s+\sqrt{-4ks+(k+s-\lambda s)^2}}{2ks}, 0)$ if $\lambda > 1$, $s \geq 2\sqrt{\frac{\lambda k^2}{(-1+\lambda)^4} + \frac{k+\lambda k}{(-1+\lambda)^2}}$. The first one is unstable for all parameter values while the second one is locally asymptotically stable if $b < \sqrt{\frac{ks}{\beta^2}}, 2\sqrt{\frac{k}{s}} + \frac{k+s}{s} < \lambda < \frac{b^2\beta^2+b\beta k+b\beta s+ks}{b\beta s}$.
(iii) two extra positive fixed points (coexistence type) which can not be found explicitly but can be written as:

$(\frac{-s-kl+\lambda sl-\sqrt{-4ksl+(s+kl-\lambda sl)^2}}{2ksl}, \frac{logl}{-b})$ and $(\frac{-s-kl+\lambda sl+\sqrt{-4ksl+(s+kl-\lambda sl)^2}}{2ksl}, \frac{logl}{-b})$

$where$ $l = e^{-bP^*}$, if $\lambda > 2\sqrt{\frac{k}{s}} + \frac{k+s}{s}$, $2\sqrt{\frac{\lambda ks^3}{(k-\lambda s)^4} + \frac{ks+\lambda s^2}{(k-\lambda s)^2}} < l < 1$

Proof. Consider the system (7). The fixed points of this system are the solutions of the following equations:

$$H = \frac{\lambda H}{1+kHe^{-bP}}e^{-bP}\frac{sH}{1+sH}$$
$$P = \beta H[1 - e^{-bP}]$$

(8)

(i) $(0, 0)$ is a solution obviously. At this point the jacobian matrix is: $\begin{pmatrix} 0 & 0 \\ 0 & 0 \end{pmatrix}$.
So it is stable [12].

(ii) Now for $H \neq 0$ (8) can be written:

$$1 = \frac{\lambda}{1+kHe^{-bP}}e^{-bP}\frac{sH}{1+sH}$$
$$P = \beta H[1 - e^{-bP}]$$

(9)

By letting $P = 0$, we find the other fixed points.
$(\frac{-k-s+\lambda s-\sqrt{-4ks+(k+s-\lambda s)^2}}{2ks}, 0)$, $(\frac{-k-s+\lambda s+\sqrt{-4ks+(k+s-\lambda s)^2}}{2ks}, 0)$ are other solutions. Both of them are positive for the parameter values $k > 0$, $\lambda > 1$,
$s \geq 2\sqrt{\frac{\lambda k^2}{(-1+\lambda)^4} + \frac{k+\lambda k}{(-1+\lambda)^2}}$

By plugging $(\frac{-k-s+\lambda s-\sqrt{-4ks+(k+s-\lambda s)^2}}{2ks}, 0)$ to the matrix J we get:

$$J* = \begin{pmatrix} 1 + \frac{\sqrt{-4ks+(k+s-\lambda s)^2}}{\lambda s} & \frac{b(-k+s-\lambda s+\sqrt{-4ks+(k+s-\lambda s)^2})}{2\lambda ks} \\ 0 & -\frac{b\beta(k+s-\lambda s+\sqrt{-4ks+(k+s-\lambda s)^2})}{2ks} \end{pmatrix}$$

The corresponding eigenvalues are $\lambda_1 = -\frac{b\beta(k+s-\lambda s+\sqrt{-4ks+(k+s-\lambda s)^2})}{2ks}$ and
$\lambda_2 = 1 + \frac{\sqrt{-4ks+(k+s-\lambda s)^2}}{\lambda s}$.
$\lambda_2 > 1$ for every values of parameters, so $(\frac{-k-s+\lambda s-\sqrt{-4ks+(k+s-\lambda s)^2}}{2ks}, 0)$ is unstable.

By plugging the other fixed point $(\frac{-k-s+\lambda s+\sqrt{-4ks+(k+s-\lambda s)^2}}{2ks}, 0)$ to the J we get:

$$J** = \begin{pmatrix} 1 - \frac{\sqrt{-4ks+(k+s-\lambda s)^2}}{\lambda s} & -\frac{b(k+(-1+\lambda)s+\sqrt{-4ks+(k+s-\lambda s)^2})}{2\lambda ks} \\ 0 & \frac{b\beta(-k+(-1+\lambda)s+\sqrt{-4ks+(k+s-\lambda s)^2})}{2ks} \end{pmatrix}$$

$\lambda_1 = 1 - \frac{\sqrt{-4ks+(k+s-\lambda s)^2}}{\lambda s}$ and $\lambda_2 = \frac{b\beta(-k+(-1+\lambda)s+\sqrt{-4ks+(k+s-\lambda s)^2})}{2ks}$.
$|\lambda_{1,2}| < 1$ if $s > 0, k > 0, \beta > 0, 0 < b < \sqrt{\frac{ks}{\beta^2}}, 2\sqrt{\frac{k}{s}} + \frac{k+s}{s} < \lambda < \frac{b^2\beta^2+b\beta k+b\beta s+ks}{b\beta s}$

(iii) Finally, if $H \neq 0$ and $P \neq 0$ letting $l = e^{-bP^*}$,

$$\left(\frac{-s-kl+\lambda sl-\sqrt{-4ksl+(s+kl-\lambda sl)^2}}{2ksl}, \frac{logl}{-b}\right) \text{ and}$$

$$\left(\frac{-s-kl+\lambda sl+\sqrt{-4ksl+(s+kl-\lambda sl)^2}}{2ksl}, \frac{logl}{-b}\right) \text{ are fixed points.}$$

To be positive $s > 0$, $k > 0$, $\lambda > 2\sqrt{\frac{k}{s}} + \frac{k+s}{s}$, $2\sqrt{\frac{\lambda ks^3}{(k-\lambda s)^4}} + \frac{ks+\lambda s^2}{(k-\lambda s)^2} < l < 1$

3.2   Numerical Simulations for the Model 7

For the parameters $\lambda = 3$, $b = 0.4$, $k = 0.2$, $\beta = 0.3$, s=5 the system will be discussed. These parameters are chosen because for these values the only stable fixed point is one of the coexistence cases. When we solve this system approximation methods there exists different non-zero results according to the given neighbourhood of the solution. Some of them are $(9.69687, -1.35339 \times 10^{-15})$, $(0.103126, 2.19998 \times 10^{-18})$ and $(8.93, 0.35)$. The first two are in fact corresponds to (H, 0), because of solving it with approximation method it gives these results with an error.

$$\begin{aligned} H &= \frac{-k-s+\lambda s-\sqrt{-4ks+(k+s-\lambda s)^2}}{2ks} = 0.103 \\ H &= \frac{-k-s+\lambda s+\sqrt{-4ks+(k+s-\lambda s)^2}}{2ks} = 9.7 \end{aligned} \qquad (10)$$

The positive fixed point, that is approximately $(8.93, 0.35)$, is locally asymptotically stable for these parameter values (Fig. 5).

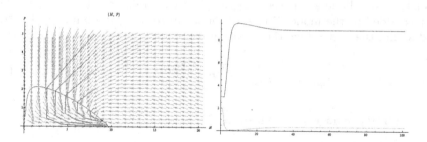

Fig. 5. Plane diagrams and time series diagrams with the parameters $\lambda = 3$, $b = 0.4$, $k = 0.2$, $\beta = 0.3$, s=5 (Color figure online)

Next, the system behavior will be discussed for $\lambda = 5$, $b = 0.7$, $k = 0.2$, $\beta = 0.3$, s=5. (H, 0) cases are unstable as well as the coexistence case $(8.11, 1.69)$ is unstable again this time (Fig. 6).

4   The Model with Host Refuge

Refuge Effect is another mechanism that makes the model more realistic. When we are talking about competitive, prey- predator or host-parasitoid models, a

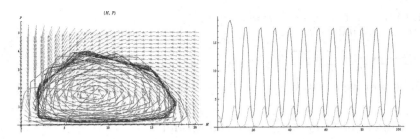

Fig. 6. Plane diagrams and time series diagrams with the parameters $\lambda = 5$, $b = 0.7$, $k = 0.2$, $\beta = 0.3$, s=5 (Color figure online)

simple question arises: "Are there some individuals protecting themselves from the damage given by other species by sheltering ?" If the answer is yes, then Refuge Effect should be added to the model. Refuge effect refers to the reality that a proportion of the prey, host or the competitors can have the state of being safe or sheltered from pursuit, danger, or difficulty. The environment is not perfectly uniform and heterogeneity of the environment is the main cause of the Refuge Effect. Part of the argument is that refuges serve as sites for maintaining vulnerable species that might otherwise become extinct. Such sites also indirectly benefit the exploiting species since a constant spillover of victims into the unprotected areas guarantees a constant food source [8].

Now, if the refuge effect is added to the model that part of the host (insect) population may be less exposed and thus less vulnerable to attack. The refuge effect is added to the model (7) as a proportion to make it simpler. A fraction $1 - d$ within the host refuge the resulting model is:

$$
\begin{aligned}
H_{t+1} &= (1 - d)\frac{\lambda H_t}{1+kH_t} + d\frac{\lambda H_t}{1+kH_t e^{-bP_t}}e^{-bP_t} \\
P_{t+1} &= \beta d H_t[1 - e^{-bP_t}]
\end{aligned}
\tag{11}
$$

4.1   Stability Analysis of Model (11)

Theorem 4. *The system (11) has*
(i) (0, 0) as non-negative fixed point for all values of parameters which is locally asymptotically stable if $\lambda < 1$.
(ii) $(\frac{\lambda-1}{k}, 0)$ as a non-negative locally asymptotically stable fixed point if $1 < \lambda < 1 + \frac{k}{\beta b}$
(iii) $(-\frac{k+ky-\lambda ky - \sqrt{4k^2y(-1+\lambda-\lambda d+\lambda dy)+(-k-ky+\lambda ky)^2}}{2k^2 y}, \frac{\log y}{-b})$ as the coexistence fixed point where $y = e^{-bP^*}$

Proof: (i), (ii) It is obvious that when P = 0 (in the absence of parasitoid the refuge effect is not meaningful. When there is no parasitoid there will be no refuge as well. As a result the model will show the same properties with the model 7.

(iii) The refuge effect has meaning and importance only in the coexistence case. Letting again $y = e^{-bP^*}$, one can find the coexistence fixed points. $(-\frac{k+ky-\lambda ky-\sqrt{4k^2y(-1+\lambda-\lambda d+\lambda dy)+(-k-ky+\lambda ky)^2}}{2k^2y}, \frac{\log y}{-b})$ is positive in the condition $\lambda > \frac{1}{1-d+dy}$ in addition to positive parameters with $0 < y < 1$ and $0 < d < 1$.

On the other hand, under the condition positive parameters and $0 < y < 1$ and $0 < d < 1$, $(-\frac{k+ky-\lambda ky+\sqrt{4k^2y(-1+\lambda-\lambda d+\lambda dy)+(-k-ky+\lambda ky)^2}}{2k^2y}, \frac{\log y}{-b})$ can not be positive.

4.2   Numerical Simulations for the Model 11

For the parameters $\lambda = 10$, $b = 0.4$, $k = 0.2$, $\beta = 0.3$, $d = 0.2$ the system will be investigated. Here, λ is chosen bigger than the other numerical simulations because the coexistence fixed point condition $(\lambda > \frac{1}{1-d+dy})$ is difficult to hold otherwise. According to these parameter values there will be coexistence point $(H^*, P^*) \approx (44.84, 0.37)$ which is locally stable sink (not spiral this time) (Fig. 7).

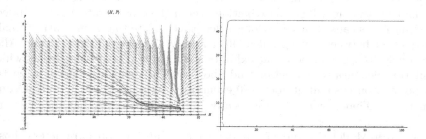

Fig. 7. Plane diagrams and time series diagrams with the parameters $\lambda = 10$, $b = 0.4$, $k = 0.2$, $\beta = 0.3$, $d = 0.2$ (Color figure online)

Next giving values $\lambda = 10$, $b = 0.7$, $k = 0.2$, $\beta = 0.3$, $d = 0.2$ the system will be discussed. Here, $(H^*, P^*) \approx (42.78, 1.88)$ which is locally stable (spiral this time) (Fig. 8).

5   Discussion

In this paper a discrete-time density dependent host parasitoid model is analyzed. Classical approach of density dependence brings intraspecific competition to the model. Because all the resources are limited, there should be capacity constraint and a decreasing growth rate function. On the other hand, when population is already small then limited resources is not a problem anymore but the problem is the individuals will have difficulties on hunting, protecting themselves, foraging and even finding mates to reproduction. We add mate limitation

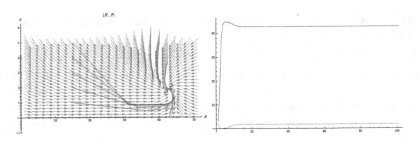

Fig. 8. Plane diagrams and time series diagrams with the parameters $\lambda = 10$, $b = 0.7$, $k = 0.2$, $\beta = 0.3$, $d = 0.2$ (Color figure online)

Allee effect to the model and analyze it in this work. Moreover, since homogeneous environment assumption is not realistic, Refuge effect is added. The further work may be about adding both effects at the same time.

With this work, we conclude that $(0, 0)$ is stable for model (3) when growth rate is small ($\lambda < 1$) while for model (7) it is locally asymptotically stable for all parameter values. In this model if the population of host is small enough at the moment, then it will go to extinction even if the growth rate λ is big. Indeed this situation is caused by the strong Allee Effect. Note that some authors make a distinction between strong Allee effect and weak Allee effect: a strong Allee effect refers to a population that exhibits a critical size or density below which population declines to extinction and above which it survives. Here, when the host population is small, it goes to 0 eventually. As a result parasitoids become extinct, too. That is why $(0, 0)$ is locally stable whatever the parameter values are.

For Both Model (3) and (7), under some conditions on parameters, hosts population can be positive when parasitoid population is 0. For this situation growth rate is neither big nor small indeed there will be survivor insects but their number is not enough to feed their parasitoids. Again for both models coexistence cases are possible but the stability changes according to the parameters.

Finally for model 11, since no parasitoid means no refuge we can simply use model 3 for both local and global stability when there is no parasitoid. That is why without doing any extra calculations we use the same results with model 3.

Otherwise mathematically we can show that there is coexistence case. But finding it numerically with approximation methods is difficult because of the condition $\lambda > \frac{1}{1-d+dy}$ that is needed to be positive. We should choose bigger λ values than the other models in order to be satisfied this condition. It is biologically meaningful because already some proportion of the host population can not be attacked because of the refuge effect, if in addition to this, growth rate is small then for parasitoid it is difficult to find host to feed on them. When there is coexistence case, since some host sheltered from the pursuit of parasitoid, we will have bigger host population in the fixed point.

References

1. Wang, Y.H., Gutierrez, A.P.: An assessment of the use of stability analyses in population ecology. J. Anim. Ecol. **49**, 435–452 (1980)
2. May, R.M., Hassell, M.P., Anderson, R.M., Tonkyn, D.W.: Density dependence in host-parasitoid models. J. Anim. Ecol. **50**, 855–865 (1981)
3. Sophia, R., Jang, J., Jui-Ling, Y.: A Discrete Time Host-Parasitoid Model. In: Proceedings of the Conference on Differential & Difference Equations and Applications, pp. 451–455 (2006)
4. Elaydi, S.: Discrete Chaos: With Applications in Science and Engineering, 2nd edn. Chapman and Hall/CRC, Boca Raton (2008)
5. Allee, W.C., Emerson, A.E., Park, O., Schmidt, K.P.: Principles of Animal Ecology. WB Saunders, Philadelphia (1949)
6. Courchamp, F., Beree, L., Allee, J.: Effects in Ecology and Conservation. Oxford University Press, Oxford (2008)
7. Scheuring, I.: Allee effect increases the dynamical stability of populations. J. Theor. Biol. **199**, 407–414 (1999)
8. Edelstein-Keshet, L.: Mathematical Models in Biology. Random House, New York (1988)
9. Guzowska, M., Lus, R., Elaydi, S.: Bifurcation and invariant manifolds of the logistic competition model. J. Differ. Eqn. Appl. **17**, 1851–1872 (2011)
10. Karydas, N., Schinas, J., Karydas, N., Schinas, J.: The center manifold theorem for a discrete system. Appl. Anal. **44**, 267–284 (1992)
11. Marsden, J., McCracken, M.: The Hopf Bifurcation and Its Application. Springer-Verlag, New York (1976)
12. Kulenovica, M.R.S., Nurkanovic, M.: Global asymptotic behavior of a two dimensional system of difference equations modeling cooperation. J. Differ. Eqn. Appl. **9**(1), 149–159 (2003)

Interpreting Heterogeneous Geospatial Data Using Semantic Web Technologies

Timo Homburg[1]([⊠]), Claire Prudhomme[1], Falk Würriehausen[1],
Ashish Karmacharya[1], Frank Boochs[1], Ana Roxin[2], and Christophe Cruz[2]

[1] Mainz University of Applied Sciences, Lucy-Hillebrand-Straße 2,
55128 Mainz, Germany
{timo.homburg,claire.prudhomme,falk.wuerriehausen,ashish.karmacharya,
frank.boochs}@hs-mainz.de
[2] Université de Bourgogne, 9 Avenue Alain Savary, 21000 Dijon, France
{ana-maria.roxin,christophe.cruz}@u-bourgogne.fr

Abstract. The paper presents work on implementation of semantic technologies within a geospatial environment to provide a common base for further semantic interpretation. The work adds on the current works in similar areas where priorities are more on spatial data integration. We assert that having a common unified semantic view on heterogeneous datasets provides a dimension that allows us to extend beyond conventional concepts of searchability, reusability, composability and interoperability of digital geospatial data. It provides contextual understanding on geodata that will enhance effective interpretations through possible reasoning capabilities. We highlight this through use cases in disaster management and planned land use that are significantly different. This paper illustrates the work that firstly follows existing Semantic Web standards when dealing with vector geodata and secondly extends current standards when dealing with raster geodata and more advanced geospatial operations.

Keywords: Heterogeneity · Interoperability · SDI · CIP · GeoSPARQL · R2RML · Semantification

1 Introduction

When working in a geospatial context, one is confronted with several historically grown data sources which may be important for the current task to be solved. Integrating heterogeneous datasets into a database ecosystem of some kind remains a necessary step before working on the actual task the data was determined for. To a certain degree, this process can be automated, yet can be seen as inflexible, as the data is often integrated and used for a limited set of usecases. This paper presents a semantic based method to simplify the process of data integration, increasing its flexibility to deal with changing requirements and possibly extracting implicit information from integrated data through usage of the semantic technologies under the Semantic Web framework [12].

© Springer International Publishing Switzerland 2016
O. Gervasi et al. (Eds.): ICCSA 2016, Part III, LNCS 9788, pp. 240–255, 2016.
DOI: 10.1007/978-3-319-42111-7_19

2   State of the Art

This section will discuss related work and the state of the art in integration of heterogeneous geospatial data.

2.1   Related Work on Geospatial Data Integration

The work on geospatial integration focuses primarily on the syntactic standardization of geospatial data. Works like the OpenGIS Geography Markup Encoding Standard (GML) that expresses the geographic features serve as an open interchange format for transactions on the internet [9]. The issue has also been addressed at a policy level. Most countries have National Spatial Data Infrastructures (NSDIs) which are SDIs at national levels that provide standardized frameworks for sharing geospatial data. Now, regional efforts are also being made to harmonize these NSDIs under a common regional SDI umbrella. An example could be the Infrastructure for Spatial Information in European Community (INSPIRE) [21] at a European level. The root problem in all these efforts is that they do not have enough constructs to express semantics [28]. However, the current upsurge of Semantic Web technologies has fuelled the implication of semantics in geospatial data. The general tendency of this implication is to combine data from different sources semantically to provide a unified semantic view [17] through mostly a global schema that maps definitions to the schema definitions at the local data sources. Projects like GeoKnow[1] apply similar strategies to combine, structure and expose geospatial data onto the web [5]. We extend this tendency by integrating distributed data sources and laying a foundation for geospatial inference.

2.2   The Semantic Web Technologies

The term "Semantic Web" is coined by Tim Berners-Lee in his work to propose the inclusion of semantics for better enabling machine-people cooperation for handling huge information that exists in the web [7]. It is basically an extension to the current web supported through the standards and technologies standardized by World Wide Web Consortium (W3C)[2]. We present a few important technologies that define the Semantic Web in this paper.

2.2.1   Web Ontology Language

Within the computer science domain, ontology is a formal representation of knowledge defined through the hierarchy of concepts and the relationships between those concepts. In theory an ontology is a formal, explicit specification of shared conceptualization [13]. OWL or the Web Ontology Language is a family of knowledge representation languages to create and manage ontologies. The World Wide Web Consortium (W3C) has standardized OWL to model

---

[1] http://geoknow.eu/Welcome.html.
[2] https://www.w3.org.

ontologies. The standardization of OWL has sparked off the development and/or adaption of a number of reasoners like Pellet[3].

In recent years a number of efforts on developing geospatial ontologies for adding semantics in spatial data have been witnessed. Geospatial ontologies take their domain and range of concepts as geospatial objects, relations and features [3]. Recent studies like [4,11], implement through geospatial semantic expressions in the Web Ontology Language[4] (OWL) and/or Resource Description Framework[5] (RDF). The Geonames ontology[6] adds possibilities to add geospatial semantics in the world wide web and also describes the relation between toponyms. Likewise, the LinkedGeoData project developed the LinkedGeoData ontology knowledge base[7] that lifts Open Street Map (OSM)[8] data to be presented in a Semantic Web infrastructure [24]. It is a lightweight ontology build through first conversion of OpenSteetMap data into RDF and then interlinking to DBPedia[9], GeoNames, and other datasets. Moreover, it supports multi-lingual class labels from various sources. The GeoKnow project applied LinkedGeoData knowledge base to set up benchmarks within its use cases. It also contributed in updating and modifying the knowledge base.

2.2.2 Query and Reasoning

The OGC GeoSPARQL standard is the spatial extension to SPARQL[10] - an RDF query language, proposed by OGC. It is the initiative taken by OGC in collaboration with the SPARQL W3C working group to define vocabularies for representing geospatial data in RDF. These standards implement the SPARQL query language to process geospatial data. Moreover, it accomodates qualitative spatial reasoning and systems based quantitative spatial computations [1]. Qualitative spatial reasoning tests the binary spatial relations between features and do not usually model explicit geometries. In the meantime, quantitative systems transform the query into a geometry based query that evaluates computational geometries between features [20].

The Semantic Web Rule Language (SWRL) is a rule language to infer ontology knowledge bases. SWRL has the form, antecedent $\rightarrow$ consequent, where both antecedent and consequent are conjunctions of atoms written a1 $\wedge$.... $\wedge$ an. Atoms in rules can be of the form C(x), P(x,y), Q(x,z), sameAs(x,y), differentFrom(x,y), or builtIn(pred, z1, ..., zn). For instance, the following rule asserts that one's parents' brothers are one's uncles where parent, brother and uncle are all individual-valued properties.

$$parent(?x, ?p) \wedge brother(?p, ?u) \rightarrow uncle(?x, ?u) \tag{1}$$

---

[3] https://www.w3.org/2001/sw/wiki/Pellet.
[4] https://www.w3.org/TR/owl-guide/.
[5] https://www.w3.org/RDF/.
[6] http://www.geonames.org.
[7] http://linkedgeodata.org/About.
[8] http://www.openstreetmap.org.
[9] http://dbpedia.org.
[10] https://www.w3.org/TR/rdf-sparql-query/.

Limited work has been conducted to include semantic on spatial geofeatures and operations into SWRL. Nevertheless, spatial extensions through spatial built-ins were proposed in the research work [16]. This work defines how spatial semantics could be used within SWRL to trigger spatial inferencing.

3   Semantification

We define semantification as the process of interpreting given geospatial data using semantic technologies. A general description of a semantification process is as follows:

1. Determining a describing set of owl classes for a corresponding data set.
2. Determining a unique identifier for instances of the describing class.
3. If possible inferring types for extracted features.
4. If possible inferring restrictions for extracted features and classes.
5. Interlinking aforementioned elements to other domains.

3.1   Schema-Assisted Semantification

A schema-assisted semantification can take place using either a database or a file that contains a certain schematic description of its content. In a GIS context this involves databases like POSTGIS and a substantial amount of geospatial data given in an XML dialect, mostly as KML or GML files. However, from a semantic web perspective it can make sense to not only integrate data directly related to a geospatial object, but rather to associate possible non-geospatial entities linked to a geospatial object. We will illustrate this thought in our usecases section.

3.1.1   Semantification of Relational Databases

Interpreting relational databases using a semantic web layer, has been extensively studied and W3C[11] recommendations such as R2RML and Direct Mapping[12] have emerged as well as in the case of SPARQLify[13] been further developed. While R2RML proposes using a manually defined mapping profile to map database tables to preexisting classes in ontologies, the Direct Mapping approach delays the interlinking part to be manually specified in the ontology after having interpreted database tables using their given designations. In our approach we are allowing currently employing R2RML and hope to add automated support for direct mapping including interlinking in the future.

---

[11] World Wide Web Consortium.
[12] https://www.w3.org/TR/r2rml/, https://www.w3.org/TR/rdb-direct-mapping/.
[13] http://aksw.org/Projects/Sparqlify.html.

3.1.2   Semantification of XML Files with a Given Schema (XSD)

A few projects like Redefer[14], Ontmalizer[15] and XS2OWL [26] have been attempting to convert given XSD schemas to RDF, in order to simplify the interpretation of given file formats. Many of those projects perform well on some XSD files, but have shortcomings in terms of completeness. This comes to little surprise as[16] points out, a general transformation process is not trivial and because of ambiguities sometimes errornous. Alternatively, rather than transforming XSD schemas to RDF, approaches like GeoKnow[17] or [14] have been providing one XSL transformation file per file format on the data files themselves. This approach compared to a schema extraction is less complicated, usually much more feasible but cannot encompass property cardinalities, the class hierarchy and available codelist references. In our approach we rely on a preextraction of XSD schemas using a modified XSL script inspired by the previously mentioned approaches to create a local ontology [8] for every file format we needed to integrate. After the creation of local ontologies, data files matching the pre-extracted schema can be imported effortlessly into the ontologies class structure. In addition to XSD transformations, codelists of XSD schemas[18] need to be taken into account. Those are often not included in the schema descriptions and therefore need to be merged independently, as they often contain the most valuable describing information (e.g. classification of an object). With our extraction mechanism we are able to gain a class hierarchy, available properties including their ranges and domains, defined restrictions on a per class basis, as well as codelist representations as enumerations along with additional information contained in comments.

3.2   Semiautomated Schemaless Semantification

File formats without a distinctive schema such as shapefiles or schemaless GML files lack relations to other classes in the to-be-built ontology. On the example of a shapefile containing data about tree locations in Germany we would like to illustrate our semantification approach for schemaless data:

Table 1. File Tree.shp

ID	Geometry	Feature1	Feature2	FeatureN
Tree1	POINT(..)	Oak	4	Mainz
Tree2	POINT(..)	Box	3	Koblenz

---

[14] http://rhizomik.net/html/redefer/.
[15] http://www.srdc.com.tr/projects/salus/blog/?p=189.
[16] http://www.ieee-icsc.org/ICSC2011/slides/XSD2OWL_Patterns_ICSC2011.pdf.
[17] http://geoknow.eu/Welcome.html.
[18] Example Codelists of INSPIRE: http://inspire.ec.europa.eu/codelist/.

In Table 1 the content of the shapefile "Tree.shp" is given. It includes the geometry column, a tree type description column, an integer column and a column including city names. This information enables us to automatically infer the representation shown in Fig. 1 of the file in RDF with the following optional pieces of information provided by the user:

– A unique identifier for individuals from one of the columns. If none is provided a unique identifier will be generated.
– An optional column defining subclasses of the class described by the files name.

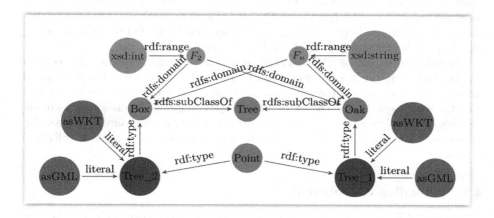

Fig. 1. Semantification example

This representation is consistent, yet far from satisfying, as we have no semantic interpretation of the integer column and the column representing the cities. Semantic interpretations of a column containing text such as the city column could be matched using two ways:

– The column description can possibly be matched to a class in a preexisting ontology using a tool like BabelNet [19].
– Attribute values may as well be matchable using BabelNet.

Each of those approaches will produce a set of possible preexisting and known concept classes a user can choose from, therefore enabling a user to semiautomatically interlink schemaless data to the ontology infrastructure. In the case of columns representing numbers only, the user should not only choose a concept but, if possible also a corresponding unit provided by the QUDT[19] ontology. On import, values of units are converted by our system to a Base SI Unit in order to guarantee a degree of comparison between values in GeoSPARQL queries.

---

[19] Quantities, Units, Dimensions and Data Types Ontology: http://www.qudt.org/.

3.2.1 Semantification and Querying of Raster Data

Raster data is represented by a (thematic) raster of pixels, each of which correspond to an associated value or the NODATA value. Raster data can therefore be interpreted as an owl:Class named after the files' designation/a user assigned concept having exactly one DataProperty describing the value or if metainformation is given an ObjectProperty respectively. Rasterdata is not natively supported by GeoSPARQL but needed in many geospatial applications.

One of our project partners' typical flood simulation cases aims to find out for concerned elements at risk of being flooded and identifying still usable rescue paths from certain areas at certain flood altitudes. In this usecase we are following the latter approach and want to find out which roads are still usable after a flood of x meters has struck the city of Cologne. We are provided with a shape file of the roads (LineStrings) in Cologne and a GEOTIFF file including the flood altitudes above ground level in our simulation.

To solve this usecase, the roads are first splitted into segments of a certain length. Following this for every segment bounding boxes of width 1 m are clipped on the raster data set to find out if the altitude of the flood increases 1m. In this case the road segment and therefore its corresponding road is considered as unaccessible. Clearly, on a semantic level there is currently no standard that provides access to both raster and vector data to accomodate a usecase as highlighted above.

3.3 Extending GeoSPARQL

The GeoSPARQL definition [6] defines a region connection calculus for the geospatial web. However, it does not include geometry constructors and geometry manipulation functions common in other spatial infrastructures. Support for raster data is also not present in its current specification. Clearly, our usecase demands such functionality to solve aforementioned problems. To our knowledge there exists no such system in the semantic web. We therefore are in the process of extending the ARQ query processor implementation using the Java Topology Suite library to support most of common functions available in comparable implementations such as POSTGIS. To support raster data we polygonize the desired region of the raster using the GDAL polygonizing library, as comparable other implementations do. On completion of this approach we want to provide a semantic platform that is capable of interpreting and dealing with both raster and vector data and combinations of them.

3.4 Interlinking

Each of the mentioned semantification methods lead to a local ontology which class tree structure is to be integrated in a global broker ontology to be accessible. For the creation of the geospatial part of the global broker ontology we follow the specifications of GeoSPARQL[20], thereby subclassing each local ontology under

---

[20] http://www.opengeospatial.org/standards/geosparql.

the class SpatialObject, as illustrated in Fig. 2b. Interlinking of generated local ontologies is a major challenge to be solved in this regard. Our goal is to create or reuse a unified vocabulary covering all concepts that have been provided us by the data sources. This will allow not only expert users familiar with the local ontologies to access geospatial data, but also expose the nowadays mainly inaccessible world of spatial data definitions to the semantic web community. In the absence of experts from our side we currently apply a semi-automatic interlinking approach to match data sources with concepts by the importing users (which we expect to have expertise on the data they import). However we intend to further investigate the quality of matching concepts in an automized fashion as described in Sect. 5.1. In contrast to [8] we would like not only to use concept matching and attribute matching but furthermore focusing on matching concepts of geometries in related data sources.

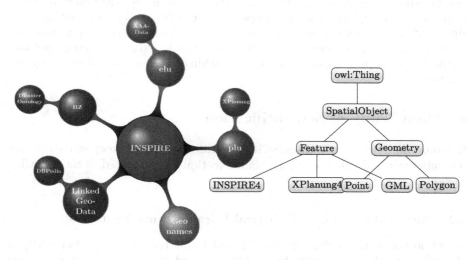

Fig. 2. (a) Current broker ontology structure, (b) Ontology with GeoSPARQL structure

However, even with manual interlinking by experts alone we were able to create an ontology infrastructure that can be seen in Fig. 2a. This ontology infrastructure currently contains ontologies of data structures near to INSPIRE and will possibly be extended in the future. Thereby INSPIRE will not be the center of our infrastructure but more likely one of many ontologies to be considered.

3.5   Management of Data Sources

Semantification is useful to interpret given data sets and to infer a class structure in OWL. However, depending on the scale of the system it might not be possible to store all datasets in one database (triple store) all of the time. We are therefore investigating caching mechanisms to be used to only store data that is needed for the current requests received by the system. To utilize caching we need up-to-date references as annotations to classes and properties in the ontology linking to the actual data sources. A source annotation contains the path to the data source, a timestamp of last access and metainformation on how to interpret the data given. The given annotations are created on import or when an update request is triggered.

3.5.1   Query Rewriting for Performance Improvements

In a previous publication, Tschirner et al. [25] proposed to use a ETL[21] process to access WFS services via a semantic layer and an annotated ontology as previously described. Depending on the SPARQL query given by the user we are taking this approach further to not only support WFS services, but rather supporting spatial databases through R2RML, as well as files queryable by OGC CQL or another query language. Depending on a (to be refined) caching algorithm to improve performance, we created our system to selectively access data by rewriting queries defined by the users to the underlying query languages and fetching data depending on demand and caching data as deemed useful by our system.

4   Case Study of Semantification

Based on the goals in the context of interpreting distributed geospatial data with semantic web, individual cases of semantification are presented in the following section.

4.1   National Strategy for Critical Infrastructure Protection

Based on the National Strategy for Critical Infrastructure Protection (CIP) in Germany [2], this case study highlights the need of integrated heterogeneous geospatial data for the purposes of environment and energy. In this context we investigate flood and disaster management cases and the protection of critical infrastructures. As described in [10], disaster management is a cycle of four steps: Mitigation, Preparedness, Response, and Recovery. In addition to the real world application concept presented in the part Sect. 3.2.1 with raster data, we will present a use case of heterogeneous data interpretation in the field of disaster management. It is based on a flood situation and aims to show how different data can be combined in order to provide relevant information for preparing the response to a disaster (corresponding to step 2 in disaster management).

Use Case: Our use case is based on a flood risk estimation of Saarland, a region in Germany. When a flood happens, people who are exposed, have to be evacuated to a safe area and transport casualties to a hospital. One of prioritary

---

[21] Extract Transfer Load.

facilities are schools. In order to organize a rescue plan of schools in a flood area, we search to determine for each school in a flood area, the hospitals which are not flooded and in a radius of 15 Km. Thanks to the INSPIRE geoportal, three GML files have been retrieved from WFS services:

- A file containing hospitals. with information about their location, provided services, and number of beds.
- A file containing the positon of schools.
- A file containing the position of flood risk areas.

Using our approach exposed previously, these files are interpreted and gathered in one semantic model which can afterwards be queried. The interpretation of these data sets allows to obtain a quick answer due to GeoSPARQL queries. The GeoSPARQL query corresponding to the research of unexposed hospitals which have a maximum distance of 15 Km from each school present in a flood area. This flood area is defined by a radius around a point of flood risk. This GeoSPARQL query is presented in Listing 1.1:

Listing 1.1. Exposed schools query

```
1   PREFIX xsd: <http://www.w3.org/2001/XMLSchema#>
    PREFIX geo: <http://www.opengis.net/ont/geosparql#>
3   PREFIX geof: <http://www.opengis.net/def/function/geosparql/>
    PREFIX school: <http://geoportal.saarland.de/arcgis/services/Internet/
    Staatliche_Dienste/MapServer/WFSServer#>
5   PREFIX uom: <http://www.opengis.net/def/uom/OGC/1.0/>
    PREFIX dangerzone:<http://geoportal.saarland.de/arcgis/services/
    Internet/Hochwasser_WFS/MapServer/WFSServer#>
7   PREFIX hospital:<http://geoportal.saarland.de/arcgis/services/Internet
    /Gesundheit/MapServer/WFSServer#>

9   SELECT DISTINCT
    ?dangerzone ?school_name ?hospital_name
11  WHERE {
    ?school a school:Schulen_SL .
13  ?school geo:hasGeometry ?school_ind .
    ?school_ind geo:asWKT ?school_geo .
15  ?school school:SCHULNAME ?school_name .
    ?dangerzone a dangerzone:Betr_EW .
17  ?dangerzone dangerzone:OBJECTID "10"^^<http://www.w3.org
    /2001/XMLSchema#int> .
    ?dangerzone geo:hasGeometry ?dangerzone_geoind .
19  ?dangerzone_geoind geo:asWKT ?dangerzone_geo .
    ?hospital a hospital:Krankenhaeuser .
21  ?hospital hospital:KRANKENHAU ?hospital_name .
    ?hospital geo:hasGeometry ?hospital_geoind .
23  ?hospital_geoind geo:asWKT ?hospital_geo .
    FILTER(geof:sfWithin(?school_geo, geof:buffer(?dangerzone_geo, 3500,
    uom:meter)))
```

```
25   FILTER(!geof:sfWithin(?hospital_geo,geof:buffer(?dangerzone_geo, 3500,
     uom:meter)))
     FILTER(geof:sfWithin(?hospital_geo,geof:buffer(?school_geo, 15000, uom
     :meter)))
27   }
```

The result of this query is presented in Table 2. It depicts the following situation on a map of the area in Fig. 3.

The interpretation of heterogeneous data is the base of a system for disaster management. This use case shows the relevance to retrieve and interpret heterogeneous data. In our future work, we will use this base to create a more complex and sophisticated system allowing to generate automatically new information thanks to rules.

4.2 Distributed Data in Spatial Data Infrastructures

4.2.1 Comprehensive Management

The need to use heterogeneous and distributed data becomes evident particularly in the current developments of SDI's and already existing geoportals on the Internet. Users should be able to create a simple search for spatial data while the usability and usage conditions should be quickly recognizable [23]. Established procedures for functional spatial data exchange among and across different levels of government are operational, such as in land use planning data management. A considerable share of planned land-use information in Germany specified in XPlanGML[22] addressed by the INSPIRE Annex III theme land-use [23]. In addition to the work in [18,27], the use of ontologies in the context of German land-use planning and INSPIRE, should be presented.

Table 2. Queryresult

Riskzone	Exposed school	Next unexposed hospital
Saubach	Erich-Kästner-Schule	Marienhausklinik Wadern
Saubach	Grundschule St. Michael	Marienhausklinik Wadern
Saubach	Berufsbildungszentrum Lebach	Marienhausklinik Wadern
Saubach	Gemeinschaftsschule ERS Lebach, Theetal Schule	Marienhausklinik Wadern

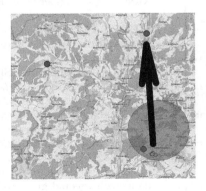

Fig. 3. Result visualized (Hospitals in green, schools in light blue, risk area in transparent green) (Color figure online)

---

[22] http://www.iai.fzk.de/www-extern/index.php?id=680&L=1.

4.2.2   Rule Management

Particularly needed are accepted rules for the provision of local government datasets, as well as for services which are able to process comprehensive spatial data based on different, even municipal, spatial data themes. Integrated data concepts and data models can help ensure cooperation even in a heterogeneous environment of organization units. In-depth analysis of data structures, therefore, is imperative. The "Technical Guidance" for INSPIRE Transformation Networking Services (TNS) [15] not only explains the functional requirements for the INSPIRE regulations but also explains the implementation rules. The needed mapping and transformation rules in this case are those between any local SDI source data model and the common INSPIRE target data model, here defined in the Annex III specification Land use. By using semantics, our mapping rules can be described with (s)ubject, (p)redicate, (o)bject in a RDF-triple, see Table 3.

Table 3. XPlanung4 to INSPIRE relations

(s)ubject	(p)redicate	(o)bject
\<http://www.xplanung.de/ xplangml/4/0#XP_Plan\>	\<http://www.w3.org/2002/07/owl# equivalentClass\>	\<http://inspire.ec.europa.eu/ schemas/plu/4.0#SpatialPlan\>
\<http://www.xplanung.de/ xplangml/4/0#name\>	\<http://www.w3.org/2002/07/owl# equivalentObjectProperty\>	\<http://inspire.ec.europa.eu/ schemas/plu/4.0#officialTitle\>
\<http://www.xplanung.de/ xplangml/4/0#BPlan_ 1000\>	\<http://www.w3.org/2002/07/owl# sameAs\>	\<http://inspire.ec.europa.eu/ schemas/plu/4.0#infraLocal\>
\<http://www.xplanung.de/ xplangml/4/0#BP_ Objekt\>	\<http://www.w3.org/2002/07/owl# equivalentClass\>	\<http://inspire.ec.europa.eu/ schemas/plu/4.0# ZoningElement\>

Table 3 shows an excerpt of transformation rules for eqivalentClass, equivalentObjectProperty and sameAs relations with attributes of the land-use in

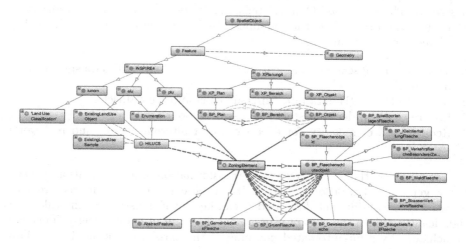

Fig. 4. Owl graph of interlinked INSPIRE and XPlanung4

national XPlanGML and European INSPIRE systems as URI. We can represent the corresponding mapping rules in an ontology graph, created with our prototype of the ontology mapping tool. It can be described in accordance with Tschirner et al. [25] as a step to integrate GML-/XSD- Source into the RDF/OWL- language. Rules are presented in two distinct forms, by using the ontology through restriction axioms and rules for the equivalence.

Figure 4 demonstrates the graphical overview of some local and INSPIRE zoning element vocabularies and how they are interlinked in the ontology. The feature types of the local-level source data model had to be mapped to the INSPIRE target data model by using semantic rule definition.

5   Conclusion

The implementation of knowledge management in this application field can help to support interoperability for the benefits of the local level. The benefits of the approach can be found in the unique flexibility of an ontology instead of some feature manipulation, in the standard-based application and the versatility of use. In that way ontologies can be used to model extensive knowledge about the data, for rule definition, and other objectives in order to reduce the increased complexity for the local government. The automated service function focuses on those data quality elements and processes that can be fully automated (e.g. logical consistency). The semi-automatic service focuses on those data quality elements and measurements that require some human intervention (e.g. data generalisation). The implementation of knowledge management in schema interlinking application fields can help to support interoperability for the benefits at the local level. The feature types of the local level source data model are interlinked to the INSPIRE target data model. Based on the model of Waters et al. [22], the steps from "Feature Transformation" to "Data Generalisation" will be described. The supported Quality steps and Rules are those between any local SDI source data model and the common INSPIRE target data model, defined in the Annex specifications.

1. Transform local schemas and features to common INSPIRE data specification framework.
2. Integrate INSPIRE data into "virtual" collaborative datasets.
3. Ensure cross-border consistency across neighbouring data.
4. Reduce data complexity, for effective use at alternate scales and real worls scenarios.

For all cases, the process results of transformation, data integration, consistency, generalisation of local sources. It can be expected that the requirement of easier usage of different data models can be met if the study of the rules can significantly reduce the cost of a user interpretation of heterogeneity. This can be achieved by semi-automated mapping of the class descriptions in the data collection process.

5.1   Future Work

Our future work will focus on extending not only the data we are about to access but also on insuring the data's quality and enriching it with inferrable knowledge.

5.1.1   Automated Concept Matching

In the future we want to fully autonize the process of matching concepts to relational geospatial datasets. This process has been described in Sect. 3.1.2. We therefore want to investigate natural language processing methods and methods to match the spatial objects we are integrating using a reference source. We hope to achieve a very accurate ranking of plausible concepts for each dataset and will test the resulting ranking using a reference data set and appropriate metrics.

5.1.2   Application of Reasoning

Through the application of reasoning techniques we will be able to infer class structures and individuals from existing data, that can be useful in several use cases. For this purpose we may need to extend an appropriate reasoning language similarly to extending GeoSPARQL to support more spatial functions. In Sect. 4.1 of disaster management we could profit from reasoning by creating a reasoning rule defining the relation between related hospitals with exposed schools through the SWRL statement presented below.

$$Hospitals(?h) \land Schools(?s) \land RiskArea(?r)$$
$$\land semGIS : buffer(?r, ?br, 2) \land semGIS : within(?br, ?s)$$
$$\land semGIS : buffer(?s, ?bs, 15) \land semGIS : disjoint(?br, ?h) \tag{2}$$
$$\land semGIS : within(?bs, ?h) \rightarrow isSuitableHospital(?h, ?s)$$

The resulting class suitableHospital could then be further reused to solve related issues.

5.1.3   Quality Assessment and Data Aggregation

Lastly, we want to focus on integrating further data sources into our system and assess the quality of data provided using to-be-defined metrics. Having evaluated our data basis in this way, our system should be able to select "good" and available data sources preferably on a request basis to give the best possible result under the current circumstances (missing, currently unaccessible data sources, etc.).

Acknowledgements. This project was funded by the German Federal Ministry of Education and Research (https://www.bmbf.de/en/index.html Project Reference: 03FH032IX4).

References

1. GeoSPARQL - a geography query language for RDF data (2016). http://www.opengeospatial.org/standards/geosparql. Accessed 17 Mar 2016

2. National Strategy for Critical Infrastructure Protection (CIP Strategy)
3. Socop, geospatial ontologies (2016). http://ontolog.cim3.net/cgi-bin/wiki.pl? SOCoP/GeospatialOntologies. Accessed 16 Mar 2016
4. Arpinar, I.B., Sheth, A., Ramakrishnan, C., Usery, L., Azami, M.: Po kwan, M.: Geospatial ontology development and semantic analytics. In: Transactions in GIS, pp. 1–15. Blackwell Publishing, Boston (2005)
5. Athanasiou, S., Hladky, D., Giannopoulos, G., Garcia-Rojas, A., Lehmann, J.: GeoKnow: making the web an exploratory place for geospatial knowledge. Eur. Res. Consortium Inf. Math. News **96**, 12–13 (2014)
6. Battle, R., Kolas, D.: GeoSPARQL: enabling a geospatial semantic web. Semant. Web J. **3**(4), 355–370 (2011)
7. Berners-Lee, T., Hendler, J., Lassila, O.: The semantic web. Sci. Am. **284**(5), 34–43 (2001). http://www.sciam.com/article.cfm?articleID=00048144-10D2-1C70-84A9809EC588EF21
8. Bizid, I., Faiz, S., Boursier, P., Yusuf, J.C.M.: Integration of heterogeneous spatial databases for disaster management. In: Parsons, J., Chiu, D. (eds.) ER Workshops 2013. LNCS, vol. 8697, pp. 77–86. Springer, Heidelberg (2014). http://dx.doi.org/10.1007/978-3-319-14139-8_10
9. Bossomaier, T., Hope, B.A.: Online GIS and Spatial Metadata. CRC Press, Boca Raton (2015)
10. Coppola, D.P.: Introduction to International Disaster Management. Elsevier, Burlington (2011)
11. Cruz, I.F.: Geospatial data integration. Department of Computer Science, University of Illinois, Chicago, ADVIS Lab (2004)
12. García-Castro, R., Gómez-Pérez, A., Munoz-Garcia, O.: The semantic web framework: a component-based framework for the development of semantic web applications. In: 2008 19th International Workshop on Database and Expert Systems Application, DEXA 2008, pp. 185–189. IEEE (2008)
13. Gruber, T.R.: A translation approach to portable ontology specifications. Knowl. Acquis. **5**(2), 199–220 (1993). http://dx.doi.org/10.1006/knac.1993.1008
14. Hacherouf, M., Bahloul, S.N., Cruz, C.: Transforming XML documents to OWL ontologies: a survey. J. Inf. Sci. **41**(2), 242–259 (2015). http://dx.doi.org/10.1177/0165551514565972
15. Howard, M., P.S.S.R.: Technical guidance for the INSPIRE schema transformation network service. version: 3.0, ec JRC contract notice 2009/s 107–153973 (2010)
16. Karmacharya, A.: Introduction of a spatial layer in the Semantic Web framework: a proposition through the Web platform ArchaeoKM. Ph.D. thesis, Le2i Laboratoire Electronique, Informatique et Image, University of Bourgogne (2011)
17. Lenzerini, M.: Data integration: a theoretical perspective. In: Proceedings of the Twenty-First ACM SIGMOD-SIGACT-SIGART Symposium on Principles of Database Systems, PODS 2002, pp. 233–246. ACM, NY, USA (2002). http://doi.acm.org/10.1145/543613.543644
18. Müller, H., Würriehausen, F.: Semantic interoperability of German and European land-use information. In: Murgante, B., Misra, S., Carlini, M., Torre, C.M., Nguyen, H.-Q., Taniar, D., Apduhan, B.O., Gervasi, O. (eds.) ICCSA 2013, Part III. LNCS, vol. 7973, pp. 309–323. Springer, Heidelberg (2013). http://dx.doi.org/10.1007/978-3-642-39646-5_23
19. Navigli, R., Ponzetto, S.P.: Babelnet: building a very large multilingual semantic network. In: Proceedings of the 48th Annual Meeting of the Association for Computational Linguistics, pp. 216–225. Association for Computational Linguistics (2010)

20. OGC: OGC geosparql - a geographic query language for RDF data. Technical report (2011)
21. Rase, D., Björnsson, A., Probert, M., Haupt, M.: Reference data and metadata position paper. Inspire RDM PP v4-3 en. European Commission, Joint Research Centre (2002)
22. Waters, R., Beare, M., Walker, R., Millot, M.: Schema transformation for INSPIRE. Int. J. Spat. Data Infrastruct. Res. **6**, 1–22 (2011)
23. Specifications, D.T.D.: D2.3: definition of annex themes and scope v3.0., European union (2008)
24. Stadler, C., Lehmann, J., Höffner, K., Auer, S.: Linkedgeodata: a core for a web of spatial open data. Seman. Web J. **3**(4), 333–354 (2012). http://jens-lehmann.org/files/2012/linkedgeodata2.pdf
25. Tschirner, S., Scherp, A., Staab, S.: Semantic access to INSPIRE. In: Terra Cognita 2011 Workshop Foundations, Technologies and Applications of the Geospatial Web, p. 75. Citeseer (2011)
26. Tsinaraki, C., Christodoulakis, S.: XS2OWL: a formal model and a system for enabling XML schema applications to interoperate with OWL-DL domain knowledge and semantic web tools. In: Thanos, C., Borri, F., Candela, L. (eds.) Digital Libraries: Research and Development. LNCS, vol. 4877, pp. 124–136. Springer, Heidelberg (2007)
27. Würriehausen, F., Karmacharya, A., Müller, H.: Using ontologies to support land-use spatial data interoperability. In: Murgante, B., et al. (eds.) ICCSA 2014, Part II. LNCS, vol. 8580, pp. 453–468. Springer, Heidelberg (2014). http://dx.doi.org/10.1007/978-3-319-09129-7_34
28. Zhao, T., Zhang, C., Wei, M., Peng, Z.-R.: Ontology-based geospatial data query and integration. In: Cova, T.J., Miller, H.J., Beard, K., Frank, A.U., Goodchild, M.F. (eds.) GIScience 2008. LNCS, vol. 5266, pp. 370–392. Springer, Heidelberg (2008). http://dx.doi.org/10.1007/978-3-540-87473-7_24

Identification of Transport Vehicle for Mobile Network Subscribers

Alexander Derendyaev[(⊠)]

Institute for Information Transmission Problems (Kharkevich Institute),
Russian Academy of Sciences, Moscow, Russia
wintsa@gmail.com

Abstract. We present a method for identification of transport vehicle via the cell phone users using the data of the mobile operator. The method is based on a model that allows to calculate the approximate speed on road sections and to estimate congestion of different transport network sections. The corresponding algorithm is implemented on the GIS platform GeoTime 3. Experimental results for the road network of Moscow city and Moscow region are discussed.

Keywords: Base stations · Transport flows · Dynamic GIS GeoTime 3

1 Introduction

Increased mobility of modern society leads to an overload of some segments of transport networks. The solution is to improve the management of such networks. This requires the development of means of determining the characteristics of a network traffic in real time, which can result in significant costs for installation and maintenance of infrastructure. For evaluating the characteristics of a network load in real-time with a good accuracy and at an affordable price it is necessary to use new monitoring technologies. Widespread mobile systems are currently considered as one of the most promising technologies for collecting such information. Mobile devices are being moved together with the vehicles through the wireless mobile fields: GPS, Wi-Fi and cellular communication. This provides anonymous monitoring of urban dynamics.

In a modern city there are a lot of overlapped transport networks, such as road network, railways and metro. Mobile phone user can travel in any network or even in several networks successively. To determine the load of different segments of networks, we have to identify which network is each mobile phone user exploiting: is he moving or not, is he moving by foot or by a transport, and what kind of transport is he using.

Recently a number of papers has been published which study the monitoring of the traffic flow velocity by means of cellular communication. Base station (BS) of the mobile network fixes the times of the cell change signals from the mobile phone. Mobile network operator fixes the BS covering the phone location. Relying on the sequence of BS switch signals it is possible with some degree of certainty to determine the route of the phone user and his speed. Lowering of the road speed below a certain level in a section of the road network may indicate congestion and traffic jams.

As far as we know, this problem wasn't yet studied. In [1, 2], the authors propose the methods for analysis of traffic jams for a separate road along which cell stations are

© Springer International Publishing Switzerland 2016
O. Gervasi et al. (Eds.): ICCSA 2016, Part III, LNCS 9788, pp. 256–264, 2016.
DOI: 10.1007/978-3-319-42111-7_20

arranged. In [3, 4], the methods for analysis of the motor transport velocity on inter-sections of several highways were considered. In [5, 6] practical approaches for assessing the urban dynamics and transport routes were suggested. SVN method for suburban traffic is discussed in [7]. In the paper [8], a statistical approach to limited mobile operator data is used to determine home and word positions.

In this paper we propose a new method of estimating vehicle speed using the mobile operator data for the entire road network of a metropolis. The corresponding algorithm is implemented on the GIS platform GeoTime 3 [9, 10] and is experimentally tested on he data of one mobile network operator for Moscow city and Moscow region transport network.

2 Method

Let there be a set of antennas, B, of BS of mobile network operator (MNO) in a certain spatial region. For each antenna of each BS the position and radiation pattern is known. Let there are a set U of all mobile phones connected to this MNO. The set U consists of phones in the cars, other vehicles and pedestrians. Mobile communication signals from base stations recorded in the database CDR (Call Data Record) of mobile operator. Additionally the geographically localized graph of the road network is known. To solve the problem of determining the velocity of road transport it is required to select a subset $U_1 \in U$ of mobile phones in cars according to the CDR, then to assess their position on the graph of the roads and define the path they traversed.

The CDR database stores data about every antenna on BS and data on binding the mobile phones to them. The CDR extraction includes the following fields:

1. The time of the event.
2. The phone ID: the hash of the phone number.
3. CID, the base station number in a subnetwork.
4. LAC, the subnetwork number. Any BS is identified from the CID/LAC pair.
5. The type of an event.
6. Geographic coordinates of the base station.
7. The type of the base station (OUTDOOR, INDOOR, METRO).
8. The direction angles of the BS antenna.

The main problem in distinction of different kinds of transport is that the road traffic can have different speeds: from as fast as the trains have, to as slow as the pedestrians do. To assess the velocities at different road segments, we construct a special method. In short, the method consists in finding maximum with respect to V of joint distribution $f(X, V)$, where V is velocities and $X = \{B, T\}$ is input data: list of events with number of base station and time of event.

$$V^* = \operatorname*{argmax}_{V} f(V|X) = \operatorname*{argmax}_{V} \frac{f(X, V)}{g(X)} = \operatorname*{argmax}_{V} f(X, V)$$

This function cannot be written in explicit form because of latent variables z - positions and states of phone users.

Let take

$$\varphi_{su}(z, X, V) = -\ln(p_{su}(z, X, V))$$

as a penalty function for user u in the state s with given z, X, V. Function $p_{su}(z, X, V)$ will be described later. As each user can change his state during observation several times, so penalty function for user u is

$$\varphi_u(z, X, V) = \sum_s \varphi_{su}(z, X, V)$$

Here s – states of user u. User can be in one state during his route, can change it once, or can change it several times. Let total penalty be the sum of penalty of all users:

$$\varphi(z, X, V) = \sum_u \varphi_u(z, X, V)$$

Considering $\varphi(z, X, V)$ as potential function of z we can find it should obey Gibbs distribution. Then it is possible to write

$$f(X, V) = \int e^{-\varphi(z, X, V)} dz$$

Summarizing all together to find velocities V^*

$$V^* = \underset{V}{\operatorname{argmax}} \int \prod_{s,u} p_{su}(z, X, V) dz$$

The standard method of optimization of such functions is EM [11]. On the E-step for each user we independently calculate all possible routes with respect to their probabilities given a priory V. On the M-step, we use only parts that associated with automobiles to get new velocities.

Function $p_{su}(z, X, V)$ depends on user state and for automobile we take it as

$$p_{su}(z, X, V) = A(z|X) B(X) \prod_i \exp\left(-\frac{(t(z_{i-1}, z_i, V) - \Delta t_i)^2}{2C^2}\right)$$

Here Δt_i – time difference between two successive events registration, $t(z_{i-1}, z_i, V)$ – transit time along the fastest path between these two events, C –parameter of methods, $A(z|X)$ – factor that doesn't depend on V and describes a priori probability for user location in the moment of BS change. It's supposed to be a 2 dimensional Gaussian with center exactly between two stations and deviations proportional to the distance between these stations. Function $B(X)$ doesn't depend on z or V and has no effect on this method. Transit time is calculated as sum of time as pedestrian to and from graph with velocity 5 km/h and transit time along the fastest path over the graph given the velocities V.

Selection of function $p_{su}(z, X, V)$ is arbitrary and requires some additional study of the form of the velocity distribution of cars on highways and urban roads.

It is rather clear how to include the other transport networks in this construction. For example, for pedestrian the function looks similar, except that the transit time $t(z_{i-1}, z_i, V)$ doesn't depend on V and is calculated as $t(z_{i-1}, z_i, V) = d/v_p$, where d is the distance between z_{i-1} and z_i, and $v_p = 5$ is supposed to be a pedestrian speed.

For those users that do not move, the function is

$$p_{su}(z, X, V) = 1/(\sigma_x \sigma_y)$$

Here σ_x and σ_y are deviations of a 2D Gaussian, which is a product of several Gaussians $A(z|X)$.

The functions for the states that are associated with metro and railways are calculated similarly to the automobile state. The main difference is that the velocities are fixed: 50 km/h for metro and 60 km/h for railways.

Because of the complexity of $p_{us}(z, X, V)$ it's not possible to check every z value, so MCMC algorithm is used. Samples are calculated from this function and are used for M-step of EM cycle. For this algorithm, function $\varphi_{su}(z, X, V) = -\ln(p_{us}(z, X, V))$ is a penalty function. For each generated route we compare with previously accepted route this penalty functions and we certainly accept new route if new value less than old one or we accept it with probability that equal to their ratio.

MCMC part of route generation and selection is still the most time and resource consuming, so we used an optimization technic such as quadtree for graphs holding, IDA* with cashing for search the fastest way over the graph and parallelization.

Complete algorithm consist of next phases:

1. Data preparation, such as graphs for automobile roads, railways and metro and corresponding quadtrees. Also user data are prepared: we take into account only the users that contain not less than 3 BS changes.
2. E-step. Do samples generation in parallel way given some velocities V:
    (a) Generate random route for current user.
    (b) Generate another random route based on last accepted route.
    (c) Check new route and accept or not it.
    (d) Return to (b) if number of accepted route is less than 100.
3. M-step. Using all samples for automobile users maximization with respect to V is performed.
4. If relative change of sum of penalty function of V is less than 1 %, EM cycle is stopped and results are prepared for visualization, else return to E-step with new velocities V.

3 Application

This algorithm has been implemented as a set of plugins for GIS GeoTime 3 [9, 10]. The GeoTime 3 system enables a user to measure interactively histograms of speeds of road sections and other techniques for analyze possible routes, such as selection and showing possible routes for special user or for special road section.

The CDR database for Moscow city and Moscow region, which is analyzed in this paper, includes 230 million events per day. It records from 44 thousand events per minute (at night) to 311 thousand (in the evening). The CDR database includes many types of events. For simplicity, in this study, we consider only LocationUpdate events, which take place on a handover of a cell, switching on/off the phone, or a timeout. The fraction of such events in the CDR is 69 %. In this work type of base stations and direction angles of antennas were not used.

The localized graph of roads was obtained from the site http://www.openstreetmap.org/ (OSM). The attributive information of the files has the following form:

- The ID in the OSM site.
- The direction of motion (one-way or two-ways traffic).
- The type of the road (pedestrian walkway; service road; 1st, 2nd, or 3rd class road; highway; etc.).
- Maximum speed.
- Number of lines.

In this paper specially prepared directed graph (taking into account directions of motion) was used. This graph contains only automobile roads and does not include pedestrian walkways and service roads. In the end, this graph was cleaned from inaccessible parts. The location of the base stations according to the CDR extraction and the road network in Moscow and the Moscow region are shown in Fig. 1.

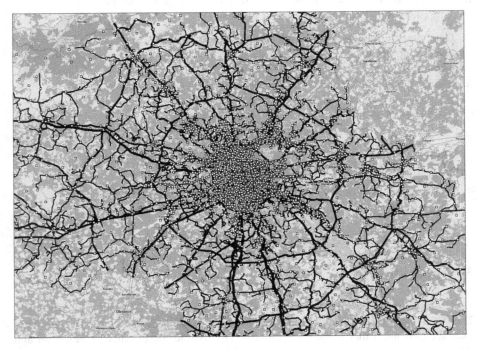

Fig. 1. The location of the BS and the road network in Moscow and the Moscow region. Line thickness corresponds to the road class.

The localized graph of metro and railways network was obtained from the same source and analogous directed graphs with analogous preparation for both metro and railways were used (Fig. 2).

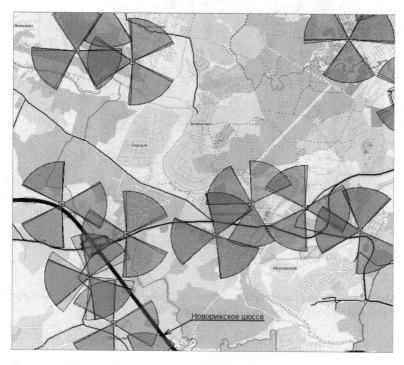

Fig. 2. The map of base station arrangement with the radiation patterns of their antennas for a section of roads in Moscow region.

For experiment, we use only 10 % of users because of slow route generation. In that case it takes about 2 h on 8 cores processor to complete calculation. Our results were compared with results of online traffic service "Yandex Jams". We use this service because of its robustness: it merge data from several sources, including GPS tracks of taxi and users of mobile application "Yandex Jams" (in Moscowthis application is popular). We constructed 2D histogram shown on Figs. 3 and 4. On this figure we have sum of road section length with some speed from Yandex (horizontal axis) and with some speed from suggested algorithm (vertical axis). Correlation between speeds with respect to road section length is about 45-47 % over several experiments.

Unfortunately, there are no other data to compare with.

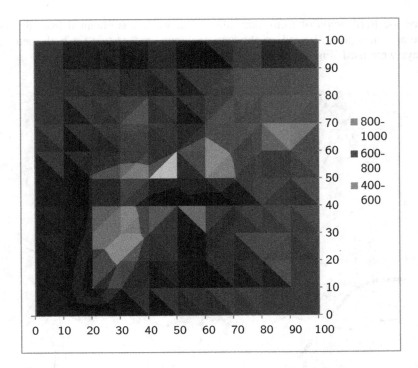

Fig. 3. The 2D histogram of road section lengths with velocities from Yandex (horizontal axis) and suggested algorithm (vertical axis). (Color figure online)

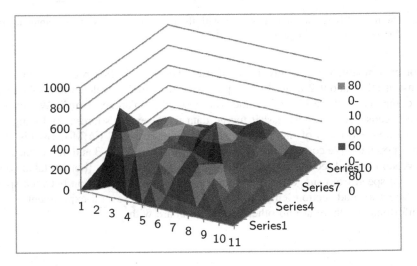

Fig. 4. The 3D histogram of road section lengths with velocities from Yandex (horizontal axis) and suggested algorithm (vertical axis). (Color figure online)

4   Conclusions

In this study, a new method for the analysis of the transport networks of a big city using the data of a mobile network operator is proposed. It is shown that this method enables one to correctly assess transport speed in different road sections and to estimate the number of users of different transport networks. The corresponding algorithm has been implemented as a set of plugins for GIS GeoTime 3. This allows to dynamically display the possible routes of individual users and visualize the transport speed mode (mode of velocity distribution on road sections) as a map.

Further studies are limited to the complexity of obtaining initial data. For example, complementing the initial data with a public transport map will allow detection of phone users using public transportation system. Using angles of antennas will allow to calculate the borders of BS sectors and these data will allow more accurately positioning the users at the moment of BS change. Complementing the initial data with a height map will allow to find out more realistic borders and user positions. To verify the model, it's necessary to have some reliable test data, for example, the speeds from video cameras. More comprehensive analysis (especially using metro loading) will allow to clarify the model.

Acknowledgments. The research is implemented in IITP RAS and supported by RSF project 14-50-00150. The author is grateful to S.A. Pirogov for helpful discussions of this work.

References

1. Caceres, N., Wideberg, J.P., Benitez, F.G.: Deriving origin–destination data from a mobile phone network. IET Intel. Transport Syst. **1**(1), 15–26 (2007)
2. Janecek, A., Valerio, D., Hummel, K., Ricciato, F., Hlavacs, H.: Cellular data meet vehicular traffic theory: location area updates and cell transitions for travel time estimation. In: UbiComp 2012, Pittsburgh, USA, 5–8 September 2012
3. Bar-Gera, H.: Evaluation of a cellular phone-based system for measurements of traffic speeds and travel times: a case study from Israel. Transp. Res. Part C **15**, 380–391 (2007)
4. Alger, M.: Real-time traffic monitoring using mobile phone data. Vodafone. http://www.maths-in-industry.org/miis/30/1/TrafficMonitoring.pdf
5. Calabrese, F., Colonna, M., Lovisolo, P., Parata, D., Ratti, C.: Real-time urban monitoring using cell phones: a case study in Rome. IEEE Trans. Intell. Transp. Syst. **12**(1), 141–151 (2011)
6. Tettamanti, T., Demeter, H., Varga, I.: Route choice estimation based on cellular signaling data. Acta Polytech. Hung. **9**(4), 207–220 (2012)
7. Lv, W., Ma, S., Liang, C., Zhu, T.: Effective data identification in travel time estimation based on cellular network signaling. In: Wireless and Mobile Networking Conference (WMNC), 4th Joint IFIP, pp. 1–5 (2011)
8. Berlingerio, M., Calabrese, F., Di Lorenzo, G., Nair, R., Pinelli, F., Sbodio, M.L.: AllAboard: a system for exploring urban mobility and optimizing public transport using cellphone data. In: Blockeel, H., Kersting, K., Nijssen, S., Železný, F. (eds.) ECML PKDD 2013, Part III. LNCS, vol. 8190, pp. 663–666. Springer, Heidelberg (2013)

9. Gitis, V., Derendyaev, A., Metrikov, P., Shogin, A.: Network geoinformation technology for seismic hazard research. Nat. Hazards **62**(3), 1021–1036 (2012)
10. Gitis, V.G., Derendyaev, A.B.: Geoinformation technology for spatio-temporal processes research. In: Proceedings of the ISPRS Workshop on Dynamic and Multi-dimensional GIS, Shanghai, China, pp. 111–115, October 2011
11. Borman, S.: The expectation maximization algorithm: A short tutorial (2004). http://www.seanborman.com/publications/EM_algorithm.pdf

Light Penetration Ability Assessment of Satellite Band for Seagrass Detection Using Landsat 8 OLI Satellite Data

Syarifuddin Misbari[1] and Mazlan Hashim[1,2,3(✉)]

[1] Faculty of Geoinformation and Real Estate (FGRE),
Universiti Teknologi Malaysia, UTM, 81310 Johor Bahru, Malaysia
syr7din@yahoo.com, mazlanhashim@utm.my
[2] Geoscience and Digital Earth Centre (INSTeG),
Research Institute for Sustainable Environment (RISE),
Universiti Teknologi Malaysia, UTM, 81310 Johor Bahru, Malaysia
[3] FGHT, Universiti Teknologi Malaysia, UTM,
81310 Johor Bahru, Johor, Malaysia

Abstract. Seagrass distribution is controlled by light availability, especially at the deepest edge of the meadow. Light attenuation due to both natural and anthropogenically-driven processes leads to reduced photosynthesis. Reliability of satellite-based seagrass mapping under different water clarity that has different attenuation coefficient value is still not fully known. Understanding the minimum light requirements for growth is crucial when light conditions are insufficient to maintain a positive carbon balance, leading to a decline in seagrass growth and distribution. By comparing the seagrass-detected pixels at two different coastal locations with the corresponding depth from the nautical chart, the assessment of seagrass map derived from Landsat 8 OLI satellite data were performed. We presented the assessment of light penetration capability of Landsat 8 OLI bands in typical tropical coastal water of Malaysia, with special attention on the different water clarity that has different amount of light deprivation on the seagrass meadow.

Keywords: Light attenuation · Seagrass · Satellite · Coastal

1 Introduction

Light availability is one of the controlling factors of seagrass distribution, especially at the deepest edge of the meadow. Pigment content, morphological structure and physical properties are regulating factors of light absorption by leaf of seagrass. While light become less intensify towards the sea bottom, chlorophyll content and morphological characteristics of leaves such as leaf thickness also can change at the deepest edge. Minimum light requirements of seagrasses (2−37 % of surface irradiance, SI) are much higher than those of macroalgae and phytoplankton (about 1−3 % of SI) [1, 2].

Globally, seagrass habitats occupy vast tracts of shallow temperate and tropical coastal waters. Over the past century, human activity has had a profound negative impact on seagrass habitats. Environmental pressure such as intensive coastal

© Springer International Publishing Switzerland 2016
O. Gervasi et al. (Eds.): ICCSA 2016, Part III, LNCS 9788, pp. 265–274, 2016.
DOI: 10.1007/978-3-319-42111-7_21

development brought dramatic changes to their spatial distribution and its meadow composition at global scale [3]. Seagrasses are particularly sensitive to reductions in light availability, where small decreases can cause significant declines in growth and distribution. Some of the best documented losses of seagrass due to light limitation have occurred in Australia [4–7].

Seagrass typically found in coastal water with good water clarity. However, it can survive on the seafloor in less water clarity, for an instance along the west coast of Malaysian Peninsular. As documented by previous studies [8], seagrass can survive to actively grow at the various depths, predominantly according to the acceptance of light intensity for them being able to survive in euphotic zone. In tropical country including Malaysia, the light availability is commonly intense for more than six hours daily where this can be benefited by the seagrass ecosystem to be more stable and has high endurance in terms of its extent and alteration of its habitat; with respect to the meteorological changes in such coastal condition. Compared to cold-weathered country and temperate coastal habitat, seagrass in tropical country is logically less vulnerable to the loss of its habitat due to light availability, except for monsoon wind and typhoons.

Radiative transfer model separates the reflectance due to substrate from of the water column. Lambert-Beers Law normally used in which the radiance in wavelength i at water depth Z is a function of radiance observed over deep water (L_s), the bottom reflectance (rBi) and the effective attenuation coefficient of the water (K_i). Seafloor mapping including seagrass has extensively conducted around the world using various methods that can be simplified into two approaches; (a) ground-based method and; (b) satellite-based sea bottom mapping; either close-range like SONAR or distance sensor like satellite and Unmanned Aerial Vehicle (UAV). For large study area, remote sensing approach using satellite data is most effective at required scale and accuracy, similar to forestry studies and mineral exploration from satellite data [9, 10]. The major focus of this study is to determine the depth light penetration that enable remote sensing approach could be performed to identify the presence of seagrass occurrences in variety of water clarity (clear to turbid water), even in the tropics which received high intensity of solar radiation. The aim of this paper is to discover how deep the seagrass habitat could be identified by the medium resolution (30 × 30 m pixel) of satellite image in the tropical coastal water. Understanding the penetration ability of selected band of remote sensing data, the potential of a satellite data specifications for seagrass mapping could be optimally known from the current operational systems or future planned systems.

2 Material

Landsat 8 OLI satellite data is the main source of data in the study. In order to test the penetration ability of the selected band, the attenuation coefficient of the study area is calculated. By extracting more than 30 random points of sand or muddy surface from the satellite image, together with the corresponding depth value, the relationship plot between these two variables is generated. The blue and red band was selected because the blue band (0.45 − 0.51 μm) has very powerful penetration ability into the water column while red band (0.64 − 0.67 μm) has high sensitivity of slight changes on the sea surfaces. The combination of these two bands is effective to detect the appearance of submerged seagrass and other shallow substrate features, especially in less water clarity. In order to assess and validate the penetration ability of satellite data in detecting submerged seagrass in water condition with low light transparency, one more seagrass habitat in ideal coastal environment was selected. The coastal environment where seagrass are found is subjected to less coastal alteration, characterize with high water quality and suitable for spawning grounds of many marine life including endangered species like sea horse and dugong. Two coastal sites (Fig. 1), namely Merambong shoal and Tinggi Island in Johor, Malaysia was selected as study area in this study.

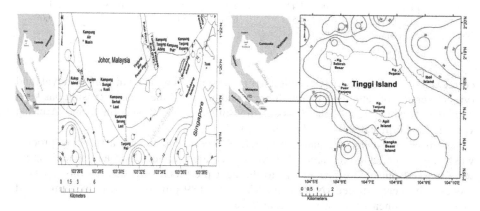

Fig. 1. Study sites: Merambong shoal (left) and Tinggi Island (right), representing less and high water clarity. Hereafter, both the area are referred to as M-area and TI-area, respectively. Depths (in meter) are shown by blue isolines. (Color figure online)

3 Method

There are two approach of techniques have been so far reported for detecting and mapping submerged seagrass from satellite remote sensing data. These, include the Depth Invariant Index (DII) [11], and Bottom Reflectance Index (BRI) [12].

In this paper, BRI technique is used as this approach is more robust in less clear water [13], indicating that it optimally work well in less clear water with high sedimentation and high light attenuation towards the seafloor. Thus, it is more suitable to test the robustness in both types of water clarities. The water quality parameter of the test area is tabulated in Table 1.

3.1 Satellite Data Processing

Only satellite data of less than 10 % cloud were chosen for experiment. Cloud-affected scenes even within thin haze are potentially degrade the data quality. Prior to the main data processing, the satellite data were subjected to data-preprocessing tasks and image preparations. This data pre-processing stage includes: (i) image subset, (ii) geometric

Table 1. Measurement of water quality parameters at 36 points around Merambong area (M) and Tinggi Island (TI).

Parameter		Temperature (°C)	pH	Conductivity (mS/cm)	Turbidity (NTU)	Dissolved oxygen (mg/L)	Total dissolved solid (mg/L)	Salinity (ppt)
Mean	M	30.03	7.96	41.01	15.09	4.72	25.14	26.17
	TI	30.87	8.21	37.37	5.32	6.41	23.54	25.44
Minimum	M	29.32	7.72	36.10	4.50	3.53	22.90	22.10
	TI	30.26	7.98	14.80	0.55	6.03	9.39	8.30
Maximum	M	31.42	8.13	43.10	46.60	8.53	26.30	27.70
	TI	31.59	8.27	39.1	25.10	12.9	23.9	24.9
*SD	M	0.47	0.08	1.44	7.41	0.88	0.64	1.17
	TI	0.40	0.05	4.51	6.21	1.65	2.70	3.08

*SD: standard deviation

correction, (iii) atmospheric correction, (iv) image masking, (v) sun glint removal, and (vi) conversion of satellite digital number to radiance. All the data processing has been carried out using digital image processing software, ENVI version 5 and ArcMap version 10. Further details can be referred to [14].

From the geometrically-and-atmospherically corrected image, a masking process was applied to exclude the areas that are not required. Land, cloud and shadow areas were masked out before proceeding to the next pre-processing step. In this masking process, the near infrared (NIR) band 5 (0.76 – 0.89 µm) was used since this band gives good delineation between land and water.

Sun glint removal was conducted according to [15], using near infrared (NIR) band. The NIR band was chosen because it exhibits maximum absorption and minimizes water-leaving radiance in clear waters. The linear relationship between NIR and visible band is performed using linear regression based on sample selected. The linear relationship was performed between visible and NIR band. When the linear relationship is known, the glint effect can be derived from NIR value and subtracted from the pixel to obtain the glint-free image. Removal of sun glint effect, R_i' is performed, such that:

$$R_i' = R_i - b_i(R_{NIR} - Min_{NIR})$$ (1)

where R_i is the pixel value in band i; R_{NIR} is the pixel value in NIR band; Min_{NIR} is the minimum pixel value in NIR band and b_i is the regression slope derived from visible and NIR band.

Next, the image was converted to radiance value (L_λ) in order to perform radiometric correction (see Table 2). The rescaling gains and biases for Landsat 8 OLI satellite data were obtained in order to proceed to image processing, as [16] stated that:

$$L_\lambda = M_L Q_{cal} + A_L$$ (2)

where L_λ is TOA spectral radiance (Watts/(m^2 * srad * µm)); M_L is Band-specific multiplicative rescaling factor from the metadata; A_L is Band-specific additive rescaling

Table 2. Conversion of digital number (DN) of Landsat 8 OLI to radiance (L_i) unit.

Band	Digital number to water radiance, (L_i) conversion
Blue (0.45 − 0.51 μm)	1.2596E-02*DN Band 2 + −62.97902
Red (0.64 − 0.67 μm)	9.7876E-03*DN Band 4 + −48.93807

Table 3. Deep water radiance (L_{si}) value of selected band is required in BRI to indicate the most clear water point that gives smallest reflected radiance to the satellite sensor due to greatest absorption of propagated signal in deep water and atmospheric perturbation.

Band	Deep water radiance, (L_{si})	Latitude (N)	Longitude (E)
Blue (Merambong)	61.8373	1°21'37.97"N	103°32'14.52"E
Red (Merambong)	22.1798	1°21'44.85"N	103°32'32.24"E
Blue (Tinggi Island)	52.3500	2°17'41.84"N	104°05'21.70"E
Red (Tinggi Island)	12.9196	2°15'30.97"N	104°06'46.27"E

factor from the metadata; and Q_{cal} is Quantized and calibrated standard product pixel values (Table 3).

3.2   Retrieval of Seagrass Features

The converted radiance (L_λ) was then further processed to retrieve the substrate-leaving radiance using BRI method. The BRI is given in Eq. (3), such that

$$BRI = \frac{(L_i - L_{si})}{[\exp(-K_i g Z)]} \tag{3}$$

where,

L_i = measured radiance in band i;
L_{si} = deep-water radiance in band i;
K_i = attenuation coefficient for band i;
g = geometric factor to account for the path length through water, and
Z = water depth (m).

4   Results

The attenuation coefficient for all the 30 points used in the light attenuation from the satellite data computed are plotted spatially as their absolute location, as depicted in Fig. 2 for both M and TI area. The average coefficients for both these area is tabulated is Table 4. The water-leaving radiance of the blue and red bands were plotted against the corresponding depths, and these are illustrated in Fig. 3 below.

The identified seagrass pixels detected for both M and TI areas using BRI technique is shown in Fig. 4 below, respectively.

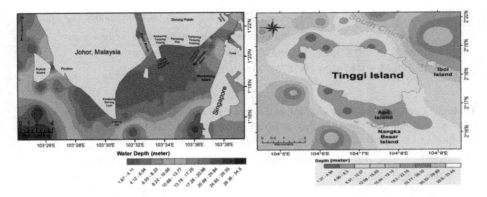

Fig. 2. Location of 30 random points in M and TI areas. Depth of water at both areas plus tidal height during satellite passes. Tidal height on 27th June 2013 at Merambong is +1.5 m; while on 13th May 2014 during satellite passes Tinggi Island, tidal height is +0.45 m. Based on the Snell's Law and satellite-earth geometry, the *g* value of Landsat 8 OLI scene of Merambong area is 2.0145, whereas Tinggi Island is 2.0994. (Color figure online)

Table 4. Computed attenuation coefficients of M and TI areas.

Attenuation coefficient	Merambong	Tinggi Island
Blue band (K_i)	0.0942	0.0834
Red band (K_k)	0.4656	0.1010

5   Discussion

The penetration ability of specific wavelength enabled the electromagnetic signal to reach submerged seagrass even through small ratio of the signal was scattered back by floating solid particles or dissolved suspended materials before reaching the seagrass physical structure. Along the M area, the seagrass can be identified in less clear water. Nevertheless, BRI robustness in detecting seagrass-dominated pixel of satellite data of 30 m × 30 m pixel size is limited only up to ≤7 m at euphotic zone where the turbidity level more than 45 NTU.

On the other hand, capability of the BRI method is proved to meet the expected penetration power where the seagrass is successfully identified at the area more than ≤15 m around the TI area that are surrounded by pristine non-polluted water as the area is gazetted as Johor Marine Park, one of 22 islands officially recognized as the protected Marine Park in Malaysia. Thus, the seagrass map accuracy around this place is far more accurate than Merambong, but BRI still relatively good in detecting seagrass in less clear tropical water where the great light attenuation occurred vertically towards sea bottom.

Using field data collection at both areas, the robustness of BRI in detecting submerged seagrass is tested. Based on the supervised classification results, the seagrass insitu points are overlaid to either fall into seagrass-detected pixel or pixel of other substrates. Based on this testing method, it is proved that BRI can work relatively well in both clear and less clear water as the depth and tidal information is important parameters considered in this method, apart from best combination of satellite band

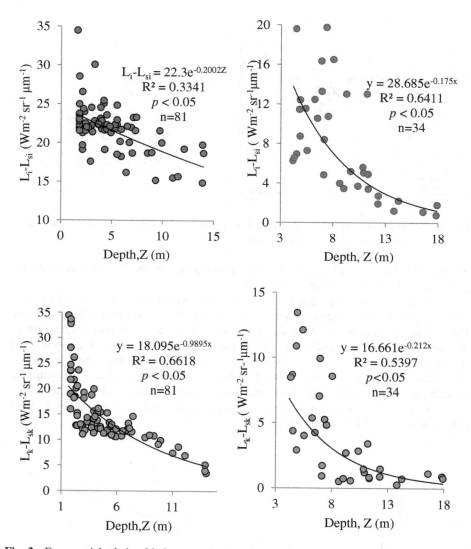

Fig. 3. Exponential relationship between depth and water-leaving radiance of blue and red band of Landsat 8 OLI at both study area. Merambong area has higher maximum radiance than Tinggi Island due to floating particles in less clear water, but Tinggi Island is relatively deeper than Merambong, close to South China Sea. (Color figure online)

used for image classification. At M-area, the maximum turbidity level that permits detection of submerged seagrass is ≤ 45 NTU and depth can reach up to ≤ 7 m. If the sea bottom is very shallow but the NTU value is more than 45, the seagrass is not detectable from Landsat 8 OLI. Seagrass around TI-area can be detected since the water has low sedimentation and good water clarity, but the bottom depth of ≥ 20 m is the limiting factor of seagrass to be detected, besides the low density of seagrass in deep water. Under low-light conditions, the leaf size is usually decreased rather than

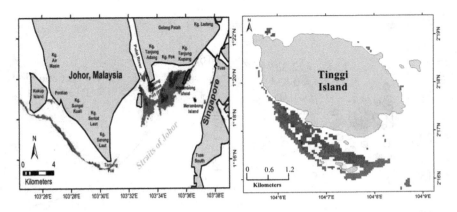

Fig. 4. Different range of BRI indicates different range of seagrass densities. Darker green indicates higher density. (Color figure online)

increased [17] which will reduce the respiratory demand of the shoot yet decrease the photosynthetic capacity of the leaves [18].

Seagrass that has longest leaf length in Malaysia, namely *Enhalus acoroides* (*Ea*) is easily found in Johor coastal water but not frequently found around Tinggi Island. Other seagrass species in Malaysia is normally small and medium length, not as obvious as *Ea*. This is an advantage of Merambong area that rich with *Ea* where the seagrass has higher potentiality to be detected by satellite image even the water is less clear than Tinggi Island. The trade-off between morphological factor of seagrass, penetration ability of satellite sensor and water clarity is remain vague, determinant factor need to be checked in next study in order to understand the limitation of depth in detecting seagrass from satellite (Fig. 5).

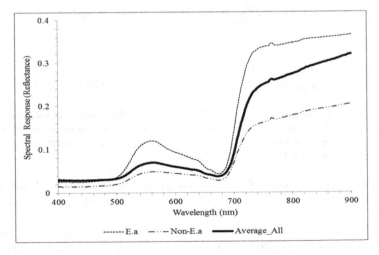

Fig. 5. Insitu reflectance of multiple species of seagrass in Malaysia. Compared to others, *Ea* is much likely easier to be identified from satellite data at deep area and low water clarity.

6   Conclusion

Seagrass is sensitive to changes in light availability across a range of spatial scales, including individual leaf responses, shoot-scale responses and alterations to the meadow structure. Using ground collection data, the limitation of seagrass detection from satellite data has been assessed at two areas with different coastal environment. Less clear water at M-area tends to have higher light attenuation coefficient, but the seagrass morphological structure helps to identify effectively the seagrass from satellite data, but limited to only area of ≤ 7 m depth and >45 NTU. In contrast, seagrass area around pristine TI-area is limited to bottom depth of ≥ 20 m with smaller seagrass physical structure compared to seagrass at Merambong.

Acknowledgements. We gratefully acknowledge a Long-Term Research Grant Scheme (LRGS)-Seagrass Biomass From Satellite Remote Sensing-R.J130000.7309.4B094, the sponsors of this study which was conducted under the network of the Asian CORE Program of the Japan Society for the promotion of Science, "Establishment of research and education network on coastal marine science in Southeast Asia", and the Ocean Remote Sensing Project for Coastal Habitat Mapping (WESTPAC-ORSP: PAMPEC III) of Intergovernmental Oceanographic Commission Sub-Commission for the Western Pacific supported by Japanese Funds-in-Trust provided by the Ministry of Education, Culture, Sports, Science and Technology in Japan.

References

1. Dennison, W.C., Orth, R.J., Moore, K.A., Stevenson, J.C., Carter, V., Kollar, S., Bergstrom, P.W., Batiuk, R.A.: Assessing water quality with submersed aquatic vegetation. Bioscience **43**, 86–94 (1993)
2. Lee, K.S., Park, S.R., Kim, Y.K.: Effects of irradiance, temperature, and nutrients on growth dynamics of seagrasses: a review. J. Exp. Mar. Biol. Ecol. **350**, 144–175 (2007)
3. Orth, R.J., Carruthers, T.J.B., Dennison, W.C., Duarte, C.M., Fourqrean, J.W., Heck Jr., K.L., Hughes, A.R., Kendrick, G.A., Kenworthy, W.J., Olyarnik, S., Short, F.T., Waycott, M., Williams, S.L.: A global crisis for seagrass ecosystems. BioScience **56**, 987–996 (2006)
4. Bulthuis, D.A.: Effects of temperature on the photosynthesis–irradiance curve of the Australian seagrass, *Heterozostera tasmanica*. Mar. Biol. Lett. **4**, 47–57 (1983)
5. Walker, D.I., McComb, A.J.: Seagrass degradation in Australian coastal waters. Mar. Pollut. Bull. **25**, 191–195 (1992)
6. Short, F.T., Wyllie-Echeverria, S.: Natural and human-induced disturbance of seagrasses. Environ. Conserv. **23**, 17–27 (1996)
7. Ralph, P.J., Tomasko, D., Seddon, S., Moore, K., Macinnis-Ng, C.: Human impact on seagrasses: contamination and eutrophication. In: Larkum, A.W.D., Orth, R.J., Duarte, C.M. (eds.) Seagrass Biology, Ecology and Conservation, pp. 567–593. Springer, Netherlands (2006)
8. Hossain, M.S., Bujang, J.S., Zakaria, M.H., Hashim, M.: The application of remote sensing to seagrass ecosystems: an overview and future research prospects. Int. J. Remote Sens. **26**, 2107–2112 (2015)

9. Okuda, T., Suzuki, M., Numata, S., Yoshida, K., Nishimura, S., Adachi, N., Niiyama, K., Manokaran, N., Hashim, M.: Estimation of aboveground biomass in logged and primary lowland rainforests using 3-D photogrammetric analysis. For. Ecol. Manage. **203**, 63–75 (2004)
10. Pour, A.B., Hashim, M.: Fusing ASTER, ALI and Hyperion data for enhanced mineral mapping. Int. J. Image Data Fusion **4**(2), 126–145 (2013)
11. Lyzenga, D.R.: Remote sensing of bottom reflectance and water attenuation parameters in shallow water using aircraft and Landsat data. Int. J. Remote Sens. **2**(1), 71–82 (1981)
12. Sagawa, T., Boisnier, E., Komatsu, T., Mustapha, K.B., Hattour, A., Kosaka, N., Miyazaki, S.: Using bottom surface reflectance to map coastal marine areas: a new remote sensing method. Int. J. Remote Sens. **31**, 3051–3064 (2010)
13. Hashim, M., Misbari, S., Yahya, N.N., Ahmad, S., Reba, M. N., Komatsu, T.: An approach for quantification of submerged seagrass biomass in shallow turbid coastal waters. In: Proceeding of IEEE Geoscience and Remote Sensing Symposium (IGARSS), Quebec City, Canada, pp. 4439–4442 (2014)
14. Misbari, S., Hashim, M.: Change detection of submerged seagrass biomass in shallow coastal water. Remote Sens. **8**(3), 200 (2016)
15. Hedley, J.D., Harborne, A.R., Mumby, P.J.: Technical Note: Simple and robust removal of sun glint for mapping shallow-water benthos. Int. J. Remote Sens. **26**, 2107–2112 (2005)
16. USGS Website (2013). http://landsat.usgs.gov/Landsat8_Using_Product.php. Accessed Nov 2015
17. Gordon, D.M., Grey, K.A., Chase, S.C., Simpson, C.J.: Change to the structure and productivity of a *Posidonia sinuosa* meadow during and after imposed shading. Aquat. Bot. **47**, 265–275 (1994)
18. Campbell, S.J., Miller, C.J.: Shoot and abundance characteristics of the seagrass *Heterozostera tasmanica* in Westernport estuary (south-eastern Australia). Aquat. Bot. **73**(1), 33–46 (2002)

Specifying the Computation Viewpoints for a Corporate Spatial Data Infrastructure Using ICA's Formal Model

Italo Lopes Oliveira[1]([⊠]), Jugurta Lisboa-Filho[1],
Carlos Alberto Moura[2], and Alexander Gonçalves da Silva[2]

[1] Department of Informatics, Federal University of Viçosa (UFV),
Viçosa, MG, Brazil
{italo.oliveira, jugurta}@ufv.br
[2] Companhia Energética de Minas Gerais (Cemig), Belo Horizonte, MG, Brazil
{camoura, ags}@cemig.com.br

Abstract. Spatial Data Infrastructure (SDI) is a concept that aids in discovering, sharing, and using geospatial data. Nevertheless, given the breadth of this concept, several ways of developing SDI have emerged. The International Cartographic Association (ICA) has proposed a formal model to describe SDIs irrespective of technologies and implementations by using three of the five viewpoints of the RM-ODP framework. However, the use of ICA's model to describe corporate SDIs had not been assessed yet. *Companhia Energética de Minas Gerais* (Cemig) seeks to develop an SDI, called SDI-Cemig, to facilitate the discovery, sharing, and use of geospatial data by its employees and partner companies. This study describes the specification of SDI-Cemig's components using the Computation viewpoint of ICA's formal model, which proved adequate. Although a single case study does not validate the model for every corporate SDI, this study showed that it can be used for this type of SDI.

Keywords: Spatial Data Infrastructure · ICA's formal model for SDI · RM-ODP · Cemig · SDI-Cemig

1 Introduction

Geospatial data are data referenced on Earth's surface and are essential in the decision-making process of an organization. However, according to [10, 13], geospatial data are a costly resource given the time and funds required to obtain them. The concept of Spatial Data Infrastructure (SDI) has emerged to help public and private organizations cut down costs in obtaining and using geospatial data.

Rajabifard and Williamson [13] define SDI as a stable environment formed by technologies and contributions, which the users employ to reach their goals. Harvey et al. [7] consider the SDI a concept that improves the sharing and use of geospatial data and services, which helps different users of a given community. Such communities may be organized hierarchically, as shown in Fig. 1, which facilitates sharing among communities.

© Springer International Publishing Switzerland 2016
O. Gervasi et al. (Eds.): ICCSA 2016, Part III, LNCS 9788, pp. 275–289, 2016.
DOI: 10.1007/978-3-319-42111-7_22

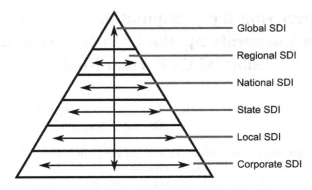

Fig. 1. SDI hierarchy – adapted from [5,13]

According to [8], the SDI concept is very broad, leading to different ways of development, both technical and organizational, which is also pointed out by [3]. In order to guarantee that the basic SDI concepts in the literature are contemplated during the specification phase, the International Cartographic Association (ICA) has developed a model to describe SDI irrespective of technologies or implementations using the concept of viewpoints from the RM-ODP (Reference Model for Open Distributed Processing) framework [8]. This model was later extended by Cooper et al. [4], Béjar et al. [1], and Oliveira and Lisboa-Filho [12].

Nonetheless, ICA's formal model for SDI has not been assessed in the development of corporate SDIs. *Companhia Energética de Minas Gerais* (Cemig) is a mixed-economy conglomerate controlled by the government of the state of Minas Gerais, Brazil, comprising over 200 companies. Cemig uses and produces large amounts of geospatial data. However, such data are currently documented and used in a non-standardized way within the company, with hinders sharing and discovering them.

The present study presents part of the SDI specification under development at Cemig (called SDI-Cemig), particularly the Computation viewpoint, using ICA's formal SDI model in order to verify whether this model is able do describe a corporate SDI. The other viewpoints approached by ICA's model (Enterprise and Information) will not be targeted by this study.

The remaining of the paper is structured as follows. Section 2 describes ICA's formal SDI model while detailing the Computation viewpoint. Section 3 presents the specification of the Enterprise viewpoint for SDI-Cemig. Section 4 discusses the results presented in this study and, finally, Sect. 5 presents the final considerations.

2   ICA's Formal SDI Model

According to [8], ICA's formal SDI model (henceforth referred to as ICA model) describes the SDI concepts irrespective of technologies or implementations using the RM-ODP framework.

RM-ODP, according to [6, 9, 14], is an architectural framework standardized by the ISO/IEC (International Organization for Standardization/International Electrotechnical Commission) used to specify heterogeneous distributed-processing systems. In order to facilitate the specification, RM-ODP uses the concept of viewpoints.

RM-ODP uses five viewpoints: Enterprise, Information, Computation, Engineering, and Technology. The use of viewpoints allows the system to be specified into smaller models, where each viewpoint answers relevant questions for different users of the system [9, 14]. Figure 2 presents the relationships among the five viewpoints in RM-ODP.

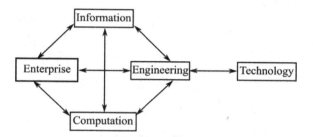

Fig. 2. Viewpoints of the RM-ODP framework and their relationships – adapted from [8]

The Enterprise viewpoint details the system's requirements and policies. The Information viewpoint details the semantics, behavior, and restrictions – which may be the policies of the Enterprise viewpoint – of the data in the system [6, 8]. According to [9], the Computation viewpoint describes the system components. However, the physical distribution of the components is not described in the Computation viewpoint, but instead in the Engineering viewpoint. Finally, the Technology viewpoint describes the technological artifacts described in the other viewpoints.

ICA's model describes only the viewpoints Enterprise, Information, and Computation. According to [3, 8], the other viewpoints heavily depend on the implementation and are not considered in ICA's model. The Computation viewpoint in ICA's model is detailed in the next subsection. The Enterprise and Information viewpoints will not be detailed since they are not relevant for this study.

2.1   Computation Viewpoint

According to [3], the Computation viewpoint "is a functional breakdown of the system modeled in a set of objects that interact through interfaces." However, during the modeling of the objects that make up the system, their physical distribution must not be considered, which will be in the hands of the Engineering viewpoint.

The RM-ODP framework determines that the interfaces of the computing objects must have a signature, a behavior, and an environment contract. To simplify the model, the behavior of the interfaces was modeled using the interfaces provided and required in the diagram and components of the UML, while the signatures and environment contracts have not been described [3].

Cooper et al. [3] identified six computational objects required in an SDI, as shown in Fig. 3: *SDI Data*; *SDI Portrayal*; *SDI Registry*; *SDI Processing*; *SDI Application*; and *SDI Management*. Each of these computational objects uses and provides a set of features through the interfaces, and their relationships are shown in Fig. 3. The connectors that have a circle are the interfaces provided, i.e., the features offered by the components. The connectors that have an arc are the required interfaces, i.e., the features the component needs to carry out one or more task and that are provided by other components.

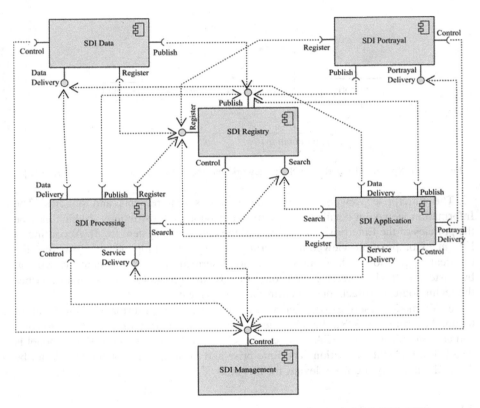

Fig. 3. Diagram of components for computational objects of the SDI – [3]

The *SDI Registry* is the component responsible for storing and registering the products (data and services), catalogs, metadata, product specifications, and SDI policies, besides allowing them to be searched. The component has a single required interface, *SDI Management::Control*, and three provided interfaces. According to Cooper et al. [3], the notation "::" indicates the interfaces of a component. Since all components use the *SDI Management::Control* interface, it will be omitted when the required interfaces of a component are described. The *SDI Registry::Register* interface is responsible for providing the features of the information an SDI may have. The *SDI Registry::Search* interface seeks the information required by the users through the catalogs registered in the SDI, while the *SDI Registry::Publish* interface publishes the information registered on the internet and other media, which allows the user to search for the information registered in the SDI.

The *SDI Data* component manages the datasets that have been shared and registered on the internet and has a single provided interface. *SDI Data::Data Delivery* is responsible for sending the data requested by the *SDI Application* and *SDI Processing* components. The data are requested after a search is performed using the *SDI Registry:: Search* interface. The *SDI Data* component requires the interfaces *SDI Registry:: Publish* and *SDI Registry::Register* since it is only able to deliver the data that are registered and published in the SDI. The component *SDI Processing* is responsible for processing the SDI data, containing services such as transformation of geographic coordinate systems, data analysis, and coordinate processing. Similarly to *SDI Data*, *SDI Processing* has a single provided interface, *SDI Processing::SDI Delivery*, which allows the other components to use the processing services offered by the component or SDI. *SDI Processing* uses the following interfaces: *SDI Data::Data Delivery* to obtain the data used by the processing services; *SDI Registry::Publish* and *SDI Registry:: Register* to automatically publish and register new geospatial data generated by the processing services; and *SDI Registry::Search*, used to search for new data whenever needed.

SDI Portrayal shows the geospatial data as static images when requested by the component *SDI Application*. It has a single provided interface, *SDI Portrayal::Portrayal Delivery* and requires the interfaces *SDI Registry::Register* and *SDI Registry:: Publish*. The component *SDI Application* is considered by Cooper et al. [3] the main component of the SDI and is the only component accessed by the user. In order to meet all the user's needs, *SDI Application* requires several interfaces and uses, at least, one interface from each of the other components, as shown in Fig. 3, besides not having any provided interface. The last component is *SDI Management*, which guarantees the other SDI components work through the interface *SDI Management::Control* as a guarantee of interoperability among the services and access rights of each user.

The authors of this study, however, disagree with some interfaces and relationships among the components proposed by [3]. The components *SDI Portrayal* and *SDI Data* use the interfaces *SDI Registry::Publish* and *SDI Registry::Register*, which do not match the features of the respective components. As described, the component *SDI Data* is the only one with direct access to the database and is responsible for delivering the data searched to the user. Since it carries out only this function, the component does not need to register or publish the data. Moreover, since it is the only component that accesses the database, *SDI Data* must provide an interface to handle (create, update, and remove) data from the SDI. Therefore, the interface *SDI Data::Data Manipulation* was added for data handling. The *SDI Portrayal* component returns the geospatial data searched such as maps, i.e., in the form of static images. In case the user wants to register or publish the map returned, it is the component *SDI Application*, and not *SDI Portrayal*, that must take up this responsibility. This way, both *SDI Data* and *SDI Portrayal* become more independent from *SDI Registry*, which facilitates its maintenance and use by other SDIs.

The component *SDI Processing*, however, kept the required interfaces *SDI Registry::Register* and *SDI Registry::Publish* for the sake of efficiency and, for the same motive, got the interface *SDI Data::Data Manipulation*. In case the services are chained, i.e., several services are executed in sequence with little or no user intervention, e.g., whether a service needs the previous service to publish the data processed to access them, that can be done with no need for going through the component SDI Application. Figure 4 presents the components adapted from the Computation viewpoint, which will be used in the remainder of this paper.

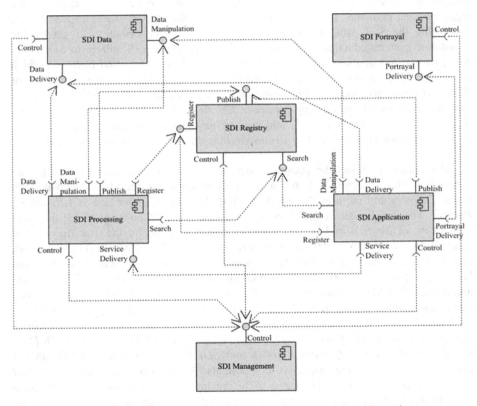

Fig. 4. Diagram of adapted components of the computation viewpoint

3 SDI-Cemig

As described in Sect. 1, the Cemig corporation seeks to standardize the use and sharing of data and geospatial services in the company conglomerate by developing and using SDI-Cemig.

The model by the ICA was used to specify SDI-Cemig so as to guarantee that the basic SDI concepts in the literature would be contemplated during the specification phase. Subsect. 3.1 describes the Computation viewpoint in ICA's model applied to the specification of SDI-Cemig. As mentioned earlier, the Enterprise and Information viewpoints are not detailed since they do not belong to the scope of this study.

Computation Viewpoint

The Computation viewpoint is detailed by comparing the components specified by Cooper et al. [3] to the computational objects in SDI-Cemig. Moreover, Subsect. 3.1 details the computational objects by specifying their interactions with other objects and their required and provided interfaces.

Figure 4 presents the basic components an SDI must have and the relationships among their interfaces. These components can be considered abstractions of the computational objects and represent a set of computational objects that have similar functionalities. The applications and services to be used by SDI-Cemig have not been defined yet, however, it has been defined that any application to be used must be compatible with the standards of the Open Geospatial Consortium (OGC). For this reason, the following generic names will be used to represent the components of SDI-Cemig: *Portrayal_SDI-Cemig*; *Data_SDI-Cemig*; and *Catalogue_SDI-Cemig*. It was verified whether these components behave similarly to those proposed in [3].

The component *Portrayal_SDI-Cemig* implements the Web Map Service (WMS) standard, responsible for generating the maps with static images from the geospatial data provided by the application. The component *Data_SDI-Cemig* is in charge of accessing the geospatial data of SDI-Cemig and is able to recover, input, change, and delete such data while implementing the standards Web Feature Service (WFS), Web Feature Service-Gazetteer (WFS-G), and Web Coverage Service (WCS). Finally, the metadata [2] and catalogs are managed by the component *Catalogue_SDI-Cemig*, which used the OpenGIS Catalogue Service standard. As well as the component *Data_SDI-Cemig*, *Catalogue_SDI-Cemig* is able to recover, input, update, and delete the catalogs and metadata from SDI-Cemig's database.

The component *SDI Application* is the only one accessed by the user and has no provided interface. However, it has several required interfaces to meet the user's needs. In SDI-Cemig, this component is equivalent to a Geoportal. Geoportals, according to Tait [15], are user access points to geographic content in the network, either the internet or intranet. Thus, in SDI-Cemig, the Geoportal aims to serve as a user access point to the features and data in the SDI, which are provided by the geospatial services.

The component *SDI Portrayal* is responsible for showing the data and results of the operations to the user when requested by the component *SDI Application*, thus providing a single interface to make this feature available. In SDI-Cemig, it is *Portrayal_SDI-Cemig* that plays this role. Just as the component proposed by Cooper et al. [3], *Portrayal_SDI-Cemig* has a single provided interface, which is responsible for providing a graphical representation, i.e., a map that presents the data provided to the component.

The access to and direct recovery of data in the SDI, a responsibility of the component *SDI Data*, corresponds, in SDI-Cemig, to the component *Data_SDI-Cemig*. *Data_SDI-Cemig* has interfaces that allow the geospatial data requested to be recovered, while the way this request is fulfilled is differentiated. Using the interfaces standardized by the WFS, the component *Data_SDI-Cemig* recovers geospatial data through spatial queries, while in the interfaces standardized by the WFS-G standard, the data are recovered using a geographic dictionary called gazetteer. However, neither standard specifies interfaces able to recover georeferenced images, a feature that is the responsibility of the interfaces standardized by the WCS standard. *Data_SDI-Cemig* implements the interfaces of the WCS standard in a similar way as those in the WFS standard, i.e., allowing the coverage data (images and matrices) to be recovered through spatial queries.

The component *SDI Registry* is responsible for registering the SDI's data and services in catalogs to facilitate searching for and recovering them. *Catalogue_SDI-Cemig* is equivalent to this component in SDI-Cemig. As in the component specified by Cooper et al. [3], *Catalogue_SDI-Cemig* has interfaces to register and search the catalogs and their records by implementing the interfaces specified by the OpenGIS Catalogue Service. However, it is noteworthy that the *SDI Registry::Register* interface guarantees the catalogs and their records are input or updated in the database, but not that they are available to the user. In order for a catalog or record to be available to the user, either via internet or intranet, the interface *SDI Registry::Publish* must be used. Nevertheless, *Catalogue_SDI-Cemig* has no specific interface for this task, which requires adapting an application or service compatible with the standard. The adaptation can be done by adding a new interface or feature, which is equivalent to the interface *SDI Registry::Publish*, or by changing the behavior of the interface *SDI Registry::Register* in a way that, when a new catalog or record in registered, it will automatically become available to the user. This last possibility was the one chosen to apply the function of the interface *SDI Registry::Publish* to the component *Catalogue_SDI-Cemig*.

Initially, SDI-Cemig will not have geoprocessing services since its focus is to facilitate sharing geospatial data relevant to Cemig's employees and clients. Moreover, SDI-Cemig has no specific component to deal with the management of data access rights and to guarantee the integrity and compatibility of the geospatial data in the exchange of messages among the interfaces. The management of access rights will be the role of the *Geoportal*, while data integrity and compatibility will be guaranteed by the components themselves. Therefore, SDI-Cemig has no component equivalent to *SDI Processing* or *SDI Management*.

3.1 Computational Objects of SDI-Cemig and Their Interfaces

According to Linington et al. [9], the Computation viewpoint is responsible for modeling the basic features of an application by specifying the services the application offers through components and their interfaces with no concern about the physical distribution of these components or which technologies will be used to implement them.

Figure 5 presents a simplified version of the components identified in SDI-Cemig and their interactions through computational objects (CV_Object), their interactions with other computational objects, and the packages that group them. According to Linington et al. [9], "computational objects encapsulate part of the system's status and functionality, thus enabling a modular system modeling."

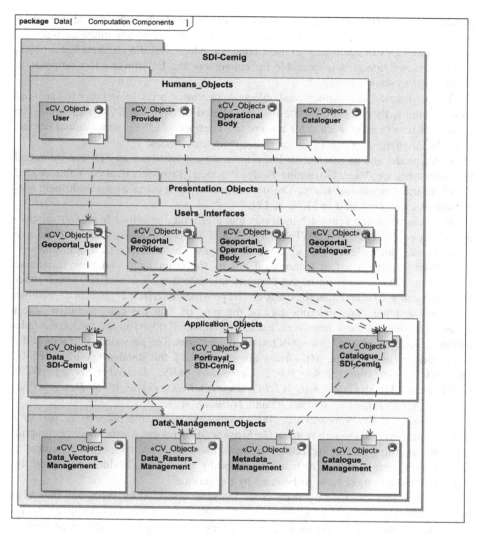

Fig. 5. Simplified view of the computational objects in SDI-Cemig

The computational objects in SDI-Cemig can be grouped into four groups: *Human_Objects*, which represent the actors specified in the Enterprise viewpoint; *Presentation_Objects*, representing the interfaces used by the actors; *Application_Objects*, whose computational objects represent the features and, consequently, the components offered by SDI-Cemig; and *Data_Management_Objects*, which access the database to recover, input, change, and delete the data requested by the computational objects of the package *Application_Objects*.

Figure 6 presents the computational objects in more details, with their provided and required interfaces and their interactions. Since the computational objects of the package *Human_Objects* (*User*, *Provider*, *Operational Body*, and *Cataloguer*) represent system actors, they have a single required interface, which interacts with the specific interfaces of each actor in the *Geoportal*. Moreover, the computational object group *Human_Objects* is responsible for connecting the Enterprise and Computation viewpoints, as shown in Fig. 2.

The actors used as computational objects were chosen because, in order to carry out their functions, they depend on the features offered through the OGC standards. The *Geoportal* interfaces of each actor require the interfaces of the computational objects from the *Application_Objects* package that meet their needs.

ICA's model and its extensions define several possible actors an SDI can have, and the main ones are: User; Governing Body; Producer; Provider; Broker; Value-Added Reseller; and Operational Body. The description of each actor, along with their specializations, can be found in [3, 4, 8, 12].

In this section, the symbol * indicates that a provided interface (e.g., *GetCapabilities_**) is equivalent for the WFS, WFS-G, and WCS standards. Catalogue Service standard was not included due to its particularities. The interface *GetCapabilities_**, when used, returns a list of features offered by the component and their respective descriptions. Thus, when an object requests this interface, it can check which other features it can access and how to access it. The provided interface *Transaction_** allows inputting, updating, and deleting data of the type of data the computational object focuses on. Besides these interfaces, each computational object in the package Application_Objects offers a specific set of provided interfaces. The provided interfaces of the computational object *Data_SDI-Cemig* are specified by the standards WFS, WFS-G, and WCS: *GetPropertyValue_WFS*; *GetFeature_WFS*; *DescribeCoverage_WCS*; *GetCoverage_WCS*; *GetFeature_WFS-G*; *DescribeFeatureType_WFS*; *DescribeFeatureType_WFS-G*; *GetCapabilities_**; and *Transaction_**.

The interface *GetPropertyValue_WFS* returns the value of a property or part of a complex property of a search feature. *GetFeature_WFS* returns one or more geographic features according to the search performed. *DescribeFeatureType_WFS* returns a schema containing the types of features offered by the service, besides specifying the standards the data must have to be used by the service.

The interfaces specified by WFS-G are the same specified by WFS, and they also behave similarly. The difference between their interfaces is the way the searches are performed. While WFS specifies that the recovery of geographic features is performed through searches that use spatial operations, such as the use of a minimum bounding rectangle, feature overlap, etc., the interfaces specified by WFS-G determine that the features are recovered by using gazetteers.

In case the user wants to recover coverage data from the SDI, the interfaces from the WCS standard implemented by *Data_SDI-Cemig* must be used. The provided interfaces related to the WCS standard are similar to those in the WFS and WFS-G standards. The interface *DescribeCoverage_WCS* returns a description of the coverage data the SDI has, including, for example, the area the data comprehends, what it represents, and its structure. The coverage data is returned to the user through the interface *GetCoverage_WCS*. As in the interface *GetFeature_WFS*, the data is recovered based on queries that use spatial concepts.

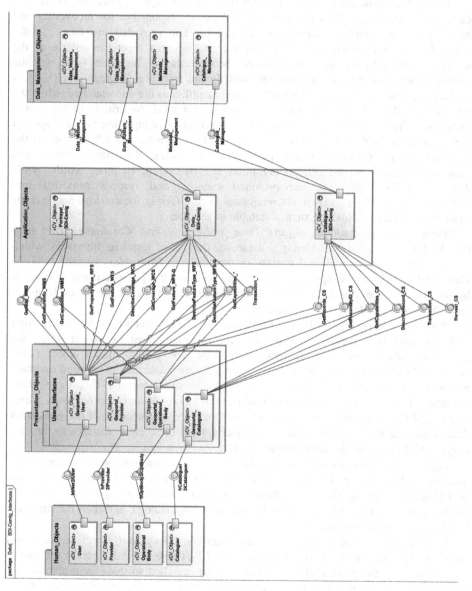

Fig. 6. Computational objects in SDI-Cemig

Since *Portrayal_SDI-Cemig* is a computational object that does not change the geospatial data in the database, it has no required interface and, therefore, does not interact with the computational object of the package *Data_Management_Objects*.

The last computational object described in the package *Application_Objects* is *Catalogue_SDI-Cemig*. This computational object is responsible for recovering and handling the geographic catalogs in SDI-Cemig. The provided interfaces *GetRecordbyID_CS* and *GetRecords_CS* return the catalogs that meet the criteria of the search performed by the user. The difference between the two interfaces is the way the catalog is searched and the amount returned. In the interface *GetRecordbyID_CS*, a single catalog is returned in case it has the same identifier as the one searched, while the interface *GetRecords_CS* will return all catalogs that meet the criteria of the search performed. The last interface, *Harvest_CS*, is responsible for inputting the catalogs and their records in SDI-Cemig, as well as the interface *Transaction_CS*. However, the interface *Harvest_CS* performs this function by "harvesting" the catalogs and records from other databases and referencing them in SDI-Cemig. In other words, while *Transaction_CS* inputs the user-provided catalogs and records provided into SDI-Cemig's database, *Harvest_CS* references in SDI-Cemig the catalogs and records from other databases, making them available to the user.

Since the computational objects *Data_SDI-Cemig* and *Catalogue_SDI-Cemig* handle data present in SDI-Cemig's database, they need required interfaces which interact with the computational objects of the package *Data_Management_Objects*. *Data_SDI-Cemig* inputs, updates, deletes, and recovers the geospatial data requested by the user. Hence, this computational object uses the provided interfaces *Management_Vectors* and *Management_Rasters*.

However, one policy in SDI-Cemig states that "the metadata must be stored along with the data they describe". Therefore, the metadata must be stored when the geospatial data are input or changed in the SDI. The computational object *Catalogue_SDI-Cemig* is responsible for maintaining the metadata and its required interface interacts with the provided interface *Metadata_Management* and *Catalogue_Management*. The interface/computational object *Metadata_Management* allows *Catalogue_SDI-Cemig* to input, update, delete, and recover the metadata of the geospatial data, while the interface computational object applies these same features to the catalogs, which maintain the metadata as records.

Figure 7 highlights the possible computational objects and interfaces the actor *User* may use in SDI-Cemig. The interface presented to the user when he or she accesses SDI-Cemig through the *Geoportal* allows the user to search for geospatial data through spatial queries, geographic names, or by browsing the catalogs, and then the data can be visualized and/or recovered.

In order to perform queries, the user interface (*Geoportal_User*) requires the interfaces *GetRecords_CS* and *GetRecordbyID_CS*. These interfaces return the metadata catalogs of the geospatial data, which can be searched so that the user finds the data that meets his or her needs. In case the user finds in the metadata the data of interest, he or she can recover them using the interfaces *GetFeature_WFS*, *GetFeature_WFS-G*, and *GetCoverage_WCS*. Besides recovering them, the user may visualize the geospatial data on the browser itself using the interface *GetMap_WMS* by forwarding to this interface the data to be visualized.

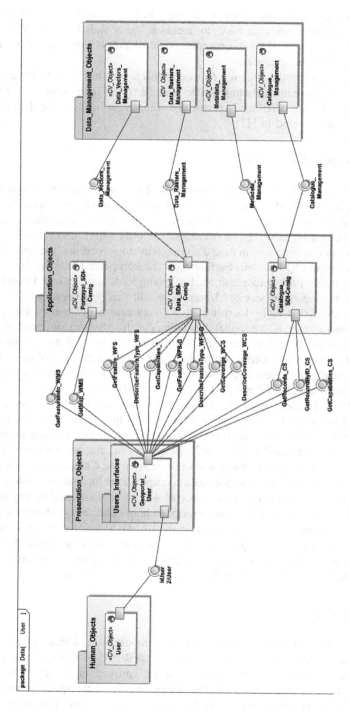

Fig. 7. Computational objects and interfaces used by the user

The interface *GetCapabilities_** is designed for users who want to use SDI-Cemig's features with no need to access it via a web browser. This interface returns the features offered by the computational object, besides the information that aids in the use of these features.

The other actors in the package *Human_Objects* will not be detailed as the actor *User* in Fig. 7 because this detailing can be inferred from Fig. 6, while Fig. 7 was used to exemplify how this detailing is performed. Finally, if needed, the detailing of the other actors can be found in [11].

4 Discussion of Results

The Computation viewpoint in ICA's model for SDI proved appropriate to describe the components in SDI-Cemig. The changes applied to the model increases the independence among the components, which facilitates specifying, maintaining, and using them. That occurs because, as each component only offers and requires interfaces directly related to its function, in case a component stops working, the operation of the components will not be affected. Furthermore, the component *SDI Application* on the ICA's model needs further discussions. For example, is it necessary to represent the component's feature chain service? If yes, how will that be represented?

Although the SDI-Cemig doesn't have a component equivalent to the component *SDI Management* of the ICA's model, the authors of this study recommend strongly the implementation of the component *SDI Management*. The *SDI Management* implementation allows that the access permissions and interoperability of data and services are guaranteed whether users and systems outside of the Cemig's environment access the web services (accessed through the component *SDI Application*).

Through the specification of the *Human_Objects*, it can be verified that, at least for SDI-Cemig, only a small set of possible actors an SDI may have compulsorily needs its computational features to carry out its role. The actor Governing Body, for example, defines the scope, policies, and funding of the SDI, which are critical aspects for any SDI to exist and does not require the use of the features presented in the Computation viewpoint.

ICA's model for SDI describes the basic concepts in the literature for SDI at all levels. Hence, ICA's model guarantees these concepts will be contemplated during the specification of the SDI. However, there is no description or orientation regarding how the model must be used. For instance, the viewpoints in the RM-ODP framework are related among themselves, as shown in Fig. 2. How should the relationship with the viewpoints Enterprise and Information be shown in the Computation viewpoint? And what elements or aspects of these viewpoints would be related with the Computation viewpoint?

5 Conclusions

The use of ICA's model for SDI allowed the components and their interfaces – considered basic for an SDI to work – to be specified in SDI-Cemig. In addition, the Computation viewpoint, after being adapted, proved appropriate to describe SDI-Cemig, with no discrepancies between the model and the specification.

Although the specification of a single corporate SDI does not guarantee that ICA's formal model is appropriate for any corporate SDI, the specification of SDI-Cemig suggests that the Computation viewpoint in ICA's formal model for SDI can be used in corporate SDIs. Moreover, the specification presented in this specification can help other designers who wish to specify an SDI, regardless of its level or of whether ICA's model is being used.

Acknowledgment. This project was partially funded by the Brazilian research promotion agencies Fapemig and CAPES, along with Cemig Enterprise.

References

1. Béjar, R., Latre, M.Á., Nogueras-Iso, J., Muro-Medrano, P.R., Zarazaga-Soria, F.J.: An RM-ODP enterprise view for spatial data infrastructure. Comput. Stand. Interfaces **34**(2), 263–272 (2012)
2. CONCAR – Comissão Nacional de Cartografia. Perfil de Metadados Geoespaciais do Brasil (Perfil MGB) (2009). http://www.concar.ibge.gov.br/arquivo/perfil_mgb_final_v1_homologado.pdf. Accessed 25 Jan 2016
3. Cooper, A.K., Moellering, H., Hjelmager, J., et al.: A spatial data infrastructure model from the computational viewpoint. Int. J. Geogr. Inf. Sci. **27**(6), 1133–1151 (2013)
4. Cooper, A.K., Rapant, P., Hjelmager, J., et al.: Extending the formal model of a spatial data infrastructure to include volunteered geographical information. In: 25th Cartographic Conference (ICC) (2011)
5. Crompvoet, J.: Spatial Data Infrastructure and Public Sector (2011). http://www.spatialist.be/eng/act/pdf/20111107_sdi_intro.pdf. Accessed 21 Oct 2015
6. Farooqui, K., Logrippo, L., De Meer, J.: The ISO reference model for open distributed processing: an introduction. Comput. Netw. ISDN Syst. **27**(8), 1215–1229 (1995)
7. Harvey, F., Iwaniak, A., Coetzee, S., Cooper, A.K.: SDI past, present and future: a review and status assessment. In: Spatially Enabling Government, Industry and Citizens (2012)
8. Hjelmager, J., Moellering, H., Cooper, A.K., et al.: An initial formal model for spatial data infrastructure. Int. J. Geogr. Inf. Sci. **22**(11–12), 1295–1309 (2008)
9. Linington, P.F., Milosevic, Z., Tanaka, A., Vallecilo, A.: Building Enterprise Systems with ODP: An Introduction to Open Distributed Processing. CRC Press, Boca Raton (2011)
10. Nebert, D.D.: Technical Working Group Char GSDI, Developing Spatial Data Infrastructure: The SDI Cookbook, vol. 2 (2004). http://www.gsdi.org/docs2004/Cookbook/cookbookV2.0.pdf
11. Oliveira, I.L.: Adequação do modelo formal da associação cartográfica internacional e sua avaliação no desenvolvimento de infraestruturas de dados espaciais corporativas: estudo de caso IDE-Cemig. Dissertation, Federal University of Viçosa (2015)
12. Oliveira, I.L., Lisboa-Filho, J.: A spatial data infrastructure review – sorting the actors and policies from enterprise viewpoint. In: Proceedings of the 17th International Conference on Enterprise Information Systems, vol. 17, pp. 287–294 (2015)
13. Rajabifard, A., Williamson, I.P.: Spatial data infrastructures: concept, SDI hierarchy and future directions. In: Proceedings of GEOMATIC's Conference, p. 10 (2001)
14. Raymond, K.: Reference model for open distributed processing (RM-ODP): introduction. In: Raymond, K., Armstrong, L. (eds.) Open Distributed Processing. IFIP, pp. 3–14. Springer, US (1995)
15. Tait, M.G.: Implementing geoportals: applications of distributed GIS. Comput. Environ. Urban Syst. **29**(1), 33–47 (2005). Elsevier

Processing and Geo-visualization of Spatio-Temporal Sensor Data from Connected Automotive Electronics Systems

Patrick Voland[✉]

Department of Geography, Geoinformation Research Group,
University of Potsdam, Karl-Liebknecht-Straße 24/25, 14476 Potsdam, Germany
patrick.voland@uni-potsdam.de

Abstract. Connected devices, paradigms of the Internet of Things and Big Data increasingly define our everyday life. In this context, modern automobiles, which are characterized by an increase of electronic components and extensive sensor devices, potentially are becoming a new kind of mobile and anytime accessible sensors. In this context "Extended Floating Car Data" (XFCD) is a rich geocoded dataset for vehicle, traffic, and environment data, augmenting more traditional geospatial databases. This paper deals with the approach to collect and use this data from automobiles for context-aware geospatial analyses by combining the sensor parameters with a spatial and temporal component. These data concern the concept of XFCD as geo-information and needs to be made available and applicable to spatio-temporal visualization. For this approach, research already conducted should be considered and findings should be used for more in-depth research.

1 Introduction

Rapidly growing technical development becomes evident through the usage of the comprehensive interconnection between electronic systems in nearly all areas of our everyday life. Modern automobiles are no longer complex mechanical, but increasingly complex electronic systems. They contain a large number of sensor devices essential for smooth technical operation (e.g. parameters like engine revolutions and vehicle speed) and environment perception (e.g. barometric air pressure and ambient air temperature). This data is a prerequisite for the growing number of "advanced driver assistance systems" (ADAS). Data from these sensors also built the base of semi-autonomous and autonomous driving systems. The Canadian Author Cory Doctorow remarked in a lecture about the consequences of electronic interconnections during the "28. Chaos Communication Congress 2011" (28C3) in Berlin: "We don't have cars anymore; we have computers we ride in" (Doctorow 2011).

According to the availability of existing techniques of communication, increasing data-bandwidth as well as decreasing cost of transmission, the collected data from automobiles for various purposes should be utilized. This includes communication models such as vehicle-to-vehicle (v2v), vehicle-to-infrastructure (v2i), or vehicle-to-x (v2x).

© Springer International Publishing Switzerland 2016
O. Gervasi et al. (Eds.): ICCSA 2016, Part III, LNCS 9788, pp. 290–305, 2016.
DOI: 10.1007/978-3-319-42111-7_23

Reacting to this situation, "Extended Floating Car Data" (XFCD) expand on the concept of "Floating Car Data" (FCD) and take on a central role in future developments. Automobiles would become interconnected and retrievable mobile sensors themselves and are high-profile within research and economy. This new massive amount of data that is valuable to many key industry players, and thus the emergence of many new applications, right up to autonomous cars, will influence our everyday life heavily.

The incentive of this research work needs the utilization of the accumulated data from the computer system automobiles for analysis. As this data from sensor-information is being generated in the light of the dimensional time and space, it can be seen as potential spatio-temporal information and consequently as geo-information or geodata. This enables a space applied process as well as a visualization and so demonstrates a fascinating field, especially from the point of view of geo-information science and cartography. Within the scope of the research work, it appears that spatio-temporal consideration, especially with the usage of cartographic visualization in combination with additional methods of visualization, provides an adequate opportunity for analysis. Consequently, this represents a basis to make the information overload manageable and evaluable for the human being. In this regard, it was possible to carry out a prototypical implementation of the data entry. The automobile had applied the application of the required by law "On Board Diagnostics" (OBD) interface, a smartphone supported procedure, as well as a web-based visualization for a usage scenario. The at hand research work is supposed to contribute to this new kind of mass information and make it available as geo-information in accordance with the concept of "Extended Floating Car Data" (XFCD) and to utilize those for spatio-temporal visualizations (visual analysis).

Therefore, the question of whether a technical implementation, through the application of open source software is solely possible, was considered. With the idea in mind, that a human is part of the automobile system (and thus creator of his own spatial mobility), the question arising is how data and its visualization can be supported, for e.g. the human as a driver or planner. Implementing methods from the field of (geo-) visual analytics, recording, processing, and visualizing data of one or more vehicles can be used to analyse driving styles and environmental behaviour. A spatio-temporal point of view – especially regarding web-based applications of cartographic visualizations in combination with other methods of representation – enables users to interpret the vehicle's data. The different applications, however, will have to take into account each specific context.

But there are still a lot of unsolved problems, which require more in-depth research. How can the data be recorded, processed, visualized, and displayed in an automated fashion? What are the most interesting facts? What data can be used in this way and how can it be combined, processed, visualized, analysed, and interpreted in a logical and productive way? Which components and strategies of visualization are most effective and how can they be defined, combined, and eventually applied? Against the backdrop of the well-known visualization pipeline, it is necessary to identify and develop suitable components and strategies for each stage of filtering, mapping, and rendering. The intention is to combine and utilize them in a pattern and rule based approach.

2 Data, Content and Application Potential

The surveillance and geolocation of the recording of traffic (specifically of automobiles) can be divided into the general recording of traffic (e.g. through induction loops, photoelectric barriers as well as camera surveillance systems), and into the individual recording of traffic or geolocation (e.g. through GPS Global Positioning System). However, general recording of traffic requires costly and local or stationary measuring sensors. The question arises, which information and potential added values during mobile measuring can be provided in terms of the individual vehicle compared to the overall traffic system and beyond.

"Floating Car Data" (FCD) and "Extended Floating Car Data" (XFCD) are suitable for the explanation of time and space applied data from automobiles and will be elaborated hereinafter (Fig. 1). In relation to a technical point of view, those specific connections between telecommunication and information are necessary and can be classified as "traffic telematics".

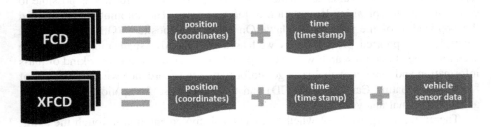

Fig. 1. Schematic representation of the composition of a FCD and XFCD information.

Floating Car Data (FCD). FCD enables the acquisition of individual geolocations, e.g. the individual data collection and transmission in regards to single vehicle with this traffic system. The more vehicles are generating information within the overall system, the more extensive the overall data system becomes and ultimately those findings and information become more reliable. The single vehicle that generates FCD for the overall data system can be recognized as samples to access and evaluate the whole traffic situation. The analogy of "corks swimming in the river" seems to be most adequate to describe this principle (Pfoser 2008a, 321).

Main components of the FCD system are called "On Board Units" (OBU), which are located (often stationary) in the vehicle and have data binding (e.g. via the mobile phone network GSM Global System for Mobile Communication) as well as a central system or server to which the information is sent through programmed algorithms, including time and space intervals or occurrence triggering. This data transfer from an automobile to a central station is usually referred as "vehicle-to-infrastructure communication" (v2i). It should be noted, that FCD is heavily related with the concept of "Floating Cellular Data" or "Floating Phone Data" (FPD) that is being used to collect and transfer data. The advantage is that there is no need for special equipment or hardware. Any mobile phone with GPS or mobile phone tracking and data connection is needed for the acquisition and transmission of data.

To summarize, FCD or FPD data sets are point data and is determined by the acquisition at the position also known as coordinates of the vehicle (spatial reference) and the timestamp (time reference). The data have spatial and temporal components that can be used for further proceedings. This way, travel time and driving progress can be reconstructed as well as information can be calculated out of this data to be used for e.g. the individual velocity of a vehicle or the behaviour of the processes of acceleration and slowing-down. The collected information out of these individual automobiles can predict situations about the overall traffic system through statistical methods. The bandwidth of potential (traffic) services on the FCD basis can therefore be versatile (according to Lorkowski 2003, 7–12):

- deflection of the average velocity,
- displaying the situation of the roads,
- automated reports for traffic jams and current status,
- dynamic route planning, on and off board navigation systems,
- generation of digital road networks.

This implies the necessity to equip a preferable large amount of automobiles with the technique to generate a reliable base of data. Automobile fleets such as cabs and public transportation are often being used for traffic recordings.

Extended Floating Car Data (XFCD). FCD already provides several possibilities to track movements of individual vehicles and to derive traffic information. Those possibilities should be extended through additional data, specifically from automotive electronic and sensor technology.

BMW investigated "Extended Floating Car Data" in 1999. Next to other automotive manufactures, this company offers automobiles with respective OBU and merges the data based on the XFCD concept through telematics services (under the brand name "BMW Connected Drive"). Without agreeing on the specifics provided by BMW, the term XFCD illustrates a fitting description for the extension of the FCD concept (as a usual geometric point object) specifically with the addition of data of the automotive electronic or sensor technology: "The location and timestamp information are enriched with vehicle status information derived from the in-vehicle bus system or additionally attached sensors" (Barceló and Kuwahara 2010, 164). XFCD so extends the range and quality of recorded (and transmitted) data, "[...] while here all information of the automotive electronics (such as ABS, ESP, rain sensors, etc.) are being used and analysed for different situations (condition of traffic, weather, condition of roads, etc.)" (Halbritter et al. 2008, 147).

According to the information and parameters that are completed through this, comprehensive and heterogeneous datasets are created as well as many potential possibilities to derive calculated new information out of this. As modern automobiles include several different sensors today, the collected data from automobiles can not only be utilized for traffic relevant considerations (e.g. Monitoring of traffic) but also for environment relevant considerations (e.g. Monitoring of the environment). This way automobiles become mobile sensors. XFCD delivers event and state data and so generates potential (traffic) service which can be characterised in general as the following (Breitenberger et al. 2004, modified):

Table 1. Content and services based on XFCD (selection).

XFCD contents	XFCD services
• sensory data	• traffic information
• data for vehicle operation	• environment and weather information
• traffic situation	• dynamic guidance
• entry and exit of traffic jams	• travel times and routed
• crossing speed at traffic jams	• individual consumption analysis (emission and fuel)
• local alerts	

The variety of possible applications can be clarified as follows: The emergence of real-time location data has created an entirely new set of location-based services from navigation to pricing property and casualty insurance based on where, and how, people drive their cars" (McKinsey 2011, 6). Fundamentally, a great number of applications for different areas are possible through XFCD. As an example (Table 1):

- traffic monitoring (e.g. traffic intensity and quality),
- environmental monitoring (e.g. weather events),
- emissions (e.g. ecological footprint and air quality),
- driving behaviour (e.g. routes, driving speeds and brake activation),
- navigation (e.g. depending on traffic or environmental situation),
- electro-mobility (e.g. route planning and range estimate).

It already becomes apparent how convincing the information of the collected data can become and how heavily privacy can be interfered through the usage of this data. Potential users or user groups can be identified, which can be considered heterogeneous:

- private user or end-user,
- planner (e.g. traffic and environment planning),
- manager (e.g. automobile fleet management),
- decision maker (e.g. politics and administration),
- economical actors (e.g. automaker, insurer).

The question is how this information, even after comprehensive (computer-aided) processing and filtering, are made analysable and controllable for the user, meaning the usage for the individual area of application. As human beings have the possibility and ability to be the generator and end user of that information and are used to read and process the information quickly, the visualization can be a suitable instrument. It needs to be emphasized that XFCD as well as FCD are spatial data. This way it is comprehensible that spatial data processing and map based or cartographic visualization is acting as a major part of the analysis and evaluation of this information. Next to the spatial component, the temporal component represents the second important condition and requires additional methods. The visualization of spatio-temporal information or spatio-temporal data can be seen as superordinate term.

This new way of data serves as a basis for technologies such as semi-autonomous and autonomous driving (however, a transmission of data from the system of the vehicle is not mandatory in this case). The possible uses of XFCD are going even further. Through technologies such as "vehicle-to-vehicle communication" automobiles have the possibility to give warnings between themselves on the basis of data of potential danger (e.g. Accidents or icy roads) by exchanging data collected from individual vehicles among themselves: "The Key aspect of XFCD is that they have the potential to indicate hazardous conditions before they turn into real incidents" (Schneider et al. 2010, 164). To enable this, general and main requirements can be formulated to the XFCD-Systems (by Stottan 2013, 52):

- low ongoing operating costs (specifically communication costs),
- scalable and usable anonymization,
- reliable and suitable OBU,
- easy and usable standardisation for provision for charitable.

Next to the transparency and data protection (specifically in terms of the effective anonymization), the standardisation of uniform and general data models turns out to be very problematic. Without those, the exchange as well as the usage of that data in scientific as well as commercial context is difficult to apply.

3 Data Logging, Filtering and Visualization

The increase of electronic components in an automobile is accompanied by a growing number of electronic control units (ECU). Taking the automobile series of model Mercedes Benz E-Class (Daimler AG) as an example, there is an increase in numbers from 17 in 1995 (Model: W210) to 67 in 2011 (Model: BR212) (cf. Schäffer 2012, 19–22). Every ECU is equipped with an intelligent mechanism in the form of a micro-controller, controlled by a software and designed for a specific task or a specific area of activity. It picks up and processes the signals of the connected sensors and generates the signals for the connected output device. The electronic interconnection within the automobile is realised through an on-board power supply. It can be thought of as a connected or allocated system of a specialised control device, whereby the single control devices can exchange data with each other through a common data bus. Therefore, data bus systems can be considered as networks for data exchange and ECUs can be considered as network nodes. The CAN-Bus (Controller Area Network) can be cited as an example of an often found bus system. It has become the common standard for automobile since its development in 1983–1987 mainly by Bosch and Intel and its introduction in 1991 and is often used in the automation engineering (cf. CiA 2016).

It arises the issue how sensor data from the automotive's electronics in terms of XFCD (and therefore as geodata) can be utilized technically. This research paper investigates the complete procedure of data acquisition, from the vehicle to the extern visualization and processing. It also develops and applies a qualified process as a technical cycle to equip the data in the vehicle's generated data with a spatial and time reference. It was specifically motivating to use a technical approach which is not exclusively covering automobiles of the new generation but also older models. No proprietary or commercial system was used to show a potential and practical approach to enable the usage of data from automobiles for every person. Also, no hardware was used that needs to be installed in the automobile. It was aimed to exclusively use complimentary and open source software as well as reasonably priced hardware.

Automakers developed a wide range of implementations such as communication protocols, diagnosis codes or connectors to diagnose the vehicle systems and its single components, specifically for fault diagnostics in repairs shops. This range is caused by the specifics of the different automakers, their technical transcription, as well as missing standards and norms. As malfunctions in the combustion system or in the exhaust gas function can be a potential danger for the driver and the environment, the compliance of statutory emission limit value for this issue needs to be defined by the legislature. In the USA, the California Air Resource Board (CARB) established measurement criteria and technical specification named "On Board Diagnostics" (OBD or OBD II). The technical details (such as form and layout of the diagnostic connector as well as defined diagnosis routines and "Diagnostic Trouble Codes" (DTC) were taken on internationally and were determined in the ISO-Norm 15031 "road vehicles – communication between vehicle and external equipment for emission-related diagnostics". Those provide the technical basis for an establishment in the regional mandatory legislation, such as Japan, USA, Europe. As far as it concerns the statuary mandatory functions, pin configurations are always determined. However, automobile manufacturers are eligible to use the not assigned pins of the diagnostic connector for own proprietary applications. It is even possible to resort to the existing OBD protocol for communication, but only with a not standardized proprietary command.

Through OBD-diagnose-hardware or with the connection to a computer and a specific software, the predefined diagnostic routine and respective diagnostic data can be obtained. The list of the OBD (II) potential possible parameters (as long as automobile specific available) is long (cf. Schäffer 2012, 205–224).

Table 2. Selected OBD PID (incl. range and und measuring units).

PID (Hex)	Initials	Description	Min	Max	Unit
0C	RPM	engine revolutions per minute	0	16.383,75	1/min
10	MAF	mass air flow	0	255	g/s
0D	VSS	vehicle speed	0	655,35	km/h
33	BARO	ambient air barometric pressure	0	255	kPa
46	AAT	ambient air temperature	-40	215	°C

By over 80 defined parameters, less than 32 parameters are regulated through the legislature for supervision of exhaust relevant components and systems. Provided that the respective automobile supports OBD, it can be used for further consumption (e.g. fuel) and supervisory sensors (e.g. fine particles or NO_x). During communication, the OBD supported information or parameters are called OBD PID (Parameter Identifier). The conclusive identification with PID enables the specific access to sensors or measured data and routines. The query of the respective supported parameter is possible (in blocks) here. Inquiries (always binary) and responses are expressed through the hexadecimal system and need to be converted for further use (cf. ib., 96f) The following table specifies the PID for further investigations as an example (Table 2).

The by law required OBD interface is therefore the physical technical interface for automotive electronics. Because of the availability of relatively reasonable hardware for diagnosis, it makes sense to gather (computer supported) (sensory) data this way. Modern smartphones (as mobile computers) provide the required technical features and are very suitable (alternative to build-in OBU) for communication with the OBD hardware (wireless or tethered), for processing and storage of the data, and for the data transfer through the combination with a web based or server-sided provided system. During the acquisition of data, the connection of space and time can be added through a GPS supported smart phone and so enables the generation of XFCD.

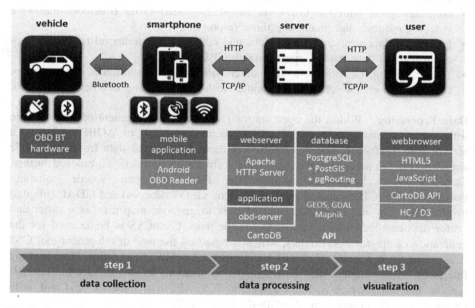

Fig. 2. Smartphone supported process for generating of XFCD from OBD data with subsequent web based process and visualisation (schematic illustration of the setup).

These discoveries made a technical process and its development and implementation possible (Fig. 2). Based on this process, three main central tasks can be identified: (1) data acquisition and transmission, (2) server-side data processing, as well as (3) the visualisation and application with the help of web browsers through the end-user (cf. Voland 2014).

Data Logging. In the first step of the process, the OBD interface of the used automobile is equipped with an OBD hardware and a connected smart phone.

Different OBD adapters with wireless Bluetooth connection are applied. A mobile application on the smartphone is being used for communication from the OBD hardware or interface to recall data (e.g. in a time interval). In this context, several software solutions are already existing, but the selection is very limited for open source software for active and advanced projects. For the open source system Android, the application, android-OBD-reader (under apache license 2.0, cf. AOBDR 2016) is an example. For the research work a constructive collaboration with the contribution of this open source project has been developed, to make the collected OBD data (XFCD) through the mobile application available. The mobile application on smart phones expands the acquisition of OBD data with the real time coordinates delivered through GPS and a time stamp. The generated XFCD are being cached or sent from the automobile (identifiable by its vehicle identification number, VIN) to a server with an existing internet connection. The tasks of the mobile applications can be summarized as:

- selection of OBD PID and configuration of the interval of inquiries,
- communication with the OBD via (ELM327-based) OBD Bluetooth hardware (sending requests and processing those responses),
- generating data sets: analysing values and adding spatio-temporal references (co-ordinates and time stamps, spec. through GPS)
- instant sending of data set to the server with an existing internet connection or caching data set for subsequent sending (with a non-existing internet connection).

Data Processing. Within the open source project, the development of a server-side software component "OBD-server" (under apache license 2.0, cf. AOBDR 2016) as a service was conducted. The possibility was created to upload data from the mobile application to the server (or the server-side software component). To proceed further, CartoDB was chosen to connect the following open source software: PostgreSQL/PostGIS as spatial database system, GEOS (libgeos) and GDAL (libgdal) to process vector and raster data, and Mapnik to generate map tiles, also raster and vector data and leaflet as a viewer to see the map. CartoCSS is being used for the generation of graphical submittals, which is based on the web development tool CSS (Cascadian Style Sheets). CartoDB encompasses an API (Application Programming Interface) for the implementation in its own applications as an interface to CartoDB specific functions, which includes a specific SQL API that enables a read and write permission to the PostgreSQL/PostGIS-database (cf. CartoDB 2016).

The server component "OBD-server" is constantly receiving data from the mobile application "android-OBD-reader". The collected data can be perceived through a soft copy on the server. Afterwards, on the server-side the readout of the "OBD-server" existing data sets occurs (of all available vehicles or VIN) and as a second step, database tables are created, as well as a quick clearing of data or databank. The interchange format is JSON, meaning the format of data sent from the mobile application, received from the server component, and then made available. This format is supported by PostgreSQL/PostGIS as well as CartoDB. Following this, further processing of data, the reprocessing of visualization (e.g. transfer of existing vector data into image and raster data), as well as the deployment of a web based application is taking place (Fig. 3).

Fig. 3. Output from the server (top), CartoDB User Interface: database table/"Table View" (middle), CartoDB User Interface: Visualisation/"Map View" (bottom).

Visualization. In the last step, the user is able to use the server provided application through a web browser (cross-platform). The ‚torque' library, provided form CartoDB, enables the use of a map-based dynamic display for movement through JavaScript, CartoCSS and HTML5 (spec. HTML5 Canvas). To allow the efficient transfer of the massive amount of data from the server to the client, a compressed ‚data cube'- format is being used in the sense of a multidimensional OLAP cube (online analytical processing) or data set. The edge of the cube is equivalent to the dimension (cf. Han 2008).

4   Additional Data Processing

The processing (specifically server-side) is an important component to increase the quality and information content of the collected data. Complex procedures are partially necessary, to evaluate realized implementation that have not taken into consideration. The following principle measures are being investigated for implementation depending on the application correlation.

Map Matching. As the determination of the location (e.g. via GPS) can result in an imprecise or insufficient outcome for several reasons, there is the need of techniques that acquire the exact location of a vehicle (or the potential location at a specific time) (cf. Pfoser 2008b). This can be done with the help of Map-Matching algorithms (specifically through Kalman filtering) that can already be applied during data acquisition as well as during later analysis data import or a subsequently processing step. It seems logical that the traffic coverage (for traffic lane analysis) requires a high accuracy for location determination. A respective method can help potentially through an open source software such as pgRouting as PostgreSQL/PostGIS, expansion of road networks (with or without the ability for routing) though OpenStreetMap-Data (OSM), as well as the implementation of a Map-Matching algorithm.

Generation of Trajectories. Since XFCD are generally geometrical and to the point data, a technique that can match the gathered points to the respective trajectories needs to be applied. This step can be implicated during the Map Matching process. Only then, a reasonable evaluation for a number of issues and use cases can be made. Processing is done through special algorithms, e.g. as the semantic method, depending on the issue or the aimed processing stage. When looking at a simple processing or derivation of trajectories, such as single tracks of the specific vehicle, data storage becomes an issue. Since a differentiation between several geometry types is done in spatial DBS (such as in PostgreSQL/PostGIS), a data base table can only contain data of determined types (e.g. point, line, polygon). As data converts from point geometry into line geometry when deriving from trajectories, the database table requires different types of geometry and therefore a modified setup of the database or the whole data base scheme.

Anonymization. Data anonymization, to protect privacy of the user, is an important requirement encompassing the data protection law. Thus, the execution should be involved in the process chain as early as possible. Depending on the technical execution, this can already be done before transmission within the vehicle or directly after transmission to the server, whereas the data anonymization for vehicle and driver should be applied before transmission already. As it is possible to derive trajectories of the single vehicle from the available data (in terms of evaluating movement profiles and to draw conclusions about the departure and final destination), the anonymization of trajectories through the alteration of individual start and target coordinates or points can form a spatial approach. In this regard, three strategies can be identified (Fig 4):

Anonymization strategies (modification of the start and end points of a trajectory)

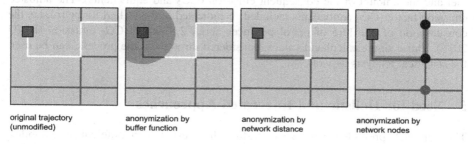

| original trajectory (unmodified) | anonymization by buffer function | anonymization by network distance | anonymization by network nodes |

Fig. 4. Anonymization strategies (modification of the start and end points of a trajectory).

A potential approach for the anonymization of trajectories can be the application of a buffer function, i.e. a removal of a trajectory's segment, which resides within a circular buffer area in the starting point or destination. Another approach can be the removal of trajectory's segments of a defined distance between starting point and destination. Since the movement and distance is looked at in the road network, it can be referred as 'network distance'. A potential third approach is the removal of trajectory's segments to the defined number of passed network nods from start point to destination. To increase security, all three named cases should preferably use random values in a defined range for the defined variables. It needs to be emphasized that the named approaches are a proposing suggestions and is likely to be inadequate for every potential use case.

Generation of Additional Data. Acquisition of data from a database is possible by several ways. One method is ‚data mining' as a "summary of different statistical procedures with the goal to filter relevant information from a massive amount of data, e.g. through intelligent identifying of relationship and distribution patterns" (Bollmann and Koch 2005). Another possibility is to calculate data from the existing one. The following shows an example of a specific calculation which elaborates the correlation between fuel consumption and CO_2 emission and to point out a potential calculation (with the use of the automotive electronics' data). Depending on available data (vehicle or fuel specific), the calculation is possible through several different formulas with different accuracy. The calculation conducted is based on the value of the air-mass sensor (MAF), the vehicles's speed (VSS) as well as the information of the kind of used fuel which also needs to be available (or can be determined by OBD PID 51 FUEL_TYPE). According to the kind of fuel (for further investigation it is petrol), specific constants need to be involved in the calculations: the needed air volume to combust one litre of fuel (14,7 kg) and the average density of petrol (740 kg/m^3).

Through this information, every single and point based data sets for the fuel consumption at a specific point in time can be calculated (so-called current fuel consumption). The described procedure shows that further statistical calculations can be made, e.g. average consumption value (based on the number of kilometres, a trajectory or route, a time frame, or aggregated sections of roads). Based on the information about the current fuel consumption, a calculation of the caused CO_2 emission can be made. It is necessary to know the specific emission factor, the relation between used amount of fuel and the amount of the consequent emission (CO_2 and equivalents). The adjusted emission factor (CO_2 equivalents included) is indicated as 2,874 kg/l for petrol, i.e. the consumption of one litre of petrol produces about 2,87 kg of CO_2 emission (BLfU 2016). Those single calculated values, considered throughout the process, can be used for statistical analysis as well.

5 Need for Holistic and Integrated Approaches

The approach presented can be successful when tested with different automobiles, smartphones, and OBD Bluetooth hardware (quoted data editing and processing excluded). It results in the conduction of the complete technical procedure, from the smart phone supported data (on board) to the transfer in a spatial database and visualization in CartoDB. A suggestion for visualizing a usage scenarios was developed, based on a conducted interview with experts. It entails the setup of a cartographical visualization as well as the setup of user interface and interaction and was implemented in form of wireframes and mock-ups. It shows to expand the addressed approaches to solve the issue of automated generation and deployment of interactive and contextual (geo-)visualization (also for XFCD). The starting point needs to be the general influencing factors and requirements for cartographic visualization.

To demonstrate the transfer of geodata as raw data to an analogue or digital graphic of a map, this pipeline shows the sequence of the process (filtering, mapping, rendering). It becomes clear that the demonstration of data is determined by the step of filtering (included in the XFCD). For mapping and rendering, there are principles and regulations for graphical processing of data, cartography, and the configuration of user interface and interaction available. A strategic approach can exist in definition and application of design patterns for approved solutions for recurring general and organizational tasks. The single design patterns (such as elements from the user interface and interaction) can be consolidated into complete libraries. Regarding cartographic media, the concept of GeoViz patterns is interesting (cf. Heidmann 2013). This shows the definition and application of policies as a guidance for a potential approach (cf. Asche and Engemaier 2012). One potential approach could be to formulate a rule-based graphical user face and interaction. It can be established on fundamental principles and requirements of graphical layouts (especially in cartography, geo-visualization, e.g. graphic variables and cartographical presentation methods, etc.) to solve the respective questions through the help of design patterns. It is to aim for merging of and applying methods and strategies of an appropriate representation for spatio-temporal visualization. A broad formalization can be achieved and consequently the automation of the process of generating (Fig. 5).

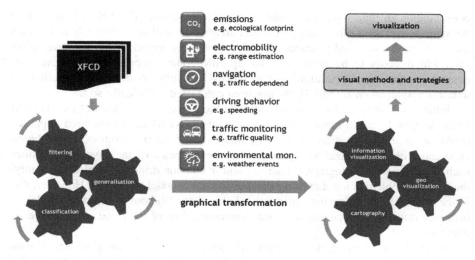

Fig. 5. Potential approach for formalization of manufacturing process (scheme).

Process modelling of knowledge-based systems, specifically rules based systems, indicate a suitable approach for automatic generation of this type of visualization. Decision diagrams (as well as the connection and combination of them) can serve to formulate and display decision rules, which are needed for the generation process. The combination of several methods of software engineering such as design patterns or libraries, service oriented architecture, and model view controller (MVC) with a rule based approach, can be a potential contribution to solve the issue. This approach would be more complex and extensive and would require an additional research work.

6   Conclusion and Outlook

The research work presented in this paper was able to demonstrate an insight and an overview of the utilization of spatio-temporal data within the automotive electronics based on the XFCD concept as well as its technical generation of data. The OBD diagnose interface enables an access to the data bus system of the automobile. This opens the possibility to have a wide but limited (by regular usage) access to specific by law required parameters. XFCD enables a variety of applications (on- and off-board), whereas the data of diverse sensor technology can be emphasized, especially for the detection of the surroundings and environment of a vehicle. As extension of the FCD concept, which is already used for traffic monitoring, XFCD or automobiles are providing an added value as mobile sensors, specifically in the field of environmental monitoring. Interesting application can be the estimation of emission values and the gathering of the concentration of surrounding air (as long as an access for respective sensors is available).

In addition, it needs to be tested whether the combination of XFCD (which show a high spatio-temporal dynamic) and data from stationary measuring sites can support the consolidation of data basis and enable a precise prediction. The field of application through visualization offers many potential areas of operation. Visualization provides the basis for decision and practices. Its elaboration is an important and comprehensive task, which needs to be referred to the specific context (application context, user context, medium for illustration) and to process carefully. A decidedly examination with the visualization of XFCD has not been conducted sufficiently so far.

Fitting and contextual solutions need to be found for this new kind of massive and spatio-temporal data to make effective visualization available. A structured approach could lie in the application of design patterns based on a cartographic rule-based approach with a visualization pipeline. The competencies of the several disciplines such as informatics, cartography, interface and interaction design is needed. Currently, there is the problem of a not existing standardized data model for XFCD. However, this is the basis for an extensive utilization of this data. Without a common model, it becomes difficult to establish an open communication of data and marketplace as platform.

Therefore, it is important to consider the security of the accruing masses of data (big data), which are generated in the vehicle and are potentially transmitted to the outside. As a minimum, the vehicle or driver specific identifying features should be removed to protect privacy. Further approaches need to be evaluated in this context. It is crucial that data is not privileged to interested industry players, e.g. automobile manufacturer or insurance companies. Instead it is necessary that the driver, as the generator of the data, is not only able to determine the utilization of data through external interested parties but also is able to gain insights in its own generated data. Here, geo-visualization is also suitable for making the information of the data legible.

Acknowledgement. This research work is funded with a PhD scholarship by the German Research Foundation (DFG) within the research training group 1539 "Visibility and Visualisation - Hybrid Forms of Pictorial Knowledge" at the University of Potsdam. This support is gratefully acknowledged. The above PhD project is supervised by Hartmut Asche (University of Potsdam) and Frank Heidmann (Potsdam University of Applied Sciences).

References

Doctorow, C.: The Coming War on General Computation. Presented at 28C3 conference in Berlin, Germany (2011). Transcribed by Joshua Wise www.joshua@joshuawise.com. https://github.com/jwise/28c3-doctorow/blob/master/transcript.md. Accessed May 2016

Pfoser, D.: Floating car data. In: Shekhar, S. and Xiong, H., (eds.) Encyclopedia of Geographical Information Sciences, p. 321. Springer, New York (2008a)

Lorkowski, S.: Erste Mobilitätsdienste auf Basis von "Floating Car Data". In: Beckmann, S. et al., (eds.) Stadt, Region, Land – 4. Aachener Kolloqium "Mobilität und Stadt" (AMUS). Institut für Stadtbauwesen der RWTH, Aachen (2003)

Halbritter, G. et al.: Strategien für Verkehrsinnovationen: Umsetzungsbedingungen, Verkehrstelematik, internationale Erfahrungen. edition sigma, Berlin (2008)

Breitenberger, S. et al.: Extended Floating Car Data – Potenziale für die Verkehrsinformation und notwendige Durchdringungsraten. In: Rohleder, M., (eds.) Straßenverkehrstechnik, 10/2014, pp. 522–531. Kirschbaum Verlag, Bonn (2004)

CiA: History of CAN technology. CAN in Automation e.V., Nuremberg (2016). http://www.cancia.org/can-knowledge/can/can-history/. Accessed May 2016

McKinsey: Big Data: The Next Frontier for Innovation, Competition, and Productivity. McKinsey Global Institute, McKinsey & Company Inc., New York (2011). http://www.mckinsey.com/business-functions/business-technology/our-insights/big-data-the-next-frontier-for-innovation. Accessed May 2016

Schneider, S., et al.: Extended floating car data in co-operative traffic management. In: Barceló, J., Kuwahara, M. (eds.) Traffic Data Collection and its Standardization, pp. 161–170. Springer, New York (2010)

Stottan, T.: XFCD als Basistechnologie für die Mobilität 3.0 - Entstehung, Entwicklung, Zukunftsanwendungen, Marktentwicklung. In: Proff, H., et al. (eds.) Schritte in die künftige Mobilität, pp. 47–59. Springer Fachmedien, Wiesbaden (2013)

Schäffer, F.: OBD-Fahrzeugdiagnose in der Praxis. Franzis Verlag, Haar (2012)

Voland, P.: Webbasierte Visualisierung von Extended Floating Car Data (XFCD) - Ein Ansatz zur raumzeitlichen Visualisierung und technischen Implementierung mit Open Source Software unter spezieller Betrachtung des Umwelt- und Verkehrsmonitoring. Master Thesis, University of Potsdam, unpublished (2014)

AOBDR: android-obd-reader and obd-server, Github Repository Android OBD-II Reader Project. Paulo Pires pjpires@gmail.com (2016). https://github.com/pires/. Accessed May 2016

CartoDB: CartoDB Github Repository. Vizzuality Inc., New York (2016). https://github.com/CartoDB/cartodb/. Accessed May 2016

Han, J.: Spatial OLAP. In: Shekhar, S., Xiong, H. (eds.) Encyclopedia of Geographical Information Sciences, pp. 809–812. Springer, New York (2008)

BLfU: Berechnung CO_2 Emissionen. Bayer. Landesamtes für Umwelt, Augsburg (2016). http://www.izu.bayern.de/praxis/detail_praxis.php?pid=0203010100217. Accessed May 2016

Pfoser, D.: Map-matching. In: Shekhar, S. and Xiong, H., (eds.) Encyclopedia of Geographical Information Sciences, pp. 632–633. Springer, New York (2008b)

Bollmann, J., Koch, W.-G.: Lexikon der Kartographie und Geomatik. Spektrum Akademischer Verlag, Heidelberg (2005)

Heidmann, F.: Interaktive Karten und Geovisualisierungen. In: Weber, W., Burmeister, H., Tille, R. (eds.) Interaktive Infografiken, pp. 66–67. Springer, Berlin (2013)

Asche, H., Engemaier, R.: From concept to implementation: web-based cartographic visualisation with cartoservice. In: Murgante, B., Gervasi, O., Misra, S., Nedjah, N., Rocha, A.M.A., Taniar, D., Apduhan, B.O. (eds.) ICCSA 2012, Part II. LNCS, vol. 7334, pp. 414–424. Springer, Heidelberg (2012)

Quality Attributes and Methods for VGI

Jean Henrique de Sousa Câmara[1], Jugurta Lisboa-Filho[1(✉)],
Wagner Dias de Souza[2], and Rafael Oliveira Pereira[1]

[1] Department of Informatics,
Federal University of Viçosa (UFV), Viçosa, MG, Brazil
{jean.camara, jugurta}@ufv.br,
rafa.oliveirap@gmail.com
[2] Engineering Department,
Federal Rural University of Rio de Janeiro (UFRRJ), Seropédica, RJ, Brazil
wagnerdiasdesouza@gmail.com.br

Abstract. The widespread use of GPS-equipped devices such as smartphones and tablets and the easy handling of online maps are simplifying the production and dissemination of volunteered geographic information (VGI) through the internet. VGI systems collect and distribute this type of information and can be used, for example, in cases of natural disasters, mapping, city management, etc. In some cases, the VGI systems must be implemented within a short timeframe, such as in response to emergencies due to natural disasters. Some tools, such as the Ushahidi and ClickOnMap platforms, have been designed to enable the quick development of VGI systems. The quality of data collected by VGI systems has been and must be questioned, but, in some situations, there are no official data or they have become outdated after the event. Hence, these platforms must provide methods to help the user produce or report data voluntarily, but with quality. This article analyzes the main quality attributes that can be applied in the VGI context and identifies, in the literature, methods that contribute to increasing VGI quality. This study resulted in a set of 19 attributes and 23 quality methods related to VGI. This paper also proposes six new methods that can help obtain data with quality assurance. These methods were implemented with success on the ClickOnMap Platform to assuring the quality of VGI.

Keywords: Quality VGI · Volunteered geographic information · Crowdsourcing · Neogeography · Collaborative mapping

1 Introduction

The number of users who provide geographic information through collaborative online systems has been growing mainly due to the widespread adoption of devices equipped with the Global Positioning System (GPS), such as smartphones and tablets. Another reason is how currently users can easily create and share maps and georeferenced photographs online. Such information with geographic location provided voluntarily has been termed by Goodchild [14] as Volunteered Geographic Information (VGI).

© Springer International Publishing Switzerland 2016
O. Gervasi et al. (Eds.): ICCSA 2016, Part III, LNCS 9788, pp. 306–321, 2016.
DOI: 10.1007/978-3-319-42111-7_24

The systems that collect, manage and distribute this type of information are called VGI systems. The OpenStreetMap is an example of this type of system. The mission of this system is to create a set of cartographic data that are free for the users to use and edit [16]. Systems that collect and distribute VGI can be used as a source of data in case of natural disasters. OpenStreetMap was used in the earthquake that shook Haiti in 2010, in the tsunami in Japan in 2011, and after the typhoon that hit the Philippines in 2013 [29].

Some tools speed up the development of these systems, such as the Ushahidi and ClickOnMap platforms. These tools can be used in emergency situations, when the data must be collected and distributed almost in real time [13]. The authors [2, 10, 17, 33] compared VGI data with data from governmental agencies and concluded that VGI data quality is nearing the level of official data, particularly in densely populated urban areas. In order to minimize this gap in quality, new methods and techniques must be identified to assess VGI quality, as suggested by Flanagin and Metzger [11].

This article analyzes the main quality attributes that can be applied in the VGI context and identifies, in the literature, methods that contribute to increasing VGI quality. Six new methods that can raise VGI quality are also proposed. Finally, the ClickOnMap Platform was extended by implementing these methods so that the quality of data collected by future VGI systems developed with it is improved, which will then contribute to the evolution of VGI quality.

The remainder of article is organized as follows: Sect. 2 presents the related works; Sect. 3 shows two tools for quickly development of VGI systems; Sect. 4 presents the main VGI quality attributes and methods, proposes the creation of six new methods to improve VGI quality and shows the relationship between the methods and the attributes this study; Sect. 5 shows how the methods proposed have been implemented and can be used on the ClickOnMap Platform; and Sect. 6 presents conclusions and future works.

2 Related Works

Leite [23] carried out a study on the attributes related to the quality of content on websites associated with healthcare units in order to develop a model to assess quality. The author studied 11 content quality models and found that the research most mentioned by these models was the one by Wang and Strong [32], who were the great pioneers in the evolution of content quality assessment based on the users' perspective.

After identifying the attributes considered by the authors of the most important researches in web content quality, Leite [23] determined the importance of each attribute using the Delphi Method [6]. After three rounds of this method, the author presented the attributes of the quality assessment model of websites related to healthcare units [Table 1].

Since no study was found involving the identification of specific quality attributes for the VGI context, the result of the study by Leite [23] was adapted for the present research since the content of VGI systems is the geographic information provided by volunteers.

Table 1. Attributes of the model proposed by Leite [23] to assess the quality of websites on healthcare.

Categories	Attributes
Intrinsic	Credibility, Exactitude, Current, Accessibility, Confidentiality, Consistency, Usable, and Completeness
Contextual	Precision, Usefulness, Valid, Efficacy, Relevance, Conformity, Surplus value, Efficiency, Specialization, and Traceable
Representational	Understandable, Concise representation, Consistent representation, Legibility, and Attractiveness

3 Tools to Develop VGI Systems

Global climate change, population growth, and the spread of infectious diseases, increased the number of natural and anthropic disasters. A method to mitigate problems related to these disasters is involving citizens of several regions through volunteered collaboration [35]. VGI systems may be an excellent alternative to help a region recover after a natural disaster, as in the aforementioned case of OpenStreetMap in Haiti's earthquake in January 2010 [35]. Therefore, tools are needed that speed up the development of VGI systems such as the Ushahidi and ClickOnMap platforms, which are described in the following subsections.

3.1 Ushahidi Platform

The Ushahidi website [18] was created in early 2007, after the electoral period in Kenya aiming to map, exhibit, and organize information on crime, violence, and peace efforts in this period. The users sent their contributions through the internet and short message system (SMS). In 2010, this website was used in response to the earthquake that destroyed much of Haiti, with a system that received information of cries for help while volunteers translated them, checked their relevance, classified them, and put them on a map for the relief organizations to be able to determine the best way to employ the limited resources [35].

Later, the site spawned the Ushahidi Platform, which can be used to quickly create a VGI system. This tool has features that enable, for example, changing the name, slogan, banner, e-mail, language, time zone, standard map location, zoom level, and map provider. The Ushahidi Platform can be used in conjunction with the four main map providers available today: Esri, Google, Bing, and OpenStreetMap. This tool also enables managing categories, subcategories, users, and contributions. The platform supports adding plugins that may be developed to expand the system's features.

The Ushahidi Platform has a tool called Crowdmap, which enables creating an online collaborative system with no need for hiring a hosting service. Thus, the user just needs to register an account to create a VGI system. However, the systems created by Crowdmap are linked to the Ushahidi company domain and the system's source code is not accessible. This feature and the possibility of receiving contributions via SMS makes this tool stand out since that minimizes the costs with hosting services and maximizes the collection of information because SMS does not require an internet connection.

3.2 ClickOnMap Platform

The ClickOnMap Platform was created to reduce the time and effort required to develop VGI systems. The user using an administrative panel can manage and customize one specific VGI-System. To use this tool does not need any technical knowledge in programming or mapping. The administrative panel enables managing settings, users, categories, types (subcategories), and the contributions made through the system. Moreover, the ClickOnMap Platform has methods that can improve the quality of the information, such as VGI validation from user ratings, definition of a user score system that creates a user ranking, and VGI documentation though the DM4VGI metadata template [31].

In addition, the systems created with the ClickOnMap Platform have VGI filters and statistics that are shown to the users as graphs, which help in decision-making processes. A comparative analysis between these platforms [3] showed that each one has its own peculiarities and that it is up to the developer to choose which best meets his or her needs.

4 Quality Attributes and Methods for VGI

Volunteered information has the advantage of being free, collected quickly, and provide data that are known by a specific group of users, such as detailed information about a community, although such content has no quality assurance as information from official agencies [15]. Thus, assessing VGI data quality remains an open challenge [12, 22, 28, 34] and new methods and tools able to estimate and increase VGI quality are needed [9, 11, 14].

Therefore, a study was carried out on which main quality attributes can be observed in VGI collection and which methods can help observe such attributes. Section 4.1 describes the survey on data quality attributes and their adaptation to the VGI context. Section 4.2 describes the methods identified in the literature. Section 4.3 presents six new methods proposed in this paper to help obtain VGI with better quality, while Sect. 4.4 shows the relationship between the methods and each quality attribute proposed in this study.

4.1 VGI Quality Attributes

Based on the study by Leite [23], a study was carried out to incorporate the attributes resulting from that study into the VGI context. Thus, the categories of the framework proposed by Wang and Strong [32] were used to organize these attributes. Assessing VGI quality may be analogous to assessing the content of websites since VGI is the content of systems that collect and distribute georeferenced volunteered information.

An analysis of the attributes found by Leite [23] shows that some attributes can be merged so that they are used in VGI related content. For example, the attributes "Understandable" and "Legibility" can be merged since, in VGI systems, they aim to facilitate reading and comprehension of the information presented in a contribution. The same can be done with the attributes "Relevance" and "Valid" since these attributes are related to the information's level of importance taking into account the context of VGI systems.

The attributes "Current" and "Surplus value" also have similar purposes regarding VGI, i.e., these attributes express the ability of modifying the content of a contribution either to correct mistakes or because of changes in environments, technologies, or specifications. The attribute "Efficiency" may be related more to the quality of the VGI systems than to the quality of the content in such systems, thus it can be ignored when dealing with VGI quality.

Therefore, 19 attributes have been proposed, which are related to VGI quality and organized according to the four categories of the framework by Wang and Strong [32], as seen in Fig. 1.

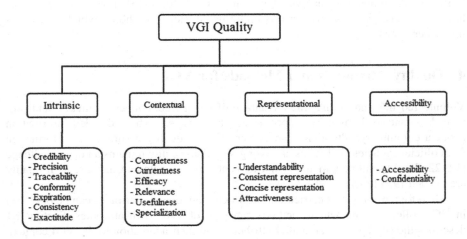

Fig. 1. VGI quality attributes

4.2   Quality Methods for VGI

The literature mentions several VGI quality methods. They were identified 23 methods in the literature, which are shown in Table 2. This table displays one code for each method, which were introduced to facilitate identifying them in the description throughout the text, and the frequency with which they were mentioned in the researchers investigated. Besides these methods, the present study proposes six new quality methods to be implemented in VGI systems. These methods are also listed in Table 2, with the code beginning with the letter N.

A user reputation system (M1) may be an important tool to assess VGI quality since it can help identify the good contributors in a system [1, 7, 11, 25]. Thus, the

Table 2. VGI quality methods and frequency with which they appear in the researchers investigated

Code	Methods	Frequency
M1	User reputation system	5
M2	Wiki review system	4
M3	Information on user identity	2
M4	Automated quality assessment tool	2
M5	Automated metadata	2
M6	Moderators	2
M7	Number of shares in social networks	2
M8	Number of users who liked the VGI	2
M9	Comments and opinions system	2
M10	Self-assessment on geographic precision	1
M11	Data classification by quality criteria	1
M12	Majority decision	1
M13	Merger of volunteered and official data	1
M14	User contribution history	1
M15	Contribution edit history	1
M16	Information on the source of data	1
M17	Number of authors	1
M18	Number of places that cite the VGI	1
M19	Zoom management system	1
M20	Rating system	1
M21	Use of geographic dictionary	1
M22	Use of fixed-field forms	1
M23	Use of step-by-step contribution	1
N1	Use of user-created keywords	*
N2	Word processing	*
N3	Interval between contributions	*
N4	Achievement seals	*
N5	Use tutorial	*
N6	Help system	*

* New VGI quality methods described in Sect. 4.3

contributions by users with good reputation tend to be truer and more acceptable. According to Flanagin and Metzger [11], the quality assessments build by collective or community efforts such as wiki review (M2) and social network applications (M7 and M8) can be new ways of improving the credibility of the information. Other features suggested by those authors are the use of tools to reveal the identity of contributors and editors (M3) through the internet protocol (IP) and the use of a system with the history of contribution editing (M15) for the user to better understand the history of the information.

Metzger et al. [26, 27] suggest including in VGI systems the possibility of the user commenting and expressing his or her opinion (M9) on a given piece of information and showing the number of authors (M17) who have edited the contribution. According to De Longueville et al. [8], a zoom management system (M19) can be used to analyze the exactitude of a contribution since, most times, if a zoom level with lots of details such as lakes, rivers, and land uses is employed at the time of the contribution, such contribution tends to offer good geographic exactitude. Those authors suggest the contributor self-evaluate the geographic precision (M10) of the information.

Cooper et al. [4] believe that information documented through metadata captured automatically (M5) adds quality to it since the user will be able to assess the quality based on its purpose and context. The authors suggest other features such as wiki review (M2) and the use of automated VGI quality assessment tools (M4) by comparing it with other sources of data. Goodchild and Li [15], when analyzing the approaches to improve VGI quality, suggest the use of the following methods: wiki review (M2), moderator users (M6), and automated quality assessment tool (M4).

According to Criscuolo et al. [5], preventing the creation of wrong contributions by instructing the contributor to provide effective data can be a good way of improving VGI quality. That can be achieved by filling out fields following a protocol (M23), using forms with fixed fields (M22), using automatically created metadata (M5), and using a geographic dictionary (M21). The authors also suggest changing faulty contributions through wiki reviews (M2), classifying data through quality criteria (M11), informing the author's experience perhaps through a reputation system (M1), and capturing information from contributors (M3) such as education, experience, age, and location.

Idris et al. [20] propose the following techniques to improve quality: publishing the identity of the data source (M16) by displaying the name, reputation, certificates, and qualification; comment and feedback system (M9); number of shares in blogs and social networks (M7); rating system for contributions (M20); number of people who liked the information (M8); and number of places linking back to the contribution (M18).

Other VGI quality improvement techniques are: employing specialists (M6) as contribution validators [21]; majority decision (M12), i.e., if many users search for and access a contribution, its quality should be good [19]; using the user's contribution history (M14) to infer the quality of the new contribution [24]; and merging volunteered data with authoritative data (M13) [30].

4.3   New Methods for VGI Quality

The present study proposes implementing six new quality methods in the VGI context: use of user-created keywords (N1); word processing (N2); interval of time between contributions (N3); achievement seals (N4); use tutorial (N5); and help system (N6).

Keywords (N1) can describe the topic or subject of a text and allow, through a search system, content to be found by a user of the system, which can help document the information. This way, the use of user-provided keywords (N1) can be a way of documenting the VGI since it allows users to describe the contribution using their own points of view.

Grammar mistakes can decrease legibility and even keep some information from being read. Therefore, as well as user-created keywords (N1), detection and removal of grammar mistakes (N2) in a contribution allows for better quality. Hence, the system can transform words written in capital letters into words written in small letters, start each sentence with capital letters, add and remove spaces from punctuation, check for punctuation that needs to be closed such as parentheses and brackets, remove unnecessary spaces and punctuation, expand abbreviated words, correct agreement mistakes, and remove offensive words.

A VGI system is vulnerable to contributions with various malicious purposes and acts of vandalism. An example of malicious contribution is related to those done through automated software aiming to overload and pollute the system. With such programs, a user can make a large number of contributions within a short period of time. Thus, preventing this type of contribution can be another efficient method to improve VGI quality. That might require setting an interval of time (N3) for the user to make a given number of contributions. For example, the system could be configured to receive at most three contributions from the same user within five minutes. In case the user goes over the number of contributions within the preset interval, the account will be blocked for some time, i.e., the user will not be able to make new contributions.

Knowing the author of a contribution can be key in inferring quality from any given information. However, some VGI systems must omit user data, such as systems that accept anonymous reports. Therefore, there must be a way of knowing the abilities of the users, i.e., what they have learned and achieved by using it. The user achievements could be a good way of presenting the abilities and, consequently, acknowledging the quality of the contributions. These achievements may be available in the user profile illustrated by seals so that other people can visualize, analyze, and assess the quality level of the contributions by that user. Besides allowing user abilities to be visualized, the achievement seals (N4) can work as a sort of bonus and display which users stand out within the system.

A use tutorial (N5) is essential at the first contact between the user and the product and allows the user to properly deal with the system. VGI systems lack user-support materials, which may lead to demotivation and abandon at the first contact with the system. Besides introducing the system to the users, the tutorial must teach the basics so that the users are able to make their first contributions.

The help system (N6) can be a good tool in VGI systems since it allows for minimizing the learning time and increasing the quality of the contributions. The help system must consist of steps, unlike the use tutorial (N5), in which the user can choose and visualize any part of the tutorial. Thus, the help system (N6) is a sequence of steps aiming to teach how to use the system and train the users.

At their first login, the users should be instructed to make their first contributions. Therefore, the system can display a dialog informing the first action to be performed so that the user begins the contribution, such as left-clicking on the desired spot in the map. These dialogs may show tips that help the user make the contribution, such as informing that the higher the zoom level used, the more details will be displayed and, consequently, the more exact the contribution. The help system may improve the level of interaction between the system and the users by playing the role of an instructor when it offers help. Therefore, the users can learn geographic information collection concepts and techniques and, consequently, make better contributions.

4.4   VGI Quality Attribute vs. Methods

This section relates each VGI quality attribute identified in Sect. 4.1 to the methods described in Sects. 4.2 and 4.3. The result is shown in Table 3. When this table is analyzed, it shows that methods M5, M13, and M15 are those related to the largest number of attributes at up to three different categories. Indeed, the use of official data in the VGI context (M13) can greatly impact the quality of a contribution since it brings more reliability and a greater level of details.

Table 3. Methods related to each VGI quality attribute

	Attributes	Methods
Intrinsic	Credibility	M1, M2, M3, M4, M5, M6, M7, M8, M10, M12, M13, M14, M15, M16, M17, M18, M20, N3, N4
	Precision	M13
	Traceable	M5, M15, N1
	Conformity	M5, M13
	Expiration	M15
	Consistency	–
	Exactitude	M1, M13, M19
Contextual	Completeness	M9, M10, M13, M15. M23
	Current	M2, N2
	Efficacy	M4
	Relevance	M4, M5, M11, M13, M18, M20
	Usefulness	M7, M8
	Specialization	M5, M13
Representational	Understandable	M13, M15, M21, N2, N5
	Consistent representation	M22, M23, N6
	Concise representation	–
	Attractiveness	M7, M8, M9, M17, M18
Accessibility	Accessibility	–
	Confidentiality	M6

The edit history in contributions (M15) can be crucial in systems that have a wiki review method (M2) since ill-intentioned users will want to damage the information. The automated metadata (M5) also stood out regarding quality attributes since this method allows VGI details such as source and author to be verified, which helps infer the quality of the information. Another fact to be noted is the number of methods related to the quality attribute Credibility. These methods verify and guarantee the veracity and acceptability of the information.

5 ClickOnMap Platform Extension

The ClickOnMap Platform was developed by students of the Computing Department of the Federal University of Viçosa and, as seen in Sect. 3.2, it can be used to quickly develop and customize VGI systems. Since its source code is available for download, the ClickOnMap Platform was used to implement the six methods proposed in this paper. All methods implemented, by default, are enabled in the systems developed with the aid of the ClickOnMap Platform, however, when the method manager is accessed, the administrator may enable or disable any method desired, as seen in Fig. 2.

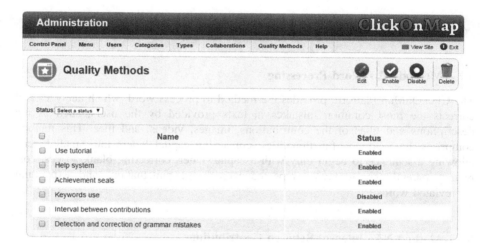

Fig. 2. ClickOnMap Platform's method manager

The following subsections show the details of each method implemented on the ClickOnMap Platform.

5.1 Method N1 - Use of User-Created Keywords

At the end of each contribution, the user may be asked to provide some keywords that describe the subject of the VGI content. For example, a contribution that reports a robbery may use the keywords "violence," "robbery," "death," "firearm," etc. That enables checking which subjects are being contributed the most and improve the VGI documentation. This method allows for a history of the keywords employed by the user and inferring the quality level of new contributions by this user, e.g., if he or she often uses the keywords "deforestation," there's a good chance that this user is knowledgeable in the subject.

The number of keywords can be defined in the administrative panel, as shown in Fig. 3. The administrator can also inform whether the use of keywords will be mandatory or optional.

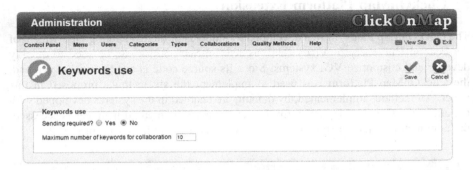

Fig. 3. Keyword manager

5.2   Method N2 - Word Processing

The ClickOnMap Platform provides the method to process word, which analyzes and corrects the most common mistakes in texts provided by the users, such as the descriptions and titles of the contributions, images, videos, and files. This method comprises the following functionalities: converting capital letters into small letters, allowing a sentence to begin only with a capital letter, correcting blank spaces, correcting parentheses, removing repeated words and punctuation, and expanding abbreviated words.

5.3   Method N3 - Interval Between Contributions

The ClickOnMap Platform employs the method proposed in Sect. 4.3, in which a user is limited to a maximum number of contributions over a certain period of time. The systems developed with this platform are set to receive at most five contributions by the same user within five minutes between the first and last contributions and a suspension for N hours, where N is a configurable value.

Thus, if a user tries to make more than five contributions to the system in less than five minutes, he or she receives a warning message and the account is blocked for N hours, i.e., a new contribution will only be possible after this suspension period. The administrator may change the settings in this module simply by informing new values, as seen in Fig. 4.

5.4   Method N4 - Achievement Seals

The platform provides some types of seals to be distributed to the VGI system's users. The system administrator can manage the achievement seals in a module of the

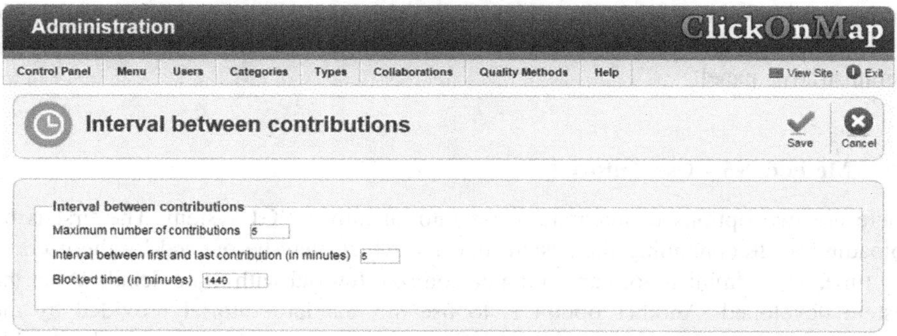

Fig. 4. Manager of the module of time interval between contributions

Table 4. Main achievement seals available on the ClickOnMap Platform

Name	Image
Expert user	
Number of contributions	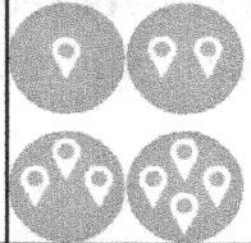
User classes	
Position in the user raking	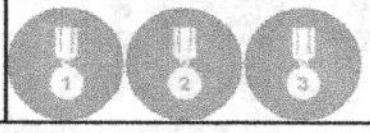

administrative panel. Thus, seals can be edited or excluded. Table 4 displays some of the seals available on the ClickOnMap Platform.

In order to use the expert user seal, the administrator must edit it to inform the time needed since the registration for the user to be able to achieve it. The platform provides four seals regarding the number of contributions by a user. Therefore, the administrator must edit and indicate the minimum number of contributions the user must make to achieve each seal. By default, these values are 10, 50, 200, and 500.

The ClickOnMap Platform has six user classes: Malicious contributor, Basic contributor, Nice contributor, Master contributor, Expert contributor, and Special contributor. These classes group the users according to their scores and a seal for each class is achieved by the users who reach the minimum score of each class. This minimum score can be changed by the system administrator by editing each user class.

The platform also provides the user ranking feature, with three seals for the top users in the ranking. When a user loses the position to another, the seal is also lost to that user, which makes these seals unique in the system. When a seal is achieved, the user may, at the administrator's discretion, receive a message by e-mail congratulating the user and informing about the achievement. Moreover, the administrator can enable or disable any achievement seal at any moment through the seal manager in the administrative panel.

5.5   Method N5 - Use Tutorial

There are two options to integrate a use tutorial into a VGI system. The first is by uploading a file containing the system tutorial, which may be opened by the users at any time. The administrator can send a customized tutorial with more details about the system developed. Another option is to use the standard tutorial provided by the platform. This tutorial can be accessed in Portable Document Format or as browsable text (HTML pages).

5.6   Method N6 - Help System

Along with the use tutorial, the system must provide the users with a help system. Thus, the system dynamically helps the users to carry out certain actions, such as contributing, uploading a picture or a video, editing the contribution, or using the wiki system to comment on a VGI. As seen in Sect. 4.3, the help system comprises steps that must be followed in order. Figure 5 shows the help system displayed when the user first logs into the system.

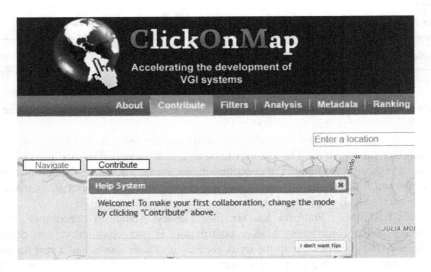

Fig. 5. Help system on the ClickOnMap Platform

As seen, the welcome message is followed by the steps to be taken for the user to make the first contribution to the system. When the users follow the message and click on "Contribute", a second message will appear informing what the next step will be. The user can leave the help system at any moment by clicking on the link "I don't want tips [x]". When the user leaves the help system, a warning can be displayed – or an e-mail can be sent – informing how to open the help system in case the user wishes to reactivate it.

All these features are optional and the administrator can enable or disable any feature in this method in the administrative panel.

6   Conclusions and Future Works

This paper proposes a set of VGI quality attributes that can be observed and a set of methods to ensure VGI quality. Based on a quality assessment model for healthcare website content, the main quality attributes that could be associated with the VGI were selected. These attributes were organized into four data-quality categories: Intrinsic, Contextual, Representational, and Accessibility. Thus, a set of 19 quality attributes was reached to be applied to the VGI context.

Besides this study, a literature survey was carried out to relate the main VGI quality methods available. That survey found 23 VGI quality methods. Besides those, six new methods are proposed to help improve VGI quality. They are: use of user-created keywords, word processing, interval of time between contributions, achievement seals, use tutorial, and help system.

Finally, the new methods were incorporated into the ClickOnMap Platform. When a VGI system is developed using this platform, administrators are able to choose which quality methods they want to use since the administrative panel allows enabling and disabling any quality methods available on the ClickOnMap Platform. That is expected to increase the quality level, make the information collected by VGI systems more reliable, and bring that quality closer to the data from official agencies, i.e., this set of attributes and the new quality methods contribute to the evolution of VGI quality.

In the future, new methods may be added to this platform, such as: displaying user information through a user profile, capturing the number of shares in social networks and the number of people who liked the information, user contribution history, and self-assessment of the geographic precision. One relevant work to be done is adding the methods proposed in this paper to OpenStreetMap, which is one of the main VGI systems.

Acknowledgments. This wok was partially financed by CAPES, CNPq, Cemig and FAPEMIG.

References

1. Arsanjani, J.J., Barron, C., Bakillah, M., Helbich, M.: Assessing the quality of open-streetmap contributors together with their contributions. In: Proceedings of the 16th AGILE International Conference on Geographic Information Science, Leuven, Belgium, 14–17 May 2013

2. Brovelli, M.A., Minghini, M., Molinari, M., Mooney, P.: Towards an automated comparison of OpenStreetMap with authoritative road datasets. Trans. GIS (2016). doi:10.1111/tgis. 12182

3. Câmara, J.H.S., Almeida, T.T., Carvalho, D.R., Ferreira, T.B., Balardino, A.F., Oliveira, G.V., Fonseca, F.J.B., Ramos, R.S., Souza, W.D., Lisboa-Filho, J.: A comparative analysis of development environments for voluntary geographical information web systems. In: Proceedings of the XV Brazilian Symposium on Geoinformatics – GeoInfo, Campos do Jordão, Brazil, 29 November to 02 December 2014, pp. 130–141 (2014)

4. Cooper, A.K., Coetzee, S., Kaczmarek, I., Kourie, D.G., Iwaniak, A., Kubik, T.: Challenges for quality in volunteered geographical information. In: AfricaGEO, Cape Town, South Africa, 31 May to 2 June 2011, p. 13 (2011)

5. Criscuolo, L., Pepe, M., Seppi, R., Bordogna, G., Carrara, P., Zucca, F.: Alpine glaciology: an historical collaboration between volunteers and scientists and the challenge presented by an integrated approach. ISPRS Int. J. Geo-Inf. 2(3), 680–703 (2013)

6. Dalkey, N.C., Bernice, B.B., Samuel, C.: The Delphi Method: An Experimental Study of Group Opinion. Rand Corporation, Santa Monica (1969)

7. D'Antonio, F., Fogliaroni, P., Kauppinen, T.: VGI edit history reveals data trustworthiness and user reputation. In: Proceedings of the 17th AGILE International Conference on Geographic Information Science, Castellón, Spain, 3–16 June 2014

8. De Longueville, B., Ostlaender, N., Keskitalo, E.C.H.: Addressing vagueness in Volunteered Geographic Information (VGI) - a case study. Int. J. Spat. Data Infrastruct. 5, 1–16 (2009)

9. Elwood, S.: Volunteered geographic information: future research directions motivated by critical, participatory, and feminist GIS. GeoJournal 72(3–4), 173–183 (2008)

10. Fan, H., Yang, B., Zipf, A.: A polygon-based approach for matching OpenStreetMap road networks with regional transit authority data. Int. J. Geograph. Inf. Sci., 1–17 (2015). doi:10. 1080/13658816.2015.1100732

11. Flanagin, A.J., Metzger, M.J.: The credibility of volunteered geographic information. GeoJournal 72(3–4), 137–148 (2008)

12. Fonte, C.C., Bastin, L., Foody, G., Kellenberger, T., Kerle, N., Mooney, P., Olteanu-Raimond, A.M., See, L.: VGI quality control. In: ISPRS Annals of Photogrammetry, Remote Sensing and Spatial Information Sciences, La Grande Motte, France, 28 September to 03 October 2015, pp. 317–324 (2015)

13. Georgiadou, Y., Bana, B., Becht, R., Hoppe, R., Ikingura, J., Kraak, M., Lance, K., Lemmens, R., Lungo, J.H., McCall, M., Miscione, G., Verplanke, J.: Sensors, empowerment, and accountability: a digital earth view from East Africa. Int. J. Digit. Earth 4(4), 285–304 (2011)

14. Goodchild, M.F.: Citizens as voluntary sensors: spatial data infrastructure in the world of web 2.0. Int. J. Spat. Data Infrastruct. 2, 24–32 (2007)

15. Goodchild, M.F., Li, L.: Assuring the quality of volunteered geographic information. In: Spatial Statistics, pp. 110–120 (2012). doi:10.1016/j.spasta.2012.03.002

16. Haklay, M., Weber, P.: Openstreetmap: user-generated street maps. In: IEEE Pervasive Computing, pp. 12–18 (2008). doi:10.1109/MPRV.2008.80

17. Haklay, M.: How good is volunteered geographical information? A comparative study of openstreetmap and ordnance survey datasets. In: Environment and Planning B: Planning and Design, pp. 682–703 (2010). doi:10.1068/b35097

18. Heinzelman, J., Waters, C.: Crowdsourcing crisis information in disaster-affected Haiti. US Institute of Peace (2010)

19. Hirth, M., Hobfeld, T., Tran-Gia, P.: Analyzing costs and accuracy of validation mechanisms for crowdsourcing platforms. Math. Comput. Model. **57**(11–12), 2918–2932 (2013). doi:10.1016/j.mcm.2012.01.006
20. Idris, N.H., Jackson, M.J., Ishak, M.H.I.: A conceptual model of the automated credibility assessment of the volunteered geographic information. IOP Conf. Ser.: Earth Environ. Sci. **18**(1), 012070 (2014)
21. Jackson, S.P., Mullen, W., Agouris, P., Crooks, A., Croitoru, A., Stefanidis, A.: Assessing completeness and spatial error of features in volunteered geographic information. ISPRS Int. J. Geo-Inf. **2**(2), 507–530 (2013)
22. Keßler, C., Trame, J., Kauppinen, T.: Tracking editing processes in volunteered geographic information: the case of OpenStreetMap. In: Identifying Objects, Processes and Events in Spatiotemporally Distributed Data (IOPE), Workshop at Conference on Spatial Information Theory, vol. 12 (2011)
23. Leite, P.I.S.T.S.: Modelo para avaliação da qualidade de conteúdos de sítios Web de Unidades de Saúde. Ph.D. Thesis, University Fernando Pessoa, Portugal, Porto (2014)
24. Mashhadi, A.J., Capra, L.: Quality control for real-time ubiquitous crowdsourcing. In: Proceedings of the 2nd International Workshop on Ubiquitous Crowdsourcing, pp. 5–8 ACM (2011). doi:10.1145/2030100.2030103
25. Maué, P.: Reputation as tool to ensure validity of VGI. In: Workshop on volunteered geographic information (2007)
26. Metzger, M.J., Flanagin, A.J., Medders, R.B.: Social and heuristic approaches to credibility evaluation online. J. Commun. **60**(3), 413–439 (2010)
27. Flanagin, A.J., Metzger, M.J.: The credibility of volunteered geographic information. GeoJournal **72**(3–4), 137–148 (2008)
28. Mooney, P., Corcoran, P., Winstanley, A.C.: Towards quality metrics for openstreetmap. In: Proceedings of the 18th SIGSPATIAL International Conference on Advances in Geographic Information Systems, pp. 514–517 (2010). doi:10.1145/1869790.1869875
29. Neis, P., Zielstra, D.: Recent developments and future trends in volunteered geographic information research: the case of OpenStreetMap. Future Internet **6**(1), 76–106 (2014). doi:10.3390/fi6010076
30. Pourabdollah, A., Morley, J., Feldman, S., Jackson, M.: Towards an authoritative OpenStreetMap: conflating OSM and OS OpenData national maps' road network. ISPRS Int. J. Geo-Inf. **2**(3), 704–728 (2013). doi:10.3390/ijgi2030704
31. Souza, W.D., Lisboa-Filho, J., Filho, J.N.V., Câmara, J.H.S.: DM4VGI: a template with dynamic metadata for documenting and validating the quality of volunteered geographic information. In: Proceedings of the Xiv Brazilian Symposium On Geoinformatics - Geoinfo, Campos do Jordão, Brazil, 24–27 November 2013, pp. 1–12 (2013)
32. Wang, R.Y., Strong, D.M.B.: Accuracy: what data quality means to data consumers. J. Manage. Inf. Syst. **12**(4), 5–33 (1996)
33. Zielstra, D., Zipf, A.: A comparative study of proprietary geodata and volunteered geographic information for Germany. In: 13th AGILE International Conference on Geographic Information Science, Leuven (2010)
34. Zielstra, D., Hochmair, H.H., Neis, P.: Assessing the effect of data imports on the completeness of openstreetmap - a united states case study. Trans. GIS **17**, 315–334 (2013). doi:10.1111/tgis.12037
35. Zook, M., Graham, M., Shelton, T., Gorman, S.: Volunteered geographic information and crowdsourcing disaster relief: a case study of the Haitian earthquake (2010). doi:10.2202/1948-4682.1069

Bacteria Foraging Optimization
for Drug Design

Sally Chen Woon Peh[1,2(✉)] and Jer Lang Hong[1,2]

[1] School of Biosciences, Taylor's University, Subang Jaya, Malaysia
chenwoonpeh@yahoo.com, jerlang.hong@taylors.edu.my
[2] School of Computing and IT, Taylor's University, Subang Jaya, Malaysia

Abstract. The bacterial foraging optimization (BFO) method has been successfully applied in a number of optimization problems, especially alongside Particle Swarm Optimization as hybrid combinations. This relatively recent method is based on the locomotion and behavior of bacteria E.coli, with modifications made over the years to increase search time, space and reduce convergence time. Regardless of changes, BFO algorithms are still based on 4 main features which are Chemotaxis, reproduction, swarming, elimination and dispersal behaviours of E.coli. A nature based algorithm, BFO has been utilized in several optimization problems such as the power loss reduction problem and in the area of PID applications. Ligand docking is another optimization problem that can potentially benefit from BFO application and this paper will focus on the methodology of BFO application and its results. We are of the opinion that the incorporation of BFO in the ligand docking problem is effective and efficient.

Keywords: Bioinformatics · Bacterial foraging optimization · Protein ligand docking · Drug design · Meta-heuristics

1 Introduction

The working mechanism of drugs heavily relies on the accurate binding of drug molecules to complementary structures or protein targets. Not only do ligands have to fit well to their binding sites, they have to bind with high affinity for increased effectiveness of drugs. The field of study is termed 'docking', which refers to the usage of computer aided drug design (CADD) in predicting binding modes between ligand and possible receptors. Over the years, this field has greatly matured with an ever-increasing database of proteins and nucleic acids contributed by X-ray crystallography and nuclear magnetic resonance. The mid-1980s also saw significant breakthrough [1] with the first revolutionary programme called DOCK. Since then, many algorithms have incorporated it for ligand conformation sampling [2].

Many factors affect CADD [3, 4] and a prime example is the equilibrium state of the ligand-receptor structure. Should complementary binding molecules have low free energy, it would be easier to have a stable structure. This binding energy is represented by a scoring or objective function for the prediction of ligand docking between two previously unpaired molecular structures. Many algorithms such as the Genetic Algorithm, Simulated Annealing and Tabu Search are examples.

© Springer International Publishing Switzerland 2016
O. Gervasi et al. (Eds.): ICCSA 2016, Part III, LNCS 9788, pp. 322–331, 2016.
DOI: 10.1007/978-3-319-42111-7_25

However before binding strength can be studied, it is crucial to first identify the protein surfaces. Functional site detection and functional site similarity [5] are 2 major geometrical aspects of ligand docking for the search of binding locations on molecules. LIGSITE, POCKET, APROPOS, vdw-fft and Drugsite are example approaches that focus on the location of binding sites [6]. The matter is further made complicated by the naturally non-static state of molecular structures, hence the need for a protein frame that is flexible. In addition, binding association takes into account the degrees of freedom involved such as hydrophobic, van der waals and electrostatic forces which may affect ligand docking. Also, the internal flexibility within ligands, receptors and solvent molecules should be considered when making predictions of binding modes and affinity [7].

Considering the many factors mentioned above, efficient docking softwares should utilise search algorithms that are more practical than thorough when searching through vast search spaces. Practical scoring functions [8, 9] will also help determine binding strength for better docking prediction whereby lower binding energies will lead to better docking processes.

The following section describes meta heuristic techniques that have found success in the ligand docking problem. Section 3 will center on BFO and its applications. Sections 5 and 6 will cover its application in ligand docking, followed by our conclusion.

2 Related Work

2.1 Genetic Algorithm

Genetic Algorithm or GA is widely incorporated in many docking softwares [10–12] or used in combination with other algorithms [13] ever since its conception by John Holland. This heuristic is based on crossover and mutation operators that determine the human genetic structure. These changes lead to formation of new genetic traits which when found essential to a particular environment, will lead to evolution of the overall population. As a result, offsprings possess 'better' qualities than the parent generation, contributing to new searching perspectives in the same search space. Rather than waiting for evolution to occur, mutation operators can be applied directly to increase diversity in the current traits of the population.

2.2 Simulated Annealing

Simulated Annealing [14, 15] works are based on the annealing process whereby glass is cooled in a non-linear method to obtain different glass formation types (different searching perspectives). Initially, the system sets off with high melting temperature before consecutive steady cooling. With every decrease in energy of the system, the next cooling step will ensue. For continuously improved solution, uphill steps are incorporated into the system. This heuristic is popular for positional and conformational searching. SA has been incorporated in many other heuristics to solve optimizational problems [16–18].

2.3   Ant Colony System

Ants tend to search for the shortest route to and fro from a food source to its nest. To mark a new path, the ant would drop pheromones for other ants to follow. As more ants choose (with high probability) to follow the same path, the pheromones count will increase, eventually becoming the preferred route for the colony. However, due to ants' constant search for the shortest pathway, the solution may change over time, introducing increasingly optimal solutions to the problem. This population based approach [19] has been used to solve the Job-shop scheduling problem (SP), Quadratic assignment problems (QAP) and similar versions of the Travelling Salesman Problem (TSP).

2.4   Particle Swarm Optimization

Kennedy and Eberhart introduced the Particle Swarm Optimization (PSO) in 1995: a population based search algorithm initialized with random solutions called particles [20]. These particles are associated with a velocity dynamically adjusted to each particles' and all particles' personal best achievement. This is to facilitate the search for better solutions. Unlike the global version, the local PSO differs slightly, with each particles' velocity adjusted to the personal best performance in a neighborhood. Although global PSO converges faster, Kennedy claimed that local PSO is better able to solve complex problems [21]. This algorithm is based on the swarm theory where birds continuously find better food sources through social cooperation with birds in the neighborhood. Quantum Binary PSO was utilized by Ghosh et al. [22] in drug design, who have determined that better results are obtained when ligand length is considered as a variable. Meanwhile, Yu Liu et al. [23] introduced FIPSDock (Fully Informed Particle Swarm) to ligand docking in 2013, and is confident that it is more suitable than conventional GA in flexible docking.

3   BFO

Bacterial foraging optimization (BFO) draws connections between the behavior of bacteria in foraging for nutrients with problem optimizations. To search for food, most bacteria are motile with the help of one or more helical flagella acting as propellers. Different bacteria has different locomotion styles, and BFO is derived from the the multiflagellated *Escherichia coli*.

Due to the wealth of knowledge on the common E.coli, its motion formed the basis of what we know of bacterial movement. E. coli is peritrichous, which means its rod shaped body is attached to 4–8 flagella, driven by a reversible rotary motor affected by proton influx [24] (Berg 2004). These motors can spin counterclockwise through hydrodynamic interactions, propelling E.coli in a forward movement –'run'at a high speed of ~ 30 µm/s; these motors can also spin in different directions to cause a change in direction and for E.coli to 'tumble' [25]. This 'run-tumble' pattern does not lead to specific angle reorientation hence the direction of movement is random, albeit only slightly different from the original run. This mechanism is common among swimming organism such as prokaryotes (i.e. S. marcescens and V. alginolyticus) and eukaryotes (i.e. C. reinhardtii) [26] (Fig. 1).

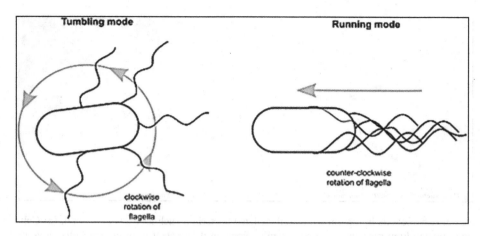

Fig. 1. Tumbling and running modes associated with swimming organisms such as *Escherichia coli and Salmonella typhimurium*. CW rotation results in tumbling or random reorientation while CCW rotation of flagellar motors produces an approximately straight-line motion. Running mode induce faster movement than tumbling mode [27].

When ideating Bacteria Foraging Optimization Algorithm (BFOA), founder Passino referred to 4 phenomena of E.coli behavior: chemotaxis, swarming, reproduction, elimination and dispersal.

3.1 Chemotaxis

'Taxes' are motion patterns of bacteria in the presence of chemical attractants or repellants, hence the term 'Chemotaxis'. Generally, this system is placed to ensure E.coli is able to find food and other necessary metabolites besides avoiding harmful substances. Of course, to search for nutrients in new search spaces, E.coli has to alternate randomly between tumbling and running. When E.coli encounters a change in concentration of a particular attractant, the ratio of running to tumbling will increase whereby the direction of the movement will be inclined towards increasing nutrients gradient [28]. Meanwhile should a repellant be present instead, mean run time will decrease and the tumbling rate increased for E.coli to effectively leave the current search space while chancing across attractants. Under constant nutrient concentration, the bacteria returns to its original rate and set of movement, however it will always be on the lookout of higher concentrations of food. This judgement is made by a constant comparison with the concentration observed the past 4 s for determination of movement direction [29] (Fig. 2).

3.2 Swarming

Swimming or moving independently is as described above. However, flagellated bacteria can also opt to move alongside other cells in a thin film of liquid, this

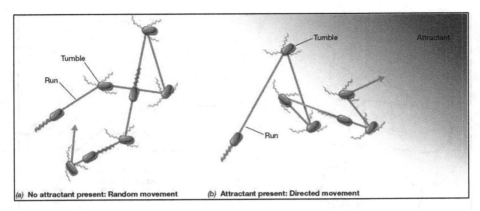

Fig. 2. Chemotaxis in a peritrichously flagellated bacterium such as Escherichia coli. (a) Cells swim randomly in runs and tumble to change direction if attractant is absent (b) Cells move up the gradient of the attractant with biased movement. (Attactant gradient is shown as shaded area) [30]

locomotion mode is known as swarming. The decision to swim or swarm depends on the availability of nutrients in the environment. Should E. coli be suspended in a homogenous, isotropic agar, it would swim independently; when it is grown on rick medium, E.coli elongate, produce more flagella and swarm [31]. A shared similarity is that both requires thrust generated by rotating helical flagella. The speed of movement of bulk swimming is similar to swarming cells' speed, but less broadly distributed. The normal run-tumble movement is largely suppressed in the sense whereby cells do not reorient as much. When moving with neighbouring cells, constant jostling will naturally reorient cells every few seconds without much need for CW rotation [32]. They move with coordination, in 4 directions of forward, lateral, reverse and stalling [31].

3.3   Reproduction

Reproduction of organisms has one main purpose, and that is to keep the population size constant. When in unfavourable environments that causes starvation and sickness, reproduction is especially important for the maintenance of swarm size (or in the case of BFOA, the search size). Healthy E.coli (or those yielding lower objective function values) asexually splits into 2 bacteria from the middle of the progeny rod [33]. The rate of reproduction is once every 30 min, and is high enough to allow adaptation to the surroundings [34].

3.4   Elimination and Dispersal

In line with the need for reproduction, environments can change either gradually or suddenly due to factors such as consumption of nutrients or abrupt change in abiotic factors. As a result, cell death or migration may ensue. Chemotaxis will either be heavily augmented or diminished, depending on the environment in which the bacteria are dispersed to [28]. Events can take place such that bacteria are liquidated at random and placed all over the search space [33].

3.5 Successful Applications of BFO

After initially developing BFO, Passino himself has applied BFO in optimization problems with comparable success [35, 36]. It is also commonly analysed and used alongside PSO [37, 38]. A review of BFO applications by Sharma et al. [39] also took note of adaptive bacterial foraging optimization (ABFO) whereby larger swim length is assigned to local global optima and has better search time and velocity modulated bacterial foraging optimization technique (VMBFO) from BF-PSO hybridization to reduce convergence time by reducing random search when in particle form. All the above mentioned modifications of the original BFO surpassed BFO in terms of search time, search space, convergence time and better ability to escape from the local optima.

4 Proposed Solution

To fully implement bacteria foraging optimization method, we first construct a 3D modelling of the protein and ligand molecular structure. Each of the geometry, atoms, and the bond between molecules are constructed based on the data available in protein database. Due to the fact that bacteria foraging method is a population based optimization method similar to the Ant Colony Systems and Genetic Algorithm, we need a parallel processor so that our algorithm could run efficiently.

Once the method is implemented, we need to test the effectiveness of the algorithm with respect to the problem given. We divided the protein ligand docking problem and its effectiveness into two stages, the first being the geometry and orientation of the molecules while the second is the binding energy generated after the docking process. We started the docking process by generating a population of bacteria. These bacteria are randomly generated by the system. It is worth noting that in some cases, even the binding energy is low, there may be some impairment between molecules, therefore making the docking process invalid. As such, we only consider the docking process successful if the protein and ligand bind to each other successfully. We use scoring function to evaluate the binding energy. Binding energy is evaluated based on the following scoring function:

$$\Delta G_{bind} = \Delta G_{vdw} + \Delta G_{H\text{-}bond} + \Delta G_{hydrophobic} + \Delta G_{rotor}$$

Where ΔG_{vdw} is the van der Waal energy, $\Delta G_{H\text{-}bond}$ is the hydrogen bonding, $\Delta G_{hydrophobic}$ is the hydrophobic bonding, and ΔG_{rotor} is the rotatable bonds.

To start the simulation process, we choose the first bacteria to bind with the protein molecular structure. We then calculate the binding energy and continue the process with the second bacteria. The binding energy is then recorded for the second bacteria and the remaining bacteria is then used to bind with the protein molecular structure where their respective binding energy is calculated and recorded. Once the process completed, we move to the next generation where new population of bacteria is generated. If the bacteria from the previous generation could bind well to the protein molecule, we keep them for the next generation where further orientation is made to its structure so that a better binding affinity can be achieved. Otherwise, the bacteria that do not binds well to the protein molecule is discarded and a new generation is produced randomly to replace the older generation. In each generation, a random mutation is applied to the bacteria population to ensure a well-diversified solution. The simulation process runs for 10000 generations. At the end of the process, we take the best solution where we assume is the most optimal solution.

5 Experimental Evaluation

We use state of the art AutoDock to evaluate and benchmark the performance of our system. AutoDock is chosen due to the fact that it uses meta heuristic method which is similar to ours. We measure the success rate of docking by calculating the binding energy of the protein ligand docking. The binding energy is only calculated after the ligand successfully docked to the protein structure. We also make sure that the respective molecules are paired between the ligand and protein structure. We test our method with respect to that of AutoDock on protein databank (PDB database at http://www.wwpdb.org/), where we randomly choose 50 random samples comprising simple and complex molecules. Simple molecules are those with lesser than 50 atoms and comprise of simple energy binding whereas complex molecules are those with more than 100 atoms with a number of energy bindings as well as hydrophobic binding. We also measure the convergence time of our method with respect to AutoDock (Fig. 3).

Table 1. The average binding energy with respect to execution time

Time (ms)	AutoDock	Our method
5	3.36	3.24
20	1.56	1.42
60	1.24	1.01
300	0.98	0.85
10000	0.76	0.64

As shown in Table 1, our method outperforms AutoDock for protein ligand docking. This is due mainly to the fact that our method employs additional variables and features compared to AutoDock which employs Genetic Algorithm. Our method uses swimming and taxing to converge to a solution quickly while uses mutation to avoid getting into local optima. We believe that our method is novel and it will be highly useful to show that bacteria foraging optimization is useful for other NP-Hard problem.

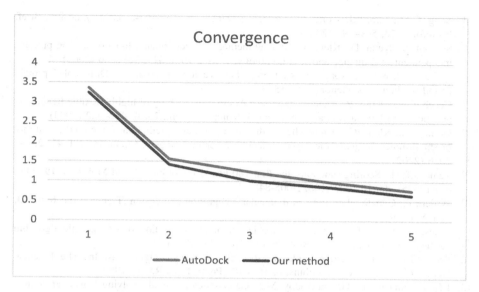

Fig. 3. Convergence graph (Color figure online)

6   Conclusions

We believe that BFO is as good as, if not better than many current meta heuristics methods available to solve the ligand docking problem Similar to past success in application to optimization problems, BFO once again proved that it is a potential answer to yet another optimization problem. Suggestions for future studies may include comparisons between different bacterial locomotion styles to be incorporated in our methodology. An example is the turning mechanism used by Vibrio alginolyticus which has been proven to surpass E.coli's ability in climbing nutrient gradients (chemotaxis) (Xie et al. 2011). We are of the opinion that further refinements on BFO will enable it to stand atop other heuristics in the field of ligand docking.

Acknowledgement. This work is carried out within the framework of a research grant funded by Ministry of Higher Education (MOHE) Fundamental Research Grant Scheme (Project Code: FRGS/1/2014/ICT1/TAYLOR/03/1).

References

1. Blundell, T.L.: Structure-based drug design. Nature **384**, 23–26 (1996). doi:10.1038/384023a0
2. Jones, G., Willette, P., Glen, R.C., Leach, A.R., Taylor, R.: J. Mol. Biol. **207**, 727 (1997)
3. Huang, S.Y., Zou, X.: Advances and challenges in protein-ligand docking. Int. J. Mol. Sci. **11**, 3016–3034 (2010)

4. Song, C.M., Lim, S.J., Tong, J.C.: Recent advances in computer-aided drug design. Brief. Bioinform. **10**, 579–591 (2009)
5. Schmitt, S., Kuhn, D., Klebe, G.: A new method to detect related function among proteins independent of sequence and fold homology. J. Mol. Biol. **323**, 387–406 (2002)
6. An, J., Totrov, M., Abagyan, R.: Comprehensive identification of "Druggable" protein ligand binding sites. Genome Inf. **15**, 31–41 (2004)
7. Cecchini, M., Kolb, P., Majeux, N., et al.: Automated docking of highly flexible ligands by genetic algorithms: a critical assessment. J. Comput. Chem. **25**(3), 412–422 (2003)
8. Gohlke, H., Klebe, G.: Approaches to the description and prediction of the binding affinity of small-molecule ligands to macromolecular receptors. Angew. Chem. Int. Ed. **41**, 2644–2676 (2002)
9. Tame, J.R.H.: Scoring functions–the first 100 years. J. Comput. Aided Mol. Des. **19**, 445–451 (2005). doi:10.1007/s10822-005-8483-7
10. Willett, P.: Genetic algorithms in molecular recognition and design. Trends Biotechnol. **13**, 516–521 (1995)
11. Jones, G., Willett, P., Glen, R.C., et al.: Development and validation of a genetic algorithm for flexible docking. J. Mol. Biol. **267**(3), 727–748 (1997)
12. Ross, B.J.: A Lamarckian evolution strategy for genetic algorithms. In: The Practical Handbook of Genetic Algorithms, p. 16. CRC Press, Boca Raton (1999)
13. López-Camacho, E., García Godoy, M.J., García-Nieto, J., et al.: Solving molecular flexible docking problems with metaheuristics: a comparative study. Appl. Soft Comput. **28**, 379–393 (2015). doi:10.1016/j.asoc.2014.10.049
14. Bertsimas, D., Tsitsiklis, J.: Simulated annealing. Stat. Sci. **8**, 10–15 (1993)
15. Rutenbar, R.A.: Simulated annealing algorithms: an overview. IEEE Circuits Devices Mag. **5**, 19–26 (1989). doi:10.1109/101.17235
16. Suman, B., Kumar, P.: A survey of simulated annealing as a tool for single and multiobjective optimization. J. Oper. Res. Soc. **10**, 1143 (2006)
17. Liu, M., Wang, S.: MCDOCK: a Monte Carlo simulation approach to the molecular docking problem. J. Comput. Aided Mol. Des. **13**(5), 435–451 (1999)
18. Friesner, R.A., Banks, J.L., Murphy, R.B., et al.: Glide: a new approach for rapid, accurate docking and scoring. 1. method and assessment of docking accuracy. J. Med. Chem. **47**, 1739–1749 (2004). doi:10.1021/jm0306430
19. Dorigo, M., Maniezzo, V., Colorni, A.: Ant system: optimization by a colony of cooperating agents. IEEE Trans. Syst. Man Cybern. Part B **26**, 29–41 (1996). doi:10.1109/3477.484436
20. Kennedy, J., Eberhart, R.C.: Particle swarm optimization. IEEE Int. Conf. Neural Netw. (ICNN) **4**, 1942–1948 (1995)
21. Kennedy, J.: Small worlds and megaminds: effects of neighborhood topology on particle swarm performance. In: Conference on Evolutionary Computation, pp. 1931–1938 (1999)
22. Ghosh, A., Ghosh, A., Chowdhury, A., Hazra, J.: An evolutionary approach to drug-design using quantam binary particle swarm optimization algorithm. In: 2012 IEEE Students' Conference on Electrical, Electronics and Computer Science, pp. 1–4 (2012). doi:10.1109/SCEECS.2012.6184776
23. Liu, Y., Zhao, L., Li, W., Yang, Y.L.: FIPSDock: a new molecular docking technique driven by fully informed swarm optimization algorithm. J. Comput. Chem. **34**, 67–75 (2013). doi:10.1002/jcc.23108
24. Berg, H.C., Berry, R.M.: E.coli in motion. Phys. Today **58**, 64–65 (2005). doi:10.1063/1.1897527
25. Stocker, R.: Reverse and flick: hybrid locomotion in bacteria. PNAS **108**, 2635–2636 (2009). doi:10.1073/pnas.1019199108

26. Polin, M., Tuval, I., Drescher, K., Gollub, J.P., Goldstein, R.E.: Chlamydomonas swims with two "Gears" in a eukaryotic version of run-and-tumble locomotion. Science **325**, 487–4890 (2009). doi:10.1126/science.1172667
27. Egbert, M.D., Barandiaran, X.E., di Paolo, E.A.: A minimal model of metabolism-based chemotaxis. PLoS Comput. Biol. (2010). doi:10.1371/journal.pcbi.1001004
28. Passino, K.M.: Bacterial foraging optimization. Int. J. Swarm Intell. Res. **1**, 1–16 (2010). doi:10.4018/jsir.2010010101
29. Segall, J.E., Block, S.M., Berg, H.C.: Temporal comparisons in bacterial chemotaxis. Proc. Natl. Acad. Sci. USA **83**, 8987–8991 (1986). doi:10.1073/pnas.83.23.8987
30. Madigan, M.T., Martinko, J.M., Stahl, D.A., Clark, D.P.: Brock Microbiology, 13th edn. Pearson, Glenview (2012)
31. Turner, L., Zhang, R., Darnton, N.C., Berg, H.C.: Visualization of flagella during bacterial swarming. J. Bacteriol. **192**, 3259–3267 (2010). doi:10.1128/JB.00083-10
32. Damton, N.C., Turner, L., Rojevsky, S., Berg, H.C.: Dynamics of bacterial swarming. Biophys. J. **98**, 2082–2090 (2010). doi:10.1016/j.bpj.2010.01.053
33. Das, S., Biswas, A., Dasgupta, S., Abraham, A.: Bacterial foraging optimization algorithm: theoretical foundations, analysis, and applications. Found. Comput. Intell. **3**(3), 23–55 (2009). doi:10.1007/978-3-642-01085-9_2
34. Stewart, E.J., Madden, R., Paul, G., Taddei, F.: Aging and death in an organism that reproduces by morphologically symmetric division. PLoS Biol. **3**, 0295–0300 (2005). doi:10.1371/journal.pbio.0030045
35. Passino, K.M.: Biomimicry of bacterial foraging for distributed optimization and control. IEEE Control Syst. Mag. **22**, 52–67 (2010). doi:10.1109/MCS.2002.1004010
36. Passino, K.M.: Biomimicry for Optimization, Control and Automation. Springer-Verlag, London (2004)
37. Biswas, A., Dasgupta, S., Das, S., Abraham, A.: Synergy of PSO and bacterial foraging optimization-a comparative study on numerical benchmarks. In: Corchado, E., Corchado, J. M., Abraham, A. (eds.) Innovations in Hybrid Intelligent Systems. Advances in Soft Computing, vol. 44, pp. 255–263. Springer, Heidelberg (2008). doi:10.1007/978-3-540-74972-1_34
38. Shen, H., Zhu, Y., Zhou, X., Guo, H., Chang, C.: Bacterial foraging optimization algorithm with particle swarm optimization strategy for global numerical optimization, pp. 497–504. ACM (2009). doi:10.1145/1543834.1543901
39. Sharma, V., Pattnaik, S.S., Garg, T.: A review of bacterial foraging optimization and its applications. In: IJCA, pp. 9–12 (2012)

Geo-Information Knowledge Base System for Drought Pattern Search

Kittisak Kerdprasop and Nittaya Kerdprasop[✉]

Data and Knowledge Engineering Research Unit,
School of Computer Engineering, Suranaree University of Technology,
Nakhon Ratchasima 30000, Thailand
nittaya@sut.ac.th

Abstract. This paper presents a novel idea of utilizing a first-order logic technique to create, search and match for drought patterns in the specific area of interest. Drought patterns in this work have been drawn from the regression analysis using weekly time period and vegetation health index (VHI) obtained from the National Oceanic and Atmospheric Administration (NOAA) satellite data. To show the devised search facility, we use drought situations in the northeast provinces of Thailand as demonstrative cases. Drought trend of each province can be inferred from the percentage of provincial area that has the value of VHI below 35; the lower VHI value, the more severe drought. According to NOAA, this VHI threshold indicates moderate-to-high drought level. The proposed method is a kind of geo-information knowledge base system that allows users to search for drought pattern in some specific area at a particular time of the year. Moreover, the system also reports other area in the same region of study that shows similar drought trend. The drought trend is recognized from a regression coefficient in which the positive coefficient is the sign of increasing drought level, whereas the negative value implies the decrease of drought level.

Keywords: Drought patterns · Logic-based pattern search · Pattern matching · NOAA satellite data · Geo-information knowledge base system

1 Introduction

It can be noticed that the use of computer in the remote sensing field is mostly for creating map with the digitization and image processing facilities. In this research, we propose a new aspect of computer technology to help analyzing the remotely sensed data and to store and search for the induced patterns. We design a logic-based method to record drought patterns of the northeast provinces of Thailand in the knowledge base called the *Geo-KB* system. Such patterns are recorded in the knowledge base as a group of facts using first-order logic representation format. Drought of each province can be inferred from the percentage of provincial area that has vegetation health index (VHI) below 35, which indicates drought level from moderate to high intensity. We define pattern of drought from the regression association between time and VHI using regression coefficient to report trend of drought level. The pattern storage and search techniques are our main contribution.

© Springer International Publishing Switzerland 2016
O. Gervasi et al. (Eds.): ICCSA 2016, Part III, LNCS 9788, pp. 332–344, 2016.
DOI: 10.1007/978-3-319-42111-7_26

The use of artificial intelligence technology in remote sensing has been done for more than 30 years [11]. The applied techniques are normally rule-based expert system and knowledge base system to help classifying the correct type of land cover from the remote images. The cover-type classification has been done through the consultation of decision rules that had been learned from the ground-based training area [4]. The rule-based expert system has been applied to study the cover type and its change in various areas including the Arabian coast [9], the semiarid and arid parts in Arizona [14], a harbor city in the northeast of Turkey [3], and the agricultural and forestry sectors in England and other countries [1, 10, 13, 15]. The classification process in these work applies the classifying facility that has been embedded in the image processing software such as ERDAS. However, we can notice some change in later work [2, 12] that remote sensing researchers start using a specialized expert system software such as JESS to generate a rule base and also to integrate JESS as a part of their geographic information system.

In this work, we propose an idea to move an ordinary expert system to the next level of intelligent rule-based system. Traditional expert and knowledge base systems are composed of two main parts: background information or facts for making decision, and the inference rules for the delivery of correct decision [5–8]. We propose in this paper that besides facts and rules, such intelligent systems should have a learning part that can analyze and learn information from the previous experiences and observations. We therefore design a knowledge base system that has three main parts: facts, patterns, and inference rules.

2   A Framework of the Geo-KB System

The knowledge base system proposed in this work is named the *Geo-KB* system and its main components are shown in Fig. 1. The four main parts in our Geo-KB system have the functionalities as follows:

- *Knowledge Classification/Consultation.* This part serves as the graphical user interface (GUI) for interactive communication with general users. The GUI used in the current version of our Geo-KB system is mostly pop-up windows and mouse-click buttons. The display of recommendation or consultation result is also responsible by this GUI part.
- *Knowledge Engineering.* This module of the Geo-KB system is the channel to communicate with domain expert and knowledge engineer for knowledge elicitation and to store specific knowledge as facts. The decision rules are also input through this module. The expert or engineer can use any kind of utility software for typing and editing rules. The rules should be saved directly to the knowledge base.
- *Knowledge Base.* The storage of information to be used for making recommendation and consultation is composed of three major sets of information: a set of facts, a set of rules, and a set of patterns. Storing patterns as additional source of knowledge is the novel idea proposed in the Geo-KB system. These patterns can actually be any kind of knowledge learnable from historical data and observations. In this paper, we store regression relationships of drought situation induced from the VHI information as patterns. Illustration of such patterns can be found in Sect. 3.

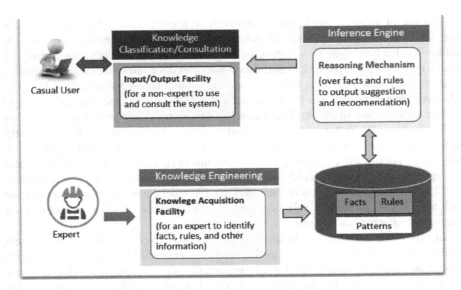

Fig. 1. Architecture of the Geo-KB system

- *Inference Engine.* The inference engine is the main component of any knowledge base and expert system. It is the reasoning mechanism to deduce answer or recommendation from the stored facts and rules. In our Geo-KB system, reasoning can also be done over patterns. We thus devise a reasoning mechanism over patterns as additional first-order logic statements.

To use the Geo-KB system, the knowledge engineer or the domain expert has to elicit facts, rules, and/or patterns of specific domain application to be stored in the knowledge base. The user can then consult knowledge from the Geo-KB system. The implementation of the system is explained in the next Section.

3   Implementation

3.1   Study Area

To demonstrate the usability of the Geo-KB system, we analyze drought patterns of the northeast provinces of Thailand. The northeast covers 160,000 km^2, which is about one-third of the country (Note that the area of Thailand is 513,120 km^2; source https://en.wikipedia.org/wiki/Geography_of_Thailand). Currently, there are 20 provinces in the northeast (Fig. 2). Agriculture is a main source of income in this region. But this area suffers from poor soil fertility and the unpredictable amount of rainfall. Average yearly precipitation varies widely from around 1,100 mm in the southwest provinces to almost 2,000 mm in the northeast provinces. A prior knowledge regarding drought situation is thus important for people in the northeast area.

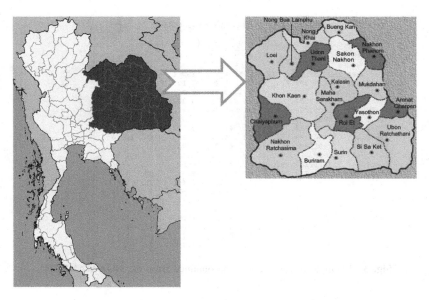

Fig. 2. Twenty provinces in northeast Thailand, which is the study area of drought pattern search   (Sources: https://en.wikipedia.org/wiki/Isan and http://www.thethailandlife.com/isaan-map)

3.2 Data Characteristics

We use the information from the remotely sensed data, which is publicly available on the website of NOAA (National Oceanic & Atmospheric Administration), U.S.A. The NOAA operational polar orbiting environmental satellites are equipped with the advanced very high resolution radiometer (AVHRR) to detect moisture and heat that are radiated from the Earth surface. AVHRR uses several visible and infrared channels to detect radiation reflected from surface objects. Collected AVHRR data are then used to compute various indices that are helpful for crop monitoring.

In this study, we are interested in the VHI index. It is a combined estimation of moisture and thermal conditions. VHI is used to characterize vegetation health. Higher VHI implies greener vegetation. VHI less than 35 indicates moderate drought condition, whereas VHI less than 15 is severe drought. The good condition is VHI > 60.

We use the data from NOAA-AVHRR that are available as time series in the ASCII format. Example of NOAA-AVHRR data is shown in Fig. 3. The last two columns (% Area VHI < 15 and %Area VHI < 35) are information for extremely drought and severe-to-moderate drought, respectively. We use only information in the last column (%Area VHI < 35) to induce regression model because they are more appropriate to our study area than the %Area VHI < 15, which occurs occasionally in some small areas of the northeast provinces.

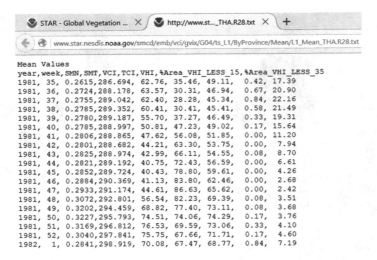

```
  STAR - Global Vegetation ...  X        http://www.st...THA.R28.txt  X   +

      www.star.nesdis.noaa.gov/smcd/emb/vci/gvix/G04/ts_L1/ByProvince/Mean/L1_Mean_THA.R28.txt

Mean Values
year,week,SMN,SMT,VCI,TCI,VHI,%Area_VHI_LESS_15,%Area_VHI_LESS_35
1981, 35, 0.2615,286.694, 62.76, 35.46, 49.11,  0.42, 17.39
1981, 36, 0.2724,288.178, 63.57, 30.31, 46.94,  0.67, 20.90
1981, 37, 0.2755,289.042, 62.40, 28.28, 45.34,  0.84, 22.16
1981, 38, 0.2785,289.352, 60.41, 30.41, 45.41,  0.58, 21.49
1981, 39, 0.2780,289.187, 55.70, 37.27, 46.49,  0.33, 19.31
1981, 40, 0.2785,288.997, 50.81, 47.23, 49.02,  0.17, 15.64
1981, 41, 0.2806,288.865, 47.62, 56.08, 51.85,  0.00, 11.20
1981, 42, 0.2801,288.682, 44.21, 63.30, 53.75,  0.00,  7.94
1981, 43, 0.2825,288.974, 42.99, 66.11, 54.55,  0.08,  8.70
1981, 44, 0.2821,289.192, 40.75, 72.43, 56.59,  0.00,  6.61
1981, 45, 0.2852,289.724, 40.43, 78.80, 59.61,  0.00,  4.26
1981, 46, 0.2884,290.369, 41.13, 83.80, 62.46,  0.00,  2.68
1981, 47, 0.2933,291.174, 44.61, 86.63, 65.62,  0.00,  2.42
1981, 48, 0.3072,292.801, 56.54, 82.23, 69.39,  0.08,  3.51
1981, 49, 0.3202,294.459, 68.82, 77.40, 73.11,  0.08,  3.68
1981, 50, 0.3227,295.793, 74.51, 74.06, 74.29,  0.17,  3.76
1981, 51, 0.3169,296.812, 76.53, 69.59, 73.06,  0.33,  4.10
1981, 52, 0.3040,297.841, 75.75, 67.66, 71.71,  0.17,  4.60
1982,  1, 0.2841,298.919, 70.08, 67.47, 68.77,  0.84,  7.19
```

Fig. 3. Example of time series data obtained from the NOAA-AVHRR

3.3 The Induction of Drought Patterns

On generating regression model to represent drought situation in each northeast province, we combine eight weekly data records to be a two-month period as the independent or predictive variable, whereas the %Area VHI < 35 is used as the dependent or target variable. The two-month period in this demonstration is for lessen the number of regression relationships because for each province, we have to analyze regression for each time period. Therefore, there exist six regression models each year for each province. Figure 4 shows the regression modelling technique with SPSS software.

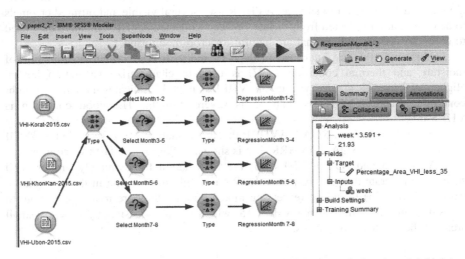

Fig. 4. Drought pattern model induction with modeler (left) and the induced model (right)

The first step in regression model building is thus the transformation of data from weekly to bi-monthly. The time and %Area VHI < 35 data for each year are then used for regression analysis. For illustration, we use the data of the year 2015. Suppose the drought situations of Nakhon Ratchasima, Khon Kaen, and Buriram provinces during the first half of the year 2015 are induced as:

Drought(Nakhon Ratchasima) $= 3.163 *$ Month 1 - to - 2 $+ 18.16$ (R $= 0.964$)
Drought(Nakhon Ratchasima) $= -0.118 *$ Month 3 - to - 4 $+ 43.40$ (R $= 0.138$)
Drought(Nakhon Ratchasima) $= 4.715 *$ Month 5 - to - 6 $- 29.93$ (R $= 0.943$)

Drought(Khon Kaen) $= 4.480 *$ Month 1 - to - 2 $- 0.683$ (R $= 0.975$)
Drought(Khon Kaen) $= 0.439 *$ Month 3 - to - 4 $+ 39.51$ (R $= 0.284$)
Drought(Khon Kaen) $= 3.116 *$ Month 5 - to - 6 $+ 4.663$ (R $= 0.948$)

Drought(Buriram) $= 1.021 *$ Month 1 - to - 2 $+ 7.616$ (R $= 0.932$)
Drought(Buriram) $= -0.131 *$ Month 3 - to - 4 $+ 11.14$ (R $= 0.235$)
Drought(Buriram) $= 1.211 *$ Month 5 - to - 6 $+ 9.183$ (R $= 0.931$)

From the drought situation in Nakhon Ratchasima province during the month 1–2 (Jan–Feb) of the year 2015, it can be interpreted from the positive coefficient value (= 3.163) that the drought area in this province is increasing. But during the month 3–4 (Mar–Apr), the drought area is decreasing. This can be noticed from the negative slope of the regression relationship (coefficient = −0.118).

It can also be compared across provinces that during the month 1 and 2, Khon Kaen province is facing the same increasing drought situation (coefficient = 4.480) as in Nakhon Ratchasima province. When the time changes to the month 3 and 4, Buriram province's drought situation (coefficient = −0.131) is more like Nakhon Ratchasima than Khon Kaen (coefficient = 0.439).

The regression coefficient (R-value) is also shown to show the strength of relationship of time against the percentage of provincial drought area. The higher R-value, the more correlation of time-drought relationship.

3.4 Geo-KB Construction

The drought patterns in each time period for each province in the northeast region of Thailand are then elicited and transformed to be first-order patterns in the knowledge base. The process of knowledge base creation is illustrated in Fig. 5. The first step is accessing satellite data for each northeast province, then selecting data attributes and transforming the data format. The second step is building the regression relationships for each time period and each province. The third step is the transformation of

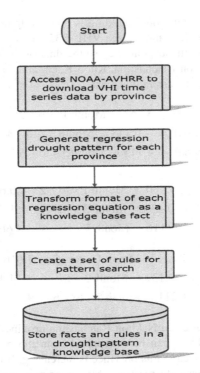

Fig. 5. The steps in drought-pattern knowledge elicitation

regression equations to be knowledge in the first-order logic format. The fourth step is the creation of inference rules for search similar/dissimilar drought patterns.

In the third step of knowledge elicitation, transformation of regression equation is necessary for the subsequent knowledge inference process. Drought patterns are to be represented as a Horn clause in the first-order logic format as follow:

$$reg\,(Province,\;Month,\;Slope,\;Y\text{-}intercept,\;Coefficient)$$

where

reg	is the predicate name to mean regression relationship
Province	is the northeast province name
Month	is the two-month time period of the year
Slope	is the slope of regression relationship for the specific time period
Y-intercept	is the point on the y-axis (amount of drought area) crossed by the regression line
Coefficient	is the regression coefficient representing the magnitude or strength of time-drought relationship for that specific time of the year.

```
1    :-ensure_loaded(system(visirule)).
2    reg(nakon_ratchasima,'1-2',3.163,18.16,0.964).
3    reg(nakon_ratchasima,'3-4',-0.1186,43.4,0.138).
4    reg(nakon_ratchasima,'5-6',4.715,-29.93,0.943).
5    reg(nakon_ratchasima,'7-8',-8.703,298.8,0.99).
6    reg(konkaen,'1-2',4.48,-0.6825,0.975).
7    reg(konkaen,'3-4',0.4387,39.51,0.284).
8    reg(konkaen,'5-6',3.116,4.663,0.948).
9    reg(konkaen,'7-8',-9.535,318.3,0.982).
10   reg(ubon_ratchatani,'1-2',3.591,21.93,0.845).
11   reg(ubon_ratchatani,'3-4',1.305,28.84,0.512).
12   reg(ubon_ratchatani,'5-6',1.659,38.98,0.719).
13   reg(ubon_ratchatani,'7-8',-8.315,267.3,0.976).
14   % ... more regression here ...
15   % -----------------------------------
16   % .... Inference rules for drought pattern search and matching
17   % .... ===================================================
18   % .... The case for positive slope
19   s1(P,N,Per):- reg(P,N,S,_,R),S>0,!,reg(PP,N,SS,_,RR),not(PP=P),
20                SS>S*(1-Per/100), SS<S*(1+Per/100),
21                flash([[PP],      ,SS ,    ,RR]),fail .
22
23   % .... The case for negative slope
24   s1(P,N,Per):- reg(P,N,S,_,R),S<0,!,reg(PP,N,SS,_,RR),not(PP=P),
25                SS<S*(1-Per/100),SS>S*(1+Per/100),
26                flash([[PP],      ,SS ,    ,RR]),fail .
27   % ........................ END OF PROGRAM ...................
```

Fig. 6. The drought patterns as facts and the inference rules for pattern search and matching

The fourth and last step of Geo-KB creation is the generation of inference rules for the pattern search and matching for provinces showing the same situations of drought for the specific time. The inference rules and some regression relationships represented as the knowledge base facts are shown in Fig. 6.

To use the Geo-KB system for drought pattern search, users have to specify three parameters:

1. The province name of interest (P).
2. Time-period of the specific year (N).
3. Threshold for similarity search (Per). This parameter constrains the search for the same drought situation in northeast provinces that it must not fluctuate beyond this threshold. For instance, if user is interested in Surin province and specifies the Per parameter to be 30, the system must search for other provinces that show the same drought situation as Surin with the difference in slope value not higher than 30 %.

The steps in using the Geo-KB system is summarized in Fig. 7.

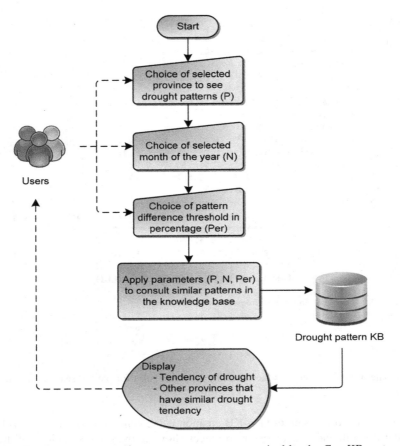

Fig. 7. The steps to specify necessary parameters required by the Geo-KB system

Fig. 8. The first screen of the Geo-KB system

4   Running Results

To test the implementation of the Geo-KB system, we demonstrate its usage through the running example. The window in Fig. 8 is the first screen welcoming user. The user can then point-and-click to command the system to search for pattern of his/her interest. In Fig. 9, user is interested in the drought situation in Nakhon Ratchsima province

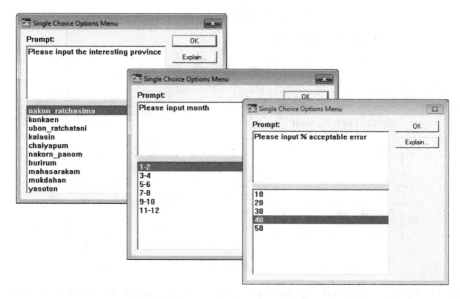

Fig. 9. Screens for inputting parameters

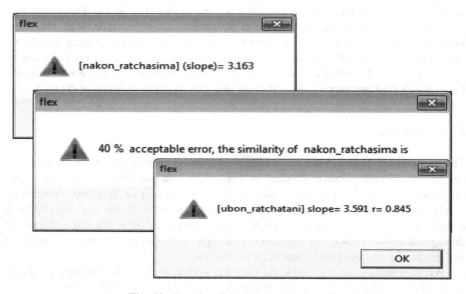

Fig. 10. Results of drought pattern search

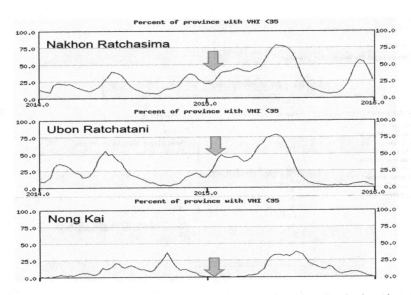

Fig. 11. Drought situation during January–February 2015 of Nakhon Ratchasima that shows similar trend as in Ubon Rtachatani, but dissimilar to the trend of Nong Kai province

during the time period January-February of the year 2015. The user also ask for similar pattern in other province with slope difference not higher than 40 %. The results that are shown in Fig. 10 identifying Ubon Ratchatani as a province with the same increasing drought situation. To confirm the correctness of drought pattern search, we also show a comparison of drought patterns of the three provinces in Fig. 11. It can be seen that during January-February of 2015 drought areas in Nakhon Ratchsima and Ubon Ratchatani are increasing, whilst the drought situation is stable in Nong Kai.

5 Conclusion

The northeast of Thailand is a large region with dense population. People in this area are mostly farmers whose planting is dependable on rainfall. Unfortunately, the northeast is not in the monsoon region as the southern part of Thailand. Amount of rainfall does not distribute normally across the rainy months. It may rain heavily for a couple of days and have a dry spell for a long period of time. Shortage of rainfall during cultivation season can cause much damage to the farmers. Drought monitoring and predicting can obviously beneficial to planters.

We thus propose in this work the Geo-KB system that can help inducing knowledge related to drought patterns in the northeast of Thailand. This system has been designed to store facts, inference rules, and drought patterns in its knowledge base. The drought patterns is the regression relationships of time and VHI index in each province of the northeast. VHI is satellite based value that is ranging from 0–100 % and can be used to represent greenness situation of each area. It can also be used as drought index.

In this work, we use the VHI value that falls below 35 % to represent moderate-to-high drought severity level. In each area, the computed VHI value changes from time to time depending on the greenness of land cover due to rainfall. Therefore, we study drought level from the VHI within six various time periods each year. Each time period spans for two months. At some specific time during the dry season, drought level may increase. But drought situation may decrease when it is during the wet season.

The devised Geo-KB system thus reports drought situation within a short period of two months in which the specific month of interest can be specified by users. The increase/decrease of drought level is reported as well as the other area showing the same drought trend.

References

1. Cohen, Y., Shoshany, M.: A national knowledge-based crop recognition in Mediterranean environment. Int. J. Appl. Earth Obs. Geoinf. **4**, 75–87 (2002)
2. Goyal, R., Jayasudha, T., Pandey, P., Devi, D.R., Rebecca, A., Sarma, M.M., Lakshmi, B.: Knowledge based system for satellite data. Int. Arch. Photogramm. Remote Sens. Spat. Inf. Sci. **XL-8**, 1233–1236 (2014)
3. Kahya, O., Bayram, B., Reis, S.: Land cover classification with an expert system approach using Landsat ETM imagery: a case study of Trabzon. Environ. Monit. Assess. **160**(1), 431–438 (2010)
4. Kartikeyan, B., Majumder, K.L., Dasgupta, A.R.: An expert system for land cover classification. IEEE Trans. Geosci. Remote Sens. **33**(1), 58–66 (1995)
5. Kerdprasop, K., Kerdprasop, N.: Integrating inductive knowledge into the inference system of biomedical informatics. In: Kim, T.-h., Adeli, H., Cuzzocrea, A., Arslan, T., Zhang, Y., Ma, J., Chung, K.-i., Mariyam, S., Song, X. (eds.) DTA/BSBT 2011. CCIS, vol. 258, pp. 133–142. Springer, Heidelberg (2011)
6. Kerdprasop, K., Kerdprasop, N.: Automatic knowledge acquisition tool to support intelligent manufacturing systems. Adv. Sci. Lett. **13**, 199–202 (2012)
7. Kerdprasop, N., Intharachatorn, K., Kerdprasop, K.: Prototyping an expert system shell with the logic-based approach. Int. J. Smart Home **7**(4), 161–174 (2013)
8. Kerdprasop, N., Kerdprasop, K.: Autonomous integration of induced knowledge into expert system inference engine. In: IMECS 2011 – International MultiConference of Engineers and Computer Scientists, vol. 1, pp. 90–95 (2011)
9. Krapivin, V.F., Phillips, G.W.: A remote sensing-based expert system to study the Aral-Caspian aquageosystem water regime. Remote Sens. Environ. **75**, 201–215 (2001)
10. Lucas, R., Rowlands, A., Brown, A., Keyworth, S., Bunting, P.: Rule-based classification of multi-temporal satellite imagery for habitat and agricultural land cover mapping. ISPRS J. Photogramm. Remote Sens. **62**, 165–185 (2007)
11. McKeown, D.M.: The role of artificial intelligence in the integration of remotely sensed data with geographic information systems. Technical report CMU-CS-86-174, Computer Science Department, Carnegie Mellon University, USA (1986)
12. Shesham, S.: Integrating expert system and geographic information system for spatial decision making. Master Theses & Specialist Projects, Paper 1216, Western Kentucky University, U.S.A. (2012)

13. Shoshany, M.: Knowledge based expert systems in remote sensing tasks: quantifying gain from intelligent inference. Int. Arch. Photogramm. Remote Sens. Spat. Inf. Sci. XXXVII, Part **B7**, 1085–1088 (2008)
14. Stefanov, W.L., Ramsey, M.S., Christensen, P.R.: Monitoring urban land cover change: an expert system approach to land cover classification of semiarid to arid urban centers. Remote Sens. Environ. **77**, 173–185 (2001)
15. Thorat, S.S., Rajendra, Y.D., Kale, K.V., Mehrotra, S.C.: Estimation of crop and forest areas using expert system based knowledge classifier approach for Aurangabad district. Int. J. Comput. Appl. **121**, 43–46 (2015)

Improving Photovoltaic Applications Through the Paraconsistent Annotated Evidential Logic Eτ

Álvaro A.C. Prado[1(✉)], Marcelo Nogueira[1,2(✉)],
Jair Minoro Abe[1(✉)], and Ricardo J. Machado[2(✉)]

[1] Software Engineering Research Group, Paulista University,
UNIP, Campus Tatuapé, São Paulo, Brazil
py2alv@gmail.com, marcelo@noginfo.com.br,
jairabe@uol.com.br
[2] ALGORITMI Centre, School of Engineering, University of Minho,
Campus of Azurém, Guimarães, Portugal
rmac@dsi.uminho.pt

Abstract. The contrast between large urban centers and other isolated locations where even the most basic resources are scarce, leads the development of self-sustainable solutions, a panorama in which the electrical power is an important demand to be supplied. Through Bibliographic and Experimental research, plus practical implementation and testing, it was possible to develop an improving solution which fits within the proposed needs. This paper aims to present a self-oriented photovoltaic system based upon the Paraconsistent Annotated Evidential Logic Eτ, its construction and practical tests, where an average yield of 3.19 W was obtained against 2.44 W from a fixed panel, representing an increase of 31.56 % in the overall power.

Keywords: Solar energy · Photovoltaic · Power optimization · Energetic sustainability · Paraconsistent Annotated Evidential Logic Eτ

1 Introduction

Nowdays, by developing new technologies and enhancements for various aspects of daily life is a constant activity, there are still common many cases of very scarce resources, particularly in locations situated far from urban centers.

One of the most important of these resources is the electricity, often unavailable because of large distances between distribution networks and the locations itself, or even because the great importance of local ecosystems [1].

The difficulties in bringing rural electrification to these places, and the need to limit the use of fossil fuels, replacing them with non-polluting and renewable energy alternatives, make urgent investments in research and development of improved alternative energy sources [2].

Fossil fuels are the main source of energy in the world and are at the center of the world's energy demands. However, its avaliability is limited, and its large-scale use is associated with environmental degradation. The negative effects known from use of these fuels include acid rain, depletion of the ozone layer and global climate changes [3].

© Springer International Publishing Switzerland 2016
O. Gervasi et al. (Eds.): ICCSA 2016, Part III, LNCS 9788, pp. 345–355, 2016.
DOI: 10.1007/978-3-319-42111-7_27

There are many different projects to obtain electricity generated from production units based on biomass, wind, solar and small hydroelectric plants (SHP), being developed for isolated communities on Northern Brazil, among other locations around the globe [2]. However, like other alternative sources, the cost of these technologies is high, creating a series of difficulties in cases of small and isolated areas, which keeps the high electricity exclusion numbers inside all Brazilian states (Fig. 1).

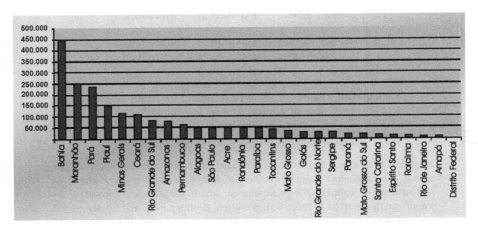

Fig. 1. Numbers of electricity exclusion inside Brazilian states [2].

Following this panorama, an important method for obtaining electricity is through a photovoltaic solar panels, where supply only implies in the cost of equipment itself, and no carbon is liberated during operation [4]. However, an important problem is related to the positioning of the solar panel, which is often fixed and does not have the ability to follow the natural movement of the sun throughout the day, which is related to the question of Maximum Power Point (MPP) of the panel [4], as seen on Fig. 2.

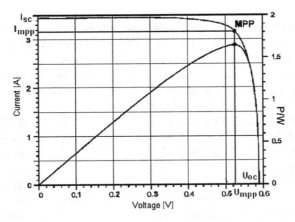

Fig. 2. Typical curve of a solar cell [5].

Figure 3 shows the operating point at which maximum solar cell power supply for a given level of irradiation. The solar cell behaves like a current source to the left of MPP as a voltage source to the right. For each solar radiation curve, there is a specific voltage point at which the cell will operate at its maximum operating point [5].

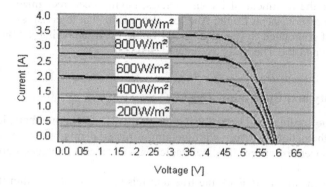

Fig. 3. Voltage versus current in a solar cell [5].

Whereas most of the devices to be powered by the cell may operate at constant voltage levels, it is necessary to find the point at which the cell is able to deliver more power, according to the requirements of the load [5].

Many authors propose different methods to circumvent this problem, by using simple timer-based traction systems or even sensors and "perturbation and observation" algorithms to move the panels, but without the ability to handle situations of inconsistency or contradiction in the collected data [6, 7].

By using embedded software, a controller board and a sample from the voltage provided by the photovoltaic panel, it is possible to obtain a correct positioning with a stepper motor mechanically attached to it [8].

This alternative combined with the use of Paraconsistent Annotated Evidential Logic Eτ on the decision-making process by the embedded software seeks to provide an optimal performance by handling situations where the signals from the panel are not conclusive or contradictory. The self-sustainable design now proposed is intended to power small devices of everyday use, by charging batteries with good performance, plus a reduced environmental impact [8].

2 Paraconsistent Logic

2.1 Historical Background

The Genesis of Paraconsistent Logic originated in 1910, by the work of logicians Vasil'év and Łukasiewicz. Although contemporaries, they developed their research independently [9].

In 1948, Jaskowski, encouraged by his professor Łukasiewicz, discovered Discursive Logic. Vasil'év wrote that "similar to what happened with the axioms of Euclidean geometry, some principles of Aristotelian logic could be revisited, among them the principle of contradiction" [6].

Going beyond the work of Jaskowski, the Brazilian logician Da Costa has extended its systems for the treatment of inconsistencies, having been recognized for it as the introducer of Paraconsistent Logic; Abe [6], also a Brazilian logician, set several other applications of Annotated Systems, specially Logic Eτ, establishing the basic study of Model Theory and the Theory of Annotated Sets.

2.2   Certainty and Uncertainty Degrees

Founded on the cardinal points, and using the properties of real numbers, it is possible to build a mathematical structure with the aim of materializing how to manipulate the mechanical concept of uncertainty, contradiction and paracompleteness among others, according to Fig. 4 [9].

Such mechanism will embark the true and false states treated within the scope of classical logic with all its consequences. To this end, several concepts are introduced which are considered "intuitive" for the purpose above:

Perfectly defined segment AB : $\mu + \lambda - 1 = 0; \ 0 \le \mu, \lambda \le 1$
Perfectly undefined segment DC : $\mu - \lambda = 0; \ 0 \le \mu, \lambda \ominus \le 1$

The constant annotation (μ, λ) that focus on the segment has completely undefined the relationship $\mu - \lambda = 0$, i.e. $\mu = \lambda$. Thus, the evidence is identical to the positive

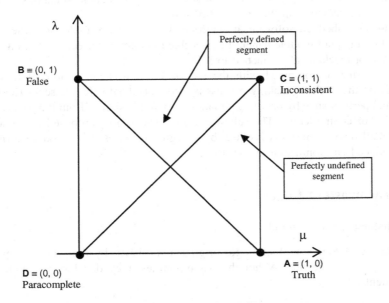

Fig. 4. τ Reticulate [10]

evidence to the contrary, which shows that the proposition $p_{(\mu, \lambda)}$ expresses a blurring. It varies continuously from the inconsistency (1, 1) until the paracompleteness (0, 0) [10].

Since the constant annotation (μ, λ) that focus on the segment has clearly defined the relationship $\mu + \lambda - 1 = 0$, i.e. $\mu = 1 - \lambda$, or $\lambda = 1 - \mu$.

Therefore, in the first case, the favorable evidence is the Boolean complement of contrary evidence and, second, the contrary evidence is the Boolean complement of favorable evidence, which shows that the evidence, both favorable and contrary 'behave' as if classic. It varies continuously from the deceit (0, 1) to the truth (1, 0) [10].

The applications are introduced as follows:

G_{ic}:[0, 1] × [0, 1] → [0, 1], G_{pa}:[0, 1] × [0, 1] → [−1, 0], G_{ve}:[0, 1] × [0, 1] → [0, 1], G_{fa}:[0, 1] × [0, 1] → [−1, 0].

Defined by:

Inconsistency Degree:	$G_{ic}(\mu, \lambda) = \mu + \lambda - 1$, since $\mu + \lambda - 1 \geq 0$
Paracompleteness Degree:	$G_{pa}(\mu, \lambda) = \mu + \lambda - 1$, since $\mu + \lambda - 1 \leq 0$
Truth Degree:	$G_{ve}(\mu, \lambda) = \mu - \lambda$, since $\mu - \lambda \geq 0$
Falsehood Degree:	$G_{fa}(\mu, \lambda) = \mu - \lambda$, since $\mu - \lambda \leq 0$

It is seen that the Accuracy Degree "measures" how an annotation (μ, λ) "distances" from the segment perfectly defined and how to "approach" of the state, and the true degree of Falsehood "measures" how an annotation (μ, λ) "distances" from the segment perfectly defined, and how to "approach" the false state. [10]

Similarly, the inconsistency degree "measures" how an annotation (μ, λ) "distances" from the segment undefined and how "close" it is from the inconsistent state, and degree of Paracompleteness "measures" how an annotation (μ, λ) "distances" of the segment undefined, and how "close" it is from paracomplete.

Is called G_{in} uncertainty degree (μ, λ) from an entry (μ, λ) to any of the degree of inconsistency or paracompleteness. For example, the maximum degree of uncertainty is in an inconsistent state, i.e. $G_{ic} (1, 1)_- = \_1$. It is called the Certainty Degree $G_{ce} (\mu, \lambda)$ of an annotation (μ, λ) to any of the degrees of truth or falsity [10].

2.3   Decision States: Extreme and Not-Extreme

With the concepts shown above, it is possible to work with "truth-bands" rather than the "truth" as an inflexible concept. Perhaps more well said that truth is a range of certainty with respect to a certain proposition. The values serve as a guide when such a proposition is considered; for example, "true" in order to make a decision positively, and so on. The extreme states are represented by Truth (V), False (F), Inconsistent (T) and Paracomplete (⊥); and the not-extreme logical states by the intermediate areas between the states. The areas bounded by not-extreme values depend on each project [10].

2.4   Para – Analyzer Algorithm

An important point for the embedded software responsible for handling the panel, the Para-analyzer algorithm reflects the paraconsistent analysis by treating the values of favorable and unfavorable evidence, resulting in certainty and uncertainty degrees, plus a logical state [6] (Fig. 5).

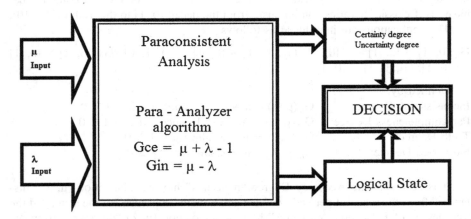

Fig. 5.   Para - Analyzer algorithm [6].

2.5   Embedded Software and Paraconsistent Controller

The embedded software to the paraconsistent controller was developed upon the spiral model, being followed by a risk analysis for all the following steps, since the adoption of software engineering and its assumptions are critical to project success [10].

Both evidence values are obtained with an interval of 500 ms between them, which allows a proper distinction and the capture of the logic states Paracomplete ($\bot$) – with low intensity and uniform λ and μ, representing a dimly lit room – and Inconsistent (T), with high-intensity and uniform μ and λ, representing an external environment with nuisances like shadows of trees, birds or other moving obstacles [8].

In order to optimize the output states of the paraconsistent controller, the non-extreme logical states were conveniently chosen (Fig. 6), according to the application requirements [11].

Taking into account that in most of the time the solar panel is exposed to light levels close to its maximum, the closest non-extreme states from truth (V) were elected [8].

These logical states, available in the output of paraconsistent controller, are properly interpreted and translated into instructions for embedded software, which will be delivered to the stepper motor through a driver circuit, as seen on Fig. 7.

Once switched on, the panel starts a 135° full scan, stopping where it finds the strongest favorable evidence (μ).

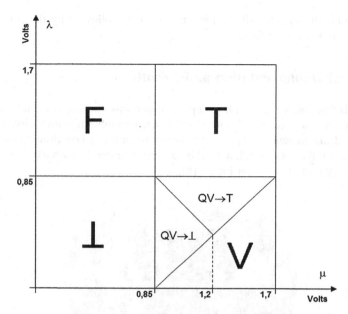

Fig. 6. Aspect of the lattice with corresponding voltage levels to the logic states [8].

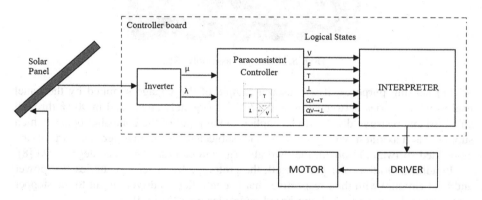

Fig. 7. Paraconsistent controller.

From this point, the corresponding logic state found in this position will indicate the following action to be taken [8].

- Logical state Truth (V): the panel stays in the same position.
- Logical states Inconsistent and Paracomplete (T and ⊥): run a second full scan of 135°.
- Logical states Almost-Truth tending to Paracomplete (QV → ⊥) and Almost-Truth tending to Inconsistent (QV → T): run a partial scan of 81°.
- Logical state False (F): after a second full scan, the system enters in Standby mode.

Every hour, the system will perform a new scan, followed by the actions corresponding to the logic state found.

3   Practical Implementation and Results

A single-axis traction system with a stepper motor was chosen, since it proved itself enough to provide a noticeable gain in performance combining simplicity, robustness and simplified maintenance [11]. The prototype assembly is constituted basically of a mobile holder for the solar panel, moved by a stepper motor through a belt system, built upon a base support, as seen on Fig. 8 [8].

Fig. 8. Prototype assembly [8].

For sensing purposes, it was used a sample of the voltage supplied by the panel itself – which varies between zero and 17 V – adequately attenuated in 90 % through an adjustable resistive divider, to be applied to the input of the controller boar – which supports a maximum voltage 5 V – and subjected to an inverter, as part of the embedded software, to obtain the favorable (μ) and contrary evidence degrees (λ) [8].

In addition to the solar panel itself, the prototype has a set of batteries (for power and load circuits) with their respective charge controllers, a driver circuit to the stepper motor and a homemade Arduino based controller board (Fig. 9).

When operating under proper conditions, the prototype remains powered only by the panel, while both batteries are charged. The power consumption of the backup battery only occurs when the stepper motor is running or when the system operates in low sunlight conditions.

In order to validate its operation, a series of daytime tests were done in a rooftop environment in November 2013, with good weather.

During three days the generated power was measured and compared with a fixed panel of the same type as the one used in the prototype. An increase of 31.56 % could be achieved in this particular days (Fig. 10) specially during the morning hours and late afternoon, when the fixed panel has its performance greatly reduced [8].

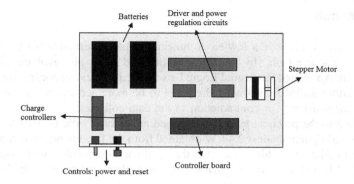

Fig. 9. Prototype components [8]

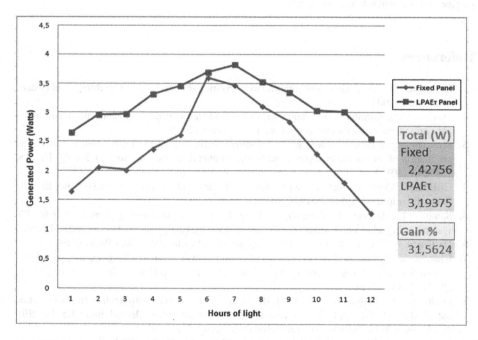

Fig. 10. Results obtained with the prototype [8].

This result is similar to others found in the literature, specially in Huang et al. [12] (35.8 %) and Salas et al. [13] (2.8 % – 18.5 %), when compared with fixed panel systems.

4   Conclusion

This paper aims to propose a low-environmental impact alternative for places where electricity is not available, by using a self-oriented solar photovoltaic panel. The Paraconsistent Annotated Evidential Logic Eτ was used in the decision-making process by the embedded software, allowing the panel to be more accurately positioned in situations of inconsistency or contradiction in the data collected.

According to the practical tests, it was found an average yield of 3.19 W provided by the proposed system against 2.44 W obtained from the fixed panel, which represents an increase of 31.56 %. This shows that the results are compatible with other similar systems [12, 13] and demonstrates that the actual implementation is perfectly feasible.

Acknowledgements. This research is sponsored by the Portugal Incentive System for Research and Technological Development PEst-UID/CEC/00319/2013 and University Paulista - Software Engineering Research Group by Brazil.

References

1. Bursztyn, M.: A Difícil Sustentabilidade: Política Energética e Conflitos Ambeintais. Garamond (2001)
2. Dantas, J.M.: Sistema Fotovoltaico para Comunidades Isoladas utilizando ultracapacitores para armazenamento de energia. Univesity of Ceará (2013)
3. Manzano-Agugliaro, F., Alcayde, A., Montoya, F.G., Zapata-Sierra, A., Gil, C.: Scientific production of renewable energies Worldwide: an overview. Renew. Sustain. Energy Rev.18, 134–143 (2013)
4. CRESESB, Centro de Referência para Energia Solar e Eólica Sérgio de Salvo Brito. Energia Solar: Princípios e Aplicações (2006)
5. Santos, J.L., Antunes, F., Chebab, A., Cruz, C.: A maximum power point tracker for PV systems using a high performance boost converter. Sol. Energy **80**, 772–778 (2005). Elsevier
6. Da Costa, N.C.A., et al.: Lógica Paraconsistente Aplicada. Atlas, São Paulo (1999)
7. Ishaque, K., Salam, Z.: A review of maximum power point tracking techniques of PV system for uniform insolation and partial shading condition. Renew. Sustain. Energy Rev. **19**, 475–488 (2012)
8. Prado, A.A.C., Nogueira, M., Abe, J.M., Machado, R.: Power optimization in photovoltaic panels through the application of paraconsistent annotated evidential logic Eτ. In: IFIP Advances in Information and Communication Technology (2015)
9. Prado, A.A.C., Oliveira, C.C., Sakamoto, L.S., Abe, J.M., Nogueira, M.: Reaching energetic sustainability through a self-oriented battery charger, based on paraconsistent annotated evidential logic Eτ. In: IFIP Advances in Information and Communication Technology (2013)
10. Abe, J.M., Silva, F., João, I., Celestino, U., Araújo, H.C.: Lógica Paraconsistente Anotada Evidencial Eτ. Comunicar. (2011)
11. Nogueira, M., Machado, R.J.: Importance of risk process in management software projects in small companies. In: Grabot, B., Vallespir, B., Gomes, S., Bouras, A., Kiritsis, D. (eds.) Advances in Production Management Systems, Part II. IFIP AICT, vol. 439, pp. 358–365. Springer, Heidelberg (2014)

12. Torres, C.R.: Sistema inteligente baseado na lógica paraconsistente anotada Eτ para controle e navegação de robôs móveis autônomos em um ambiente não estruturado. Doctoral thesis, University of Itajubá (2010)
13. Huang, B.J., Ding, W.L., Huang, Y.C.: Long-term field test of solar PV power generation using one-axis 3-position sun tracker. Sol. Energy **85**(9), 1935–1944 (2011). Elsevier
14. Salas, V., Olías, E., Lásaro, A., Barrado, A.: Evaluation of a new maximum power point tracker (MPPT) applied to the photovoltaic stand-alone systems. Sol. Energy Mater. Sol. Cells **87**(1), 807–815 (2004)

Remote Sensing Based Model Induction for Drought Monitoring and Rainfall Estimation

Kittisak Kerdprasop and Nittaya Kerdprasop[(✉)]

Data and Knowledge Engineering Research Unit,
School of Computer Engineering, Suranaree University of Technology,
Nakhon Ratchasima 30000, Thailand
{kerdpras,nittaya}@sut.ac.th

Abstract. Droughts are natural phenomenon threatening many countries around the globe. In this work, we study regional drought in the northeastern area of Thailand. Since 1975 droughts in the northeast, especially Nakhon Ratchasima province in the south of this region, have occurred more frequently than the past with stronger intensity. We firstly investigate the relationship of regional drought to the cycle of El Nino Southern Oscillation (ENSO) and find that the cool phase of ENSO (or La Nina) shows positive effect to the increase of rainfall, whereas the warm phase (or El Nino) has no clear relationship to the decrease of rainfall in Nakhon Ratchasima province. We then further our study by inducing a model to monitor drought situation based on the historical remotely sensed data. Data in the past six years had been selected from both the excessive rain fall years due to the La Nina effect and the drought years with unclear cause. We also draw a model to estimate the amount of rainfall from the lagged two and three months of remotely sensed data. The drought monitoring model and the rainfall estimation model are built by the decision tree induction algorithm and the models' accuracy tested with 10-fold cross validation are 68.6 % and 72.2 %, respectively.

Keywords: Remotely sensed data · Drought monitoring · Rainfall estimation · Decision tree model · ENSO effect

1 Introduction

Drought is obviously a disaster not only for farmers, but also for the entire community including humans and wild animals. Generally, the physical drought can be categorized as meteorological drought, hydrological drought, and agricultural drought [4, 5]. Meteorological drought is defined as the deficit in precipitation for an extensive period of time. Hydrological drought is defined as the deficiency of runoff. It is the consequence of meteorological drought in that the shortage of rainfall for some long period will lead to the decrease below critical level of stream flow, groundwater, and reservoirs. Meteorological drought also the cause of agricultural drought, which has been defined as the deficit in soil moisture such that plant growing is almost impossible.

O. Gervasi et al. (Eds.): ICCSA 2016, Part III, LNCS 9788, pp. 356–368, 2016.
DOI: 10.1007/978-3-319-42111-7_28

From its various categories, drought can thus be assessed using various indices such as standardized precipitation index (SPI), standardized soil moisture index (SSI), Palmer drought severity index (PDSI), and many more [5]. The computation of these indices is mostly based on climatic variables such as precipitation, humidity, runoff, and soil moisture. Some researchers invented instruments and techniques for collecting accurate variables to estimate rainfall [15, 16]. Several works on rainfall estimation have employed data from environmental satellite to generate rainfall estimation model [2, 10, 14, 20]. Remotely sensed data used in these research works are mostly direct precipitation monitoring and indirect interpretation based on cloud characteristics.

Recently, to characterize rainfall sufficiency or deficiency that leads to drought, many researchers propose the idea of using remotely sensed vegetation health or stress as substituting information to the amount of actual rainfall measured from the ground based station. Advantages of remote sensing is its low price in obtaining data, real-time access, and the ability to collect data in some unreachable areas. Therefore, more and more researchers are interested in using remotely sensed data such as normalized difference vegetation index (NDVI) as a substitution for precipitation data [3, 7, 8, 19]. Besides NDVI, other indices such as vegetation health index (VHI), vegetation condition index (VCI), standardized thermal condition (SMT), and temperature condition index (TCI), generated from environmental satellite can also be used to analyze sufficient/deficient rainfall condition [1, 9, 11–13, 22].

Precipitation estimation models computed from the remotely sensed data are mostly based on the statistical methods such as linear and non-linear regression techniques. Theses statistical-based models can be too complicate to comprehend is there are so many predictors. We thus present in this paper the decision tree model constructed from satellite-based indices including SMN (smoothed NDVI), SMT, VCI, VHI, and TCI to monitor regional drought. The decision tree induction used in this study is a data mining technique that aims at producing a concise model for easy interpretation but accurate enough for explaining current situation as well as estimating the future event.

We also construct the tree model to estimate precipitation level as a categorical value based on these satellite indices. The data used in this study are selected from some La Nina years. This selection is based on our ENSO effect analysis in that the lower northeast region of Thailand is affected by the La Nina stage, whereas the El Nino effect is unstable. The study of ENSO effect in the north region of Thailand [21] also identified that such effect can be recognized only for the strong and medium ENSO phase. The detail of our ENSO effect analysis is explained in Sect. 2. The models for drought monitoring and precipitation level estimation are illustrated in Sects. 3 and 4, respectively. We then conclude our work in Sect. 5.

2   The Analysis of ENSO Effect Over the Study Area

We focus our drought analysis over Nakhon Ratchasima area, which is the largest province of Thailand covering 25,494 km^2. Nakhon Ratchasima is in the lower northeastern region situated along the southwest rim of the plateau with mountainous area on the far west and south borders of the province (Fig. 1). Population in this province is around 2.6 million and the majority of them is in the agricultural sector.

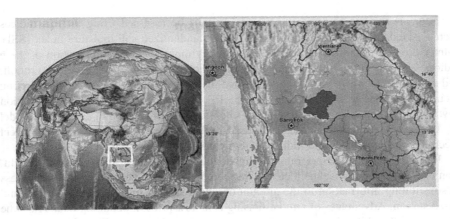

Fig. 1. The study area: Nakhon Ratchasima province with geographical location 14.957 N and 102.104 E (source: www.maphill.com)

Major crops in Nakhon Ratchasima are rice, sugar cane, tapioca, corn, and fruits. Planting of these crops is mainly based on precipitation and irrigation. Estimating correctly the period and sufficiency of rainfall is thus very important to the farmers. Unlike the southern and eastern parts of Thailand that have plenty of rain during the monsoon season, the plateau and mountainous terrain in the northeast has less amount of rain during the rainy season. This part of the country is thus experiencing drought more often than other parts of the country.

Prior to building a model to monitor drought and a model to estimate amount of rainfall, we have to firstly specify years of interest for accessing satellite data to be used in our next phase of model induction. The time selection is based on the analysis result of ENSO effect over the northeast region of Thailand.

The El Nino-Southern Oscillation (ENSO) is a phenomenon of streams and wind shifts in the equatorial pacific between the west coast of South America and the eastern equatorial Pacific including Southeast Asia and Australia. The streams and wind shifts cause fluctuation of the Pacific ocean temperature, called the sea surface temperature (SST), to be warmer or cooler. The warm phase of ENSO causes the SST to be higher than average and is called El Nino. The cool phase is the reverse phenomenon and is called La Nina. If the fluctuation of SST is not higher or lower than 0.5 °C, it is the neutral phase.

To identify the El Nino or La Nina events, the National Oceanic & Atmospheric Administration (NOAA) uses the Oceanic Nino Index (ONI) as the indicator. The SST for every three overlapping month period will be observed at the Nino 3.4 region over the Pacific area located at 0.5°N–0.5°S and 120°–170°W. The two ENSO phases can be classified as follows:

- El Nino: the warm event occurs when 5 consecutive overlapping 3-month periods showing SST at or higher than +0.5 °C.
- La Nina: the cool event occurs when 5 consecutive overlapping 3-month periods showing SST at or lower than −0.5 °C.

The intensity of each ENSO phase can be categorized with the following threshold:

- Weak: with SST anomaly between 0.5 to 0.9
- Moderate: with SST anomaly between 1.0 to 1.4
- Strong: with SST anomaly between 1.5 to 1.9
- Very Strong: with SST anomaly 2.0 or higher

From the ONI index to classify El Nino/La Nina/neutral phase, we thus analyze the relationship of these ENSO phases to the amount of rainfall in Nakhon Ratchasima province during the years 1951–2015. According to the NOAA, the ONI-based ENSO phases may span as long as three years. But in this work, we analyze the deviation of amount of rainfall on a calendar year basis. We thus set the threshold that a specific year is classified as the El Nino (or La Nina) year if the SST is above (or below) 0.5 °C for five consecutive overlapping 3-month periods within the same calendar year. From this setting criteria, during the years 1951 to 2015, there are 19 El Nino and 20 La Nina events. The horizontal axis in Fig. 2 is ENSO events and the left-hand-side vertical axis is amount of rainfall in Nakhon Ratchasima that is above (positive value) or below (negative value) the 65-year average. The fluctuation of SST is also plotted on the right-hand-side vertical axis to show intensity of each ENSO event.

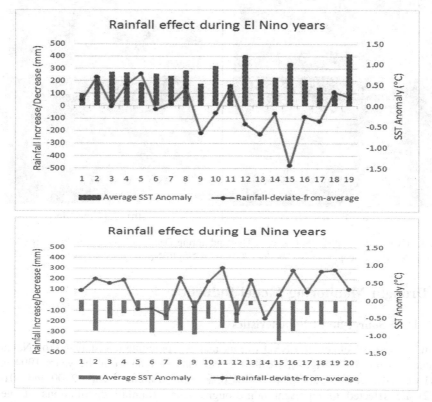

Fig. 2. The deviation of rainfall from 65-year average value (1103.14 mm) during the El Nino years (above) and the La Nina years (below) (Color figure online)

It can be seen from Fig. 2 that from the total 19 El Nino events, 10 of them (or 53 %) cause the decrease in rainfall in Nakhon Ratchasima province, whereas 9 of them (or 47 %) increase rainfall. The effect of El Nino over this area in terms of rainfall increasing or decreasing is thus unpredictable. The phenomenon of La Nina is however different. During the 20 La Nina years, rainfall has increased for 70 % of the time. The decrease of rainfall has occurred only 30 % of the events.

The effect of increasing rainfall during La Nina episode in Nakhon Ratchasima area follows the patterns anticipated by NOAA (Fig. 3). From the global impact pattern of El Nino during December-February and June-August, northeast of Thailand (the white star in the figure) is expected to experience higher temperature than normal situation. The relation to either wet or dry is not identified. This is in contrast to the La Nina effect that the wet pattern can be expected.

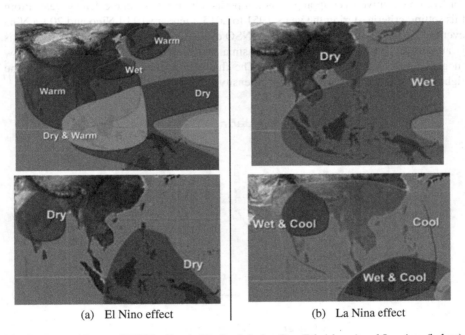

(a)   El Nino effect   (b)   La Nina effect

Fig. 3. Expected area of ENSO effect in Thailand during Dec-Feb (above) and Jun-Aug (below) (source: www.cpc.ncep.noaa.gov/products/CWlink/ENSO/ENSO-Global-Impacts/)

3   Drought Monitoring Model Induction

3.1   Data Source and Characteristics

From the rainfall effect of ENSO events and the precipitation statistics of Nakhon Ratchasima province (Fig. 4), we thus select satellite data from the years 1994 and 2001 to study drought pattern, while the data during La Nina years (2000 and 2010–2012) are selected to represent non-drought years. Rainfall distributions of these drought and non-drought years are illustrated in Figs. 5 and 6, respectively.

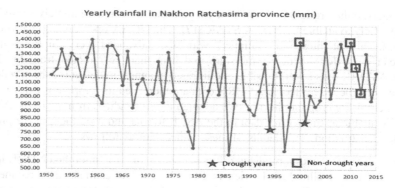

Fig. 4. Yearly rainfall in Nakhon Ratchasima during the year 1951–2015 with the dotted trend line showing a decline of average rainfall. Years 1994 and 2001 are selected as representatives of drought years, while the year 2000 and 2010–2012 are representatives of non-drought years. (Color figure online)

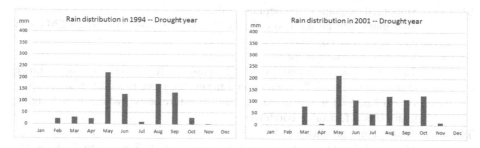

Fig. 5. Rainfall distribution during the drought years (Color figure online)

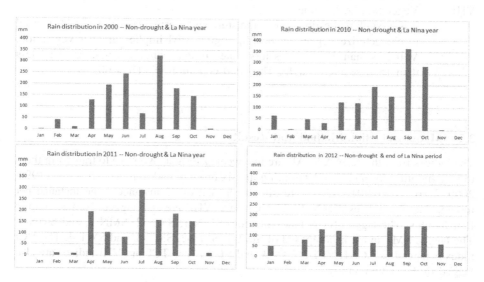

Fig. 6. Rainfall distribution during the non-drought years due to the La Nina effect (Color figure online)

To analyze the drought and non-drought patterns, we use the remotely sensed data that is publicly available by NOAA STAR [17]. The NOAA operational polar orbiting environmental satellites are equipped with the advanced very high resolution radiometer (AVHRR) to detect moisture and heat that are radiated from the Earth surface. AVHRR uses several visible and infrared channels to detect radiation reflected from surface objects. Collected AVHRR data are then used to compute various indices that are helpful for crop monitoring. We adopted the five indices from NOAA-AVHRR that are available as time series data in the ASCII format and used them in the process of model construction. The details of these indices are as follows:

SMN =   Smoothed and normalized difference vegetation (or smoothed NDVI) index. It is a satellite-based index that can be used to estimate the greenness of vegetation. The value is in the range +1.0 to −1.0. Low value (0.1 or less) indicates rock or sand surface. Moderate value (0.2 to 0.5) means sparse vegetation area such as shrubs or agricultural area after harvesting. High value (0.6 to 0.9) implies dense vegetation such as forest or crops at the peak growth state.

SMT =   Smoothed brightness temperature index. This index can be used to estimate the thermal condition. High SMT can lead to the dry condition of vegetation.

VCI =   Vegetation condition index. The VCI was expressed as NDVI anomaly. It can be used to estimate moisture condition. The higher VCI means the more moisture detected from vegetation. VCI less than 40 indicates moisture stress, whereas VCI higher than 60 is a favorable condition.

TCI =   Temperature condition index. TCI can be used to estimate thermal condition. TCI less than 40 indicates thermal stress, whereas TCI higher than 60 is a favorable condition.

VHI =   Vegetation health index. It is a combined estimation of moisture and thermal conditions. VHI is used to characterize vegetation health. Higher VHI implies greener vegetation. VHI less than 35 indicates moderate drought condition, whereas VHI less than 15 is severe drought. The good condition is VHI > 60.

3.2   Model Induction Method and Result

The indices from the NOAA STAR are in a weekly format. We have to transform them to be a monthly format. The data during the six drought and non-drought years (1994, 2000, 2001, 2010, 2011, 2012) are thus composed of 72 instances: one instance per month. The five indices (SMN, SMT, VCI, TCI, VHI) are also lagged from zero up to four months. The lagged SMN index includes SMN-1, SMN-2, SMN-3, and SMN-4. The lagged SMT index is composed of SMT-1, SMT-2, SMT-3, and SMT-4. The VCI index is lagged as VCI-1, VCI-2, VCI-3, and VCI-4. The TCI and VHI indices are also lagged in the same manner.

These data instances are then labelled as either drought or non-drought according to their categorization, that is, drought years in 1994 and 2001 and non-drought years in 2000 and 2010–2012. We then apply the C5 algorithm, which is the tree induction algorithm [18] for classification task available in IBM SPSS Modeler software [6].

Decision tree induction [18] is a popular method for inducing knowledge from data. Popularity is due to the fact that mining result in a form of decision tree is interpretability, which is more concern among practitioners than a sophisticated method but lack of understandability. A decision tree is a hierarchical structure with each node contains decision attribute and node branches corresponding to different attribute values of the decision node. The goal of building decision tree is to partition data with mixing classes down the tree until the leaf nodes contain pure class.

In order to build a decision tree, we need to choose the best attribute that contributes the most towards partitioning data to the purity groups. The metric to measure attribute's ability to partition data into pure class is *Info*, which is the number of bits required to encode a data mixture. To choose the best attribute, we have to calculate information gain, which is the yield we obtained from choosing that attribute. The information gain calculates yield on data set before splitting and after choosing attribute with two or more splits. The gain value of each candidate attribute is calculated. Then choose the maximum one to be the decision node. The process of data partitioning continues until the data subset has the same class label.

The result of drought/non-drought classification based on remotely sensed data in a format of decision tree is shown in Fig. 7. Performance of a tree is tested with 10-fold cross validation method and its average accuracy is 68.6 %.

The meaning of each variable appeared in the tree pattern is:

VCI-2 = lagged two months of vegetation condition index (or VCI value of the month preceding the last month)
SMN-3 = lagged three months of smoothed NDVI index (or SMN from the month preceding the last two months)
SMN-1 = lagged one month of smoothed NDVI index (or SMN of previous month)
VHI-2 = lagged two months of the vegetation health index (or VHI from the month before the last month)

The use of this tree as decision rules for drought classification can be explained as follows:

1. If VCI-2 ≤ 21.236,
   then drought occurs (100 % confidence).
2. If VCI-2 > 21.236 and SMN-3 ≤ 0.156,
   then drought occurs (85.71 % confidence).
3. If VCI-2 > 21.236 and SMN-3 > 0.156 and VHI-2 ≤ 50.405 and SMN-1 > 0.258,
   then drought occurs (100 % confidence).
4. If VCI-2 > 21.236 and SMN-3 > 0.156 and VHI-2 > 50.405,
   then not_drought (100 % confidence).
5. If VCI-2 > 21.236 and SMN-3 > 0.156 and VHI-2 ≤ 50.405 and SMN-1 ≤ 0.258,
   then not_drought (73.91 % confidence).

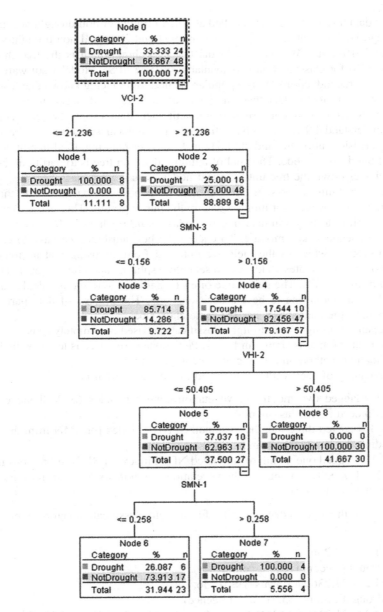

Fig. 7. A tree-based model induced from remotely sensed data for drought monitoring (Color figure online)

4   Rainfall Estimation Model Induction

Besides drought monitoring, we are also interested in estimating the amount of rainfall over a specific month. We thus use binning method to transform the six-year precipitation data in the continuous form to be an interval form. The 5 bins are firstly created using the ±2 standard deviations (Fig. 8). We further collapse the lowest two bins to be in the same interval. This is due to the fact that the negative value of the lowest bin is meaningless. The final four bins are thus referred to the following levels of monthly precipitation:

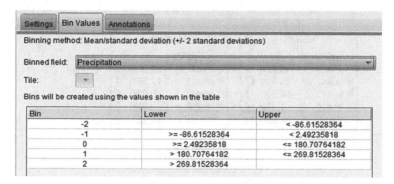

Fig. 8. Binning method for setting precipitation level prior to the building of rainfall estimation model (Color figure online)

Level 1: precipitation 0–2.498 millimeters (mm).
Level 2: precipitation 2.492–180.707 mm.
Level 3: precipitation 180.708–269.815 mm.
Level 4: precipitation more than 269.815 mm.

The monthly remotely sensed data with annotated precipitation level are then used for tree pattern induction. The result is shown in Fig. 9. Its accuracy tested with 10-fold cross validation is 72.2 %. The decision tree can also be interpreted as a set of conjunctive decision rules. Therefore from the tree in Fig. 9, the following remote sensing based rules can be applied to estimated amount of rainfall:

- If SMN-2 ≤ 0.290,
  then precipitation in current month should be 2.492–180.707 mm.
- If SMN-2 > 0.290 & VHI-3 ≤ 61.362,
  then precipitation in current month should be 0–2.498 mm.
- If SMN-2 > 0.290 & VHI-3 > 61.362,
  then precipitation in current month should be 2.492–180.707 mm.

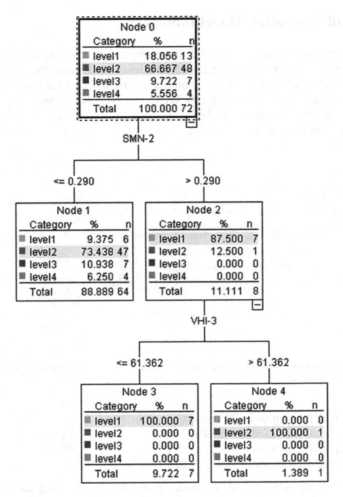

Fig. 9. A tree model with lagged SMN and VHI decision attributes to estimate precipitation level (Color figure online)

5 Conclusion

We have performed a series of experiments with the main objective of applying the remotely sensed data to monitor meteorological drought, which is the fundamental event for other kinds of droughts such as agricultural drought and hydrological drought. Advancement in remote sensing makes satellite data widely available and generated timely to facilitate rapid analysis of climate and environmental situation. We obtain climate satellite data from the NOAA-AVHRR. Index data related to vegetation monitoring have been used in this research to construct a drought monitoring model. We also extend our drought study based on remote sensing data to the estimation of amount of rainfall by constructing a model to guess precipitation level. Both drought monitoring and precipitation level estimation models are a tree-based model automatically induced by the C5 algorithm.

Prior to constructing the tree-based models, we analyze the ENSO effect to the increase or decrease of rainfall around Nakhon Ratchasima area. We have found that 70 % of the La Nina events coincided with the increase of rainfall, but the El Nino years did not show a clear pattern related to the rainfall effect. We thus selected satellite-based data during the La Nina years for tree model construction. The data from the drought years in the past during the neutral phase of ENSO are also selected for the efficiency of classification that needs both positive and negative examples.

The drought monitoring model shows the classification accuracy of 68.6 % with the tree depth at four levels. The precipitation level estimation model has a shallow depth of two levels and a 72.2 % classification accuracy. The accuracy has been assessed with 10-fold cross validation method.

References

1. Bayarjargal, Y., Karnieli, A., Bayasgalan, M., Khudulmur, S., Gandush, C., Tucker, C.: A comparative study of NOAA-AVHRR derived drought indices using change vector analysis. Remote Sens. Environ. **105**, 9–22 (2006)
2. Bellerby, T., Hsu, K., Sorooshian, S.: LMODEL: A satellite precipitation methodology using cloud development modeling. Part I: algorithm construction and calibration. Journal of Hydrometeorology **10**, 1081–1095 (2009)
3. Boken, V., Hoogenboom, G., Kogan, F., Hook, J., Thomas, D., Harrison, K.: Potential of using NOAA-AVHRR data for estimating irrigated area to help solve an inter-state water dispute. Int. J. Remote Sens. **25**(12), 2277–2286 (2004)
4. Farahmand, A., AghaKouchak, A., Teixeira, J.: A vantage from space can detect earlier drought onset: an approach using relative humidity. Sci. Rep. **5**, 8553 (2015). doi:10.1038/srep08553
5. Hao, Z., Singh, V.: Drought characterization from a multivariate perspective: a review. J. Hydrol. **527**, 668–678 (2015)
6. IBM Corporation: IBM SPSS Modeler 15 Algorithms Guide (2012)
7. Jalili, M., Gharibshah, J., Ghavami, S., Beheshtifar, M., Farshi, R.: Nationwide prediction of drought conditions in Iran based on remote sensing data. IEEE Trans. Comput. **63**(1), 90–101 (2014)
8. Karnieli, A., Agam, N., Pinker, R., Anderson, M., Imhoff, M., Gutman, G., Panov, N., Goldberg, A.: Use of NDVI and land surface temperature for drought assessment: merits and limitations. J. Clim. **23**, 618–633 (2010)
9. Kerdprasop, K., Kerdprasop, N.: Rainfall estimation models induced from ground station and satellite data. In: The 24th International MultiConference of Engineers and Computer Scientists, pp. 297–302 (2016)
10. Kidd, C.: Satellite rainfall climatology: a review. Int. J. Climatol. **21**, 1041–1066 (2001)
11. Kogan, F.: Operational space technology for global vegetation assessment. Bull. Am. Meteorol. Society **82**(9), 1949–1964 (2001)
12. Kogan, F.: 30-year land surface trend from AVHRR-based global vegetation health data. In: Kogan, F., Powell, A., Fedorov, O. (eds.) Use of Satellite and In-situ Data to Improve Sustainability, pp. 119–123. Springer, Dordrecht (2011)

13. Kogan, F., Guo, W.: Early detection and monitoring droughts from NOAA environmental satellites. In: Kogan, F., Powell, A., Fedorov, O. (eds.) Use of Satellite and In-situ Data to Improve Sustainability, pp. 11–18. Springer, Dordrecht (2011)
14. Levizzani, V., Amorati, R., Meneguzzo, F.: A review of satellite-based rainfall estimation methods. Technical report MUSIC-EVK1-CT-2000-0058, European Commission under the Fifth Framework Programme (2002)
15. Marzano, F., Cimini, D., Ciotti, P., Ware, R.: Modeling and measurement of rainfall by ground-based multispectral microwave radiometry. IEEE Trans. Geosci. Remote Sens. **43** (5), 1000–1011 (2005)
16. Marzano, F., Palmacci, M., Cimini, C., Giuliani, G., Turk, F.: Multivariate statistical integration of satellite infrared and microwave radiometric measurements for rainfall retrieval at the geostationary scale. IEEE Trans. Geosci. Remote Sens. **45**(2), 1018–1032 (2004)
17. NOAA STAR Center for Satellite Applications and Research. STAR-Global Vegetation Health Products, USA (2015).http://www.star.nesdis.noaa.gov/smcd/emb/vci/VH/
18. Quinlan, J.: Induction of decision trees. Mach. Learn. **1**, 81–106 (1986)
19. Quiring, S., Ganesh, S.: Evaluating the utility of the Vegetation Condition Index (VCI) for monitoring meteorological drought in Texas. Agric. For. Meteorol. **150**, 330–339 (2010)
20. Tapiador, F., Kidd, C., Levizzani, V., Marzano, F.: A neural networks-based fusion technique to estimate half-hourly rainfall estimates at 0.1° resolution from satellite passive microwave and infrared data. J. Appl. Meteorol. **43**, 576–579 (2004)
21. Ueangsawat, K., Jintrawet, A.: The impacts of ENSO phases on the variation of rainfall and stream flow in the upper Ping river basin, northern Thailand. Environ. Natural Resour. J. **11**(2), 97–119 (2013)
22. Zarei, R., Sarajian, M., Bazgeer, S.: Monitoring meteorological drought in Iran using remote sensing and drought indices. Desert **18**, 89–97 (2013)

From Bicycle to Bites: Indoor vs. Outdoor User Testing Methodologies of Georeferenced Mobile Apps

Letizia Bollini[(✉)] and Giulia Cicchinè

Department of Psychology, University of Milano-Bicocca,
Piazza dell'Ateneo Nuovo 1, 20126 Milan, Italy
letizia.bollini@unimib.it

Abstract. The paper is aimed to explore and discuss the way we evaluate, asses and test with users mobile applications—which main interaction modality is GPS and geo-referenced data—both in-lab and *en plein air*. The research intends to asses user experience evaluating methodologies to have better insight to understand how to design and plan spatial interactions among people, mobile devices, the physical environment and the digital space of geo-located information. The study adopts user test task-based methodologies coming from the user-centered design qualitative methods comparing infield research and usability lab conditions. The paper proposes experimental evidences coming from the indoor experiences—where geo-localization is simulated, but other research parameters are in control—with outdoor situation—where geo-localization is the real driver of interactions, but many variables interferes with some parameters and measurement observation—to understand experimental variables and bias to prevent them in the design process, using the field of cycling as a case study.

Keywords: Space-based interaction design · Geo-based experience design · Outdoor user experience testing · Infield usability test · Contextual usability test

1 Indoor vs. Outdoor Geo-located Experience Testing

User research methodologies raised in the field of user-centered design have been borrowed from many *exo*-disciplines such as ethnography, cognitive ergonomics, social science and so on. Many of them adopt field research approaches to people observing them in their real context and *natural* interactions. On the other hand most of the research practice in the field of digital media have developed and conducted in the laboratory according to Meyer and Tognazzini [1] and Nielsen work [2] among other pioneers of the field. The usability test labs—widely described and discussed by Krug [3]—guarantee an optimal control on external influence on the user experience with the digital navigation, content and interactions and give to research observers (and clients) a good and *live* insight of the process and problems emerging during the evaluation sessions.

But as already pointed out in a previous research about user-centered design methodologies applied to geo-referenced interactive ecosystems "the mobile revolution,

O. Gervasi et al. (Eds.): ICCSA 2016, Part III, LNCS 9788, pp. 369–382, 2016.
DOI: 10.1007/978-3-319-42111-7_29

the so-called *Web 3.0*—eo-referred, semantic and social—and the environmental pervasiveness of digital technology in its various forms (*pocketable* web, Internet of the Things, ubiquitous, Augmented Reality, intelligent environments, wearable devices etc.) is opening up new scenarios of research and development both in the field of geo-referenced information and in user-centered design". [4]

This turning point—both technological and cultural—implies new tools and approaches to develop user research of contextual interactions. It requires moving back to *in field research approaches* or, at least, to explore and discuss problems, limits and bias generated by outdoor user tests. Significant theoretical considerations and practical experience have been held in the last few years such as the contributions of Kjeldskov in 2004 [5], Kjeldskov and Skov ten years later [6], Brown et al. [7] and the workshop held by Korn and Zander during NordCHI in Reykjavik in 2010 [8]. Nielsen himself with the contribution of Budiu has revised and extend user testing methodologies and heuristic in 2012 [9].

Moreover, when geo-referenced information and data are intended to be the main part of app interaction the variable typically controlled in the lab context are significant part of the evaluation process of the service and/or device performances in the user experience.

The question is about how to identify, classify, cluster and manage the variables so that they become significant in the user experience evaluation process although they are not completely controllable to give significant insight to designer and planners involved in geo-defenced or spatial interaction projects. In a qualitative research methods perspective they offers a *real* insight from the people point of view and help not to underestimate contextual conditions and environmental interference. The inability to see the real-time data because of poor Wi-Fi connection, the lack of touch-interaction due to atmospheric humidity or perspiration have a bad impact on user experience as well as an ambiguous button-label or a too complex interaction with the information architecture or navigation through the app.

2 Research Hypothesis

The research goal of this first experimental contribution to research in the field of mobile and geo-referenced information interactions, is to understand how different an indoor evaluation and an outdoor one, could be applied to geo-referenced app using the field of cycle-tourism and bike competition as a case study.

That could be promising because of the nowadays application in sport's market related to maps, position and geo-location aimed to measure time, kilometres physical performances during the training according to the way we do and track sport and fitness activities through digital devices such as GPS, smartphones, smart watches or fit bands. In the past few years, healthcare and wellness markets are growth more than other services and even mobile application market moved in this direction, giving customers more and more sport and wellness apps. Starting from this point, the research tries to understand if the interaction of sport's and wellness' geo-located apps, is different, if used in an indoor setting, or in an outdoor one.

The first step is to discern the two typologies of fitness activities but, there are sports areas in which a mobile app, could be useful both in an indoor training and in an outdoor one. The case study touches one of the sports activities that could be done in the two manners: cycling, moreover cycling training related to races.

When a professional cyclist trains, he always tries to go out to ride *en plain air*, first of all because cycling is a sport that was born on the street and second because, in this way it is possible to recreate all the race's situations such as air friction or grip of the road. Unfortunately, this is not always possible because of the weather conditions so it may be preferable to stay and train at home, to avoid illnesses. That give the opportunity to observe the *same* activity in two *different* contexts.

As Kjeldskov says "a recent literature study showed that most mobile HCI research projects apply lab-based evaluations" [5]. Nevertheless, it is important to say that, in particular for applications that are born to be used outdoor, it would be proper to evaluate these mobile systems on the field, because these projects are highly context-dependent. On the other hand, field-based usability studies are difficult to conduct, time consuming, sometimes even expensive and it is unknown if they worth the time. The research tries, therefor, to understand if these kinds of evaluation counts, especially for mobile projects that should be used on the field and how it impacts on designing geo-based interactions between people and mobile devices in the real environment.

3 Research Method

To understand the validity of research hypothesisl—if these researches aloud really "worth the hassle"—two kinds of evaluations of cycling mobile applications were conducted: first indoor, trying to simulate a training, and second, on the field, giving the tester the possibility to use a mobile application to track his/her workout.

All the activities were video-recorded to understand what testers do and what they were thinking about—often using thinking aloud methodology—what they were doing, using a *GoPro action camera* to film clips that could mark the actions done to revise and discuss. A first camera was put on the helmet and it had to film all the fingers interactions with the smartphone and the mobile application. A second one, was on the handlebar turned on the tester's face to monitor his/her expression while using the app. Although a direct analysis has not been planned and conducted on facial expressions interpretation or on eyes tracking data this kind of experimental materials could be included in a further development of the project.

Two different mobile apps on an Apple iPhone 5S were used and compared: *Strava Cycling* and *Cyclometer*. The first one to track the train and see the map, the second one to see some statistics data that in Strava are accessible just payng a fee.

3.1 Indoor Evaluation

The idea of an Indoor evaluation was to evaluate cycling mobile app, in a controlled and selected environment where it could be possible to closely monitor the use of the system. In addition to this, that allows to understand how much the context would have mattered in an on-the-field evaluation, in order to compare the results.

A standard scenario has been used to conduct the user test including the *typical* indoor training setting called *Home Trainer* (see Fig. 1).

Fig. 1. Indoor setting: *Home trainer.* The picture comes directly from the GOPro Action Camera on the user helmet to see the tester perspective on the device's display

Setting: The standard setting in which a cyclist uses to do indoor training.

His/her bicycle is fixed on the home trainer, a *LeMond Home Direct Drive Trainer,* a particular home trainer that can repeat as best the situation outside, thanks to little windmill blades that can simulate air friction. Each of the tester has his/her own bicycle, that means that all the measures and the position on the bike was personal, to guarantee the comfort of each one's back and to pedal in a fluid way.

Artificial lights or sun light seem not to be critical in trying to recreate the setting, cyclist were simply requested where they would place his/her bicycle and the home trainer at best. The choice was mediate by the space available, by the size of each bicycle and—the funniest—by the proximity to a television.

Data Collection: There were set two action cameras, one on the helmet facing the smartphone, and the other one on the handlebar turned on the tester's face. Action cameras are small and lightweight so the scenario was as similar as the real indoor workout situation and they wouldn't interfere with a standard training session.

The Subjects: At the very beginning the experiment started with two subjects aged 20 and 24 years both agonistic cyclist to asses the overall experimental framework. Having high-competitive-level cyclists was a plus, because they train everyday and they are versatile, so they are able to use home trainers without any difficult and they could contribute giving an *expert walk through* insight to the cycle experience.

Tasks: All subjects were given a series of tasks to perform while using the system.

The tasks were derived from a previous theoretical and practical study by *there-searchers* and the tasks were related to standard way of cycling training.

This involved

(1) recording the activity,
(2) check the GPS signal,
(3) check the map and the route,
(4) check the statistic data.

Procedure: Before the evaluation session started, it was asked the subject to pedal for a while, to place each tool in the right way, without creating troubles and obstacles not only for the research but also for the whole workout.

The evaluation sessions were structured by the task assignment; these tasks involved the interaction of the cyclist with the mobile app that was recording his indoor training. They were not encouraged to *think aloud* or to say something about the research, because all their expressions were filmed in a clip, to understand what they thought about what they were doing by the analysis of their expressions. The evaluation lasted between 10 and 15 min.

Roles: Each evaluation session involved three people, a *researcher*, a *subject* using the system and an *observer and facilitator* that controlled all the situation and helped in technical problems with video equipment.

3.2 Outdoor Evaluation

The second evaluation took place in Lombardy, in the streets of Bergamo's province.

The aim of this study was to test the importance of the context in mobile applications that were born to be used on the field and if and how much laboratory/indoor testing are restrictive, because indoor testing could have gaps and could even not consider some particular elements strictly related to outdoor practising.

The focus is to compare these scenarios to check if indoor evaluation is enough or if researchers, in general, should evaluate geo-located application on the field, that is their real use-case.

The field evaluation is described in detail below.

Setting: Some quiet and not congested streets of Bergamo were chosen as test scenario and testers were asked to take their own bicycle. As said before, having a personal bike is something important, even in our research because all the measures are personal and the position of the smartphone and of the cameras, is related to these measures in order to have comfort while pedalling.

Data Collection: There were set two action cameras, one on the helmet facing the smartphone, and the other one on the handlebar turned on the tester's face. Action cameras are small and lightweight so the scenario was as similar as the real outdoor workout situation. Unlike the indoor evaluation, the image of the outdoor setting, was a little more bothered because of the wind and the weather changes (see Fig. 2).

The Subjects: At the very first beginning two subjects were selected aged 20 and 24 years, they were both agonistic cyclist. We thought that having high-competitive-level cyclists was a plus, because they train everyday and they can carefully move on the streets even if they have cameras and a mobile phone with them.

Fig. 2. Outdoor setting. The picture comes directly from the GOPro Action Camera on the user helmet to see the tester perspective on the street and the mobile device.

Tasks: All subjects were given a series of tasks to solve while using the system. The tasks were derived from a previous theoretical and practical study by the researchers and the tasks were related to standard way of cycling training.

This involved

(1) recording the activity,
(2) check the GPS signal,
(3) check the map and the route,
(4) check the statistic data.

Procedure: Before the evaluation session started, it was asked the subject to pedal for a while, to place each tool in the right way, without creating troubles and obstacles not only for the research but also for the whole workout. All the tools were positioned at best trying different speeds to see if everything was ok.

As we said before, the evaluation sessions were structured by the task assignment, these tasks involved the interaction of the cyclist with the mobile app that was recording his outdoor training. They were not encouraged to think aloud or to say something about the research, because all their expressions were filmed in a clip, so we tried to understand what they thought about what they were doing by the analysis of their expressions.

The evaluation lasted between 10 and 15 min.

Roles: Each evaluation session involved three people, a *researcher*, a *subject* using the system and a *observer and facilitator* that controlled all the situation and helped in technical problems with video equipment.

4   Analysis and Discussion

The data analysis aimed at creating two lists of usability problems identified on the two scenarios. The usability problems were classified as aesthetic, relevant and critical based on the guidelines by Molich [10]. The analysis of the video involved the department of Communication and Psychology of The University of Milano-Bicocca. Videos have been analysed and problems have been identified and ranked according to their impact or "relevance" category, all these usability problems should have been reviewed. Problems emerging from the user-test activities were listed, classified and prioritized (Fig. 3):

	Indoor	Outdoor
Critical – 4	3	3
Relevant – 2	1	2
Aesthetic – 1	1	1
Total – 7	5	6

Fig. 3. Ranking of user tests evidences

- Battery life;
- Wet weather;
- Screen Glare;
- Altitude;
- Temperature;
- Connectivity.

In the first user test setting and assessment conducted with two subjects, 7 problems have been identified, 1 from the indoor sessions, 3 from the outdoor ones and 3 problems that are in common between the indoor and the outdoor workouts. From this 7 problems, 4 of them were classified as *critical*, so problems that could stop the use of the system; 2 were classified as *relevant*, so they have to be solved as soon as possible with high priority and 1 problem was assessed to be *aesthetic*.

Just with only two testers, it was impossible to identify some critical problems such as the weather conditions, the battery, the light exc. that could deeply impact on the app performances and the user experience.

According to Krug [3] quantification about the numerosity of user test research method and its validity—5 to 8 testers are enough to find the 80 % of problems and as stated by Nielsen "The curve clearly shows that you need to test with at least 15 users to discover all the usability problems in the design" [11]—a second experimental session was conducted both in- and out-door involving 15 subjects (10 man and 5 women) which were asked to complete same tasks of the pre-test and to give bot a quantitative evaluation—ranked with a 5 grade Lickert scale—and a qualitative one collected through a questionnaire (Fig. 4).

	Indoor	Outdoor
Challenges	/	1.5
Map	2.5	2.2
Segments	/	1.4
Friends	/	2.4
Performance Comparison	/	1.6
Ranking	/	2.4
Navigation	1.6	1.2
Texts	3	2.4
Images	3.2	3
Going Back	1.6	1.4
Task Solving	2	2.2
TOT	2.36	2.01

Fig. 4. Results of the questionnaire and the Likert Scale

In general it seems that the problems mean is bigger in the indoor condition but this is probably due to the fact that there is a lack of connectivity dimension in indoor evaluation so there are less data.   Then problems have been classified in three categories called: Workflow; Navigation; Comprehension.

Workflow dimension is strictly related to app's operation, so its use as a cyclo computer where the app has to measure parameters like kilometres and average speed. *Navigation* is connected to primary and secondary menus and eventual gestures ambiguity. The *Comprehension* is related to legibility and readability of the interface's elements (See Fig. 5).

Workflow		Navigation		Comprehension	
Outdoor	Indoor	Outdoor	Indoor	Outdoor	Indoor
1,5		1,2	1,6	1,4	1,6
2,2	2,5	2,2	2,5	2,2	2
1,4				2,4	3
2,4				3	3,2
1,6					
2,4					
1,916666667	2,5	1,7	2,05	2,25	2,45
TOT= 2		TOT= 1,8		TOT= 2,35	

Fig. 5. Calculation of the means of indoor/outdoor evaluation of each category.

If we consider that the biggest part of the critical (CR) and serious (S) problems is related to workflow (W), so to check the cycling performance and so there are difficulties related to the data recording or functionally to GPS connection, it is important to notice that the gap between outdoor (7) and indoor (3) is big. Surely the indoor evaluation can't provide the same connection data because, generally, WI-FI connection or even 4G/LTE connection in a laboratory setting is stable but, on the other hand, the purity of location-based-data and GPS- based-data is not so clear (Fig. 6).

	Indoor (N=12)	Outdoor (N=9)
Critical (N=7)	3	7
Serious (N=6)	2	6
Cosmetic (N=2)	2	2
TOTAL (N=15)	7	15

Fig. 6. *Molich distinction of problems.* On the left, the critical, serious and cosmetic problems identified at the beginning of the research compared to user-test results.

This big difference is a great sign in theory, because it shows that the most important problems are just related to outdoor workouts, as we expect.   This gap also shows that obviously GPS problems can't be studied in an indoor evaluation, but if the problems were studied in the outdoor condition before the app's release, why are there so many problems and especially related to performance measurement (Fig. 7)?

	Molich	Category
Kilometers	CR	W
Average Speed	CR	W
Difference in Altitude	CR	W
Stop-n-go	CR	W
Challenges	S	W
Map	CR	W-N
Segments	S	W
Friends	S	W
Performance Comparison	S	W
Ranking	S	W
Navigation	CR	N
Texts	C	C
Images	C	C
Going Back	CR	C
Task Solving	S	C

Fig. 7. Molich distinction of problems compared to the categories of usability.

Finally 3 big and critical different problems related to connectivity, touch screens and battery life have been carried out by the tests. Statistics shows that the difference between what it was supposed and what really happens is not significant (p = .5; p = .7). That's important because some problems are structural and that emerges both in indoor training and in outdoor one (Figs. 8, 9, 10 and 11).

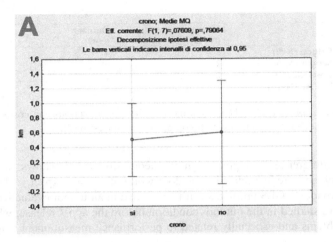

Fig. 8. Results: Kilometres. When the stopwatch, stops, the confidence bar is over the 0, so there's a significant relevance. When the app does its normal workflow, data recorded are not correct.

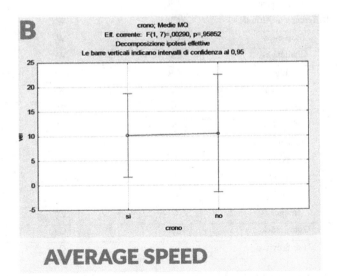

Fig. 9. Results: Average Speed. When the stopwatch, stops, the confidence bar is over the 0, so there's a significant relevance. When the app does its normal workflow, data recorded are not correct.

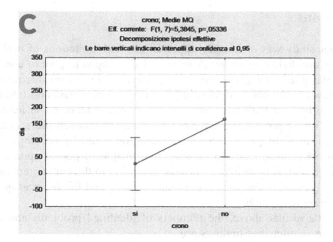

Fig. 10. Results: Drop. When the stopwatch has stopped, there isn't a huge difference, there is just a 25 m overestimation but it is impossible to say that it is statistically significant. On the other hand, when the stopwatch doesn't sop, it is visible from the table above that the difference is relevant because there's a 150 m overestimation.

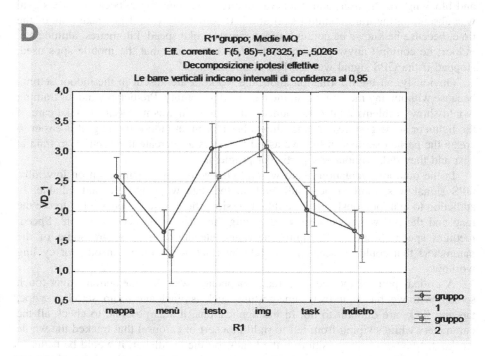

Fig. 11. Results: comparison Indoor vs. Outdoor. The comparison between the outdoor and the indoor group. The red group, the outdoor one, in general has lower opinion that the indoor group and it reaches the limit of significance. (Color figure online)

5 Conclusions

The aim of our study was to identify opportunities and limitations of usability evaluation in indoor and field conditions. A primary follow up is to give to user experience and interface designer *scientific* criteria to evaluate the *en plain air* interactions that the people experience with mobile devices exploring geo-based data and information and consequently to be able both to conduct user research outdoor or, at least to provide guide line for further project that have the relationship between users and the spatial environment as main driver. A second consideration could be developed in order to provide cognitive studies, research methods and insight of people approach to urban space to planner, architects and technologies according to the convergent development of *smart city's' culture*—that means the macro-scale—and the *Internet of Thins*—the micro-scale—next revolution.

Based on the results above, the numbers of identified problems and considering their nature, the summarized findings are:

Added value of on-the-field evaluation: At the beginning of the research, some general problems that could have been related to mobile phones' geo located applications, which could be battery life, connectivity, the reaction of the smartphones to atmospheric conditions, the light etc. have been identified.

The results of the indoor workout confirmed that some problems were really critical and blocking for the user, and it all was due to the connectivity, because GPS signal was absent, so the train couldn't start at all. If somebody would just take note of the time, because he knows he couldn't use parameters like speed, kilometres, altitude etc. indoor; he couldn't anyway. It was noticed in both cases that the mobile apps used, stopped if the GPS signal was too bad or anyway lost.

Obviously, all the information about the workout are absent in the indoor setting, because without any movement, all the data field are empty. Predictably indoor training would show problems related to connectivity, neither it was necessary to take care of the lights or some gestures or other things because in an indoor training, it is easier to create the perfect setting, in fact we asked to testers to recreate their setting to train at best and they did, without any particular problem.

In the outdoor evaluation, there weren't connectivity problem, even on low hills, GPS signal was, more or less full, and all the data were updated and reliable. In addiction to that, obviously, in the outdoor session, there were the possibility to see the map and the pin was always moving, giving the chance to check the route. Speed, medium speed, best speed, climbs, descents, elevations, slopes, are some of the parameters that could be seen in a mobile interface while doing an outdoor cycling workout.

A critical point, expected, and really happened was the interaction finger-touch screen. With wet hands, in fact, some gestures such as pinch and zoom, scroll or swipe, tap and flick, are difficult to do. In the applications, it was possible to check all the parameters while swiping from left to right in a sort of carousel that tracked the whole workout. Maybe, for the position of the hands, on the handlebar, it would be better to scroll vertically to change menus; the hands have to be moved anyway from the handlebar if a cyclist wants to track the train because of the position of the smartphone at the middle of the handlebar, so it would be better to move the entire hand in the position where the smartphone is. Index and middle fingers are always exposed to air, because these are the fingers apt to change bicycle gears, so they probably would be drier than the others and the vertical scroll is easier with index finger. In addiction to all this things, both the two testers found out that, in an application, swiping horizontally it could be seen a map, without tapping anything. Once a cyclist reaches the map, he couldn't swipe back or forth anymore, because all the movement are related to the map, so it is only possible to navigate into it. To get back? Just a tap at the top, but unfortunately it is not so easy because of the size of the buttons.

In the end, the general assumption is that the evaluation of mobile geo-located applications should be conducted in their natural habitat in order to generate appropriate findings. This means that thanks to both the studies and thanks to the both setting created, it has been possible to understand how important is evaluation on-the-field. It could bring more data, clearer information; it could underline things that not necessarily are seen about apps' usability in their right environment of use.

Indoor evaluations have less problems than the outdoor ones, as the study confirms but, they have a lot of limitations and blocks that it couldn't be considered totally reliable and trusty.

Finally it could be interesting to increase the numbers of the subjects in order to collect more information and statistical data and enlarge the study to other sports and geo-based application for training, waking or out door dynamic interactions.

Other methods for understanding the use and the interaction, could provide different perspectives on context-aware mobile systems use. This could then be a supplement or it could contradict the findings of the present study.

Acknowledgments. Although the paper is a result of the joint work of all authors, Letizia Bollini is in particular author of parts 1, 2 and 5 and Giulia Cicchinè is author parts 3 and 4.

We acknowledge Professor Natale Stucchi, University of Milano-Bicocca for is precious advice and supervision on data analysis of experimental results.

References

1. Meyers, J., Tognazzini, B.: Apple IIe Design Guidelines. Apple Computer (1982)
2. Nielsen, J.: Usability Engineering. Academic Press Inc., Orlando (1994)
3. Krug, S.: Rocket Surgery Made Easy: The Do-It-Yourself Guide to Finding and Fixing Usability problems. New Riders Press, Berkley (2010)
4. Bollini, L.: Orienteering and orienteering yourself. User centered design methodologies applied to geo-referenced interactive ecosystems. In: Murgante, B., Misra, S., Rocha, A.M. A., Torre, C., Rocha, J.G., Falcão, M.I., Taniar, D., Apduhan, B.O., Gervasi, O. (eds.) ICCSA 2014, Part II. LNCS, vol. 8580, pp. 642–651. Springer, Heidelberg (2014)
5. Kjeldskov, J., Skov, M.B., Als, B.S., Høegh, R.T.: Is it worth the hassle? exploring the added value of evaluating the usability of context-aware mobile systems in the field. In: Brewster, S., Dunlop, M.D. (eds.) Mobile HCI 2004. LNCS, vol. 3160, pp. 61–73. Springer, Heidelberg (2004)

6. Kjeldskov, J., Skov M.B.: Was it worth the hassle?: ten years of mobile HCI research discussions on lab and field evaluations. In: Proceedings of the 16th International Conference on Human-Computer Interaction with Mobile Devices and Services (MobileHCI 2014), pp. 43–52. ACM, New York (2014)
7. Brown, B., Reeves, S., Sherwood, S.: Into the wild: challenges and opportunities for field trial methods. In: Proceedings of the SIGCHI Conference on Human Factors in Computing Systems (CHI 2011), pp. 1657–1666. ACM, New York (2011)
8. Korn, M., Zander, Pär-Ola: From workshops to walkshops: evaluating mobile location-based applications in realistic settings. In: Workshop on Observing the Mobile User Experience at NordiCHI 2010, 16–20 October 2010, Reykjavik, Iceland (2010)
9. Budiu, R., Nielsen, J.: Mobile Usability. New Riders Press, Berkeley (2012)
10. Molich, R. Usable Web Design. Ingeniøren, Bøger (2000). In Danish
11. Nielsen, J.: Why You Only Need to Test with 5 Users, Nielsen Norman Group Blog, 19 March 2000. https://www.nngroup.com/articles/why-you-only-need-to-test-with-5-users/

Digital Tom Thumb: A Digital Mobile and Geobased Signage System in Public Spaces Orientation

Letizia Bollini[✉]

Department of Psychology, University of Milano-Bicocca,
Piazza dell'Ateneo Nuovo 1, 20126 Milan, Italy
letizia.bollini@unimib.it

Abstract. The paper presents an experimental project of a digital mobile and geobased signage system applied to a public space—the University of Milano-Bicocca Campus—and to a digital information tool aimed to guide the choices of students in choosing their university career and to support them along the experience (called the *Students' Guide*). The project approaches all the information, services, tasks and the interaction both with the external—environmental findability and orientation supported by GPS and geobased data—and internal spaces —localization and information inside the building supported by Beacon technology—according to a digital ecosystem design approach. The app *MoBi Moving in Bicocca. Digital Pollicino (Moving in Bicocca. Digital Tom Thumb)* has been designed, prototyped, tested, revised and validated thorough the user-centered approach, co-design and qualitative research methods. Although applied to a specific case study—the whole campus and the building U6 of Milano-Bicocca University—the research approach and the systemic perspective are intended to be replicable as a design frame-work.

Keywords: Geobased interaction design · Geobased mobile app · Mobile app design · Mobile app user experience · Beacon · Digital signage

1 Social Geographies and Cognitive Dimensions

According to the studies of La Cecla [1] it is possible to classify two kind of *orientation* approaches developed to interact with a place: the *topographic* orientation that uses quantitative and absolute measurement units to define spatial relations, and a *relative* orientation—a sort of egocentric point of view—that has the subject in itself as the epicentre of spatial distribution.

On the other hand places themselves have a *story-telling* vocation and *intention* (see the work of Fiorani about the *Grammar of Communication* [2]) that allows people— both as individuals and as groups—to build collective mental models and shared maps of the space they interact with according to Lynch's work *The image of the city* [3].

To surf through and unknown space—both physical and cognitive has underlined by Wurman [4] in his information architecture studies about urban and web environment —activates two different relational modalities: orientation and navigation. The first need is to understand were we are and how the surrounding is related to us, where we come

© Springer International Publishing Switzerland 2016
O. Gervasi et al. (Eds.): ICCSA 2016, Part III, LNCS 9788, pp. 383–398, 2016.
DOI: 10.1007/978-3-319-42111-7_30

from and in which directions we can further proceed. The second is to explore, to discover, to know and understand the place for building a mental model to refer to in the next interactions, the so-called *cognitive mapping*.

In return the environment could have good topological and eidetic categories or a visual and perceptual homogeneity, on one hand, or a deep *environmental overload* that interferes with the creation of a clear image of it as expressed by Augé in his well-known philosophical essay about the *non-lieu* [5].

Identity, structure and meanings determinate the ability of people to move or—better to say—to interact with the space in a sort of aporetic process in which experience has been acquired through experience. The process is a progressive addition in terms of space and time that implies to generate both perceptual and cognitive dynamics according to Ittelson's theory [6].

If the interaction with a physical space implies so many individual and social construction that means individual and collective images, shared knowledge and symbolic culture of the community that inhabit in it a further exploration can give significant insight on geobased information and the connection built among them, people and their spatial relation.

Intercepting these shared images could give a significant insight both individual and collective of the mental models that people share which are the cognitive maps and semiotic meanings emerging from the space with/in which they interact.

Consequently—to you this images as a driver of planning and communication activity—it is important to identify a way to discover how they are socially and individually built. The culture of *psychological geography* suggest some direct and implicit methodologies to use in the field research and directly when working with users. The work of Bagnara and Misiti [7] collects and presents some theoretical positions and an experimental research based on a qualitative, hybrid approach to people relational and cognitive connection with city's environment.

2   The Cognitive and Spatial Perspective on Orientation

Representative test

This social representation of the space is well emerged by a first experimental approach and user testing activities on different segmentation of students population to identifies possible differences among them.

One of the first issues, according with the whole research plan, has been to map and analyse the user needs, the strategies adopted and the tools considered useful in such a circumstance that involves a long-term decision with huge consequences in the future.

A first qualitative recognition has been conducted defining the main *user personas*, archetypical users arise from research. During this recognition phase volunteers were asked to respond to an interview aimed to understand the level of knowledge both of the academic offers and the Bicocca Campus. Then their were asked to use their memory and previous experience to draw a neighbourhood map according to a qualitative representative method presented in the work of Francescato and Mebane [8] and already adopted and discussed at *ICCSA Conference* in 2011 [9]. Finally they were asked—

according to a user test task-based approach—to find critical places and services describing their cognitive strategies and choices expressing aloud their thought (thinking aloud method).

All the results, answers and paths were recorded, analysed and compared to identify common behaviours, problems and recursive patterns. The main tasks were designed or adapted according to a previous experiment conducted in the same area in 2006—as presented at the *VIII Colour Conference* in 2012 [10]—to have a comparative benchmark.

Spatial user test task-based and persona

The collected data were used to build persona and user-scenarios. A set of task-based user test were conducted according to the hypothesis on target segmentation beside questionnaire and representative methods were used to collect more insight on mental models built by people and difficult emerging in exploring and finding way in a complex environment.

A first group is represented by high-school students, which come to the Bicocca Campus for the first time during the *open days*. In spring time the universities organize some activities to promote master and degree courses to involve potential *clients* in a marketing and advertising campaign to directly *grab* the best and more motivated students before the admission tests to be held in early September.

This group has been clustered and described by *the newbies* label. They are 18/19 years old, not yet familiar with some of the structural and formal requirements and fulfilments of the university world. They do not know the Bicocca Campus and need some extra support in orientation both in the physical space and in the log-term decision about university career.

A second group is the so-called *intermediate:* age 22, they have already a first level degree and are aware of university procedures and bureaucracy, but unfamiliar with the Bicocca Campus because coming from another athenaeum or from another city.

The third segment is represented by *Gurus:* people between 22 and 24 years old, that have been attending the Bicocca Campus from the beginning of their career. They are well informed both on academicals fulfilments and of the space of the neighbourhood, not only the campus.

According to recent literature—see the studies conducted by Perkins [11] and in more recent time by Schmidt and Hawkins [12] about *Millenials*—they are already considered *digital natives*. Newbies, Intermediate and Gurus have a massively use of mobile devices (98,5 % of involved subjects)—at least a smartphone—social networks —Facebook and Snapchat—and instant messaging—whatsapp—and often helps themselves with georeferenced applications: Maps or Google Maps.

A first in field research *iBi*—conducted in 2014 to understand problems and potential targets of the *Students' Guide* of the *Department of Psychology* to promote its courses —gave a first insight of their information retrieval strategies and the use of technological infrastructure in orientation activities both when they are intended to take decisions about their university and professional futures and in a way-finding activity in the space of the Campus.

According to user-testing qualitative methodologies developed in user-centered design —see the work of Budiu and Nielsen [13], and Krug [14]—and their declination in geographical psychology research investigated by Bagnara and Misiti [7]—an experimental activity was conducted on a segmented panel of users. Subjects—9 for each category—were previously asked to prepare themselves to the spatial tasks finding information (see Figs. 1 and 2 for indoor and outdoor orientation strategies) and then—after a couple of days—to complete a list of spatial activities shaped on their Campus or daily students' life experience out- and in-doors (see Fig. 3 on data retrieval strategies).

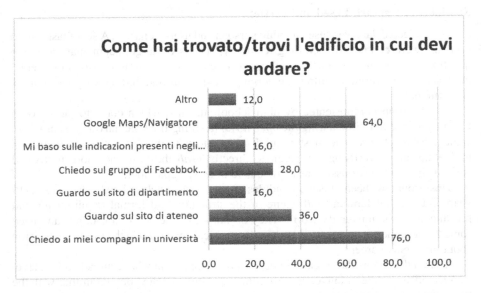

Fig. 1. *Outdoor orientation:* Strategies used to identify and get to a specific building

The given tasks to complete were:

1. to get to building U6 from train station or other interconnection points to reach the Aula Magna where courses will be presented;
2. to identify the students' secretariats where to register for the admission test;
3. to visit a professor in his/her room;
4. to find the bar and the canteen where to have coffee-break and lunch;
5. to find the eco-bus stop to go to U16 for English test;
6. to discover facilities linked to students life: copy services and book-shops;

What emerges from this first task-based spatial user tests sessions is the difficult to identifies a clear mental model of the space around them—bot outdoor and inside the single building. The place has an overall lack of *figurability*—or environmental layout —small and not well-differentiated landmarks and visual anchor-points according to the *legibility* described by Weisman [15].

Signage elements are not enough to identify the correct or shorter pathway to the final destination and not always present in crucial point where clearer information were

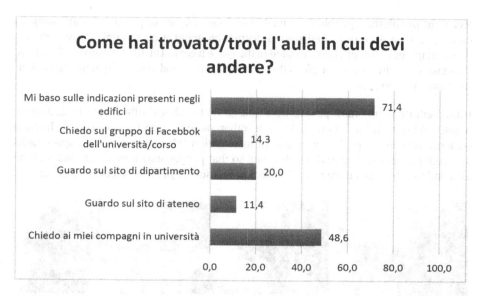

Fig. 2. *Indoor orientation:* Strategies used to identify and get to a specific classroom

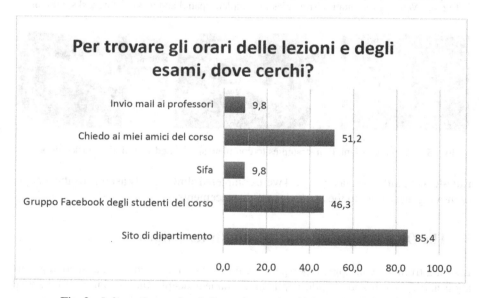

Fig. 3. *Information retrieval:* Strategies used to find practical information

needed. Sometimes they are neither updated nor corresponding to real configuration and location of offices and students' services.

Clustering the tasks in order of criticality to define a clear design goals ranking.

Relevant problems: people were unable find the eco-bus stop—task 5—and subjects took more than 40 min to find some information about it on the university web site. Signs, maps and other elements located in the space have not at all or incorrect directions. This task was critical also for gurus that already have a good familiarity with the campus structure and services.

Intermediate problems: people were able to find students' offices—tasks 2, 3 and 6 —just asking to information desks or to other students (see Figs. 4 and 5). Internal signage and wall-maps do not give such information. Moreover, the inner space lacks of visual anchors or orientations elements so that people start moving around with no idea and find by chance the right destination in a sort of *pervert* serendipity effect.

Fig. 4. *Newbies:* orientation strategies to complete spatial and way-finding tasks outdoor.

Fig. 5. *Newbies:* orientation strategies to complete spatial and way-finding tasks indoor.

Low-level problems: tasks 1 and 4 were completed almost easily asking to other people or moving according to the *critical mass* displacements.

3   The Digital Dimension of Spatial Interactions

Looking from a technological perspective, it is important to remark that second generation and 3.0 web and mobile application connect deeply the spatial and the social dimension of interactions.

First generation was mainly aimed to give operative GPS *instructions* such as where the users were, how to move, or to arrive in a specific place showing this date on a map as well exemplified bay iOS Maps® and Google Map® for mobile. Second generation is represented by the use of GPS date and geolocated information to connect people to each other according to a spatial and *in presence* criteria: the archetype is Four Square®. In this case people are able directly interact with the place thanks to the *check in* task, to *see* other person on a map, interact with them personally or using social dynamics—such as the *major* challenge—leaving traces, suggestions, information (user-generated) or notes through the *tips* features.

Open air and indoor orientation digital supports

Among the many technological solutions available on the market an analysis on pros and cons has been conducted to understand and choose a best practice applicable to the project. For a deeper discussion on the issues Manh Hung presents an interesting comparison table about indoor navigation system, although not so updated to the contemporary state of the art. [16].

Following this categorization the research has been expanded to two main issues: the *indoor navigation apps*—Google Indoor, Apple e Wi-Fi SLAM and Nexton have been identified, analysed and evaluated—and Internet of Things and Smart Tagging technologies.

In particular QR-Code, RFid/NFC and Beacon were considered.

QR-codes—then used as a second choice support in some of the digital signage system—were originally discarded. It seemed that they are *not so digital* to be included as an efficient support in an IoT approach to spatial interactions. Nevertheless they have been restored and integrated in the final version to support non-Android devices—in this first version of the app—or older devices non compliant with contemporary solutions. Moreover the project was intended to break down barriers between *physical* and *virtual* giving a totally transparent experience to the users.

On the other hand only the Android mobile devices and Windows phone are equipped with Bluetooth 4.0 required by Beacon. This latter technology enable a *one-to-many* experience generating a direct interaction and feed-back on the user device when in proximity. The notification push system is, surely, very efficient, but, at the same time, very invasive and, potentially, disturbing, giving the user a *bad feeling* to be under control. However it allows people mot to miss information, help and support in very complex, crowded or spatially confused situations.

NFC, nevertheless, are spreading in commercial context and other public spaces such as museums and other cultural institutions. The communication is *one-to-one* and directly controlled by the user who must be very closed the NFC tag and intentionally decides to interact with the digital environment. The potentiality of this system consists mainly in the possibilities to be used as payment technologies directly connected to banking on line services.

4   MoBi-Digital Tom Thumb: A Mobile Context-Aware App

The spatial scenario and the context aware approach

Mobile technologies can be, then, a support in facilitating way-finding strategies thanks to the direct relation with geobased data and the physical environment.

At the same time in a complex spatial system such as a university campus related to a peripheral and peculiar neighbourhood—for a wider comprehension of the specificity of the Milano-Bicocca suburb *the urban island* see the work presented at ICCSA 2011 in Santander [9]—a *context-aware* application could help and support efficiently the users in their exploration activities.

Among the many definition and theoretical elaboration of the concept, the *context-aware system* is, here, intended as described by Dey: "Context is any information that can be used to characterize the situation of an entity. An entity is a person, place, or object that is considered relevant to the interaction between a user and an application, including the user and applications themselves" [17]. Besides, Dey underlines how this means that *every* situation is relevant for the system and its users. Consequently, information and context are *dynamic* factors. Furthermore Chen and Kotz [18] describe the context as "the set of environmental states and settings that either determine an application's behavior or in which an application event occurs and is interesting to the user."

A final refinement of the idea of the relationship between the space ant the interaction with users comes from the work of Schmidt et al. *There is more to context than location* [19] suggesting that *context* is made from human factors—the social environment, user's identity and tasks—and physical landscape intended as the system of places, environmental conditions and infrastructures. A way-finding approach, according to the conceptual framework of spatial *findability* developed by Morville [20], therefore could be successfully applied both to this spatial and digital environmental model.

The Beacon plan: contextually and proximity interactions

The challenge of the project is to develop a user experience completely transparent in the transition between inside and outside space so that people remain focused on the orientation interactions focused on spatial and personal decision-making.

As emerged in the preparatory study, inside the buildings people do not have the support of GPS functionalities and are forced to interact with not-updated printed signs or asking to information desk or other students. Although Google makes available outdoors and indoors API for Google Maps, university campus are considered sensible targets so that taking photos or publishing interoperable internal floor-plans of buildings is not allowed.

According to previous consideration about the support of technologies for indoor-interactions and facing the constrains given by the specificity of the place and its security issues, the application uses Beacon and Bluetooth as *micro-localization facilitators.* They were used both to enable the users' interaction with contextual information connected with their positioning inside the building and to activate or deactivate the out/in-door navigation flow switching from GPS and Wi-Fi + Beacon.

A background location layer is provided by Google Maps® API and its digital cartography support to create, map and align the personalized floor-plans with the image of the building itself. The next operation consists in the mapping activity give place to *makers* and relative Wi-Fi hot spots and Beacons to have an performing repeaters network validated through *Google Maps® Floor Plan Marker* [21], to have an effective geolocated-data map (see Fig. 6 where the experimental infrastructure is spatially described and positioned).

The services and experience delivered by *MoBi* are not only geolocated, but also context aware and *situated,* relevant for the user and the tasks that he/she is intending to complete, according to Coutand researches on *Contextual Personalised Applications*

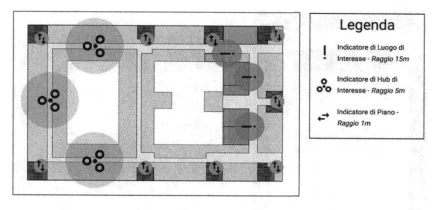

Fig. 6. *Beacon plan:* positioning hypothesis inside 3rd floor of building U6.

[22] where context itself is part of the information and part of the fruition modalities of data and functions of the application.

According to Schmidt's working model [19] the contest is made by the Physical Environment—the sum of the place, environmental conditions and the infrastructure—and by the Human Factors—that means the social environment, the identity and the tasks to be done by the user—and a *temporal* structure. Therefore the MoBi app and the digital located infrastructure supports students in the way-finding processes both orienting people through the *physical space*—the whole Campus, the single building and its inner space—, the information offer according to *users' needs*—mainly common, but with different levels of access according to the persona: seniority, degrees and majors—organized in a timeline—news, reminders and alerts of the university life-cycle.

Point of interest: interacting with the digital signage plan

Following the context-aware approach, Beacon where placed along the inner space of the building—paying attention that they signals do not interfere one to another—in the *point of interests* emerged during the first research phase.

They are: *access point* to the building, *access point* to single floor, single *point of interests* and *hub of interest* where many spatial functions cross or overlap.

Every point of interest signal is composed by three modules (as shown in Fig. 7), each one using a specific technological support:

1. *contextual element* (beacon): information give are the name of the place—office, number of the room etc.—and the Beacon showing the *Point of interest sign.* It uses both the analogic and the digital modality.
2. *information element*: dynamic, specific and time-based information are displayed chronologically—such as news, time schedule, updated information—scrolling on the display of a tablet
3. *documents element*: a grid of NFC tag and QR-code is displayed connecting users to URL where to download documents, files and forms.

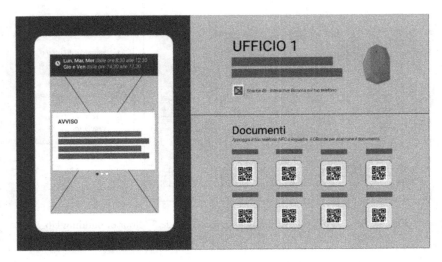

Fig. 7. *Point of interest:* horizontal configuration of the 3 modules (from left to right counter wise) and related technologies: information element (tablet); contextual element (beacon), documents element (QR-Code).

Designing the geobased interface of the MoBi App

To approach the information architecture structure and to define services and contents to deliver through the application different activities of user-centered and co-design where developed.

First of all an on-line questionnaire was launched early in the design phase to understand how people of the different targets where used to search information and which expectation they could have on a future mobile application aimed to help people both as a *student guide* and as a orientation tool in the campus. 67 people took part spontaneously to the survey and 10 of them where involved in the next research activity.

A card-sorting workshop was organized to understand which services were important for students and how organize and label them according to the user experience and expectation.

This approach to mobile navigation was crucial to prioritize functions according to the different persona needs—newbies were less confident and they need a wider and deeper support, gurus prefers shortcuts and abbreviations—to cluster services and information and to correctly label them. This activity was revised during the iterative test-assess-redesign cycle that brought some modification in the previous design choices (the item: "collections" becomes "to-do list" and some icons have been changed following the user test results).

In addition to user space tests already described, 15 subjects took parte to an A/B test on the Lo-Fi mockups to asses some of the interface spatial organization of the wire-frame and some structural components disposition.

In the next few images the interface design adopted to represent some of the spatial tasks of the application and their relation with the signage system are shown (Figs. 8, 9, 10 and 11).

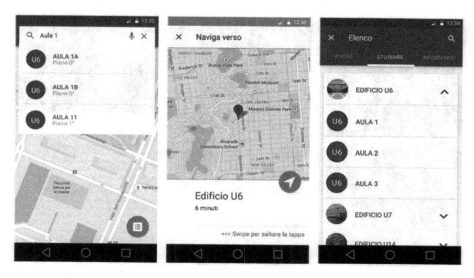

Fig. 8. *MoBi app:* how to locate: a. the single classroom c. inside a building and b. where the building is inside the Bicocca-Campus area

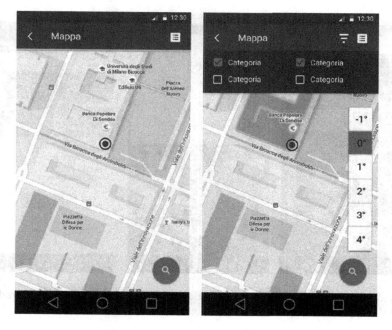

Fig. 9. *MoBi app:* comparing a standard georeferenced visualization (on the left) and the iBi/MoBi multilayer, filterable georeferenced visualization (on the right).

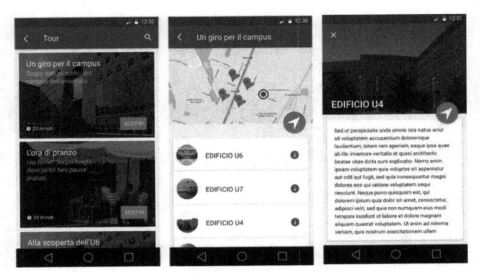

Fig. 10. *MoBi app:* the *Tour* feature designed to help newbies in the campus discover

Fig. 11. *MoBi app:* offices' location, basic function and information, opening days/hours directly connected to module 2 of the *Point of interest* signal.

Interface elements have been designed according to *Material design guide lines* developed by Google for resident and custom applications development. The application, based on Beacon infrastructure is manly intended for Android devices—mainly used by students as emerged by the questionnaires—although further development with Beacon can be implemented for iOS target.

Other visual values and choices—such as the colour language—are intended to be coherent with the corporate image of the University institution, Further more the purple tint is the declination of the academic symbolic palette for the Psychology Department. Yellow has been chosen as accent colours thanks to the high contrast level able to guarantee a good discrimination with the primary one, its variables and neutral grey scale although in the code of *gogliardia* it is associated to Economical Degrees.

The font used—Google font Roboto—is a third generation screen font able to guarantee a high level of legibility of the text also in bad visibility conditions (movement, dazzling lights, small dimension or display resolution and quality). Text, contents, labels and titles are black or white not to be confused with other significant elements of the interface or interactive actions triggers. They remain neutral and with the highest contrast with the interface backgrounds' colours.

Physical vs. digital: comparing the way-finding tasks

The whole project has been tested with two different finality.

Firstly it has been validated with users through a intense user-testing activity: 20 people took part to the evaluation sessions.

Task analysis: 12 tasks were proposed and tested to understand the navigation, services and an overall assessment of the application.

System usability score: users were asked to score the application, their experience and the perceived usability using a 0–5 Likert Scale.

Semi-structured interviews and co-design workshop: critical navigation elements and screen-shot have been discussed and redesigned directly involving the users in a co-design session.

A second type of evaluation activities has been conducted to compare the previous wayfinding and orientation experience with the one mediated and supported by the MoBi app. 12 people were involved in a dynamically session out- and in-door.

Task analysis: 3 tasks—from the easiest to the hardest according to the first phase results—were proposed and tested to understand the navigation, services and an overall assessment of the application.

System usability score: users were asked to score the application, their experience *live* using a 0–5 Likert Scale.

Semi-structured interviews: overall navigation and qualitative feed-back.

As Fig. 12 shows the comparison between smartphone + signage support + peer help experience and the MoBi support are significantly different also in quantitative terms. Time to complete the way-finding tasks are simply lower when people use the digital application. The time to complete the activity is almost halved and—as emerges from the qualitative and semi-structured interviews—the overall perception is an easy to use and helpful support.

			Tempo (s)	Errori	Hanno Utilizzato iBi?
Newbies	Con iBi	Soggetto 1	184	0	Sì
		Soggetto 2	197	0	Sì
		Media	190,5		
	Senza iBi	Soggetto 3	408	1	
		Soggetto 4	513	1	
		Media	460,5		
Interme-diate	Con iBi	Soggetto 5	185	0	Sì
		Soggetto 6	188	0	Sì
		Media	186,5		
	Senza iBi	Soggetto 7	350	0	
		Soggetto 8	372	1	
		Media	361		
Guru	Con iBi	Soggetto 9	192	0	Sì
		Soggetto 10	183	0	Sì
		Media	187,5		

Fig. 12. *MoBi app:* comparing persona-based test executions with and without MoBi (iBi) App support: here are exemplified the task 3 results.

5   Conclusions

After the different validation sessions and the design assessment the project should still face some further possible improvement and development.

The experimental pilot of the *MoBi* application should be extended to the whole Campus—buildings, services, infrastructure, departments and sub-campus—to verify the hypothesis and the complexity of the information architecture embedded in the system. A cross platform app could be tested according to iOS and windows phone requirements or a responsive web site, although this solution implies a huge unification effort of all the university digital frameworks both on line and intranet services.

Even if the project is developed and applied on a specific case study, the methodological approach proposed is intended to be replicable on other spatial scenarios.

The idea of a diffused spatial interaction with digital signage systems embedded in the physical environment is one of the more promising evolution of mobile, pocketable and wearable devices. On the other hand Internet of Things used both in commercial and cultural location will probably have a huge development in the next few years. This converging infrastructures—personal and widespread—are the next transparent interface able to connect us and our *situated* interactions and information needs. Geobased technologies and user-centered design are the driver to plan and shape the potentiality of such a scenario.

A participatory approach aimed to understand social images of the space and cognitive mental models can give design tools and guideline to develop application based on the spatial *aware*.

Acknowledgments. The project was conducted with Matteo Cesati who developed and tested the application prototype and the (e)beacon settings for his Master Degree in *Theory and Technologies of Communication* dissertation: *iBi – Interactive Bicocca app – progettare un'esperienza contestuale per il wayfinding spaziale e digitale* and the contribution of Giulia Busdon, Matteo Cesati and Annalisa Mazzola involved in the *MoBi research* explorative phase in 2014.

References

1. La Cecla, F.: Perdersi; l'uomo senza ambiente. Laterza, Bari (1988)
2. Fiorani, E.: Grammatica della comunciaizone. Lupetti, Milano (1992)
3. Lynch, K.A.: The Image of the City. The MIT Press, Cambridge (1960)
4. Wurman, R.S.: Making the City Observable. The MIT Press, Cambridge (1971). Design Quarterly No. 80
5. Augé, M.: Non-Lieux. Introduction à une anthropologie de la surmodernité. Le Seuil, Paris (1992)
6. Ittelson W.H.: Percezione dell'ambiente e teorie percettive. In: Ittelson, W.H. (ed.) La psicologia dell'ambiente. Franco Angeli, Milano (1973)
7. Bagnara, S., Misiti, R. (eds.): Psicologia ambientale. Il Mulino, Bologna (1978)
8. Francescato, D., Mebane, W.: How citizens view two great cities: Milan and Rome. In: Downs, R., Stea, D. (eds.) Images and Environment. Aldine, Chicago (1973)
9. Bollini, L.: Territories of digital communities. Representing the social landscape of web relationships. In: Murgante, B., Gervasi, O., Iglesias, A., Taniar, D., Apduhan, B.O. (eds.) ICCSA 2011, Part I. LNCS, vol. 6782, pp. 501–511. Springer, Heidelberg (2011)
10. Bollini, L.: Comunicare con il colore spazi e percorsi: Aspetti metodologici, ergonomici e user-centered. Campus bicocca: Un caso studio. In: Rossi, M., Siniscalco, A. (eds.) Colore e colorimetria. Contributi multidisciplinari, vol. VIII, pp. 431–438. Maggioli Editore, Santarcangelo di Romagna (2012)
11. Perkins, M.: Digital natives, digital immigrants part 1. Horizon **9**(5), 1–6 (2001)
12. Schmidt, L., Hawkins, P.: Children of the tech revolution. In: Sydney Morning Herald (2008)
13. Budiu, R., Nielsen, J.: Mobile Usability. New Riders Press, San Francisco (2012)
14. Krug, S.: Rocket Surgery Made Easy: The Do-It-Yourself Guide to Finding and Fixing Usability Problems. New Rider, San Francisco (2009)
15. Weisman, J.: Evaluating architectural legibility: way-finding and the built environment. Environ. Behav. **13**(2), 189–204 (1981)
16. Manh, H.V., et al.: Indoor Navigation System for Handheld Devices. Worcester Polytechnic Institute, Worcester (2009)
17. Dey, A.K.: Understanding and using context. Pers. Ubiquit. Comput. **5**, 4–7 (2001)
18. Chen, G., Kotz, D.: A survey of context-aware mobile computing research. Dartmouth Computer Science Technical Report TR2000-381, Dartmouth College (2000)
19. Schmidt, A., Gellersen, H.W., Beigl, M.: There is more to context than location. Comput. Graph. **23**(6), 893–901 (1999)

20. Morville, P.: Ambient Findability: What We Find Changes Who We Become. O'Reilly Media, San Francisco (2005)
21. Play Store, Google Maps Floor Plan Marker. https://play.google.com/store/apps/details?id=com.google.android.apps.insight.surveyor&hl=it. Accessed 28 July 2015
22. Coutand, O.: A Framework for Contextual Personalised Applications, p. 21. Kassel University Press, Kassel (2009)

Land Suitability Evaluation for Agro-forestry: Definition of a Web-Based Multi-Criteria Spatial Decision Support System (MC-SDSS): Preliminary Results

Giuseppe Modica[1(✉)], Maurizio Pollino[2], Simone Lanucara[1,3],
Luigi La Porta[2], Gaetano Pellicone[1,4], Salvatore Di Fazio[1],
and Carmelo Riccardo Fichera[1]

[1] Dipartimento di Agraria, Università degli Studi Mediterranea
di Reggio Calabria, Loc. Feo di Vito, 89122 Reggio Calabria, Italy
{giuseppe.modica,salvatore.difazio,
cr.fichera}@unirc.it
[2] DTE-SEN-APIC Lab, ENEA National Agency for New Technologies,
Energy and Sustainable Economic Development,
Via Anguillarese, 301, 00123 Rome, Italy
{maurizio.pollino,luigi.laporta}@enea.it
[3] Institute for Electromagnetic Sensing of the Environment,
National Research Council (IREA-CNR), Via Bassini, 15, Milan, Italy
lanucara.s@irea.cnr.it
[4] Institute for Agricultural and Forest Systems in the Mediterranean
(ISAFOM-CNR), Via Cavour 4-6, Rende, CS, Italy
gaetano.pellicone@isafom.cnr.it

Abstract. Land suitability evaluation (LSE) is a widespread methodology that supports environmental managers and planners in analysing the interactions between location, development actions, and environmental elements. In the present paper, we discuss on a web-based multi-criteria spatial decision support system (MC-SDSS) implemented to accomplish LSE for agro-forestry. We propose a MC-SDSS developed on a free open source software for geospatial (FOSS4G) environment, accessible through a user-friendly geographical user interface (GUI) that allows to perform geospatial analyses. To this end, the MC-SDSS has been conceived as a multi-tier architecture, able to manage processes executable via OGC (Open Geospatial Consortium) web processing services (WPSs) and produce output maps and data available via the largely used OGC services: web feature service (WFS), web coverage service (WCS) and web map service (WMS). In this first application, we chose the weighted linear combination (WLC) as decision rule to aggregate data, weighted by judgements provided by experts following the analytical hierarchy process (AHP).

Keywords: Free open source software for geospatial (FOSS4G) · Webgis platforms · Web-based Multi-Criteria spatial decision support system (MC-SDSS) · Land suitability evaluation (LSE) · Analytical hierarchy process (AHP)

© Springer International Publishing Switzerland 2016
O. Gervasi et al. (Eds.): ICCSA 2016, Part III, LNCS 9788, pp. 399–413, 2016.
DOI: 10.1007/978-3-319-42111-7_31

1 Introduction

Today, landscape planning and management involve the identification of consensus-based solutions that address multiple societal needs and demands [1]. To manage agro-forestry resources according to the economic, environmental, and social dimensions of sustainability, Decision-Making (DM) approaches and procedures that examine trade-offs between often competing/conflicting objectives/alternatives should be implemented. Considering that soil can be fairly recognisable as an ecosystem structure and a non-renewable resource, studies for the assessment of land take phenomenon and actions for its mitigation have been encouraged at European level [2, 3].

GIS-based technologies and approaches can play a fundamental role both in suitability assessment and in complex DM, in the course of environmental impact analysis [4]. Recently, geo-scientific community has been focusing on the use of GIS technologies and techniques for supporting DM in numerous fields of application. The need for related standard and effective spatial interfaces, geo-visual analytic tools, integrated geographic platforms (SDIs, Spatial Data Infrastructures) has been recognised by several researchers as shown by their research works [5–10]. It is universally recognised the capacity of maps to offer an overview of and insight into spatial patterns and relations [11–15].

Land suitability evaluation (LSE) can effectively support environmental managers and planners in analysing the interactions between location, development actions and environmental elements [16]. The LSE, today widespread, can be defined as a method for identifying and assessing the suitability of an area for a specific land use and is based on the explicit identification of constraints and opportunities either for conservation of present land uses or for planning and location of new ones [17–19]. In implementing a LSE two different components can be distinguished: the so-called physical LSE (suitability is an intrinsic characteristic of the examined area); a subsequent usability evaluation (usability is the current and actual possibility to use the available resources).

Today, in a LSE procedure the principles of sustainable development should explicitly be taken into account when choosing and weighting criteria and/or alternatives. In this direction, multi-criteria decision analysis (MCDA), that can be synthesised as a complex and dynamic technique supporting DM in combining several criteria to make an optimal decision, can support Decision Makers (DMs) in taking into account multiple criteria explicitly [20]. Coupling MCDA techniques with geographical information systems (GIS) in GIS-MCDA procedures [21] represents a key element in the implementation of a multi-criteria spatial decision support system (MC-SDSS) [22]. As a matter of fact, the introduction of GIS-MCDA in physical LSE, which was concomitant with the scientific and technological advance of such tools, has enabled to see the whole spatial planning approach from a new perspective [5].

Referring to the issue of the present paper and as highlighted in other researches [23], GIS-MCDA procedures are widespread in spatial planning and DM processes, since they allow to consider, at the same time, the objectives and the different criteria influencing a specific land suitability analysis. The integration of web-based GIS and MCDA techniques allows to obtain a MC-SDSS which provides appropriate analytical tools for direct involvement of people in a collaborative spatial planning process [24]. Thanks to the use of web applications, MC-SDSSs, traditionally developed for stand-alone users, can be implemented on a web server (WebGIS), in order to allow, through the internet, the interaction with thematic maps and data associated with it. WebGIS platforms [25] are exploitable by common browsers (Mozilla Firefox, Google Chrome, Internet Explorer, etc.), thus they extent and improve the utilisation of these systems, that can be referred as web-based MC-SDSSs. The main benefits of using WebGIS technology are: (i) global sharing of geographic information and geospatial data, (ii) ease usability by the client, (iii) data network dissemination and ability to reach a wider community of users [7]. The WebGIS tool can be used, therefore, as an information consultation tool enriched by the geospatial component, for querying and analysing geographic data and thematic maps. In this way, it is possible to exploit all the advantages of being able to display and manage data in a spatial way, adding greater new significance on environmental analyses, jointly with the possibility to combine this information with other types of data characterizing an area of interest by different aspects (territorial, geo-morphological, socio-economic, etc.). Using the mapping power, a GIS-based environment allows visualizing, querying and analysing data to understand relationships, patterns and trends in order to improve communication and DM processes [26]: this implies the adoption of specific and suitable GIS and SDI architectures, that in the framework of the study here described have been developed using free/open source (FOSS) packages.

Moreover, while originally MC-SDSSs were proposed as systems that accomplish single-user evaluations, today an increasing number of applications in different research and operational fields propose MC-SDSS within a framework of group (e.g., collaborative) DM.

Over the last 50 years, a wide literature has proposed several multi-criteria methods and techniques dealing with theoretical and practical issues on DM [27]. Among these, one of the more widespread methods to derive the relative criteria weights according to an appropriate hierarchical system is the analytic hierarchy process (AHP) proposed by Saaty [28].

The main goal of our research is the implementation of a web-based MC-SDSS, capable to provide a user-friendly geographical user interface (GUI) for perform spatial analyses (geoprocessing): in particular, via a specific WebGIS application, users are allowed to set and tune MCDA parameters according to their preferences (based on their knowledge), in order to produce scenarios and identify most suitability locations according to a specific use. To this end, the MC-SDSS has been conceived as a server-side architecture, able to manage processes executable via OGC (Open Geospatial Consortium, www.opengeospatial.org) web processing service (WPS) and to produce output maps and data available via the largely used OGC services: web feature service (WFS), web coverage service (WCS) and web map service (WMS). In this first application, we propose a web-based MC-SDSS implemented as a general schema to perform LSE for agro-forestry in which geo-processes implemented as WPS are those related to criteria weighting and aggregation.

2 Materials and Methods

To implement the MC-SDDS and provide users with an effective MCDA accessible online, a suitable architecture has been designed and implemented. The logical and physical components of such architecture, as well as the data workflow, are depicted in the general schema reported in Fig. 1. A detailed description of each component (and of the related functionalities/capabilities) is provided in the following sub-paragraphs.

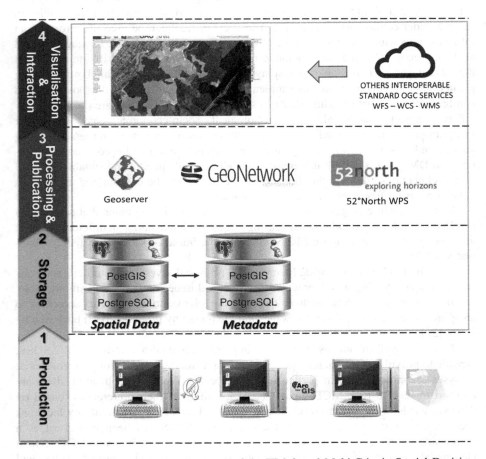

Fig. 1. Data workflow and general schema of the Web-based Multi-Criteria Spatial Decision Support System (MC-SDSS)

2.1 Spatial Data Infrastructure (SDI)

In order to facilitate the access to all the geospatial data and analysis produced, and to share them with stakeholders and decision makers (DMs), so as to support the planning process, a spatial data infrastructure (SDI) was developed in a FOSS4G (Free and Open Source Software for Geospatial) environment [29].

In general terms, SDIs can be referred as common platforms for management, discovery, publication and use of geospatial data and metadata. A SDI provides an ideal environment to connect applications to data, at the same time influencing the creation of data and the development of applications based on standards and appropriate procedures. Various and widespread are the solutions commonly adopted to design and implement SDI architectures [30]: in particular, the implementation of applications in FOSS4G environment is actually supported by the large availability of software suites and packages [31], which make possible - among other things - to effectively share and publish the resulting maps and data by means of suitable (and customisable) web interfaces. The architecture here adopted has been properly designed to allow the interchange of geospatial data over the Web and to provide the beneficiaries with a user-friendly application, characterised by accessibility and versatility. Following that, we implemented the SDI with a multi-tier architecture composed by three layers with different functions:

1. Data repository stratum to store data and metadata in a geospatial database implemented in PostgreSQL with PostGIS extension (http://postgis.net);
2. Server stratum, composed by GeoServer (http://geoserver.org) and GeoNetwork Opensource (http://geonetwork-opensource.org), to manage stored data and metadata and publish them on the web using OGC standard interfaces, 52° North WPS (http://52north.org/communities/geoprocessing/wps/) to enable the infrastructure to the deployment of geo-processes using OGC WPS standard interface;
3. Client front end stratum (i.e., the WebGIS client).

All above components are implemented in a virtual environment, (i.e., a virtual machine), based on the Ubuntu operating system, and managed by a VMware® vCenter server.

The Data Repository identifies the storage area containing the data-set and metadata used. The access is allowed only to devices physically defined at the level of the storage area network (SAN), in order to ensure the complete integrity and consistency of the data themselves.

The Server stratum represents the software environment that allows to organise information making it accessible from the network. Moreover, it provides maps, data and geo-process, from a variety of formats, to standard-OGC interfaces, so they are available to clients such as web browsers and Desktop-like GIS suites that manages OGC standards. This makes it possible to store spatial data in almost any format that users prefer. In our case, we adopted the GeoServer suite, a widespread used open source application server, which plays a key-role within the SDI as server for sharing geospatial data. It allows sharing and managing (by means of different access privileges) the information layers stored in the repository and is also interoperable (e.g., the capability of two or more systems to exchange data and information among them meaningfully and accurately, and then be able to use them independently from the operating system or the software used).

From the technical point of view, GeoServer is the reference implementation of the W * S standards defined by OGC. In this way, for instance, users are allowed to access geospatial data using any client (Desktop GIS software and/or web application) enabled to manage W * S connections. Concerning the front-end stratum, by accessing the WebGIS interface through its own web browser, the user (not necessarily with specific GIS skills) can gather and view thematic maps, charts and other results available (e.g., scenarios). In particular, to visualise the data available, the WMS standard is commonly and widely exploited, by means of a map-server approach that allows producing thematic maps of geo-referenced data and responding to queries about the content of the maps themselves. Other than the OGC standards related to geospatial data access and sharing, it is also possible to exploit another type of service specifically devoted to the processing task: the web processing service (WPS, www.opengeospatial.org/standards/wps), that we implemented by means of the 52° North WPS software. Such standard provides rules in order to define how inputs (requests) and outputs (responses) can be processed in the frame of a Web service. The WPS standard also defines how the client can request the execution of a process, and how to manage the output resulting from the process. To this end, a specific interface is defined in order to support the publishing of geospatial processes and clients' discovery of and binding to those processes. The data required by the WPS during the processing can be provided via Web or, alternatively, can remain available at the server. By using the WPS, it is possible to implement any calculation (including all its inputs and outputs) and trigger the execution as a Web service. Such standard is able to support simultaneous processes (via HTTP and SOAP), allowing the client to choose the most appropriate interface mechanism. The WPS processes execution is invoked by submitting a request (in XML or URL-encoded) to a specific URL (identifying the inputs, the type of process to be executed and the output format to be provided). A WPS is usually not invoked directly, but generally it is requested by a client application (preferably, web-based), which provides the user with interactive controls.

2.2 WebGIS Platform Implementation

The publication of geographic data through the web has stimulated the development of tools for the implementation of WebGIS platform and applications. The WebGIS is not a simple extension of a GIS Desktop suite, but it is part of the larger category of web-oriented software [32]. The network is the way for the interchange of data through the Web browser and the communication is based on client-server architecture in which two independent modules interact to perform a task. In such a way, it is possible to build geographic web-oriented tools for publishing, consulting and analyse data, useful for planning purposes. The main the advantages of using a WebGIS client are:

- the effective usability (WebGIS applications are accessible and exploitable through common internet browsers);
- the spread on the network and the capability to reach a wider audience of users.

Fundamental, therefore, is the role of a similar approach/architecture as a support tool in DM and planning, as part of a more articulated and complex system like the MC-SDSS subsequently described. The WebGIS client was developed in heron mapping client MC (http://heron-mc.org), a free and open source framework based on

GeoExt (http://geoext.org). GeoExt is a JavaScript powerful toolkit that combines the web mapping library OpenLayers with the user interface of Ext JS to help build powerful desktop-style GIS apps on the web based on JavaScript language. Heron MC provides to developers prebuild widgets to view, query, print, as well as advanced tools to data editing and filtering. Also Heron MC developers are generating new functions continuously, so it is possible to add new functions and menus to the web interface. OpenLayers is an open source JavaScript library, used to visualise interactive maps in web browsers: it provides the so-called application programming interface (API) allowing the access to various sources of cartographic information on the Web, such as OGC protocols, commercial maps (i.e., Google and Bing Maps, etc.), different GIS formats, maps from the OpenStreetMap (OSM) project, etc. Ext JS is a JavaScript framework for building feature-rich, cross-platform web applications for desktops, tablets, and smartphones. Ext JS leverages HTML5 features on modern browsers while maintaining compatibility and functionality for legacy browsers. It features hundreds of high-performance UI widgets that are designed to fit the needs of the most complex web applications.

To gather a good degree of interactivity, the WebGIS application allows user to get the maps and work with them, according to different functions. The set of functions available encompasses, first of all, the basic ones: the classic pan and zoom on a map, which involve not only the scaling of the image, but also the identification of geographic objects in the archive and related to the specific request. Another capability resides, then, in the selection mode, which can be graphical and spatial. In such WebGIS environment, in fact, the selection is made on a set of geographical objects characterised by mutual spatial relationship (such as, for example, contiguity, adjacency, intersection, etc.) and by specific descriptive attributes (qualitative/quantitative). Objects (also named as "features") can be selected on the basis of a request made in WFS standard.

As stated in the previous section, one of the common ways to invoke a WPS process is to exploit a web-based application: this is another specific capability of the current WebGIS implementation. The WebGIS, thus, provides an interface mechanism to: (i) identify/select the geo-referenced data required by the process, (ii) launch the calculation and (iii) manage the output obtained by means of the calculation in order to make it available to the client.

2.3   The Multi-Criteria Spatial Decision Support System (MC-SDSS) Module

The MC-SDSS module of our platform has been implemented according to four main phases, as reported in the following flow-chart (Fig. 2) and subsequently described.

Phase 1 - Criteria Definition and Selection. As in previous activities of the research group [23], and following the approach proposed by Eastman et al. [33], we consider evaluation criteria as factors and constraints. In more details, factors are the measured decisional variables on the basis of which the suitability of an area for a specific use can be assessed, while constraints represent the limitation or, as in our case, the exclusion

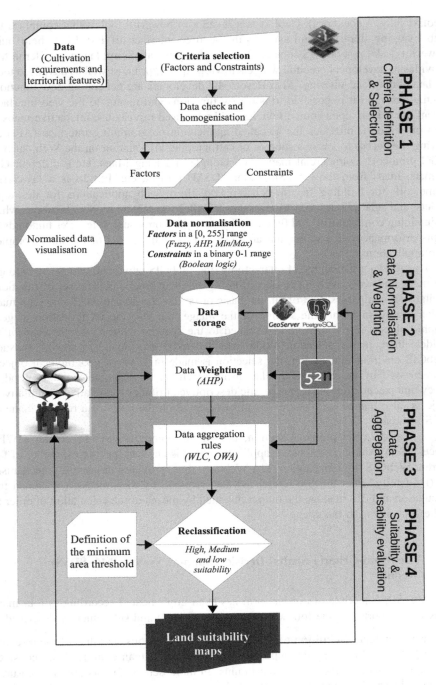

Fig. 2. Flow-chart showing the main four phases of the multi-criteria spatial decision support system (MC-SDSS) module

of an area for that use. In general terms, the criteria reported in Table 1 should be considered in implementing a LSE for agroforestry.

Phase 2 - Data Normalisation and Weighting. Criteria are physical variables measured in different scales. A procedure of normalisation, i.e. a standardisation procedure to refer all criteria in a common range, must be performed. Factors are normalised in a closed [0, 255] 8 bit interval, through Min/Max, AHP and fuzzy logic approaches;

Table 1. Criteria (Factors and constraints) to be considered in implementing a land suitability evaluation (LSE) for agroforestry

Factors		Constraints	
Name	**Description**	**Name**	**Description**
Land Use/Land Cover (LU/LC)	Referring to the considered agro-forestry use of the LSE, judgement is expressed on the productive suitability of the considered kinds of vegetation	*Core areas of Natural parks*	In Italy, in accordance with the general policy law on protected areas (Framework law 394/1991), every kind of human activity is forbidden in the integral reserves (so-called Zone A)
Elevation	Micro-climate variations depending on this parameter (air temperature gradient in particular), significantly influencing the phenology of vegetation	*Urbanized areas*	A technical constraint to exclude urban areas from the evaluation
Slope	Slope directly influences soil loss, thus the limits of practicability of agro-forestry activities, especially referred to the mechanisation of operations	*Areas with steep slopes*	Areas whose use for agro-forestry would lead to a high risk of erosion. Chosen limits depend on the considered agro-forestry use
Aspect	This factor directly influences the amount of solar radiation to the soil surface during the growing season	*Landslide and earth-flow areas*	Constraint based on the hydrogeological hazard & risk assessment plan (in Italian known as PAI, *Piano di AssettoIdro-geologico*) that defines the areas of the territory where the risk of damage to people and environmental heritage is relevant
Pedology	Pedological factors category includes the chemical and physical characteristics of soils that directly influence the definition of the optimum growing conditions of vegetation	*Flood expansion fields*	Riverbed of the so-called *fiumare* (typical Calabrian torrents) serving as flood expansion fields
Climate	Climate factorsmainly accomplish for temperature and precipitations. Depending on the considered agroforestry use, they are normalised as single parameters as well as specific	*Archaeological areas and Sea coastline*	Pursuant to the ICCLH (Italian Code of the Cultural and Landscape Heritage, Italian Legislative Decree n° 42 of 22/01/2004) these areas are not suitable for crop production
	climate indices		

these last, in order to best fit the physical quantities examined, by choosing membership functions with two (monotonic) or four (symmetric) control points. Constraints are normalised trough Boolean algebra in a binary [0-1] map: 0 for the areas excluded from the evaluation and 1 for those included.

Among MCDA techniques and procedures, to obtain the factors' weights basing on judgments provided by experts and DMs, we implemented an AHP-based sub-module. Despite relevant criticism highlighted by scholars, the AHP procedure remains widely applied in many research and operational fields [27, 34, 35] for several reasons. Among others, (i) the possibility to allow relationships between factors and to overcome the human difficulty in simultaneously judging the importance with regard to all the factors inserted in the model; (ii) a relative simplicity compared to other MCDA techniques; (iii) the possibility, through a specific index (i.e., the consistency ratio, CR), to check the inconsistencies in judgments provided by experts [35].

Following the AHP method, to obtain a ratio scale of measurement (i.e., the factors' weights), judgments provided by the experts through iterative pairwise comparisons (PCs) are organised as numeric data in a positive reciprocal matrix (the pairwise comparison matrix, PCM). Therefore, in a PCM if the priority of element i compared to element j is w_{ij} (relative weights), the priority of element j compared to element i is $1/w_{ij}$. To obtain factors' weights, each expert makes a judgment w_{ij} of all pairs of the n elements that include in the PCM as a number (a_{ij}) according to the Saaty's fundamental scale of absolute numbers [28]. Values range from 1 (indifference) to 9 (extreme importance, preference or likelihood); when compromise is needed, the following intermediate values 2, 4, 6, 8 must be used [28].

The AHP is also useful when many interests are involved and a number of people participate in the judgement process. Following our previous experiences [35], we provided the possibility to supports group-DM by aggregating the individual PCMs. To do that, we adopted the aggregating individual judgment (AIJ) approach suggested by Forman and Peniwati [36], in which each individual preference/judgment provided by experts is aggregated by means of a geometric mean. Following what suggested by Saaty [28], only PCMs with CR < 0.1 are considered consistent, thus can be considered in following phase 3 (data aggregation), while PCMs with CR > 0.1 should be omitted.

Phase 3 - Data Aggregation: Decision Rules. In this first conceiving, aggregation of results is performed by means of weighted linear combination (WLC). Once obtained, the results of the WLC are masked with the normalised constraints by Boolean intersection (AND type operator). In mathematical terms, the implemented decision rule to derive the land suitability index (LSI) of the j-th area can be expressed as reported in the following formula (Eq. 1).

$$LSI_j = \left(\sum_{i=1}^{n} w_{ji} \cdot x_i \right) \cdot \prod_{k=1}^{m} c_{jk} \qquad (1)$$

where: n is the total number of considered factors x_i; w_i is the weight of the i-th factor; m is the total number of constraints; c_{jk} are the k constraints present in the j-th area (i.e., the raster cell).

Phase 4 - Data Reclassification, Suitability and Usability Evaluation. Results coming from the aggregation phase, are reclassified in three suitability classes: High, Medium and Low suitability. A class with zero suitability is also included in the reclassification of the results to include those areas having no specific constraints but which prove completely unsuitable for the considered agro-forestry use. Furthermore, the selection and localisation of best areas in terms of suitability and minimum area thresholds is provided.

3 Results and Discussion

In accordance with Zhao [37], we implemented a SDI that uses web services to manage, analyse, and distribute geospatial data connecting electronic devices (desktop and mobile) through the web, therefore it shares the so-called geoservices or geospatial web services (GWS). The data workflow includes the following steps: data and metadata obtained in GIS and remote sensing software environments (1), are uploaded and stored into the PostgreSQL-PostGis geospatial Database (2), managed by Geo-Server (data), GeonetWork OpenSource (metadata) and 52° North WPS (geo-process) servers (3), then provided as GWSs to thick clients (i.e. QGIS, www.qgis.org), thin clients and also integrated in the implemented WebGIS that allows for browsing, zooming, identifying, querying the data, maps and execute geo-process (4). All produced geospatial data (orthophoto mosaics, Land Use/Land Cover maps, etc.) as well as reference base data (orthophotos, cadastral maps, road networks, etc.) are exposed as OGC standard interfaces (WMS, WFS, WCS) and in a GUI of the implemented WebGIS platform.

Data can reside locally or in different remote servers (accessible via the above mentioned OGC-compliant services). The use of such standards allows to provide heterogeneous geospatial data as geospatial services. Thus, the WebGIS platform interface efficiently allows to retrieve data in geospatial format from GeoServer (e.g., via WMS service) and to display them as map layers.

Thanks to the architecture previously pictured for the WebGIS, the user (not necessarily endowed with specific GIS skills), through a normal web browser, can display basic geospatial data and maps, execute geo-processes and all the scenarios representing the results produced within the MC-SDSS module. In particular, to display the data of interest, the WMS standard is exploited: such a type of map-server interface provides a simple HTTP interface, allowing the client to request and get a map from one or more distributed spatial databases. In response to that request, one or more map images (i.e., JPEG, PNG, etc., formats) are returned, so that they can be displayed within the browser (or, alternatively, by a desktop application client).

The MC-SDSS here described has been conceived and designed to offer an effective solution, accessible also to non-experts, and based on specific online tools (e.g. web-based processing and multi-criteria assessment). While in this first application, aggregation rules are conceived only by means of a WLC procedure, in a next development ordered weighted averaging (OWA) operators [38] will be implemented to better manage trade-off among evaluation factors.

To this end, the WPS has been configured to offer GIS geoprocessing capabilities to clients via web, including access to pre-programmed calculations and/or computation models, operating on geo-referenced data. The WPS, thus, allows to execute simple calculations, such as adding one or more sets of spatially referenced layers (raster maps representing territorial features or environmental characteristics), or as multiplying a single layer by a scalar (values representing indices and/or a weights). In such a way, it is possible to perform the WLC, which assigns weights for each criterion considered (Land Cover, Elevation, Slope, Aspect, etc.) and sums these to an overall score for searching suitable locations.

The data required by the WPS are delivered via web, and the service is targeted at processing both vector and raster data: for the specific purposes of this work, raster data (in GeoTIFF format) have been selected and used as input for the processing tasks implemented. Considering the general framework described and the processing steps outlined, the WebGIS application represent the front-end interface of the MC-SDSS, through which the user is able to access data, perform calculation and assess the results obtained (land suitability maps).

The final product consists of maps of land suitability for each specific agro-forestry use under consideration, giving a current picture of the areas suitable for that use. In that way, the implemented Web-based MC-SDSS can represent a powerful tool to support sustainable planning and management actions.

4   Conclusion and Final Remarks

Digital mapping and WebGIS, jointly with spatial analysis, are largely considered as effective tools in providing a comprehensive overview of environmental phenomena [39]. Through appropriate descriptions and thematic maps, it is easier to understand environmental features and characteristics, and to point out patterns and interactions. The geo-referencing of information is always a basic element for planning and management of resources and often this process puts various subjects around the same table in order to contribute, each in their skills and their missions, an immersive common or otherwise taking advantage of the mutual collaboration and cooperation [40]. Moreover, it must be highlighted their potentiality in informing people, promoting transparency in choices, thus providing local authorities with the possibility to collect and use valuable knowledge [41].

In this paper we have described the design of a Web-based MC-SDSS, relying on a specific SDI architecture, having the target to perform spatial and environmental analyses, supporting DM processes related to the agro-forestry LSE [42, 43].

Thanks to WebGIS interface and tools specifically implemented, information sharing is able to improve the effectiveness of analyses carried out, in order to support the DM, and of the scenario evaluation tasks (e.g., what-if analysis). Moreover, in perspective, WebGIS coupled with Web 2.0 technologies will allow to overcome most of the limitations and critical aspects of traditional methods of public participation inherent to their synchronous and place-based nature [42].

From the practical point of view, the web-based MC-SDSS has been pictured using FOSS4G environments, which encompass a set of application solutions suitable for our purposes and implementable in a well-integrated and easy-to-use platform. This solution, by providing on-line tools in order to perform multi-criteria spatial analyses via WPS protocols, allows to publish on the Web geospatial information following the standard required by the OGC (through a series of specific features for viewing and consulting of thematic maps in an advanced framework), and consequently to share the results with DMs, in order to support LSE processes.

While in this first application we implemented our web-based MC-SDSS platform without any limitation to potential users, in next development we are planning to manage three different type of users that can access the system via the WebGIS client: guest, expert, DM. To each of them, corresponds a different level of privileges in using the MC-SDSS platform.

Acknowledgements. This research has been funded by projects PONa3_00016-RI-SAF@MED (Research Infrastructure for Sustainable Agriculture and Food in Mediterranean area) and PON03PE_00090_3, in the framework of National Operational Programme (NOP) for Research and Competitiveness 2007–2013 of the Italian Ministry of Education, University and Research (MIUR) and Ministry of Economic Development (MiSE), and co-funded by the European Regional Development Fund (ERDF).

References

1. Antognelli, S., Vizzari, M.: Ecosystem and urban services for landscape liveability: a model for quantification of stakeholders' perceived importance. Land Use Policy **50**, 277–292 (2016)
2. Attardi, R., Cerreta, M., Sannicandro, V., Torre, C.M.: The multidimensional assessment of land take and soil sealing. In: Gervasi, O., Murgante, B., Misra, S., Gavrilova, M.L., Rocha, A.M.A.C., Torre, C., Taniar, D., Apduhan, B.O. (eds.) ICCSA 2015. LNCS, vol. 9157, pp. 301–316. Springer, Heidelberg (2015)
3. Colombo, L., Palomba, I.G., Sannicandro, V., Torre, C.M.: Geographic data infrastructure and support system to the evaluation of urban densification. In: Gervasi, O., Murgante, B., Misra, S., Gavrilova, M.L., Rocha, A.M.A.C., Torre, C., Taniar, D., Apduhan, B.O. (eds.) ICCSA 2015. LNCS, vol. 9157, pp. 330–341. Springer, Heidelberg (2015)
4. Hamilton, M.C., Nedza, J.A., Doody, P., Bates, M.E., Bauer, N.L., Voyadgis, D.E., Fox-Lent, C.: Web-based geospatial multiple criteria decision analysis using open software and standards. Int. J. Geogr. Inf. Sci. **30**, 1667–1686 (2016)
5. Malczewski, J.: GIS-based multicriteria decision analysis: a survey of the literature. Int. J. Geogr. Inf. Sci. **20**, 703–726 (2006)
6. Caiaffa, E., La Porta, L., Pollino, M.: Geomatics in climate services and local information: a case study for Mediterranean area. In: Gervasi, O., Murgante, B., Misra, S., Gavrilova, M.L., Rocha, A.M.A.C., Torre, C., Taniar, D., Apduhan, B.O. (eds.) ICCSA 2015. LNCS, vol. 9157, pp. 540–555. Springer, Heidelberg (2015)
7. Pollino, M., Caiaffa, E., Carillo, A., La Porta, L., Sannino, G.: Wave energy potential in the Mediterranean Sea: design and development of DSS-WebGIS "Waves Energy". In: Gervasi, O., Murgante, B., Misra, S., Gavrilova, M.L., Rocha, A.M.A.C., Torre, C., Taniar, D., Apduhan, B.O. (eds.) ICCSA 2015. LNCS, vol. 9157, pp. 495–510. Springer, Heidelberg (2015)

8. Boroushaki, S., Malczewski, J.: ParticipatoryGIS: a web-based collaborative GIS and multicriteria decision analysis. URISA J. **22**, 23–32 (2010)

9. Steiniger, S., Hay, G.J.: Free and open source geographic information tools for landscape ecology. Ecol. Inform. **4**, 183–195 (2009)

10. Jolma, A., Ames, D.P., Horning, N., Mitasova, H., Neteler, M., Racicot, A., Sutton, T.: Chapter ten free and open source geospatial tools for environmental modelling and management. In: Jakeman, A.J., Voinov, A.A., Rizzoli, A.E., Chen, S.H. (eds.) Developments in Integrated Environmental Assessment, pp. 163–180. Elsevier, Amsterdam (2008)

11. Caiaffa, E., Pollino, M., Marucci, A.: A GIS based methodology in renewable energy sources sustainability. Int. J. Agric. Environ. Inf. Syst. **5**, 17–36 (2014)

12. Pollino, M., Modica, G.: Free Web mapping tools to characterise landscape dynamics and to favour e-participation. In: Murgante, B., Misra, S., Carlini, M., Torre, C.M., Nguyen, H.-Q., Taniar, D., Apduhan, B.O., Gervasi, O. (eds.) ICCSA 2013, Part III. LNCS, vol. 7973, pp. 566–581. Springer, Heidelberg (2013)

13. Baban, S.M., Parry, T.: Developing and applying a GIS-assisted approach to locating wind farms in the UK. Renew. Energy **24**, 59–71 (2001)

14. Fichera, C.R., Modica, G., Pollino, M.: GIS and remote sensing to study urban-rural transformation during a fifty-year period. In: Murgante, B., Gervasi, O., Iglesias, A., Taniar, D., Apduhan, B.O. (eds.) ICCSA 2011, Part I. LNCS, vol. 6782, pp. 237–252. Springer, Heidelberg (2011)

15. Vizzari, M.: Spatio-temporal analysis using urban-rural gradient modelling and landscape metrics. In: Murgante, B., Gervasi, O., Iglesias, A., Taniar, D., Apduhan, B.O. (eds.) ICCSA 2011, Part I. LNCS, vol. 6782, pp. 103–118. Springer, Heidelberg (2011)

16. Collins, M.G., Steiner, F.R., Rushman, M.J.: Land-use suitability analysis in the United States: historical development and promising technological achievements. Environ. Manage. **28**, 611–621 (2001)

17. FAO: A framework for land evaluation. FAO Soils Bulletin, vol. 32. Food and Agriculture Organization of the United Nations (FAO), Rome (1976)

18. FAO: Guidelines for land-use planning. FAO Development Series, vol. 1. Food and Agriculture Organization of the United Nations (FAO), Rome (1993)

19. Steiner, F.: Resource suitability: methods for analyses. Environ. Manage. **7**, 401–420 (1983)

20. Spruijt, P., Knol, A.B., Vasileiadou, E., Devilee, J., Lebret, E., Petersen, A.C.: Roles of scientists as policy advisers on complex issues: a literature review. Environ. Sci. Policy **40**, 16–25 (2014)

21. Carver, S.J.: Integrating multi-criteria evaluation with geographical information systems. Int. J. Geogr. Inf. Syst. **5**, 321–339 (1991)

22. Malczewski, J.: GIS and Multicriteria Decision Analysis. Wiley, New York (1999)

23. Modica, G., Laudari, L., Barreca, F., Fichera, C.R.: A GIS-MCDA based model for the suitability evaluation of traditional grape varieties: the case-study of "Mantonico" grape (Calabria, Italy). Int. J. Agric. Environ. Inf. Syst. **5**, 1–16 (2014)

24. Jelokhani-Niaraki, M., Malczewski, J.: A group multicriteria spatial decision support system for parking site selection problem: a case study. Land Use Policy **42**, 492–508 (2015)

25. Kraak, M.-J.: The role of the map in a Web-GIS environment. J. Geogr. Syst. **6**, 83–93 (2004)

26. Dragićević, S., Balram, S.: A Web GIS collaborative framework to structure and manage distributed planning processes. J. Geogr. Syst. **6**, 133–153 (2004)

27. Vizzari, M., Modica, G.: Environmental effectiveness of swine sewage management: a multicriteria AHP-based model for a reliable quick assessment. Environ. Manage. **52**, 1023–1039 (2013)

28. Saaty, T.L.: The Analytic Hierarchy Process: Planning, Priority Setting, Resource Allocation. McGraw-Hill, New York (1980)
29. Willmes, C., Kürner, D., Bareth, G.: Building research data management infrastructure using open source software. Trans. GIS **18**, 496–509 (2014)
30. Steiniger, S., Hunter, A.J.S.: Free and open source GIS software for building a spatial data infrastructure. In: Bocher, E., Neteler, M. (eds.) Geospatial Free and Open Source Software in the 21st Century, pp. 247–261. Springer, Heidelberg (2012)
31. Ballatore, A., Tahir, A., McArdle, G., Bertolotto, M.: A comparison of open source geospatial technologies for web mapping. Int. J. Web Eng. Technol. **6**, 354 (2011)
32. Anderson, G., Moreno-Sanchez, R.: Building Web-based spatial information solutions around open specifications and open source software. Trans. GIS **7**, 447–466 (2003)
33. Eastman, R., Jin, W., Kyem, P.A.K., Toledano, J.: Raster procedures for multi-criteria/multi-objective decisions. Photogramm. Eng. Remote Sens. **61**, 539–547 (1995)
34. Girard, L.F., Torre, C.M.: The use of Ahp in a multiactor evaluation for urban development programs: a case study. In: Murgante, B., Gervasi, O., Misra, S., Nedjah, N., Rocha, A.M.A., Taniar, D., Apduhan, B.O. (eds.) ICCSA 2012, Part II. LNCS, vol. 7334, pp. 157–167. Springer, Heidelberg (2012)
35. Modica, G., Merlino, A., Solano, F., Mercurio, R.: An index for the assessment of degraded Mediterranean forest ecosystems. For. Syst. **24**, 13 (2015)
36. Forman, E., Peniwati, K.: Aggregating individual judgments and priorities with the analytic hierarchy process. Eur. J. Oper. Res. **108**, 165–169 (1998)
37. Zhao, P., Yu, G., Di, L.: Geospatial Web services. In: Hilton, B. (ed.) Emerging Spatial Information Systems and Applications, pp. 1–35. IGI Global, Hershey (2007)
38. Yager, R.R.: On ordered weighted averaging aggregation operators in multicriteria decisionmaking. IEEE Trans. Syst. Man. Cybern. **18**, 183–190 (1988)
39. Goodchild, M.F.: Spatial thinking and the GIS user interface. Procedia Soc. Behav. Sci. **21**, 3–9 (2011)
40. Modica, G., Zoccali, P., Di Fazio, S.: The e-participation in tranquillity areas identification as a key factor for sustainable landscape planning. In: Murgante, B., Misra, S., Carlini, M., Torre, C.M., Nguyen, H.-Q., Taniar, D., Apduhan, B.O., Gervasi, Osvaldo (eds.) ICCSA 2013, Part III. LNCS, vol. 7973, pp. 550–565. Springer, Heidelberg (2013)
41. Murgante, B., Tilio, L., Lanza, V., Scorza, F.: Using participative GIS and e-tools for involving citizens of Marmo Platano-Melandro area in European programming activities. J. Balk. Near East. Stud. **13**, 97–115 (2011)
42. De Longueville, B.: Community-based geoportals: the next generation? Concepts and methods for the geospatial Web 2.0. Comput. Environ. Urban Syst. **34**, 299–308 (2010)
43. Stirling, A.: Analysis, participation and power: justification and closure in participatory multi-criteria analysis. Land Use Policy **23**, 95–107 (2006)

A Multicriteria Assessment Model for Selecting Strategic Projects in Urban Areas

Lucia Della Spina$^{(\boxtimes)}$, Claudia Ventura, and Angela Viglianisi

Mediterranea University of Reggio Calabria, Reggio Calabria, Italy
{lucia.dellaspina, claudia.ventura,
angela.viglianisi}@unirc.it

Abstract. The main goal of this paper lies in the need to define an assessment methodology to search for the feasibility of strategic programs aimed at achieving the urban sustainable development. The methodology utilizes a multidimensional approach for selecting the strategic actions, in which the choice and the feasibility are conditioned by the multiplicity of involved interests.

To achieve an efficient allocation of the available resources, equity for different social groups involved and transparency in the evaluation process, the evaluation methodology here shown was built on the conceptual basis of the Community Impact Evaluation. This provides the establishment of a social budget matrix, in order to facilitate the participation of the involved social groups to the decision-making process within the project, highlighting for each of them advantages and disadvantages subsequent to the implementation of the various design alternatives.

Keywords: Strategic programs · Urban regeneration · Sustainable development · Community Impact Evaluation · Stakeholder analysis · Financial analysis · Multicriteria assessment

1 The Evaluation Context, Objectives and Methodology

Since the 90s in Italy the interest in urban and regional strategic programs has been growing, introducing a more strategic approach to the practice of urban planning.

These programs were created to implement urban renewal and territorial transformation through the integration of public and private financial resources. Recently they have been addressed on achievement of sustainable development of the territory (Fusco Girard and Nijkamp 1997) The latest generation of integrated complex programs is in fact merged in programs for urban regeneration and sustainable development of the territory (Calavita et al. 2014; Calabrò and Della Spina 2014a), which have undoubted innovative aspects and for whose setup is required the contribution of assessment procedures, both during wording of the proposal, and in the subsequent definition phase of the program.

The ministerial tender inviting municipalities and local authorities to promote the formation, explicates the criteria that will be used to evaluate the proposals and in this way it provides guidance and rules for their processing. (Decree of President of the Council Ministers 2015/10/15. Official Journal of the Italian Republic n. 249 of 2015/10/26).

© Springer International Publishing Switzerland 2016
O. Gervasi et al. (Eds.): ICCSA 2016, Part III, LNCS 9788, pp. 414–427, 2016.
DOI: 10.1007/978-3-319-42111-7_32

When the selected programs will enter into the realization phase, the projects that form it will be classified according to an order of priority, and then it will possible to select for each project the more efficient and effective implementation methods.

In this second phase projects must be identified where to convey, as a priority, public and private activatable funding.

The success of this phase requires that the interests involved reach a proper settlement through preventive explanation of the advantages and disadvantages - in terms of social, economic and environmental impact of public and private stakeholders (Calabrò and Della Spina 2014b).

The programmatic evaluation of ex-ante intervention plays a very important role, especially in the context of spatial planning. It allows to test the feasibility and sustainability of the interventions motivating public choices by integrating technical and policy considerations analysis. These components are able to create a basis for discussion and interaction between the various stakeholders, having them different skills and interests, considering also different roles of those involved in decision-making (Fusco Girard and Nijkamp 1997). Participation is here understood as the involvement of more actors within the decision-making process through the adoption of an inclusive process.

Thus the pre-feasibility of the interventions is critical, since it determines which of a finite number of possible transformation alternatives, it should be preferred in relation to a specific decision context. Therefore, it is a very delicate phase and of great relevance as the decisions taken in this step will determine the final outcome of redevelopment, recovery or transformation programmes. As indicated by the regulations in both national and Community level, this stage requires the use of multi-attribute or multi-goals evaluation methodologies (Multiple-Criteria Decision Analysis - MCDA), to allow to deduct a priority ranking of alternatives of possible intervention.

The multicriteria evaluation methods have proven to be the most appropriate technical instrument to take account of the values and goals of all relevant social partners involved. These evaluation methods have the goal of producing decisions in which are involved the possible largest number of actors and they are able to take simultaneously account of a plurality of aspects of the decision problem, bringing out the different points of view the involved actors (Fusco Girard and Nijkamp 1997).

Of course, the end result which is reached (ranking) is not absolute and unique, but relative to the considered elements (in particular, referring to criteria and their respective weights) and in any case differently depending on the scenario.

The need to define an evaluation methodology to research the feasibility and sustainability of strategic programs is therefore the main objective of the case study illustrated below, which combines a financial analysis within this approach (Calabrò and Della Spina 2014c; De Mare et al. 2015). In fact, according to the criteria of the tender for the institution about programs for urban regeneration and sustainable development, the modalities - to define the priorities for the implementation of intervention projects under the program - should also take into account the economic, environmental and social impacts of the projects (Tajani and Morano 2015; Morano et al. 2015).

In addition, the feasibility of programs is conditioned by the multiplicity of interests that are put into play. The most appropriate evaluation methodology for this purpose was therefore built on the conceptual basis of the Community Impact Evaluation (CIE) proposed by Lichfield in the early '60s (Lichfield 1996), which provides for the preparation of a social report matrix for all social groups involved in the project, showing for each of them advantages and disadvantages that are subsequent to the implementation of the different design alternatives.

Therefore general problem facing the CIE becomes to verify whether and to what extent the community, vaguely understood not as a whole, but in the various groups that compose it, will improve or not their own after the project intervention wellbeing (Lichfield 1996). Or, in other words, to what extent the project under discussion enables the extension of the territory oriented towards sustainability.

The CIE basically allows to build an integrated assessment, summarizing economic analysis, financial analysis, environmental analysis and social analysis, showing the distribution of net benefits among the different groups involved. It allows to detect the socio-economic efficiency and socio-environmental analysis using multidimensional impact (De Mare et al. 2015), resolving the problem of identification of priority over multiple objectives, heterogeneous and conflicting on the basis of a solution that minimizes the total cost-opportunity direct users, indirect, potential and future in a sustainability perspective (Lichfield 1996).

In accordance with the criteria of economic efficiency, environmental enhancement and social equity, advocated by the tender, the economic financial analysis is in fact combined with the assessment of the impacts on the community sectors (Nesticò and Pipolo 2015).

This methodology is then applied on a trial basis in a program for urban regeneration and sustainable development of the territory (Della Spina et al. 2015), promoted by some local authorities in the region of Calabria, to order the projects relating to the redevelopment of abandoned areas in the city of Reggio Calabria in accordance with the preferences expressed by various sectors of the local community.

2 Description of the Alternatives and Involved Groups

The first step for the application of the valuation model, includes a series of phases ranging from the description of the alternative projects to the identification of sectors of the community involved and their sectorial objectives.

The three areas in question are attributable to the adjacent hospital Morelli (Alternative A), to the former Italcitrus (Alternative B) and the former Fiera Pentimele (Alternative C).

For the different alternatives are listed in a table the values that express the main urban characteristics, in terms of activities to be set up and public spaces (Table 1).

The definition of the sectors of the community involved in the project are coated from the division into producers/consumers (Table 2) used by the Lichfield Community Impact Evaluation (Lichfield 1996), which is enriched with additional subjects and divided into sub-groups following a more targeted Stakeholders analysis conducted for the study areas (Valenti et al. 2015).

The recognition of sectoral objectives was carried out following different techniques. For manufacturers and the local government, seen as active and economically involved in the subject of urban transformation process, sectoral preferences were identified with the help of social survey techniques (Rosenberg 1968; Marradi 1980), using direct interviews with privileged witnesses, such as representatives of particular economic categories, associations and interest groups.

The objectives were defined through an interview, while for consumers it was considered more appropriate to use a questionnaire to a sample of the resident population, adjacent to the three considered areas. The sample used in the survey was constructed from the census (National Institute of Statistics census 2011) of permanent residents in the three areas, totalling 2.696 people, stratified compared to "age group" variable. 284 subjects were interviewed, 10.5 % of the resident population of the universe, observing the proportions of sex and age. The questionnaire consists of 23 questions except two open-ended questions.

Table 1. Land use for the three areas of intervention

Land use	Alternatives		
	A	B	C
Land area (m^2)	104.00	155.600	92.800
Territorial use capacity (m^2/m^2)	0,25	0,28–0,60	0,55
Green Park, public spaces, pedestrian and cycle paths	60.000	49.100	13.200
Parking areas and roads (m^2)	16.700	15.250	36.200
Public equipment (cultural, recreational, etc.)	1.100	19.500	1.900
Healthcare equipment (m^2)	–	15.100	–
Free residential activities (m^2)	7.780	12.000	20.000
Rent-controlled housing activities (m^2)	6.890	2.100	–
Directional and craft services (m^2)	4.300	3.100	2.450
Commercial business (m^2)	2.400	5.100	2.640

Table 2. Description of community sectors and involved alternatives

Community sectors			
Producers/operators	Alternatives	Users/consumers	Alternatives
A1. Local governement	A,B,C	B1. New employees	A,B,C
A2. Transport company	C		
A3. Public company	A	B2. New residents	A,B
A4. Owners	A,B,C		
A5. Developers	A,B,C	B3. Park's visitors	A
A6. Constructors	A,B,C		
A7. Industrial	A,C	B4. Users	A,B,C
A8. Commercial operators	A,B,C		

3   Analysis of the Economic, Environmental and Social Impacts

To evaluate economic, environmental and social impacts, each project alternative has been divided into two stages of the process, due to the phase of implementation and management phase. The purpose of this division is to analyze the impacts of the interventions in the short and long term (Calabrò and Della Spina 2014a; Trovato and Giuffrida 2014).

The economic effects were analyzed using as indicators the cash flow discounted to the public operator and the private operator and the cost of maintenance and management of public spaces (e.g.: green areas, open spaces, etc.) (Realfonzo 1994; Prizzon 1995; Simonotti 1997).

The cash flow generated with a year by year difference between revenues and costs, was calculated to current value using a discount rate of 8 %. The net present value obtained is positive for all three alternatives and maximum in the case of Alternative C (Table 3).

The required financial information has been obtained by crossing the information obtained from indirect sources (Real estate consultants, Gabetti Agency, etc.) with those obtained by direct survey of local construction companies and real estate experts (Simonotti 1997).

Table 3. Summary of social financial analysis

Alternatives	Cost-Revenue	Local government	Housing cooperatives and services	Private developers and society for transformation
A	Cost	−3.481.952,41	−5.035.454,77	−14.616.246,70
	Revenue	1.196.114,18	72.975.359,84	21.410.753,67
	Financial NPV	−2.140.197,39	11.114.152,47	1.887.133,51
B	Cost	−10.582.718,32	17.662.825,95	21.944.253,64
	Revenue	1.895.396,82	22.569.166,49	33.714.306,37
	Financial NPV	−6.627.174,93	981.268,11	3.501.577,78
C	Cost	−7.106.446,93	0	28.715.003,59
	Revenue	2.509.980,53	0	41.533.463,82
	Financial NPV	−4.462.187,61	0	4.043.857,52

From the social point of view, it has been estimated the employment impact of new equipment planned within each field (Soderstrom 1981; Taylor et al. 1995), distinguishing the temporary occupation, induced by the material construction of the works, and the durable, linked to the management of public facilities (recreation centers and health social housing for the elderly) and private (service industries) (Calabrò and Della Spina 2014a). For this purpose were used the index established by the National

Association of Builders, which expresses the number of jobs generated by construction investment (both direct employment in related sectors) and indices derived from official publications of the National Institute of Statistics.

National Association of Builders on their 1998 data leads the multiplier to get the number of jobs compared to the billions invested to 8.4 (million investments for employees) in the construction sector and 4.6 (million investments for employees) in the areas connected. The surfaces/staff ratios based on National Institute of Statistics data was then cross-checked with data provided by the companies concerned and the study "Urban Transformations related to the redevelopment of areas of complex system", developed by Oikos Searches. The National Institute of Statistics shows, for different size classes of surfaces, the data relating to the number of local units and employees. From this, it can be deduced the average number of employees required for classes of areas, and then an average value sqm/clerk. These data are serving a certain approximation increases as the size of local units, but that seemed acceptable in reason of the purpose of the present study.

The temporary employment and the lasting so estimated are reported in Table 4. Considering that the new structures host companies in part transferred from less functional premises, it was considered reasonable to estimate the number of jobs created from scratch 65 % of those needed for the operation of new facilities.

From an environmental and town planning point of view, the indicators employed were: the traffic induced by future users of the areas, the noise barriers to protect the residence and the equipment to be set up and finally the proportion of green areas and public spaces compared to settled activities (Rickson et al. 1990).

Table 4. Synthetic scheme for the estimation of employment induced by the alternative

	Total area (m²)	Area/worker (m²)	Investiment (€ million)	Temporary jobs		Permanent jobs	
				Costruction sector (n. jobs)	Supply sector (n. jobs)	n. employees	n. new jobs
Alternative A				*208*	*122*	*134*	*86*
Residential	13.725		9.146,45	142	81		
Directional	3.300	50	2.386,03	33	21	66	42
Commercial	2.300	40	1.662,99	25	14	57	37
Pubb. services	1.100	100	738,53	8	6	11	7
Alternative B				*469*	*246*	*363*	*235*
Residential	14.000		10.432,43	168	92		
Directional	5.000	50	3.615,20	58	32	100	65
Commercial	2.800	40	2.024,51	33	17	70	45
Pubb. services	19.300	100	12.957,90	210	115	193	125
Alternative C				*321*	*180*	*130*	*83*
Residential	20.000		15.493,71	252	138		
Directional	2.500	50	1.807,60	25	16	50	32
Commercial	2.500	40	1.807,60	25	16	62	40
Pubb. services	1.800	100	1.208,51	19	10	18	11

Table 5. Synthetic scheme for estimate the traffic generated by the three alternatives

Alternatives	Total car	Peak times at daytime (n. car)		Peak times at nighttime (n. car)	
		Incoming	Outgoing	Incoming	Outgoing
A	1.421	168	133	278	283
B	3.555	409	136	495	981
C	1.625	204	195	364	307

Notes: Hours 7:30 to 8:30 and 17:30 to 18:30. The daily attendance (residents, employees, users, etc.) In each area and the hourly concentration of users of shops, offices and various functions of public interest were drawn from the study undertaken by OIKOS research on "Urban Transformations related to redevelopment of the areas of complex system". Modal choice of the transport vehicle by users is drawn from the Census 2011 National Institute of Statistics data recalibrated to take account of the operator and the users of public transport changes. Finally, the number of trips (round trip) generated per day per person/user and by type of operation, the load factor of the car to cause displacement and concentrations of private traffic for the peak time were derived from studies carried out by the traffic Research Centre as part of preparing the General Plan of Reggio Calabria traffic.

The data on the presumed traffic for the three settlement scenarios examined here have been derived from a study aimed at drawing up of the urban mobility plan. These data are summarized in Table 5.

In this study, the demand for mobility has been quantified on the basis of traffic generated by urban forecasts for an average weekday, further developed through the application of appropriate parameters to define more precisely the extent of the means of transport in hours of peak traffic.

4   The Assessment of Sectorial Preferences

The valuation method continues with the second phase, with assessment of the preferences and objectives for the different sectors of the community involved, assessment of impacts and normalization (standardization) of the obtained values.

Once you find the information on the sectorial assessments, these have been reported in a J order of evaluation matrix E (criteria) x I (alternative), which reports the eij values of the effects for each alternative (Table 6).

The analysis was followed by the definition of the objective functions for the different social groups (MacArthur 1997), and their subsequent optimization. The analysis shows that the Local Government (A1) has an interest in maximizing the positive effects in the social sphere generated by the three alternatives, such as the increase of permanent and temporary jobs and social housing than the free building. In urban and environmental context, it has an interest in minimizing the private vehicular

Table 6. Matrix evaluation for economic, urban plannig, environmental and social effects

Effects		Alternative		
		A	B	C
Economic	Measure unit			
E1 – Public operating costs for 8 years	€	43.788	150.787	91.913
E2 – Private Net flow (financial NPV)	€	1.887.134	3.501.578	4.043.858
E3 - Public Net flow (financial NPV)	€	−2.140.197	−6.627.175	−4.596.466
Environmental-Urban planning				
EU1 - Car traffic caused by future users in the areas	n. car (daily mean)	431	1010	535
EU2 - Acoustic and atmospheric barriers to protect axis protection	Linear metres green protection	4.400	2.800	0
EU3 - Green areas and open spaces for the resident pop. and employees to be set up	m^2 Green areas/Resident pop. and employees	157	67	27
Social				
S1 - Right to housing for weakest categories	m^2 Rent-controlled housing	6.855	2.000	0
S2 - Permanent jobs	New employees in activities planned	86	287	83
S3 – Temporary jobs	New employees in construction industry	330	752	501

traffic and noise barriers, while in the economy to minimize the operating costs of policy interventions and maximize its economic and financial return achieved by the implementation of the projects.

Consumers/users (group B) do not have strong interests in the economic, but with one voice they have as aim the reduction of traffic, the increase in noise barriers and green spaces, the right to housing for the most disadvantaged classes and the increase of long term employment. A slight conflict of interest within this group is relatively temporary employment: in fact, the residents (B2) are opposed because they often attract non-EU workers and the presence of foreigners in the neighbourhood.

The producers/workers (group A) have as their main objective the maximization of revenues obtained from the implementation of interventions, followed by the increase of employment related to the establishment of new management activities, sales and production.

The impacts related to the project variables were then weighed concerning sectorial targets: the assignment of weights has tried to reflect the relative importance of the various types of impacts considered for the main social groups, re-aggregated in local government, producers and consumers.

For the estimation of the weights, it was used a rating method often applied in the practice of planning (Voogd 1983). In this type of methods, you are called the interlocutor to assign a set amount of points (for example 100) between the criteria identified in such a way that the number of points assigned to each criterion reflects its relative importance.

Since in the analysis are used quantitative data, it was necessary to proceed to a standardization. In the present case we have been used two well-known methods, obtained from the following formulas:

$$Eij = (eij - \min ej)/(\max ej - \min ej) \qquad (1)$$

$$Eij = eij/\max ej \qquad (2)$$

Where Eij is the data corresponding to j and to the standardized criteria, eij is the data before standardization and *max ej* and *min ej* represent the maximum value and the minimum value observed for the j criterion among all other alternatives i (i = 1, 2... I).

With the first method of standardization, Alternative B it will be preferable for the Local Government attributing high importance to the criterion of social equity, protecting the underprivileged classes in search of housing (at discounted prices), and temporary and permanent employment (Table 7).

Adverse to this alternative, instead will be residents of the middle classes who opt for the best environmental conditions, in terms of green space equipment, less pollution and noise barriers. For this group, and more generally for consumers (in the broad sense as the users of the facilities of the area, as residents, employees or users of the assets or frequenters of green spaces) the preferable alternative is the A.

Finally, the alternative C, which maximizes the criterion of economic efficiency is preferable for manufacturers/operators.

It is difficult from this analysis to make an unambiguous conclusion about the preference among the alternatives, because each of them meets certain community groups rather than others in terms of economic efficiency, environmental and social solidarity enhancement (Nesticò and Pipolo 2015) (Table 7).

Using the second method of standardization, leaving unchanged the scoring by the different sectors of the community system, it appears a different situation and of considerable interest. In fact, the alternative B is preferred by manufacturers and the local government, while consumers prefer alternative A, which is still the second option also for the local government, with little difference compared to the first option, that is the B.

Table 7. Priority matrix of the alternatives, standardization criterion min/max

Criterion min/max				Local government				Producers				Consumers			
Effects	**A**	**B**	**C**		**A**	**B**	**C**		**A**	**B**	**C**		**A**	**B**	**C**
Economic				25%				65%				10%			
E1	0.00	1.00	0.45	10%	0,00	10,00	4,50	0%	0,00	0,00	0,00	5%	0,00	5,00	2,25
E2	0.00	0.75	1.00	10%	0,00	7,49	10,00	65%	0,00	48,66	65,00	0%	0,00	0,00	0,00
E3	1.00	0.00	0.45	5%	5,00	0,00	2,26	0%	0,00	0,00	0,00	5%	5,00	0,00	2,26
Environmental / Urban planning				30%				10%				45%			
AU1	1.00	0.00	0.66	15%	15	0	9,91	5%	5,00	0,00	3,30	20%	20,00	0,00	13,22
AU2	1.00	0.64	0.00	5%	5	3,18	0	0%	0,00	0,00	0,00	10%	10,00	6,36	0,00
AU3	1.00	0.31	0.00	10%	10	3,08	0	5%	5,00	1,54	0,00	15%	15,00	4,62	0,00
Social				45%				25%				45%			
S1	1.00	0.29	0.00	15%	15,00	4,38	0,00	5%	5,00	1,46	0,00	15%	15,00	4,38	0,00
S2	0.01	1.00	0.00	20%	0,29	20,00	0,00	5%	0,07	5,00	0,00	25%	0,37	25,00	0,00
S3	0.00	1.00	0.41	10%	0,00	10,00	4,05	15%	0,00	15,00	6,08	5%	0,00	5,00	2,03
					50,29	58,13	30,72		15,07	71,66	74,38		65,37	50,36	19,76

5   Final Ranking of Alternatives

As part of the interventions of urban transformation of the territory, the MCDA techniques constitute a supporting tool to the ex-ante evaluation stage. This phase is becoming increasingly important, because it is a moment aimed at identifying possible alternatives and benefits of transformation and disadvantages that could result from their implementation (Calabrò and Della Spina 2014b).

Within the MCDA family, now also recognized by regulations, it is possible to identify a number of which are derived from different scientific theories and methodologies methods and concepts. Such a set of methods allows to aggregate the various evaluation criteria with the aim to select one or more actions. The individual methods are different from each other, in function of the theoretical background to which are linked, and lending themselves to the solution of various problems. This heterogeneity and function corresponds to the same diversity of multifarious problems encountered in reality.

Depending on the content and the ultimate purpose of the assessment, for MCDA are possible classifications, however, characterized by a common methodological process (Voogd 1983; Figueira et al. 2005a; Figueira et al. 2005b).

So the choice of which method to use between the different analysis techniques Multicriteria is not trivial.

In the specific case illustrated here, in order to arrive at the final ranking of alternatives, given that the valuation problem is a set of "discrete" (a finite number of alternatives) and that the information available relating to the performance of the alternatives are of cardinal type, it was decided to opt for a technique that uses hard information - i.e. data quantities - which include, among others, of the weighted summation method and ELECTRE methods, better referred to as "outranking" or outranking methods.

The known ELECTRE methods (acronym for Elimination Et Choix Traduisant the Réalité, which means "Deleting and choice that express the reality") are actually a technical Analysis Multicriteria family developed by the French mathematician Roy (1968; 1985). As part of the MCDA, the so-called methods outranking (Mousseau and Roy 1996; Munda 2005) they employ a procedure based on the construction of binary relations (of concordance and discordance) between pairs of elements.

Therefore, now, in participatory planning processes in order to reach the more consistent choice, "justified", a "compromise" between the three alternative scenarios, the last step of the methodology involves the analysis of the distributional asymmetries through the analysis of discrepancy (Roy 1968; 1985; Roy and Bouyssou 1993).

For verification to the ordinances on the basis of discrepancy absolute index, it must first calculate the matrix of discrepancy from the standardized assessment matrix.

The discordance matrix represents a measure of "regret" in choosing an alternative rather than another, because it gives an idea of the maximum value (environmental, economic and social) that is lost by making this choice (Mousseau and Roy 1996; Roy and Bouyssou 1993; Bresso et al. 1992).

Table 8. Absolute discordance indexes for the alternatives

	Id (criterion min/max)	Id (criterion 0/max)
A	0,00	−0,57
B	−0,34	−0,33
C	0,34	0,90

The discordance of absolute index represents an overall measure of regret, in the case where the final decision is that of realizing the alternative under consideration: much lower is its value, the more so the alternative is satisfactory.

The ranking of alternatives is therefore made from lower value. For the criterion of standardization min/max, the preferable alternative will be the alternative B, followed by the alternative A and to last from option C, while the criterion for 0/max the preferable alternative is the A followed by B (Table 8).

This arrangement is similar to that obtained with the technique of weighted sum (Table 7) and highlights that the alternatives preferable - according to the old system by the local government and consumers - are also less conflicting.

This last step allows to extend the analysis to the conflicts between different social groups concerned, as it highlights asymmetries in the distribution of costs and benefits, pointing out the alternatives that generate more conflicts of interest.

In this case, a preferred alternative will therefore B, since it expresses a low index of social conflict and also through the technique of weighted sum it reflects the preferences of the local government and producers and is the second option for consumers.

This step is essential to take into account the needs of different groups and it is a prerequisite to the identification of forms of compensation that can be activated to prevent the opposition of disadvantaged individuals leading to paralysis of the program during the implementation phase.

6 The Main Results

The construction of an evaluation methodology for the feasibility of programs for urban regeneration and sustainable development of the territory must take responsibility for reconciling hypothesis of transformation with the objectives of protection, according to the principle of sustainable development. Moreover the assessment should help the decision maker to verify the compatibility of the envisaged changes to those objectives and to seek socially shared solutions and yet economically feasible.

The proposed methodology was, therefore, developed to achieve efficient allocation of available resources, equity in the distribution of advantages and disadvantages to the affected groups and transparency in the evaluation process to facilitate social participation to decision-making.

A further requirement is flexibility, defined as the capacity to adapt to the specificities of the present case, to accommodate an increase or a variation of the available information and to be permeable with respect to the program, to facilitate the ongoing process of adjustment between the design phase and the evaluation.

Compared to this requirement, further testing and refinement of the methodology are required, as the application to a single case study makes it impossible to use "up to speed" in the context of these programs.

Although it was formalized in relation to programmes for urban regeneration and sustainable development of the territory, the assessment methodology can be implemented on all complex programs and easily adapted to the evaluation of projects within other negotiated programming tools. In fact, it can be used along the different phases of the program development, both as ex ante evaluation tool of design alternatives during the project definition to be implemented as a priority, both as an ongoing evaluation tool for: processing projects identified; the implementation of which, in parts, it must be constantly monitored; and, finally, as an ex-post evaluation tool, to check the gap between goals and results and to help the public decision maker to refine new programs.

Acknowledgements. The paper reflects the opinion and the serious commitment of its authors who all contributed to its writing. However, Lucia Della Spina wrote paragraphs 1, 4, 5 and made the assessment model; Claudia Ventura wrote paragraph 3 and Angela Viglianisi wrote paragraphs 2.

References

Bresso, M., Gamba, G., Zeppetella, A.: Valutazione ambientale e processi decisionali. NIS, Roma (1992)

Calabrò, F., Della Spina, L.: The cultural and environmental resources for sustainable development of rural areas in economically disadvantaged contexts: economic-appraisals issues of a model of management for the valorisation of public assets. In: Advanced Materials Research, vols. 869–870, pp. 43–48. Trans Tech Publications, Switzerland (2014a)

Calabrò, F., Della Spina, L.: The public-private partnerships in buildings regeneration: a model appraisal of the benefits and for land value capture. In: Advanced Materials Research, vols. 931–932, pp. 555–559. Trans Tech Publications, Switzerland (2014b)

Calabrò, F., Della Spina, L.: Innovative tools for the effectiveness and efficiency of administrative action of the metropolitan cities: the strategic operational programme. In: Bevilacqua, C., Calabrò, F., Della Spina, L. (eds.) Advanced Engineering Forum, vol. 11, pp. 3–10. Trans Tech Publications, Switzerland (2014c)

Calavita, N., Calabrò, F., Della Spina, L.: Transfer of development rights as incentives for regeneration of illegal settlements. In: Bevilacqua, C., Calabrò, F., Della Spina, L. (eds.) Advanced Engineering Forum, vol. 11, pp. 639–646. Trans Tech Publications, Switzerland (2014)

Della Spina, L., Scrivo, R., Ventura, C., Viglianisi, A.: Urban renewal: negotiation procedures and evaluation models. In: Gervasi, O., Murgante, B., Misra, S., Gavrilova, M.L., Rocha, A. M.A.C., Torre, C., Taniar, D., Apduhan, B.O. (eds.) ICCSA 2015. LNCS, vol. 9157, pp. 88–103. Springer, Heidelberg (2015)

De Mare, G., Granata, M.F., Nesticò, A.: Weak and strong compensation for the prioritization of public investments: multidimensional analysis for pools. Sustain. 7(12), 16022–16038 (2015)

Fusco Girard, L., Nijkamp, P.: Le valutazioni per lo sviluppo sostenibile delle città e del territorio. Angeli, Milano (1997)

Figueira, J., Greco, S., Ehrgott, M.: Multiple criteria decision analysis: state of the art surveys, vol. 78. Springer, New York (2005a)

Figueira, J., Mousseau, V., Roy, B.: ELECTRE methods. In: Figueira, J., Greco, S., Ehrgott, M. (eds.) Multiple Criteria Decision Analysis: State of the Art Surveys, pp. 133–153. Springer, New York (2005b)

Lichfield, N.: Community Impact Evaluation. UCL press, London (1996)

MacArthur, J.: Stakeholder analysis in project planning: origins, applications and refinements of the method. Proj. Appraisal 12(4), 251–265 (1997)

Marradi, A.: Concetti e metodi per la ricerca sociale. La Giuntina, Firenze (1980)

Morano, P., Tajani, F., Locurcio, M.: Land use, economic welfare and property values: an analysis of the interdependencies of the real-estate market with zonal and socio-economic variables in the municipalities of Apulia region (Italy). Int. J. Agric. Environ. Inf. Syst. (IJAEIS) 6(4), 16–39 (2015)

Mousseau, V., Roy, B.: A theoretical framework for analysing the notion of relative importance of criteria. J. Multi-Criteria Decis. Anal. 5, 145–159 (1996)

Munda, G.: Measuring sustainability: a multi-criterion framework. Environ. Dev. Sustain. 7(1), 117–134 (2005)

Nesticò, A., Pipolo, O.: A protocol for sustainable building interventions: financial analysis and environmental effects. Int. J. Bus. Intell. Data Min. 10(3), 199–212 (2015)

Prizzon, F.: Gli investimenti immobiliari. Celid, Torino (1995)

Realfonzo, A.: Teoria e metodo dell'estimo urbano. NIS, Roma (1994)

Rickson, R.E., Western, J.S., Burdge, R.J.: Social impact assessment: knowledge and development. Environ. Impact Assess. Rev. **10**(1–2), 1–10 (1990)

Rosenberg, M.: The Logic of Survey Analysis. Basic Books, New York (1968)

Roy, B.: Classement et choix en présence de points de vue multiples. Revue française d'automatique, d'informatique et de recherche opérationnelle. Recherche opérationnelle **2**(1), 57–75 (1968)

Roy, B.: Méthodologie Multicritère d'Aide à la Décision. Economica, Paris (1985)

Roy, B., Bouyssou, D.: Aide multicritère à la decision: méthodes et case. Economica, Paris (1993)

Simonotti, M.: La stima immobiliare. Utet, Torino (1997)

Soderstrom, E.J.: Social Impact Assessment. Praeger publishers, New York (1981)

Tajani, F., Morano, P.: An evaluation model of the financial feasibility of social housing in urban redevelopment. Property Manage. **33**(2), 133–151 (2015)

Taylor, N.C., Goodrich, C., Bryan, H.: Issues-oriented approach to social assessment and project appraisal. Proj. Appraisal **10**(3), 142–154 (1995)

Trovato, M.R., Giuffrida, S.: The choice problem of the urban performances to support the Pachino's redevelopment plan. Int. J. Bus. Intell. Data Min. **9**(4), 330–355 (2014). Inderscience Enterprises Ltd., Genève

Valenti, A., Giuffrida, S., Linguanti, F.: Decision trees analysis in a low tension real estate market: the case of troina (Italy). In: Gervasi, O., Murgante, B., Misra, S., Gavrilova, M.L., Rocha, A.M.A.C., Torre, C., Taniar, D., Apduhan, B.O. (eds.) ICCSA 2015. LNCS, vol. 9157, pp. 237–252. Springer, Heidelberg (2015)

Voogd, H.: Multicriteria evaluation for urban and regional planning. Pion Ltd., London (1983)

Augmented Reality Applications in the Transition Towards the Sustainable Organization

Radu Emanuil Petruse[1], Valentin Grecu[1(✉)],
and Bogdan Marius Chiliban[2]

[1] Department of Industrial Engineering and Management,
Lucian Blaga University of Sibiu, Sibiu, Romania
{radu.petruse,valentin.grecu}@ulbsibiu.ro
[2] Department of Machines and Industrial Equipment,
Lucian Blaga University of Sibiu, Sibiu, Romania
bogdan.chiliban@gmail.com

Abstract. Business companies are increasingly aware that the world where they operate in is facing increasing environmental, economic and social challenges, thus there is a trend in most companies to align their processes and services with a sustainability agenda. Customers, shareholders, financial partners, governmental agencies, NGOs, globalization and internationalization of environmental and social standards exert new pressures on them to do so. Industry is widely recognized as being essential for wealth creation and development, although it is at times considered a source of environmental degradation and social concerns. Thus, industry must play an important role for the transition towards the sustainable society and a better future, as it is such an important social actor These factors, combined with genuine self-commitment have led to important contributions from the business sector to the development of sustainable energy and transport infrastructures, and in designing and implementing new sustainable technologies. This paper proposes Augmented Reality (AR) applications as a pertinent solution to the challenges that organizations face in their transition towards sustainability. AR allows companies to improve the efficiency of the product development, shorten the manufacturing and delivery times at a cost which is compatible with the demand. By using AR team members with different skills and roles from a product development team can view and interact with the required digital content without having to develop new specialized skills or purchase new software or hardware.

Keywords: Sustainable organization · Augmented reality · Knowledge management · Flexible manufacturing

1 Introduction

The way in which modern society produces and consumes is linked with many of the challenges associated with sustainable development. The natural resources are affected both quantitatively and qualitatively by the production, supply and distribution of

© Springer International Publishing Switzerland 2016
O. Gervasi et al. (Eds.): ICCSA 2016, Part III, LNCS 9788, pp. 428–442, 2016.
DOI: 10.1007/978-3-319-42111-7_33

goods and services which require material and energy consumption, generating waste, pollution and disrupting ecosystems (Petry et al. 2010).

Business companies are increasingly aware that the world where they operate in is facing increasing environmental, economic and social challenges, thus there is a trend in most companies to align their processes and services with a sustainability agenda. However, the sustainability concept is interdisciplinary, complex and contested. There is a wide range of approaches to sustainable businesses, from increasing efficiency through new technologies to reframing technological uses and pursuing more fundamental changes through integrating sustainability in the organizational culture. Furthermore, the ability to make a difference is different depending on the size of the organization or the sector where it activates, but nonetheless all business enterprises can make a contribution towards sustainability (Grecu and Denes 2012).

Many businesses focused their strategies on a good environmental services management, aspiring to 'greening' their infrastructures and product deliveries. Major new markets have emerged due to the growing demand for "green" products and visionary entrepreneurs already reap the rewards of approaching sustainability (IISD 2010). Therefore, companies can increase their market share, gain competitive advantage and increase shareholder value by adopting sustainable practices.

The perception of each individual impacts how we work to improve prospects for sustainability. Collective social desires will emerge from individual perceptions and therefore these desires will expand to political and legislative actions that can institutionalize what might begin as a lifestyle. Local, national and international organizations, the governance systems and the beliefs of key political decision makers, play each an important role in pointing towards sustainability. Often people are urged to "act fairly" by incentives, as it's visible regarding usage of cars. However there are also tougher policy instruments that are applied for the same issue, such as limiting fuel consumption. Policies and governance must generate and support these incentives and prohibitions.

2 The Sustainable Organization

2.1 What is a Sustainable Organization

Businesses that have no negative impact on the society, community, environment and economy are considered sustainable or green businesses. A business that aims to satisfy the triple approach of sustainability, namely its social, environmental and economic components is a sustainable business, according to Matthews (2009). Often, companies have progressive policies for environmental sustainability and human rights (Statman 2005). According to Cooney (2009), a business is considered to be green if it meets the following four criteria:

- it incorporates sustainability in all of its business decisions;
- it offers environmentally-friendly products and /or services which replace the demand for traditional products and /or organic services;
- it is greener than its traditional competition;

- it made a commitment to the principles of environmental sustainability in its business operations.

Any organization that ensures that all its products, production activities and processes are addressing properly the current social and environmental issues, while maintaining a profit is considered a sustainable business. In other words, it is a business that "meets the needs of the present world without compromising the ability of future generations to meet their own needs" (WECD 1987). According to Rennie (2008a), a sustainable organization implies the design of products and services that take advantage of current environment, while an indicator of success is how well a company behaves when products are made from renewable resources.

Sustainability is a three legged chair, as stated in the Brundtland Report: people, planet and profit (WECD 1987). A balance between these three components is the aim of sustainable businesses, who try to implement the sustainable development concept in the supply chain in order to reduce its impact on the environment, social welfare and economic growth (Galvao 2008).

The planet and the market sustainability are affected by everyone in one way or another. However, an enterprise that chooses to implement the principles of sustainable development can create value for customers, investors, and the environment. Customer needs can be addressed by a sustainable business, while it is treating the environment well at the same time (Rennie 2008b).

Another pertinent definition of the sustainable business is given by Gerard Keijzers in his paper "The transition to the sustainable enterprise" (Keijzers 2002). He argues that a sustainable business implies not only processes that reduce polluting emissions, use renewable energy, recyclable resources or re-uses resources, but it also allows for the preservation of the natural capital, while at the same time it assures economic and social development, both locally and globally.

Many conceptions and practical ideas have been associated with "greening" businesses. Authors like Keijzers (2002), Rennie (2008a, b), Galvao (2008), and Matthews (2009), attempted to provide definitions of green or sustainable business as business that leave no negative impact on the society, community, economy and the environment, satisfy customer needs while preserving natural resources, using recyclables, re-using renewables and reducing pollution.

Hart (1995) defined a model for the transition towards sustainable businesses, identifying the steps that an enterprise needs to undertake in order to become sustainable:

- adapting their strategies to prevent pollution and waste,
- implement product stewardship schemes and improve the collaboration in the supply chain,
- incorporate strategic environmental management in all levels within the business organization.

Companies are constrained to take on ecological and social responsibilities as governments, NGOs, stakeholders, customers, internationalization of environmental and social standards, financial partners and globalization of economies exert new pressures on them to do so. These factors, combined with genuine self-commitment have led to important contributions from the business sector to the development of sustainable energy and transport infrastructures, and in designing and implementing new sustainable technologies (Keijzers 2002).

2.2   Benefits of the Transformation into the Sustainable Organization

Issues such as environmental protection, workplace conditions, human rights, corporate social responsibility or business ethics, are among others typically included into the sustainable and socially responsible agenda (Petrini and Pozzebon 2009). According to Raynard and Forstarter (2002), in order to assess the increasing importance of sustainability and CSR, the context of deregulation, privatization and globalization should be considered as its complexity is given by social, environmental and economic inequalities that increase continuously (Grecu 2015). Therefore, managers need to shift their perspective that is focused mainly on the decrease of costs and/or on the growth of profits and sales, towards the sustainable development of the business itself and of the society and the environment where it operates. As a result, there is an increasing number of companies that aligned their business practices with the sustainable development principles, having undergone considerable efforts to do so (Jones 2003).

The implementation of sustainable business practices faces serious obstacles, even if in the last years the interest and concern of companies towards these aspects has increased considerably. According to Petrini and Pozzebon (2009) until recently, sustainability indicators and "purely economic" performance indicators have been considered separately. However, in addition to defining environmental and social performance indicators, it is crucial to address the question of how they can be tracked, visualized, operated and monitored in terms of added value to corporate social responsibility management, since accessibility of corporate information and transparency have become important issues (Clarkson 1995).

Corporate sustainability has become an important tool for many corporations to explore ways to create new products, implement internal changes in culture and structure, manage risks or reduce costs. Nonetheless, integrating sustainability in the organizational structure, culture and practice is not easy. It requires a vision, commitment and leadership. A systems approach with an adequate management framework in order to enable design, management and communication of corporate sustainability policies is also required (Azapagic 2003).

Industrial systems are a fundamental component of the human economy as they determine and cause flows of material and energy in society. Industry is widely recognized as being essential for wealth creation and development, although it is at times considered a source of environmental degradation and social concerns. Thus, industry must play an important role for the transition towards the sustainable society and a better future, as it is such an important social actor (Azapagic and Perdan 2000).

Any business that aims to be a sustainable one faces the challenge to ensure that it contributes to a better quality of life today without compromising the quality of life of future generations. Industry needs to respond to this challenge by improving its environmental, social and economic performance, within new and evolving governance systems (Azapagic 2003).

The pressure for companies to commit to sustainability more seriously is increasing due to the environmental (e.g. climate change) and social (e.g. accountability) demands from shareholders and stakeholders. However, industries and companies face the challenge to prove their actual contributions to the society, while keeping the potential to deliver improvements for the generations to come. Thus, in order to meet the sustainability challenges a company needs to implement sustainability management practices allowing the management board to align its corporate and business strategy with the principles of sustainable development (Schaltegger and Burritt 2005).

Despite all the pressure from external factors, there is another factor that justifies incorporating sustainability into business practices and its relevance is increasing: it makes business sense to be more sustainable. The benefits of addressing sustainability include among others the following (IISD and WBCSD, 2002):

- Cleaner production methods leads to cost reduction. Product efficiencies, better material and energy consumption are among the advantages of investing in innovation and technology for sustainable development;
- Health and safety costs will decrease if companies commit to assuring a safe and healthy working environment. This means increased productivity, reduced costs for medication and social services and less compensation and damage suits;
- Innovative solution and decreased labor costs are to be expected, as better working conditions can improve productivity and motivation and will result in fewer union disputes and less work absenteeism;
- Implementation of sustainable practices will lower insurance costs and loan rates and will ease the access to insurers, lenders and preferential loans and insurance rates;
- Following best practices give companies much better chances to influence how standards are set and the direction of regulatory change, than their competitors;
- Being committed to sustainable development will increase the company's reputation and will attract the best to join the company;
- A better and deeper relationship with the customers can be built by moving towards an integrated supply chain management and capturing more value by providing service rather than selling products only;
- Investors are more concerned about unacceptable environmental and social performance, thus the commitment to sustainability will increase the chances of companies to attract funds from ethical investors.

3   Knowledge as a Key Resource in Developing a Sustainable Organizational Paradigm

Knowledge is important, in today's highly flexible and increasingly demanding market place in which organizations compete, because it is becoming increasingly clear that the only sustainable competitive advantage that can be achieved is through its correct utilization and leverage (Halal 1999), (Alavi and Leidner 2001). The strategic importance of knowledge is beautifully emphasized by Lew Platt, ex CEO of Hewlett Packard who stated: "If only HP knew what it knows it would make three times more profit tomorrow".

Knowledge as a resource is difficult to pinpoint and measure within the organizational paradigm. Although it is recognized by stakeholders as being the key to organizational performance and increased consumer satisfaction it has eluded a clear definition and optimum codification. Knowledge has been divided into three main types: tacit knowledge, explicit knowledge and some experts also argue the existence of implicit knowledge.

Polanyi is the first to make, a much debated distinction between two types of knowledge: "tacit" and "explicit" (Polanyi 1967). He famously characterizes tacit knowledge by the following expression: "we know more than we can tell".

Tacit Knowledge is knowledge present in the mind of a person that cannot be fully explicated by the possessor. It is the kind of knowledge that is difficult to transfer to another person by means of writing it down or verbalizing it. It consists of beliefs, ideals, values, schemata and mental models which are deeply ingrained in us and which we often take for granted. While difficult to articulate, this cognitive dimension of tacit knowledge shapes the way we perceive the world.

Polanyi noted that human action is often based on what to the observer seems inexplicable reasoning. Polanyi found an explanation in the deeply, often culturally ingrained, beliefs and understandings which we carry with us but of which we are not consciously aware. At best as observers we may be able to infer at least some aspects of this knowledge from the behavior of the subject (Land 2009).

In contrast to tacit knowledge, explicit knowledge is easily codified and transmitted between individuals. It resides within a plethora of forms, such as books, reports, procedures, works of art, videos etc. Explicit knowledge constitutes the basis for every educational system implemented by humanity to this point.

There is another form of knowledge that is supported by academics in the field of knowledge management. This is the so called implicit knowledge, which Wilson (Wilson 2002) considers as the true replacement of tacit knowledge. Others in the community see implicit knowledge as a grey area existent between explicit and tacit knowledge. In this perspective implicit knowledge is seen as knowledge that can be transformed into an explicit state but has not been thus far. Tacit knowledge is considered as impossible to express in any form outside the mind of the owner.

Regardless of the lack of a universally accepted definition and consensus as to its nature, knowledge is a present force in our everyday lives which molds society itself. It can be defined as a "green", inexhaustible, unbound by conventional limitations, resource, that can be easily created within the confines of the original virtual space, the human mind. Although it can be argumented that knowledge can be used for both good and evil purposes, history has proven that ultimately human beings uncannily recognize its merits and strive to use it for the betterment of their lives. Thus it has a profound transformative and positive effect on individuals by changing their perspectives on life; on the planet as a whole by enabling smart utilization of finite resources and on organizational wellbeing by improving products and processes effectiveness and efficiency leading to stakeholder satisfaction. This symbiotic relationship is the key to a sustainable paradigm not only at the organizational level but also at a societal level.

Although this process of continually developing and improving knowledge at all levels of abstraction is a highly desired outcome, in practice there are countless hurdles to overcome.

O'Dell and Grayson also state that a commonly occurring problem in today's organizations is the lack of transparency which can lead to an unpleasant situation in which "the organization does not know what it knows" (O'Dell and Grayson 1998).

Knowledge is difficult to codify and transfer, especially the most valuable of it, tacit (implicit) knowledge. This is the highly interconnected situational based resource whose availability is directly correlated to successfully achieving the sustainable organizational paradigm. Thus, there has been significant investments in setting up technological platforms that support business processes and increased efficiency of operational structures in many organizations and therefore most of them have reached a point where the use of said tools, emerges as a key factor (Petrini and Pozzebon 2009). One of the most important and technologically advanced tools utilized for knowledge integration within the decision making processes related to product and process development is the implementation and utilization of augmented reality.

4   Implications on Manufacturing Technologies

Over the last few years the society has slightly changed its demands, requesting environmentally friendly, highly customizable products. Simultaneously, the time between the clients' order and the finished product is in a constant decrease. Such being the case, products and manufacturing processes are becoming more and more complex. This challenges companies to improve the efficiency of the product development, shorten the manufacturing and delivery times at a cost which is compatible with the demand. Together with the market globalization and the increase in international competitivity, leads to a complete review of the classic product design and manufacturing principles. Turning the production paradigm from a linear model to an interactive model characterized by mass customization of the product (Drăghici 2015). Forcing the companies to change the product development and manufacturing processes from a linear, sequential model to a simultaneous one distributed between different global actors, each sharing their own know-how in a specific domain in order to obtain a greater reactivity.

In order to improve the product development process in the collaborative work environment computer-aided technologies such as CAD, CAM, CAPP, FEA, PDM, etc. are used. These are solutions that assist the user in the stages of design, engineering, manufacture, inspection and maintenance of the product's lifecycle.

The latest challenge is to gather all the required resources into a centralized management that can be used in a geographically dispersed collaborative environment. Obtaining a networked manufacturing based on decentralization.

Significant attention from both academia and industry is oriented on cloud-based manufacturing (CBM) which can be a solution for these challenges. CBM should be viewed as a whole system which has the goal to gather all the required resources for a product development into a centralized management (Petruse and Bondrea 2016).

Together with the benefits of implementing the concept of CBM, there are also disadvantages. The main disadvantage is created by the high costs of the necessary hardware and software. More than that, the degree of digitization of the fabrication technologies has known a sustained progress due to the desire to reduce both costs and time, and also to increase the product development efficiency. To be able to compare the fast evolution of the digitalization degree, we have compared a graph from (Xiumin Fan 2011) and the current state, represented in Fig. 1.

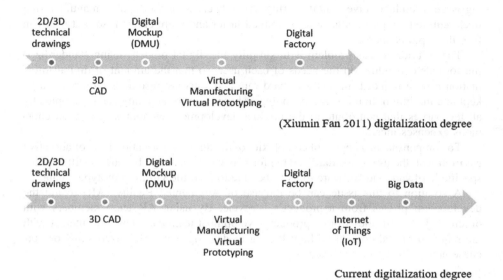

Fig. 1. 2010 vs. current digitalization degree

The deficiency produced after applying this complex manufacturing concept, based on virtual reality, is that the digital information of a product becomes very complex and the resources needed requires developing a whole informatics system.

Additional costs are incurred for the administration and maintenance systems that have to be implemented. Besides the costs of acquisition and the maintenance of the equipment required to enable the implementation of CBM, there are extra expenses with employees training and costs generated by the time lost during the implementation and verification of the new technologies.

Global corporations are using external suppliers for more than 50 % of their products which implies that their suppliers, the smaller companies, are constrained to adopt the same technologies that are used by their clients in order to meet the demands, thus being forced to increase the price of products supplied to amortize their investments.

In order to be able to cope with these demands, a solution for the micro companies is to introduce modular production systems. A sustainable company must be able to easily adapt its manufacturing technologies to the client's needs. This implies having an interdisciplinary know-how in process development and flexible manufacturing systems. The flexible manufacturing systems are a suitable link between the market requirements and the current state of development in micro production technologies.

Another disadvantage of this CBM concept is that it requires specific software tools to read all the digital information. In most cases these tools are very expensive and require the same software version with which the original digital information was created. Considering the large amount of software suppliers on the market, this can be a big issue in collaborative manufacturing. Also, because of the flexible manufacturing lines required, employees have to be trained faster and more efficient so that they can face the rapid changes.

The solution to this problem is to simplify the digital models, using standardized file formats depending on the needs of each user, so that the amount of digital information can be reduced. In this way hardware and software resources required can be kept to a minimum. In this case, technologies based on virtual reality can be adopted by all the parties that contribute to the product development without incurring an enormous expenses increase.

To compensate this degree of complexity of the digital information in a collaborative environment, the use of standardized file formats it is not enough because it still requires specific hardware and software tools to be able to read the virtual prototype's data.

A solution to this issue can be the use of Augmented Reality (AR) to aid the development phases from a product's lifecycle. By using AR team members with different skills and roles from a product development team can view and interact with the required digital content without having to develop new specialized skills or purchase new software or hardware.

5 Augmented Reality Applications in a Sustainable Organization

In order to verify this hypothesis, we have applied AR a FESTO Flexible Manufacturing system. This is a modular production line that is easily reconfigurable which provides compatibility between the modules. The FESTO assembly line composed out of 7 modular workstations as follows: a NC manufacturing centre, a MITSUBISHI robot, an assembly module, a sorting module, a testing module, a handling module and a conveyer (Fig. 2).

It is a highly scalable approach that enables the user to customize the production system. The flexible modular production lines are the ideal solution for a sustainable micro company because it offers the possibility of customizing and/or changing its products with minimal costs.

The customized production systems are 3D modelled in Catia CAD software and simulated virtually in Tecnomatix Plant Simulation 12. In this simulation different production scenarios can be tested without disrupting the real manufacturing process. Information regarding production times, bottlenecks, logistics but also a digital model of the production system are obtained from the virtual simulation as presented in Fig. 3.

In this scenario we have added another CNC milling machine and a Mitsubishi robotic arm to the manufacturing process in order to improve the production time and to reduce the bottlenecking.

After the digital model is verified we are using AR to superimpose the 3D digital models over a real environment in order to verify if it is feasible. The superimposed 3D

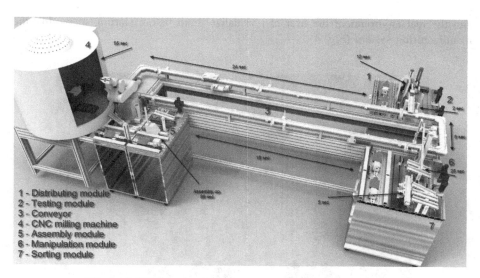

Fig. 2. FESTO flexible manufacturing line modeled in CATIA

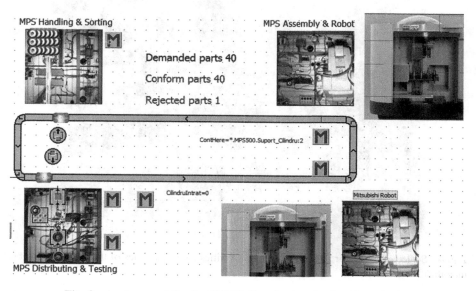

Fig. 3. An improved for the FESTO line simulated using Tecnomatix

models are simplified representations of the CAD models so that any user can use them with a smartphone or a tablet without having to develop specialized skills.

With the aid of AR, we can test if the new model fits to the designated location and if it is still an ergonomic workplace. If collisions are detected between the 3D models and the real environment the user can visualize them before starting to modify the manufacturing system (Fig. 4).

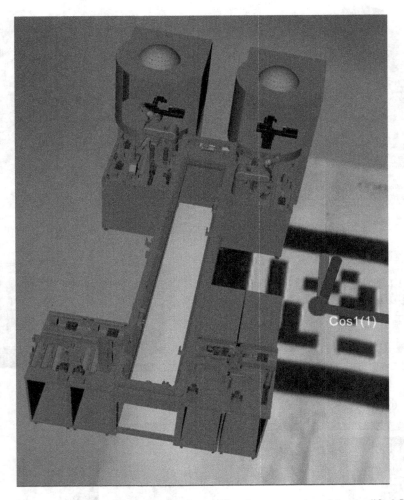

Fig. 4. Scaled example for AR application that verifies the ergonomics of the modified flexible manufacturing system

AR is also used to provide information for each separate module from the flexible production system. This information consists in maintenance data, installation info and production data thus reducing the required worker qualification.

Further, one of the most complex AR applications from the FESTO manufacturing line is presented. We used AR to aid the programing of the RV-2AJ Mitsubishi robotic arm which performs an assembly task on the FESTO flexible manufacturing line. AR is used to teach the position required for the assembly process after the robot loses its origin.

By default, the robot's coordinate system origin is not used as a functional origin in the robot's assembly program. This is due to the fact that an origin for the FESTO assembly program is defined using a caliber which is not included when buying the FESTO module, requiring an expensive maintenance intervention each time the robot's batteries get depleted and the origin is lost. Based on this origin all the positions required for the assembly process are calculated. In order to avoid this, we reprogrammed the robot based on its mechanical origin.

AR instructions are created to guide the user in defining the robot's mechanical origin for each of the 5 joints. These AR instructions are based in the robot's user manual images and text superimposed over the real environment.

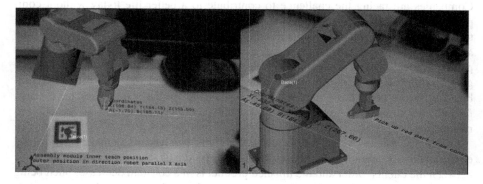

Fig. 5. AR instructions for setting the robot's positions

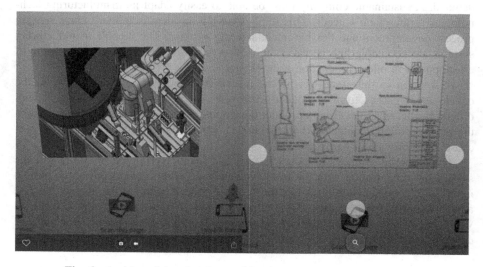

Fig. 6. A video of the virtual assembly process superimposed using AR

After setting this mechanical origin all the coordinates for positions required in the assembly process must be changed in relation to the new origin. Here is where AR is used again to provide the 3D model of the required robot's posture for each position and the informative text containing the coordinates (Fig. 5).

In order to give the user, the possibility to verify the robot's program we also superimposed a video of the virtual assembly process (Fig. 6) over the RV-2AJ robot's technical drawing. The implications of these AR instructions are beneficial, as shown in our previous papers (Bondrea and Petruse 2015) the AR instructions are much easier to understand and to follow that classic instructions.

6   Conclusion

Companies are facing the challenge to contribute to a better life today without compromising the chances of future generations to satisfy their own needs. Industry needs to demonstrate commitment and continuous improvement in its social, environmental and economic performance, given the increased pressure from customers, business partners, governmental and non-governmental organizations, internationalization and globalization of the economy and homogenization of standards. These trends, combined with the benefits of implementing sustainable business processes have transformed the transition towards the sustainable organization into a generalized movement.

Over the last few years the society has slightly changed its demands, requesting environmentally friendly, highly customizable products. Simultaneously, the time between the clients' order and the finished product is in a constant decrease. Such being the case, products and manufacturing processes are becoming more and more complex. This challenges companies to improve the efficiency of the product development, shorten the manufacturing and delivery times at a cost which is compatible with the demand. A sustainable company must be able to easily adapt its manufacturing technologies to the client's needs. This implies having an interdisciplinary know-how in process development and flexible manufacturing systems. The flexible manufacturing systems are a suitable link between the market requirements and the current state of development in micro production technologies.

A solution to this issue can be the use of Augmented Reality (AR) to aid the development phases from a product's lifecycle. By using AR team members with different skills and roles from a product development team can view and interact with the required digital content without having to develop new specialized skills or purchase new software or hardware.

The proposed solutions to reducing costs and advancing the transition towards the sustainable organization with the use of augmented reality applications show that an interdisciplinary approach leads to a much easier understanding of the instructions and a reduced implementation time. AR applications have an immediate impact on reducing waste and resource consumption, thus contributing to the challenge of companies to create a better future.

References

Alavi, M., Leidner, D.: Knowledge management and knowledge management systems: Conceptual foundations and research issues. MIS Q. **25**(1), 107–136 (2001)

Azapagic, A.: Systems Approach to Corporate Sustainability: A General Management Framework. Trans. IChemE, vol. 81, pp. 303–316 (2003). http://www.ingentaselect.com/titles/09575820.htm. Accessed 17 Apr 2014

Azapagic, A., Perdan, S.: Indicators of sustainable development for industry: a general framework. Trans. IChemE Part B Proc. Safe Env. Prot. **78**(4), 243–261 (2000)

Clarkson, M.B.E.: A stakeholder framework for analyzing and evaluating corporate social performance. Acad. Manag. Rev. **20**(1), 92–117 (1995)

Cooney, S.: Build A Green Small Business. Profitable Ways to Become an Ecopreneur. McGraw-Hill Publishing, New York (2009)

Drăghici, G.: Infrastructure for integrated collaborative product development in distributed environment. Appl. Mech. Mater. **760**, 9–14 (2015)

Galvao, A.: The Next Ten Years: Energy and Environment. Crossroads 2008 presentation, MIT TechTV beta, 55 min., 51 sec (2008). http://techtv.mit.edu/videos/107

Grecu, V.: The global sustainability index: an instrument for assessing the progress towards the sustainable organization. Acta Univ. Cibiniensis **67**, 215–220 (2015)

Grecu, V., Denes, C.: A decision support system for the transition towards the sustainable university. In: Proceedings of the International Conference on Engineering & Business Education, Innovation and Entrepreneurship, Sibiu, Romania, 18 - 21 October, 2012, pp. 319–324 (2012a)

Halal, W.: The infinite resource: mastering the boundless power of knowledge. In: Halal, W., Taylor, K. (eds.) 21st century economics, pp. 53–75. Macmillan, London (1999)

Hart, S.A.: A natural-resource-based view of the firm. Acad. Manag. Rev. **20**(4), 986–1014 (1995)

International Institute for Sustainable Development (IISD) (2010). http://www.iisd.org/business/. Accessed 07 Sept 2011

Bondrea, I., Petruse, R.E.: Augmented reality CAD/CAM training application version 2.0. In: Advances in Computers and Technology for Education, pp. 147–150 (2015)

Jones, T.M.: An integrating framework for research in business and society. Acad. Manag. Rev. **8**(4), 559–564 (2003)

Keijzers, G.: The transition to the sustainable enterprise. J. Clean. Prod. **10**(4), 349–359 (2002)

Land, F.: Knowledge management or the management of knowledge? Knowledge Management and Organizational Learning. Annals of Information Systems, vol. 4, pp. 15–25. (2009)

Matthews, R.: What constitutes a green business? (2009). http://globalwarmingisreal.com/2009/09/16/what-constitutes-a-green-business/. Accessed 10 Sept 2011

O'Dell, C., Grayson, J.: If Only We Knew What We Know. Free Press, New York (1998)

Petrini, M., Pozzebon, M.: Managing sustainability with the support of business intelligence: integrating socio-environmental indicators and organisational context. J. Strateg. Inf. Syst. **18**(4), 178–191 (2009)

Petry, R.A., et al.: Educating for sustainable production and consumption and sustainable livelihoods: learning from multi-stakeholder networks. Sustain. Sci. **6**(1), 83–96 (2010)

Polanyi, M.: The Tacit Dimension. Routledge and Kegan Paul, London (1967)

Petruse, R.E., Bondrea, I.: Augmented reality aided product life cycle management collaborative platform. In: IATED Academy, pp. 4652–4659 (2016). ISBN: 978-84-608-5617-7

Raynard, P., Forstarter, M.: Corporate social responsibility: implications for small and medium enterprises in developing countries. United Nations Industrial Development Organization, Viena (2002). http://www.unido.org/doc/5162

Rennie, E.: Growing green, boosting the bottom line with sustainable business practices. APICS Mag., 18(2) (2008a)

Rennie, E.: Painting a green story, APICS Extra, 3(2) (2008b)

Schaltegger, S., Burritt, R.: Corporate sustainability. In: Folmer, H., Tieten-Berg, T. (Eds.) The International Yearbook of Environmental and Resource Economics. Edward Elgar, Cheltenham, pp. 185–232 (2005)

Statman, M.: The Religions of Social Responsibility, SSRN (2005). http://ssrn.com/abstract= 774386. Accessed 21 Oct 2011

Wilson, T.: The nonsense of knowledge management. Inf. Res. 8(1), 144–183 (2002)

World Commission on Environment and Development (WCED): Our Common Future (The Brundtland Report). Oxford University Press, Oxford (1987)

Xiumin Fan, R. Y.: Virtual assembly environment for product design. In: Virtual Reality & Augmented Reality in Industry, pp. 147–161 (2011)

Urban Solar Energy Potential in Europe

Federico Amato[1](✉), Federico Martellozzo[2], Beniamino Murgante[1],
and Gabriele Nolè[3]

[1] University of Basilicata, 10,Viale Dell'Ateneo Lucano, 85100 Potenza, Italy
fdrc.amato@gmail.com, beniamino.murgante@unibas.it
[2] University of Rome "La Sapienza",
Via Del Castro Laurenziano 9, 00161 Rome, Italy
f.martellozzo@hotmail.com
[3] Italian National Research Council, IMAA,
C.da Santa Loja, Tito Scalo, Potenza 85050, Italy
gabriele.nole@imaa.cnr.it

Abstract. Among the objectives of the Sustainable Development Goals by United Nations, "Affordable and Clean Energy" aims at ensuring access to affordable, reliable, sustainable and modern energy for all. However, in Europe there is not a precise understanding of the unleashed potential that could be achieved through the exploitation of solar and wind resources. This study presents an application to retrieve spatial explicit estimates of Direct Normal Irradiance (DNI) through the use of data from geo-stationary satellites. The energetic demand of large metropolitan areas in Europe is then retrieved and compared with the potential production of energy for domestic use through solar panels. Results of this comparison are presented based on the assumption that only the 1 % of the built up area could be covered with solar panels, and hence devoted to energy production. Outcomes suggest that even such a little coverage, if spread systematically over urban areas can in most of the cases satisfy urban population domestic needs.

Keywords: Renewable energy · Energy supply · Sustainable development goals · Direct normal irradiance · Meteosat

1 Introduction

The demand of energy from renewable sources has increased significantly at the global scale. In September 2015 the United Nations unveiled the Sustainable Development Goals (SDGs); a list of tangible targets to be reached within 20130 [1]. The seventh objective is named "Affordable and Clean Energy" and it aims at ensuring access to affordable, reliable, sustainable and modern energy for all. More in details, target 7.2 wants to reach by 2030 a substantial increase of the share of renewable energy in the global energy mix. The success in reaching this target relies, above all, on accurate precise and up-to-date data about solar and eolic energy availability and potential. However, although several relevant studies [2] and progress have been conducted in this regard, there is not a precise understanding of the unleashed potential that could be achieved through the exploitation of solar and wind resources. The amount of solar

© Springer International Publishing Switzerland 2016
O. Gervasi et al. (Eds.): ICCSA 2016, Part III, LNCS 9788, pp. 443–453, 2016.
DOI: 10.1007/978-3-319-42111-7_34

radiation hitting Earth's surface depends heavily on the orographic profile of our planet's surface and on atmospheric factors. Regarding the former, surface's aspect and slope substantially influence the amount of solar energy that reaches Earth's surface. Usually, the amount of radiation measured at the ground is assessed by interpolation of data retrieved with meteorological stations on the ground (e.g. through splines or Kriging methodologies) [3, 4]. Nevertheless, utilizing data from meteorologic geo-stationary satellites is also a technique widely applied [5]. However, this sort of analysis on the one hand produce results that are sufficiently accurate although less precise; but on the other they allow to obtain a larger time and spatial coverage that cannot be achieved with data from meteorological stations on the ground. Accurate and frequent measurements of solar irradiance is of fundamental importance to investigate the potential productivity of thermal energy from solar energy, because the amount of incident solar radiation is largely affected by microclimatic variations [6].

This study presents an application to retrieve spatial explicit estimates of Direct Normal Irradiance (DNI) for continental Europe and United Kingdom through the use of data from geo-stationary satellites. A time series covering a period of 24 years (from 1990 to 2013) was utilized to derive an average of the daily DNI estimate (given in $KWh/m^2/day$). The DNI is then used to elaborate potential estimates of energy production from solar irradiation.

The energetic demand of large metropolitan areas in Europe is then retrieved (Larger Urban Zones, Eurostat), and it is compared with the potential production of energy for domestic use through solar panels. Results of this comparison are presented based on the assumption that only the 1 % of the built up area could be covered with solar panels, and hence devoted to energy production. Results suggest that even such a little coverage, if spread systematically over urban areas can in most of the cases satisfy urban population domestic needs.

2 Materials and Methods

2.1 Direct Normal Irradiance

DNI is defined as the amount of solar radiation received per unit area by a surface that is always held perpendicular (or normal) to the rays that come in a straight line from the direction of the sun at its current position in the sky. Typically, DNI is intended as the amount of solar irradiance that can be converted into energy through Concentrated Solar Technologies (CST) or Concentrated Photovoltaic Technologies. Buie and Monger define DNI as the non-scattered radiant flux collected by a surface normal to the direction of the Sun, within the extent of the solar disk only [7].

The estimate of DNI from satellite geostationary data (in this study we used METEOSAT data) involves the calculation of the effective cloud albedo (CAL) though the Heliosat algorithm [8] to be used in combination with the clear sky radiation model MAGIC [9]. CAL defines the amount of sun irradiance that is reflected for all sky relative to the amount of reflected irradiance for clear sky. The Effective Cloud Albedo can be estimated as per Eq. 1 [10]:

$$CAL = \frac{R - R_{sfc}}{R_{max} - R_{sfc}} \tag{1}$$

where:

- R_{max} is a measure of the maximum cloud reflection;
- R_{sfc} is the clear sky reflection, dominated by the surface albedo;
- R is the observed irradiance.

Hence, to determine CAL only the value of reflection are needed, because R_{max} and R_{sfc} are statistically determined from the values of R.

CAL is fundamental to determine the Surface Incoming Solar Radiation (SIS), which is defined as the irradiance reaching a horizontal plane at the Earth surface in the 0.2-4 μm wavelength region. It is expressed in W/m². According to the Heliosat method, based on the principle of conservation fo energy, SIS can be calculated as per Eq. 2:

$$SIS = SIS_{CLS}(1 - CAL) \tag{2}$$

where:

- SIS_{CLS} is the clear sky irradiance.

The latter can be derived at the aid of the Radiative Transfer Model (RTM) [11]. Hence, it is possible to derive DNI at this point, because it can be defined as the irradiance at the surface normal to the direction of the sun in the 0.2-4 μm wavelength region. Thus, it is expressed in W/m² and it derives from a normalization of the direct irradiance (SID) with the cosine of the solar zenith angle (SZA). In other words, SID is the irradiance reaching a horizontal plane in the 0.2-4 μm wavelength region at the surface without scattering.

Mueller *et al.* proposed an algorithm to determine SID in clear sky conditions [12]. In this algorythm the atmospheric input and interpolation routine is identical for SIS and SID, with exception of a background surface albedo map, which is not needed for SID.

In order to account also for cloud presence, the direct irradiance for all sky is derived as per Eq. 3:

$$SID_{all\,sky} = SID_{clear}((1 - CAL) + 0.38\,CAL)^{2.5} \tag{3}$$

where:

- $SID_{all\,sky}$ is the all sky irradiance;
- SID_{clear} is the clear sky irradiance.

Thus far, DNI can be calculated as per Eq. 4:

$$DNI = \frac{SID}{\cos(SZA)} \qquad (4)$$

where:

- SZA is the solar zenith angle.

2.2   Urban Solar Energy Potential

In this paper we want to estimate the potential of energetic production from the solar renewable source based on a simplistic scenario. The scenario adopted allows for only the 1 % of the urban rooftop cover of the largest European urban areas to be allocated for hosting solar panels for energy production, and that all panels are equal.

To this end, the energetic Per Capita Consumption (PCC) by country in Europe is determined as per Eq. 5:

$$PCC = \frac{Total\,Electricity\,Consumption}{Total\,Country\,Population} \qquad (5)$$

It is expressed in KW/m^2, thus an average of the daily value is assumed as per Eq. 6:

$$PCC_{per\,day} = \frac{PCC}{365} \qquad (6)$$

Once DNI for the entire study area expressed in KWh/m^2/day is known, is possible to estimate its average value for the urban areas within the study region. Consequently, the authors propose to estimate the Potential Daily Solar Production (PDSP) as the product between the Average DNI and the surface that is assumed to be possibly allocated for hosting solar panels (1 % of urban areas), as per Eq. 7:

$$PDSP = Average\,DNI * (0.01 * Urbanized\,Areas) \qquad (7)$$

To this end, the Urban Solar Energy Potential (USEP) can be determined as per Eq. 8:

$$USEP = \frac{PDSP}{Urban\,Area\,Population * PCC_{per\,day}} \qquad (8)$$

it is a percentage estimate expressing the ratio between the energy produced by a surface covered with solar panels (i.e. 1 % of urban areas) and the energetic demand for the study area.

2.3   Data Preparation

DNI was derived through elaborations of freely distributed data from Satellite Application Facility on Climate Monitoring (CMSAF) [13, 14]. CMSAF distributes free raster maps of DNI in W/m^2. In particular, CMSAF releases mean of hourly, daily and monthly data derived from 1st and 2nd generation Meteosat. The use of 1st and 2nd generation satellites is necessary to provide a high quality monitoring of climatic dynamics. In fact, climatic analysis require long and high frequency time series data to ensure robust results.

Meteosat Visible-InfraRed Imager (MVIRI) installed on board of the Meteosat 1st Generation carrier is a passive imaging radiometer with three spectral channels: a visible channel covering 0.5–0.9 microns, and infra-red channel covering 5.7–7.1 microns and 10.5–12.5 microns. MVIRI comes with a spatial resolution of 2.5 km for the visible and 5 km for the IR channels, sub-satellite point respectively. 2nd Generation Meteosat satellites are equipped with a Spinning Visible and InfraRed Imager (SEVIRI) [15] and with the Geostationary Earth Radiation Budget (GERB). GERB is a visible-infrared radiometer extremely useful to observe Earth radiation. CMSAF develops DNI maps by applying the equation elicited in paragraph 2.2 to MVIRI and SEVIRI data [16].

The Average DNI value in KWh/m^2/day was derived using the monthly mean DNI values of a 24 years timespan (from 1990 to 2013).

With the DNI was possible to derive the USEP and calculate its average value within the main LUZs of continental Europe and UK. The LUZs dataset consists of core urban areas and corresponding commuting zones. The commuting zone is identified based on three commuting patterns:

- If 15 % of employed persons living in one city work in another city, these cities are treated as a single city;
- All municipalities with at least 15 % of their employed residents working in a city are identified;
- Municipalities surrounded by a single functional area are included and non-contiguous municipalities are dropped.

LUZs boundaries and population data residing in each zone are distributed by Eurostat [17–19]. For LUZs is also available the extent of artificial surface within each zone, and it is considered as the sum of continuous urban fabric, discontinuous urban fabric, industrial, commercial, public, military and private units, and other artificial areas. Nevertheless, Eurostat distributes also the data of the domestic energetic consumption in GWh per country (except for, Switzerland, whose energetic data were derived form the World Factbook, 2012 [20]).

3   Results and Discussion

The proposed methodology was applied to analyse 266 LUZ in 22 countries in the EU area. The calculation of the average of DNI for the 1990–2013 period considering CMSAF data shows how South European countries have significantly higher values of index than North European countries, with medium DNI variations that can reach 5 KWh/sq.mt/day (Fig. 1). Among the LUZ, 88 would be able to produce, through the use of photovoltaic panels, an amount of energy higher than their total electricity needs, reaching USEP values greater than 100 % and in 11 LUZ USEP even exceed 200 %.

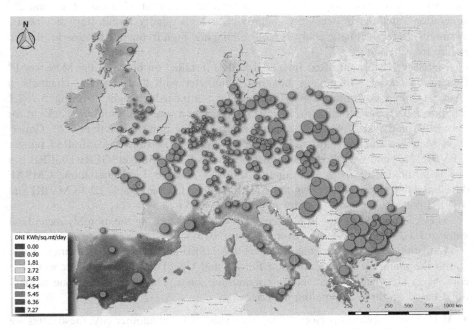

Fig. 1. Average of DNI in the countries of the object of study. The circles indicate the position of the analyzed LUZ. Their size is proportional to the USEP value calculated for the LUZ.

USEP particularly high levels are found in the countries of Southeast Europe, where at the same there are energy needs not excessively high and DNI with values always higher than 3.5 kWh/sq.mt/day. Liepaja (LT) and Piotrkw Trybunalski (PL) LUZ reach high values of USEP (respectively 226.5 % and 220.2 %) compared with an Average DNI respectively 2.66 and 2.83 kWh/sq.mt/day. The potential for high energy production is justified, in quite obvious way, considering the low per capita values of $PCC_{per\ day}$, equal to 8.66 for Liepaja and 9.07 for Piotrkw Trybunalski.

Considering 20 LUZ with the highest population (Table 1), only in Budapest (HU) and Sofia (BG) is calculated an USEP value greater than 100. Rome (IT) measures the highest value of the Average DNI, 93.1 %. The percentages of Paris (FR, 31.6 %) and Ruhrgebiet (DE, 39.2 %) are significantly lower, while Berlin (DE) reaches 58.8 %.

Particularly interesting is to observe as among the 20 most populous LUZ, to date the percentage of electricity produced from renewable sources in their own countries never exceed 30 %, with the exception of Rome and Turin (IT, 33.4), Bucharest (RO, 41.7) and Stockholm (SE, 63.3). Particularly interesting is the virtuous case of

Table 1. Results obtained in the analysis of 20 LUZ with highest population.

Country	LUZ	LUZ population	Artificial land [%]	Average DNI KWh/sq. mt/day	Electricity consumption (Country) [GWh]	USEP [%]	% of Electricity from renewable resources (Country)
FR	Paris	11800687	16,4	3,22	415325	31,6	18,3
DE	Ruhrgebiet	5045784	28,5	2,70	512835	39,2	28,2
DE	Berlin	5005216	10,1	2,88	512835	58,8	28,2
IT	Roma	4370538	17,7	5,07	281498	93,1	33,4
DE	Hamburg	3173871	13,6	2,63	512835	47,9	28,2
HU	Budapest	2915426	15,8	3,71	34704	133,4	7,3
UK	West Midlands	2909300	30,9	2,39	303563	41,2	17,8
UK	Manchester	2815100	30	2,05	303563	31,1	17,8
DE	Stuttgart	2668439	15,3	3,13	512835	37,9	28,2
BE	Bruxelles/Brussel	2607961	23,4	2,85	80561	42,7	13,4
DE	Frankfurt am Main	2573745	13,1	2,99	512835	37,8	28,2
NL	Amsterdam	2502189	18,5	2,78	101630	36,4	10,0
RO	Bucharest	2395515	27	3,94	41905	82,1	41,7
CZ	Praha	2204730	10,6	2,99	56203	68,8	13,9
SE	Stockholm	2091473	12,4	2,90	122191	35,5	63,3
FR	Lyon	1934717	17,8	3,98	415325	78,5	18,3
DE	Köln	1930036	26,3	2,80	512835	36	28,2
IT	Torino	1801729	19,7	4,65	281498	71,4	33,4
BG	Sofia	1543377	7,9	3,82	27674	106,5	18,9
DE	Dusseldorf	1515921	28,9	2,78	512835	36,8	28,2

Stockholm, that despite the DNI Average of 2.90 KWh/sq.mt/day and 35.5 USEP, has a domestic production of energy from renewable equal to 63.3 %.

Figure 2 compares DNI Average with the calculated values of USEP. It is evident that there is no a clear relationship, so we can not say that the growth of the DNI increases the capacity of a LUZ to approach to self-sufficiency energy. This is because USEP is also a function of the energy needs of the LUZ and their populousness.

In this article USEP was calculated assuming to cover 1 % of the urbanized areas of the LUZ with photovoltaic panels. The results showed that even with this low percentage of photovoltaic many LUZ could achieve a self-sufficient energy conditions, producing in some cases even more energy than used. This result stimulates several considerations.

First, not all energy produced by the photovoltaic is used. The panels, obviously, produce energy from sunrise to sunset, with a production peak in the middle of the day. However, very often the photovoltaic panels are installed on buildings by the owners. In these cases, it is rare that all daily energy produced should be exploited, especially considering that the average daytime energy consumption is typically less than the evening consumption, since the typical user spends the day outside home for work. Consequently, generally, there are two possibilities: that the user decides to transfer the surplus of produced, but not used, energy to the national grid (According to specific regulations that vary from country to country), exchanging a slight reduction of the cost

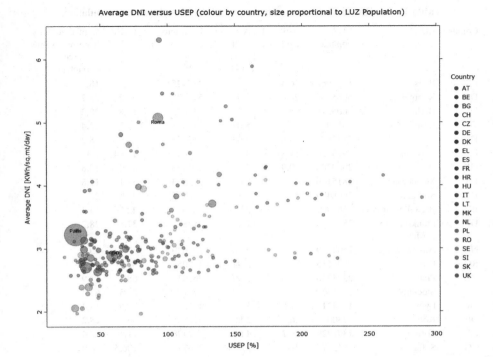

Fig. 2. Dot-scatterplot using DNI and USEP for all LUZ areas. Color indicates the Country, while the size is proortionla to population.

of the energy used in the evening or that the energy produced and not used is stored in special batteries purchased and installed within the single building. This second solution, which for private citizens seems to be more convenient, it is still very uncommon due to the high cost of the batteries. However, recent studies by the Massachusetts Institute of Technology indicate that the cost of the batteries will be reducing in the next few years to reach a price of about $ 100 in 2030 [21].

It is also important to understand the conditions required so that LUZ achieve self-sufficiency energy. In this case, it is necessary to modify the paradigm adopted in this paper. In other words, in this case will be necessary to calculate the surface area to be covered with photovoltaic panels so that for a specific LUZ is assessable a USEP equal to 100 %. This kind of analysis must be counterbalanced by a greater and more careful evaluation of electricity consumption, for example by separating the consumption due to residential and industrial uses. Future studies have to be focussed more on these aspects.

4 Conclusions

In this study an indicator of Urban Solar Energy Potential calculation has been proposed considering average values of Direct Normal Irradiance. The calculation of this indicator for the European 266 Larger Urban Zones has shown that the increased use of solar resource can ensure the achievement of sustainability goals proposed by the United Nation.

The topic of energy renewal of existing buildings is currently of great relevance. Models for energy rehabilitation and upgrading of existing buildings, such as the nearly zero energy building or the net zero energy building, are the subject of discussion of the international scientific community [22, 23].

The average energy performance of buildings in European cities are low and not suitable to achieve the most recent industry standards. The need to intervene on existing buildings is underlined also by the need to reduce urban growth and the phenomena of land take [24, 31]. It is important that the scientific community tackles with increasing intensity the energy issues, in order to build a complete and articulated knowledge framework, considering the potential deriving from sun and wind resources exploitation.

References

1. United Nations. Transforming Our World: The 2030 Agenda for Sustainable Development. Seventieth Session, Agenda Items 15 and 116 (2015). http://www.un.org/ga/search/view_doc.asp?symbol=A/RES/70/1&Lang=E. 21 October 2015
2. Schillings, C., Pereira, E.B., Perez, R., Meyer, R., Trieb, F., Renne, D.: High resolution solar energy resource assessment within the UNEP project SWERA. In: World Renewable Energy Congress VII, Cologne, Germany, 29 June – 5 July 2002
3. Hulme, M., Conway, D., Jones, P.D., Jiang, T., Barrow, E.M., Turney, C.: A 1961–1990 climatology for Europe for climate change modelling and impact applications. Int. J. Climatol. **15**(12), 1333–1364 (1995)
4. Zelenka, A., Czeplak, G., D'Agostino, V., Josefson, W., Maxwell, E., Perez, R.: Techniques for supplementing solar radiation network data. Technical Report, International Energy Agency, # IEA-SHCP-9D-1, Swiss Meteorological Institute, Switzerland (1992)
5. Noia, M., Ratto, C.F., Festa, R.: Solar irradiance estimation from geostationary satellite data: I. Stat. models, Solar Energ. **51**(6), 449–456 (1993)
6. Broesamle, H., Mannstein, H., Schillings, C., Trieb, F.: Assessment of solar electricity potentials in north Africa based on satellite data and a geographic information system. Solar Energ. **70**(1), 1–12 (2001)
7. Buie, D., Monger, A.G.: The effect of circumsolar radiation on a solar concentrating system. Solar Energ. **76**(1), 181–185 (2004)
8. Hammer, A., Heinemann, D., Hoyer, C., Kuhlemann, R., Lorenz, E., Müller, R., Beyer, H. G.: Solar energy assessment using remote sensing technologies. Remote Sens. Environ. **86**(3), 423–432 (2003)
9. Mueller, R.W., Matsoukas, C., Gratzki, A., Behr, H.D., Hollmann, R.: The CM SAF operational scheme for the satellite based retrieval of solar surface irradiance–a LUT based eigenvector approach. Remote Sens. Environ. **113**, 1012–1024 (2009)

10. Cano, D., Monget, J.M., Albuisson, M., Guillard, H., Regas, N., Wald, L.: A method for the determination of the global solar-radiation from meteorological satellite data. Solar Energ. **37**(7), 31–39 (1986)
11. Mayer, B., Kylling, A.: Technical note: the libRadtran software package for radiative transfer calculations-description and examples of use. Atmos. Chem. Phys. **5**(7), 1855–1877 (2005)
12. Mueller, R., Behrendt, T., Hammer, A., Kemper, A.: A new algorithm for the satellite-based retrieval of solar surface irradiance in spectral bands. Remote Sens. **4**, 622–647 (2012)
13. Data. http://www.cmsaf.eu/EN/Home/home_node.html
14. Posselt, R., Müller, R.W., Stöckli, R., Trentmann, J.: Remote sensing of solar surface radiation for climate monitoring the CM-SAF retrieval in international comparison. Remote Sens. Environ. **118**, 186–198 (2012)
15. Blasi, M.G., Serio, C., Masiello, G., Venafra, S., Liuzzi, G.: SEVIRI cloud mask by cumulative discriminant analysis. J. Phys: Conf. Ser. **633**, 012056 (2015). doi:10.1088/1742-6596/633/1/012056
16. Müller, R., Pfeifroth, U., Träger-Chatterjee, C., Cremer, R., Trentmann, J., Hollmann, R.: Surface Solar Radiation Data Set-Heliosat (SARAH) - Edition 1, Satellite Application Facility on Climate Monitoring (2015). doi:10.5676/EUM_SAF_CM/SARAH/V001, http://dx.doi.org/10.5676/EUM_SAF_CM/SARAH/V001
17. Eurostat, Geographical Information System of the Commission (2010)
18. Dijkstra, L., Poelman, H.: Cities in Europe – the new OECD-EC definition, Regional Focus (2012)
19. Eurostat: Eurostat regional yearbook 2014, Publications Office of the European Union (2014). doi:10.2785/54659
20. https://www.cia.gov/library/publications/the-world-factbook/rankorder/2233rank.html
21. Schmalensee, R., et. al.: The future of solar energy, Energy Initiative Massachusset Institute of Technology (2015). ISBN: (978-0-928008-9-8)
22. Buonomano, A., De Luca, G., Montanaro, U., Palombo, A.: Innovative technologies for NZEBs: an energy and economic analysis tool and a case study of a non-residential building for the mediterranean climate. Energ. Buildings **121**, 318–343 (2016)
23. Brinks, P., Kornadt, O., Oly, R.: Development of concepts for cost-optimal nearly zero-energy buildings for the industrial steel building sector. Appl. Energ. **173**, 343–354 (2016)
24. Amato, F., Martellozzo, F., Nolè, G., Murgante, B.: Preserving cultural heritage by supporting landscape planning with quantitative predictions of soil consumption. J. Cultural Heritage Elsevier (2016). doi:10.1016/j.culher.2015.12.009
25. Amato, F., Pontrandolfi, P., Murgante, B.: Using spatiotemporal analysis in urban sprawl assessment and prediction. In: Murgante, B., et al. (eds.) ICCSA 2014, Part II. LNCS, vol. 8580, pp. 758–773. Springer, Heidelberg (2014)
26. Amato, F., Martellozzo, F., Murgante, B., Nolè, G.: A quantitative prediction of soil consumption in Southern Italy. In: Gervasi, O., Murgante, B., Misra, S., Gavrilova, M.L., Rocha, A.M.A.C., Torre, C., Taniar, D., Apduhan, B.O. (eds.) ICCSA 2015. LNCS, vol. 9157, pp. 798–812. Springer, Heidelberg (2015). doi:10.1007/978-3-319-21470-2_58
27. Amato, F., Pontrandolfi, P., Murgante, B.: Supporting planning activities with the assessment and the prediction of urban sprawl using spatio-temporal analysis. Ecol. Inf. **30**, 365–378 (2015). doi:10.1016/j.ecoinf.2015.07.004
28. Amato, F., Maimone, B., Martellozzo, F., Nolè, G., Murgante, B.: The effects of urban policies on the development of urban areas. Sustainability **8**(4), 297 (2016). doi:10.3390/su8040297

29. Martellozzo, F.: Forecasting high correlation transition of agricultural landscapes into urban areas. Diachronic case study in north-eastern Italy. IJAEIS Anal. Model. Vis. Spat. Environ. Data **3**(2), 22–34 (2012)
30. Martellozzo, F., Clarke, K.C.: Urban sprawl and the quantification of spatial dispersion. In: Borruso, G., Bertazzon, S., Favretto, A., Murgante, B., Torre, C.M. (eds.) Geographic Information Analysis for Sustainable Development and Economic Planning, pp. 129–142. Hershey, PA, USA, IGI Global (2013)
31. Martellozzo, F., Clarke, K.C.: Measuring urban sprawl, coalescence, and dispersal: a case study of pordenone. Italy. Environ. Plan. B **38**, 1085–1104 (2011)

Computational Environment for Numerical Modeling of the Results of Cloud Seeding

Elena N. Stankova[1,2(✉)], Vladimir V. Korkhov[1], Natalia V. Kulabukhova[1],
Arkadyi Yu. Vasilenko[1], and Ivan I. Holod[2]

[1] Saint-Petersburg State University, 7–9, Universitetskaya Nab., St. Petersburg 199034, Russia
{lena,vladimir}@csa.ru, kulabukhova.nv@gmail.com,
vsvasilenko1@mail.ru
[2] Saint-Petersburg Electrotechnical University "LETI" (SPbETU),
5, Ul. Professora Popova, St. Petersburg 197376, Russia

Abstract. The paper describes numerical environment for forecasting the results of cloud seeding. Cloud seeding is a form of weather modification intended to change the amount or type of precipitation or to suppress fog and hail appearance. The effect of such a modification is crucially depended on the moment of seeding and the stage of evolution of the cloud under modification. Precise determination of these parameters and forecasting of the results of cloud seeding is impossible without using of numerical models of convective clouds. The results of simulations often need to be validated by simultaneous usage of several models of different dimension with a different degree of detail in the description of the microphysical processes. All the models should be involved into distributed computational environment, intended for integration of meteorological data about the state of the atmosphere, realization of the numerical calculations, visualization and comparison of the model output results. In this paper we describe the Virtual Cloud environment for modeling the results of cloud seeding on distributed computational resources. We present the concepts and the design of the Virtual Cloud and describe the prototype implementation.

Keywords: Distributed computing · Problem solving environment · Numerical modeling · Convective cloud · Cloud seeding

1 Introduction

Cloud seeding is a form of weather modification intended to change the amount or type of precipitation or to suppress fog and hail appearance.

At present, the mechanism of such modification is reduced to changing the phase state of a cloud by "seeding" it with some reagents, such as dry ice, smoke of silver iodide or lead iodide [1–3].

In recent years, numerical models of clouds are increasingly used to identify the effect of seeding [4, 5]. Using a numerical model allows, without expensive field experiments to analyze the development of clouds after modification, to trace the impact of different methods of seeding in different weather conditions on the evolution of the

© Springer International Publishing Switzerland 2016
O. Gervasi et al. (Eds.): ICCSA 2016, Part III, LNCS 9788, pp. 454–462, 2016.
DOI: 10.1007/978-3-319-42111-7_35

processes of cloud and precipitation, to identify the parameters most suitable for monitoring of seeding effects.

The implementation of cloud seeding is done by using of crystallizing reagents which are delivered either by aircrafts or by "shooting" a cloud with artillery guns. The success of the actions is determined by the right choice of time and place of seeding. Right time of seeding is determined by the stage of cloud evolution. Cloud should be on the stage of development. It this case injection of reagent will facilitate the formation of condensation or ice forming nuclei. Place of seeding is determined by the height where the maximum amount of so-called "supercooled" water is observed, the rapid freezing of which under the influence of reagent will facilitate the most effective process of sedimentation.

The time and the place of seeding can be determined most precisely by using numerical simulation. Ideally, the airfield from which planes fly, or artillery guns should be the nodes of the distributed environment, designed to predict and implement the cloud seeding.

This environment must include a distributed information system for the integration of meteorological information about the state of the atmosphere, as well as the number of compute nodes designed for the calculations with the help of various cloud models, as well as units for visualization and interpretation of the results of the calculations.

In order to forecast the effect of seeding several types of cloud models should be used. The models should differ by dimension and degree of detailed description of the microphysical processes.

Note that the calculations using the "simple" 1-D and 1.5-D cloud models could be implemented on a desktop computer equipped with multi-core processors. The calculation time is 2–3 min.

Meanwhile each calculation run for 2-D or 3-D models can last for several hours especially if the microphysical block is represented in the detailed version. These models require powerful multiprocessor systems.

Such environment using high-performance systems and computations have been developed, for example, for the calculation of complex chemical systems [6], for the numerical modeling of plasma chemical deposition reactors [7], as well as the simulation of the dynamics of elementary particle beams [8].

We intend to use the approach proposed in [8, 9], where the so-called "Virtual Reactor" (VR) is discussed. VR is a software environment for the modeling of the dynamics of particle beams.

By analogy with the "Virtual Reactor" such an environment can be called a "Virtual Cloud" (VC).

2 Approach

The main idea of the concept of the computing environment is the modelling of the results of cloud seeding using several numerical models differs by degree of description of dynamical and microphysical characteristics of clouds. It is supposed to use 1-D [10–14], 2-D [15] and 3-D [16, 17] models.

The main use of the VC is simulation of the results of cloud seeding by different numerical models with the opportunity to compare the results and to use the output of one model as the input to another model. Thus the most simple 1-D cloud model can be used for obtaining preliminary results very quickly (in 2–3 min of calculations).

The preliminary results in turn can be used for input data modification in order to prepare properly input data for 2-D and 3-D models, which are much more "slowly", requiring essential computational resources and several hours of computer running to obtain the results.

The VC in "hard", full variant can be considered only as information and computing environment designed for determination the best time and location of cloud seeding in different atmospheric conditions. However, in "light" variant, using only 1-D cloud model it can be used in real-time systems intended for operational works for rainfall enhancement or hail suppression.

The general idea of the software implementation is based on the Service-Oriented Architecture (SOA) that allows using Grid and Cloud computing technologies and enables remote access to the information and computing resources. Distributed services establish interaction between mathematical models, sources of meteorological information and operational cloud seeding centers.

The VC user interface allows getting both research simulations and real recommendations concerning the time and place of seeding. This approach gives researchers ability for system identification, parameter optimization, and result verification, which is impossible without computational models. Similar approach to develop virtual laboratory based on WS-VLAM workflows [18] is discussed in [8, 19].

The LEGO paradigm [8, 20] is used for the VC design. In terms of information technology it corresponds to object oriented design and component programming. Each object is represented as an independent component with own parameters and behavior. Development of distributed computing systems based on this concept is examined in [21] in more detail.

The Virtual Cloud is considered as a set of services and tools enabling transparent execution of numerical cloud models for simulating the optimal parameters and results of cloud seeding on distributed computing resources. Users will get the access to VC resources by unified interface including GUI on different platforms.

High-level scheme of VC environment similar to presented in [8] is shown in Fig. 1.

3   Design and Realization of the Virtual Cloud

The main purpose of a virtual cloud is to conduct computational experiments to simulate optimal parameters for cloud seeding using various cloud models with the ability to compare the results of calculations (different models simulate cloud features with different degree of detail). In addition, preliminary results obtained by one model can be used as input data for another, more elaborated model.

Technology for parallel computing and massively parallel computing systems can be used efficiently in case of complex 2-D and 3-D cloud models. In this case special

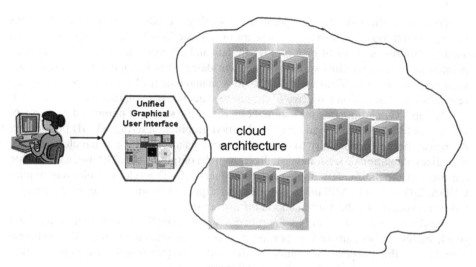

Fig. 1. High-level scheme of Virtual Cloud environment

investigations can be provided for choosing the optimal configuration of computational resources designed for problem solution.

The user has access to the resources of the virtual cloud through a "single window" [8] – a portal or a desktop-based interface. In this interface the user selects the sources of meteorological information and the type of cloud model, set the input data and parameters specific for each model, and the task is run on the available computing resources, see Fig. 2.

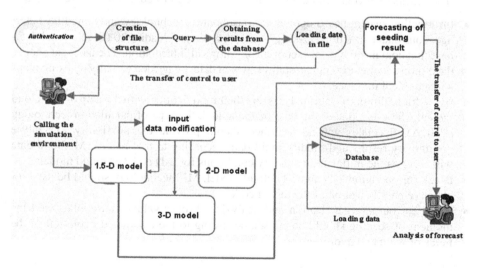

Fig. 2. Computational experiment in Virtual Cloud

To provide the sequence of operations smoothly we need workflow management system, which should be also responsible for the conversion of data between the formats used by different sources of meteorological data and different cloud models. As it was pointed in [8] this requires special experiment description language that can provide such conversion. The Virtual Cloud provides management of the tasks workflow by realization of description language of the entire sequence of computational experiment.

During VC operation the user should control all the stages of numerical experiment via access to all Virtual Cloud services. Unified graphical interface (UGI) provides to the researcher the opportunity to choose data sources and the type of the cloud model, to realize calculation process and visualization of the output results. The user can indicate the method for the models implementation. Thus 1.5 –D model calculations require CUDA, 2-D and 3-D – MPI technology. User's requirements are matched to the computational resources by the Virtual Cloud infrastructure.

We illustrate this matching by the process of cloud seeding result forecasting. First of all, the researcher must define appropriate meteorological data sources. Virtual Cloud contains for this purpose the complex information system based upon consolidation technology and aimed for producing model input.

Model input data represent vertical distributions of atmospheric temperature and humidity (radiosonde sounding) and several surface characteristics. In some specific cases of the state of atmosphere they should be modified before using in 2-D and 3-D cloud models. Modification should be done with the help of 1-D model, the simplest and the fastest in calculation sense.

After the user selects a proper numerical technology for calculation, the Virtual Cloud will build the computational experiment. This experiment can be executed on different resources via grid and cloud services.

The blocks of Virtual Cloud environment can be described as follows, see Fig. 3:

- Information system based upon data consolidation technology and aimed for heterogeneous data extraction [22–24], transformation and loading to the relational database that contains the whole set of meteorological information. The user can choose the region and meteorological stations which data are needed actually by using maps and graphical interface.
- Block for preliminary data analyses and their modification with the help of 1-D cloud model. The user pointed the type of the model, the type of calculation technology (CUDA). Special graphical interface should provide the possibility to analyse preliminary results and modify input data according to that results. Modified data will serve as the initial and boundary conditions for 2-D and 3-D cloud models.
- Block for operation 2-D and 3-D cloud model. MPI technology should be used for effective parallelization of the model codes.
- Block for output visualization and analyzing. Proper values of the place and the moment of seeding should be chosen according to the calculated evolution of the fields of wind and hydrometeors.

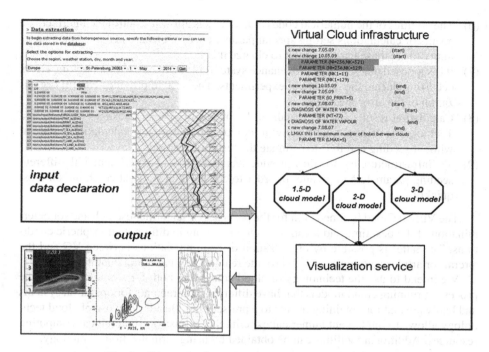

Fig. 3. Schematic view of the Virtual Cloud environment

The researcher does not need to use all types of the models available in VC. For example, in small meteorological centres located in airports, where high performance computational resources are not available, using 1-D model will be sufficient.

The Virtual Cloud usage scenario is the following:

(1) The user provides authentication and authorization to get access to the Virtual Cloud user interface.

(2) The user switches to special integrated information system and selects meteorological stations numbers in the region of cloud seeding. The system downloads automatically meteorological data about the state of the atmosphere and the characteristics of the ground surface and transforms the heterogeneous meteorological data into the format of the model input data.

(3) The user starts working with the 1.5-D model to verify the quality of input data. If necessary, input data may be modified using the GUI

(4) The user instructs the system to run calculations using 2-D and 3-D cloud models and modified initial data. This is done using either a dedicated resource (e.g. cluster or multicore processor) manually or automatically selected resource based on information about requirements of the application.

(5) The user sets the initial parameters of the cloud seeding, i.e. the amount of reagent and the moment of its injection into the cloud. Afterwards it is necessary to repeat the whole set of calculation with all the models in order to estimate the result of seeding.

(6) The abilities of the numerical codes of the models allow to control the intermediate results of calculations setting special parameters of viewing. VC gives the opportunity to set these parameters or to change them.
(7) It is supposed that so called provenance systems [25] will be used for tracking errors occurring during the numerical experiments. The system provides careful organization of the error messages.
(8) Calculation results can be visualized by means of special packages GrADS and Tecplot. The Virtual Cloud allow to compare the results of calculation with and without seeding and thus to estimate the result of seeding
(9) VC infrastructure has to allow provide sets of numerical experiment with different seeding parameters in parallel in order to choose their optimal combination most quickly.

The Virtual Cloud is aimed both for the purposes of scientific research and for determination of the best time and location of cloud seeding in different atmospheric conditions. The latter is possible by organization of communications between VC and the aircraft or artillery gun units, which provide real seeding in natural conditions.

We intend to use the technology of cloud computing both for simplification of the process of running cloud models that have different requirements for operation systems and hardware and for obtaining results to a pre-set time. Besides, the use of cloud technology allows to implement numerical experiments in case of limited local computing resources. Additional facilities can be obtained by using virtualization technology.

4   Conclusions

This paper presents a work-in-progress prototype of the Virtual Cloud environment intended to be used for modeling the results of cloud seeding with the help of numerical cloud model differ by the dimension and method of microphysical process description. We discuss the approach to the development of computing environment and describe design and realization of the Virtual Cloud. Some modules such as complex information system for consolidation heterogeneous meteorological information, block for input data formation and modifications, parallel versions of cloud models are completely developed, others are in progress.

The VC in "hard", full variant can be considered only as information and computing environment designed for determination the best time and location of cloud seeding in different atmospheric conditions. However, in "light" variant, using only 1-D cloud model it can be used in real-time systems intended for operational works for rainfall enhancement or hail suppression.

Acknowledgment. The paper was prepared in SPbETU and is supported by the Contract No. 02.G25.31.0149 dated 01.12.2015 (Board of Education of Russia).
The authors wish to thank the anonymous referees for their useful remarks and comments.

References

1. Weather Modification Association. http://www.weathermodification.org/
2. Dennis, A.S.: Weather Modification by Cloud Seeding, p. 284. Academic Press, New York (1980)
3. Breuer, G.: Weather Modification. Cambridge University Press, Cambridge (1980)
4. Dovgalyuk, Y., Dracheva, V.P., Egorov, A.D., Kachurin, L.G., Ponomarev, Y.F., Sinkevich, A.A., Stankova, E.N., Stepanenko, V.D.: Results of complex investigation of a thick cumulus cloud after seeding. Meteorologia i Gidrologia, N11, pp. 20–29 (1997)
5. Krauss, T.W., Sin'kevich, A.A., Veremey, N.E., Dovgalyuk, Y.A., Stepanenko, V.D.: Estimation of the Results of the Cumulonimbus Cloud Modification Aiming at Hailstorm Mitigation in Alberta (Canada) on the Radar and Numerical Modeling Data. Meteorologia i Gidrologia, N4, pp. 39–53 (2009)
6. Schuchardt, KL., Myers, JD., Stephan, EG.: Open data management solutions for problem solving environments: application of distributed authoring and versioning to the extensible computational chemistry environment. In: Proceedings HPDC-10 2001 (2001)
7. Krzhizhanovskaya, V.V., Korkhov, V.V., Tirado-Ramos, A., Groen, D.J., Shoshmina, I.V., Valuev, I.A., Morozov, I.V., Malyshkin, N.V., Gorbachev, Y.E., Sloot, P.M.A.: Computational engineering on the grid: crafting a distributed virtual reactor. In: Proceedings of Second IEEE International Conference on e-Science and Grid Computing (e-Science 2006) Amsterdam, The Netherlands, p. 101 (2006)
8. Korkhov, V., Ivanov, A., Kulabukhova, N., Bogdanov, A., Andrianov, S.: Virtual accelerator: distributed environment for modeling beam accelerator control system. In: 2013 Proceedings of 13th International Conference on Computational Science and Its Applications (ICCSA 2013), Ho Chi Minh City, Vietnam, 24–27 June 2013
9. Kulabukhova, N., Korkhov, V., Andrianov, S.: Virtual accelerator: software for modeling beam dynamics. In: Computer Science and Information Technologies Proceedings of the Conference, Yerevan, Armenia, 23–27 September 2013
10. Raba N.O., Stankova E.N.: Research of influence of compensating descending flow on cloud's life cycle by means of 1.5-dimensional model with 2 cylinders. In: Proceedings of MGO, vol. 559, pp. 192–209 (2009)
11. Raba, N., Stankova, E., Ampilova, N.: One-and-a-half-dimensional model of cumulus cloud with two cylinders. Research of influence of compensating descending flow on development of cloud. In: Proceedings of the 5th International Conference "Dynamical Systems and Applications" Ovidius University Annals Series: Civil Engineering, vol. 1, Special Issue 11, pp. 93–101, June 2009
12. Raba, N., Stankova, E.: On the possibilities of multi-core processor use for real-time forecast of dangerous convective phenomena. In: Taniar, D., Gervasi, O., Murgante, B., Pardede, E., Apduhan, B.O. (eds.) ICCSA 2010, Part II. LNCS, vol. 6017, pp. 130–138. Springer, Heidelberg (2010)
13. Raba, N., Stankova, E., Ampilova, N.: On investigation of parallelization effectiveness with the help of multi-core processors. Procedia Comput. Sci. 1(1), 2757–2762 (2010)
14. Raba, N.O., Stankova, E.N.: On the problem of numerical modeling of dangerous convective phenomena: possibilities of real-time forecast with the help of multi-core processors. In: Murgante, B., Gervasi, O., Iglesias, A., Taniar, D., Apduhan, B.O. (eds.) ICCSA 2011, Part V. LNCS, vol. 6786, pp. 633–642. Springer, Heidelberg (2011)
15. Khain, A., Pokrovsky, A., Pinsky, M.: Simulation of effects of atmospheric aerosols on deep turbulent convective clouds using a spectral microphysics mixed-phase cumulus cloud model. part i: model description and possible applications. J. Atmos. Sci. 61, 2963–2982 (2004)

16. Dovgalyuk,Y.A., Veremey, N.E., Vladimirov, S.A., Drofa, A.S., Zatevakhin, M.A., Ignatyev, A.A., Morozov, V.N., Pastushkov, R.S., Sinkevich, A.A., Stasenko, V.N., Stepanenko, V.D., Shapovalov, A.V., Shchukin, G.G.: A conception of the numerical three-dimensional convective cloud model development I. The model structure and main equations of hydrothermodynamical block. In: Proceedings of MGO, vol. 558, pp. 102–142 (2008)

17. Dovgalyuk, Y.A., Veremey, N.E., Vladimirov, S.A., Drofa, A.S., Zatevakhin, M.A., Ignatyev, A.A., Morozov, V.N., Pastushkov, R.S., Sinkevich, A.A., Stasenko, V.N., Stepanenko, V.D., Shapovalov, A.V., Shchukin, G.G.: A conception of the numerical three-dimensional convective cloud model development. II. Microphysical block. In: Proceedings of MGO, vol. 562, pp. 7–39 (2010)

18. Wibisono, A., Vasyunin, D., Korkhov, V.V., Zhao, Z., Belloum, A., de Laat, C., Adriaans, P.W., Hertzberger, B.: WS-VLAM: a GT4 based workflow management system. In: Shi, Y., van Albada, G.D., Dongarra, J., Sloot, P.M. (eds.) ICCS 2007, Part III. LNCS, vol. 4489, pp. 191–198. Springer, Heidelberg (2007)

19. Korkhov, V., Vasyunin, D., Belloum, A., Andrianov, S., Bogdanov, A.: Virtual laboratory and scientific workflow management on the grid for nuclear physics applications. In: Proceedings of the 4th International Conference on Distributed Computing and Grid-Technologies in Science and Education, Dubna, Russia, pp. 153–158 (2010)

20. Andrianov, S.: LEGO-technology approach for beam line design. In: Proceedings of the Eighth European Particle Accelerator Conference. Paris. France, pp. 1607–1609 (2002)

21. Kulabukhova, N., Ivanov, A., Korkhov, V., Lazarev, A.: Software for virtual accelerator designing. In: 13th International Conference On Accelerator And Large Experimental Physics Control Systems: Proceedings Of Icalepcs 2011, Grenoble, France, WEPKS016, p. 816 (2011)

22. Petrov, D.A., Stankova, E.N.: Use of consolidation technology for meteorological data processing. In: Murgante, B., Misra, S., Rocha, A.M.A., Torre, C., Rocha, J.G., Falcão, M.I., Taniar, D., Apduhan, B.O., Gervasi, O. (eds.) ICCSA 2014, Part I. LNCS, vol. 8579, pp. 440–451. Springer, Heidelberg (2014). doi:10.1007/978-3-319-09144-0_30

23. Petrov, D.A., Stankova, E.N.: Integrated information system for verification of the models of convective clouds. In: Gervasi, O., Murgante, B., Misra, S., Gavrilova, M.L., Rocha, A.M.A.C., Torre, C., Taniar, D., Apduhan, B.O. (eds.) ICCSA 2015. LNCS, vol. 9158, pp. 321–330. Springer, Heidelberg (2015). doi:10.1007/978-3-319-21410-8_25

24. Stankova, E.N., Petrov, D.A.: Complex information system for organization of the input data of models of convective clouds. Vestnik of Saint-Petersburg University. Series 10. Applied Mathematics. Computer Science. Control Processes. Issue 3, pp. 83–95 (2015) (in Russian)

25. Gerhards, M., Sander, V., Matzerath, T., Belloum, A., Vasunin, D., Benabdelkader, A.: Provenance opportunities for WS-VLAM: an exploration of an e-science and an e-business approach. In: Proceedings of the 6th Workshop on Workflows in Support of Large-Scale Science, pp. 57–66, ACM, New York (2011)

Using Technologies of OLAP and Machine Learning for Validation of the Numerical Models of Convective Clouds

Elena N. Stankova[1,2(✉)], Andrey V. Balakshiy[1], Dmitry A. Petrov[1],
Andrey V. Shorov[2], and Vladimir V. Korkhov[1]

[1] Saint Petersburg State University,
7-9, Universitetskaya nab., St. Petersburg 199034, Russia
{lena,vladimir}@csa.ru, andreyspbsu94@gmail.com,
g_q_w_petrov_dm_alex@mail.ru
[2] Saint Petersburg Electrotechnical University "LETI", (SPbETU),
ul. Professora Popova 5, St. Petersburg 197376, Russia

Abstract. The paper is a continuation of the works [1–3] where complex information system for organization of the input data for the models of convective clouds is presented. In the present work we use the information system for obtaining statistically significant amount of meteorological data about the state of the atmosphere in the place and at the time when dangerous convective phenomena are recorded. Corresponding amount of information has been collected about the state of the atmosphere in cases when no dangerous convective phenomena have been observed. Feature selection for thunderstorm forecasting based on Recursive feature elimination with cross-validation algorithm is provided. Three methods of machine learning: Support Vector Machine, Logistic Regression and Ridge Regression are used for making the decision on whether or not a dangerous convective phenomenon occurs at present atmospheric conditions. The OLAP technology is used for development of the concept of multidimensional data base intended for distinguishing the types of the phenomena (thunderstorm, heavy rainfall and light rain).

Keywords: Numerical modeling · Weather forecasting · Machine learning · OLAP technology · Multidimensional data base

1 Introduction

Nowadays the problem of validation of numerical models is very acute especially for solution of practical problems such as forecasting of dangerous convective phenomena (thunderstorms, squalls, hail and heavy rainfall). Validation assumes that the model describes the simulated phenomenon adequately. In order to prove the adequacy it is necessary to make sure that the cloud model is able to distinguish the cases with and without the phenomenon and to distinguish the type of the phenomenon, for example, to differentiate a thunderstorm from hail or heavy rainfall. We use the technology of machine learning [4, 5] in order to solve the first problem and OLAP (OnLine Analytical Processing) for the second one.

© Springer International Publishing Switzerland 2016
O. Gervasi et al. (Eds.): ICCSA 2016, Part III, LNCS 9788, pp. 463–472, 2016.
DOI: 10.1007/978-3-319-42111-7_36

Modeling of a convective cloud has been conducted by using 1.5D cloud model with the detailed description of microphysical processes [6–10]. We use complex information system [1–3] for obtaining statistically significant amount of meteorological data about the state of the atmosphere in the place and at the time when a dangerous convective phenomenon takes place. Corresponding amount of information has been collected about the state of the atmosphere in cases when no dangerous convective phenomena have been observed. The collected data are used as the initial data for the cloud model for conducting series of model calculations. The results of numerical experiments are processed afterwards using machine learning and OLAP technologies.

Machine learning is a mathematical discipline which allows extracting knowledge from existing data using mathematical statistics, numerical optimization methods, probability theory and discrete analysis.

Machine learning is used for the solution automation of various professional tasks in different fields of human activity, such as computer vision, medical diagnostics and speech recognition. The range of machine learning applications is constantly expanding due to widespread computerization resulted in accumulation of big amounts of data in all kinds of industry, science, business and health. We use three methods of machine learning: Support Vector Machine, Logistic Regression and Ridge Regression on making the decision for whether or not a dangerous convective phenomenon occurs. An algorithm is provided before feature selection for thunderstorm forecasting based on Recursive feature elimination with cross-validation.

The OLAP technology is used for development the concept of multidimensional data bases intended for distinguishing the types of the phenomena (thunderstorm, heavy rainfall and light rain).

Multidimensional data model integrates information from several sources for its further online analytical processing [11–14]. This technology allows obtaining answers to the queries, comprising large data volumes and thus detecting common trends of changing. A multidimensional analysis is intended for knowledge evolvement via semiautomatic search of unknown patterns and relationships.

Multidimensional data storages and multidimensional analysis are usually used in the interest of business in order to improve the quality of decision making, to quickly produce all kinds of reports on sales of various products in different outlets. In this paper an attempt is made to apply the advantages of a multidimensional approach to the analysis of specialized meteorological information for the identification of a dangerous convective phenomenon via the cloud parameters, calculated by the numerical model.

2 Model Settings and Usage

In order to use the machine learning technology and multidimensional analysis, the model running must be automated. For example, we do not need a graphical user interface for setting the input and viewing the output data. All we need is numerical values of calculated cloud parameters at any moment of the cloud evolution. In addition a group of soundings (vertical distributions of atmospheric temperature and humidity) is to be processed, serving as input model data, instead of working with each sounding separately, as it is set in the original version of the model.

We have to set the values of such input model parameters as vertical eddy exchange coefficient, time step, height increase step and some others, despite the possibility of correcting them in every numerical experiment in original version of the model. They are prerequisites for the application of the machine learning methods which require provision of both learning and implementation of the model under the same conditions (otherwise we cannot be sure of the objective reasons for algorithm processing results). It is possible to change the parameter values, but it requires training data to be set anew and restart of the learning process.

The following values of the parameters are used: time step $\Delta t = 20$ s, height increase step $\Delta h = 150$ m, mixing coefficient $\alpha 2 = 0.08$, vertical eddy exchange coefficient $Kv = 100$. Computational domain H equals 10 km, radiuses of inner and outer cylinders equal 3 and 10 km correspondingly. All computations are performed within a period of 1 h, which is the typical time frame for a convective cloud evolution.

After the implementation of the model automation a representative sample of soundings for the cases both with and without dangerous convective phenomena is to be obtained. The integrated information system [1–3] is used for this purpose. 289 soundings have been collected in the place and at the time when dangerous convective phenomena are recorded. 326 soundings have been collected in cases when no dangerous convective phenomena have been observed.

All the numbers of the collected input data need preprocessing due to the simulation problems connected to the presence of the atmospheric layers, where inverse or constant vertical distribution of temperature is observed. The problem is solved by correction of soundings by finding condensation level using Ippolitov's formula [15] and providing dry adiabatic temperature gradient from the surface up to this level. Experiments show that after modification 117 soundings from 128 control ones used as the input data allow reproducing the actual cloud development, the simulated cloud appears when it must appear and does not appear when it must not. Similar experiments provided for the cases without phenomena show the ratio of 102 from 104 soundings.

Corrected soundings serve as the model input; the output is presented by the data in CSV (Comma-Separated Values) format. All output data are displayed in a table that contains columns for the time, the altitude, the name of the cloud parameter and the value of the cloud parameter, see Fig. 1. The following cloud parameters are calculated: the vertical and radial components of the velocity, pressure, air density, ambient temperature, temperature excess in the cloud, relative humidity, vertical height of the cloud, mixing ratio of water vapour, mixing ratios of aerosols, water drops, ice particles, graupel and hail.

One sounding is processed for an average of 2–3 min, using Intel Core i5 processor and NVidia GeForce 840 M video card.

3   Using the Methods of Machine Learning

As a result of numerical experiments training data are obtained that consist of simulated cloud parameters. Our training data are hand-labelled. The next step is the feature selection. We need to set a certain moment of cloud evolution and select the most representative cloud parameters (features). It is decided to set the moment of the

time   height   name   value

1160.0,4050.0,velocity,13.067016

1160.0,4050.0,velocityU,9.343779

1160.0,4050.0,temperature,272.22715

1160.0,4050.0,deltaTemperature,2.2229649

1160.0,4050.0,relativeHumidity,1.0090644

1160.0,4050.0,vapor,0.0058138833

1160.0,4050.0,pressure,61652.788

1160.0,4050.0,density,0.7889759

1160.0,4050.0,aerosol,0

1160.0,4050.0,drop,0.002514255

1160.0,4050.0,ice,1.7051359E-0013

1160.0,4050.0,hailAndGrits,2.2671957E-0006

1160.0,4200.0,velocity,11.32452

1160.0,4200.0,velocityU,39.735083

1160.0,4200.0,temperature,270.61552

1160.0,4200.0,deltaTemperature,1.595726

1160.0,4200.0,relativeHumidity,1.0096239

1160.0,4200.0,vapor,0.0052634661

1160.0,4200.0,pressure,60491.81

1160.0,4200.0,density,0.77872896

Fig. 1. Fragment of the CSV file with cloud parameters.

maximum difference between the height of the upper and lower height of the cloud (cloud thickness). The value of the cloud thickness allows obtaining the most realistic estimation of convection intensity and consequently of the probability of a dangerous convective event. In addition to the time moment we set the altitude selection of cloud parameters on the level of maximum liquid water content. Then all the selected data are normalised.

All of the cloud parameters enumerated above are used as the features. Feature selection has been provided by using recursive feature elimination algorithm with automatic tuning of the number of features selected with cross-validation.

Prediction accuracy as a function of the number of the considered features is presented on Fig. 2.

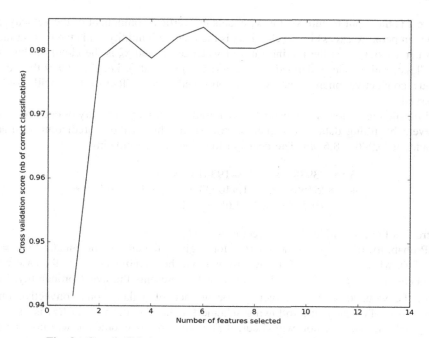

Fig. 2. Cross validation score versus number of the selected features.

Usage of the following 6 features appears to be optimal in our case: the vertical component of the velocity, temperature excess in the cloud, relative humidity, mixing ratio of water vapour, mixing ratio of water drops and cloud thickness.

The following three methods of machine learning are used: Support Vector Machine (SVM), Logistic Regression and Ridge Regression.

Cross validation method is used in order to estimate the quality of the implemented machine learning methods.

Support Vector Machines (SVM) belongs to the family of linear classifiers. It is one of the most popular methods of machine learning. The key idea of the method suggests transfer of the original vectors to a space of higher dimension and the search for the separating hyperplane with the maximum gap in this space. Two parallel hyperplanes are constructed on both sides of the hyperplane, dividing classes of investigated features. Separating hyperplane is the hyperplane that maximises the distance between the two parallel hyperplanes. The algorithm works on the assumption that the greater the difference or distance between the two parallel hyperplanes, the smaller the average error of the classifier.

The SVM with linear kernel is used. The accuracy of the prediction equals 0.977 (97.7 %). The resulting hyperplane

$$\begin{aligned} f(\vec{x}) = {}&2.33572643 * x_1 + 0.83159455 * x_2 \\ &+ 0.59938731 * x_3 + 0.7031227 * x_4 + 0.65584503 * x_5 \\ &+ 2.77717664 * x_6 + 1.31997495 \end{aligned} \qquad (1)$$

where $\vec{x}$ is the input feature vector, x_1 equals the vertical component of the velocity, x_2 is the temperature excess in the cloud, x_3 is the relative humidity, x_4 is the mixing ratio of water vapour, x_5 is the mixing ratio of water drops and x_6 is the cloud thickness.

Classification rule is defined as follows: $G(x) = \text{sign}(x)$, i.e. if $f(x) > 0$ then dangerous convective phenomena will be observed, and if $f(x) <= 0$ it will not be observed.

Logistic regression is a statistical model used to predict probability of occurrence of an event by fitting data to a logistic curve. Using this method, prediction accuracy amounts to 0.986 (98.6 %). The decision function is the following:

$$z = 3.88523988 * x_1 + 1.84193614 * x_2$$
$$+ 2.8254668 * x_3 + 1.94093366 * x_4 + 1.05776969 * x_5 \qquad (2)$$
$$+ 2.90365228 * x_6 + 3.68298532$$

where $\vec{x}$ is the same input feature vector as in (1).

Probability of the phenomenon (Pr) for a given feature vector equals: $\Pr\{y = 1|x\} = f(z)$, where $y = \{0, 1\}$, 0 is the case without the phenomenon, 1 is the case with the phenomenon, $f(z) = \frac{1}{1+e^{-z}}$. The function f(z) presents the event probability. The larger the value of f(z) the larger the occurrence of a dangerous convective phenomenon is. Typically, the following scheme for the classification of f(z) is used: if $f(z) \geq 0,5$ the phenomenon will occur, if $f(z) < 0,5$ the phenomenon will not occur.

Ridge regression addresses some of the problems of Ordinary Least Squares by imposing a penalty on the size of coefficients. The ridge coefficients minimize a penalized residual sum of squares: $\min_{\omega} \|X\omega - y\|_2^2 + \alpha\omega_2^2$. Here, ω is the weight of the predicate, $\alpha \geq 0$ is a complexity parameter that controls the amount of shrinkage: the larger the value of α the greater the amount of shrinkage and thus the coefficients become more robust to collinearity.

The accuracy of the prediction equals to 0.981 (98.1 %). The decision function f(z) is defined as follows:

$$f(\vec{x}) = 1.24574578 * x_1 - 0.19004467 * x_2$$
$$+ 1.3471659 * x_3 - 0.03611449 * x_4 + 0.65584503 * x_5 \qquad (3)$$
$$+ 0.74499218 * x_6 - 1.40857864$$

where $\vec{x}$ is the same input feature vector as in (1). Classification rule is defined as follows: $G(x) = \text{sign}(x)$, i.e. if $f(x) > 0$ then dangerous convective phenomena will be observed, and if $f(x) <= 0$ it will not be observed.

So, taking into account the results of machine learning, the following scheme of, for example, thunderstorm forecasting can be suggested.

1. Operational data of radiosonde sounding are used as an input for the cloud model.
2. The data are modified by using Ippolitov's formula and dry adiabatic temperature gradient.
3. Model calculation is performed and used features (selected cloud parameters) are obtained
4. Decision functions are calculated.

5. Conclusions of a thunderstorm probability are drawn if two of three methods provide the same result.

4 The Concept of a Specialized Multidimensional Database

In this paper an attempt is made to apply the advantages of a multidimensional approach for the type definition of dangerous convective phenomena, for example, to distinguish thunderstorm from heavy rainfall or hail using cloud characteristics obtained as a result of numerical simulation.

The numerical model that we used allows calculating vertical distributions and time evolution of dynamical and microphysical characteristics of clouds, such as vertical velocity, liquid water content, mixing ratios of water drops, ice crystals, hail and graupel particle. We can also calculate the total amount of precipitation on the ground and radar reflectivity. A study of the development of all these characteristics in space and time is of great interest from a scientific point of view. However, these characteristics cannot uniquely specify the type of dangerous convective phenomena, which is accompanied, or is a consequence of the development of the model cloud.

The model does not have the so-called "electrical" block, which allows simulating the development of the electrical field in the cloud and to fix the electrical discharge, the occurrence of which clearly indicates a thunderstorm. Such phenomena as hail or heavy rain can be determined by the size of the ice particles on the surface and the amount of precipitation fallen on it. However, as a rule, thunderstorms are accompanied by heavy rain and hail, so we are not able to determine clearly what kind of dangerous convective phenomena may occur under given initial and boundary conditions.

However, forecasting of the type of convective phenomena is crucial for the practical use of the model in operational practice. Meteorologists in the airport must know whether the thunderstorm will occur in their airport and on the route of the airplane.

Our multidimensional database is a 3-D cube, the axes of which represent the following dimensions: "Dates," "Phenomena," and "Cloud features." The dates of the specific cases when the specific dangerous convective phenomenon has been observed or has not been observed are represented along the "Dates" axis. The types of phenomena (storm, hail, heavy rain, with no events) are located along the "Phenomena" axis. Cloud features selected above are located along the corresponding third axis. The view of the cube is presented on Fig. 3.

We fill in our multidimensional data base with the help of integrated information system, which allow to integrate information about the dates and types of different convective phenomena, about vertical distributions of temperature and relative humidity observed on these dates and cloud features, obtained as a result of numerical experiments by using the cloud model with corresponding input parameters. The values of the calculated cloud features are filled in the cells of the cube at the intersection of corresponding axes.

Technical realization of the development of a multidimensional database, its filling and multidimensional analysis is carried out using PHP API (Application Program

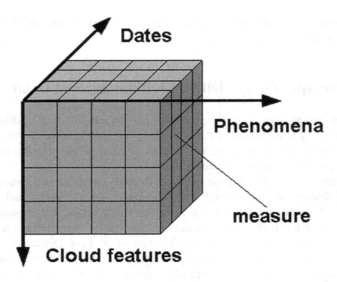

Fig. 3. 3-D cube

Interface written on PHP language). PHP API is provided by the developers of the Jedox Palo BI Suite system produced by a German company Jedox AG [16]. It has been chosen because Jedox Palo is an open source OLAP-system allowing becoming a client for any outer program or system (e.g. ERP or CRM).

It is possible to provide appropriate analyses when the cells are filled in. In particular, it is possible to define the range of cloud features which corresponds to the specific convective phenomenon.

Thus, in the future, using the model for operational forecasting, it will be possible to draw conclusions about the type of convective phenomena after analyzing the range of values of the calculated cloud parameters.

5   Conclusions

Machine learning and OLAP technologies are used for validation of 1.5 numerical model of a convective cloud.

Integrated information system is applied for gathering of specific meteorological data used as the model input parameters. 289 radiosonde soundings were collected in the place and at the time when dangerous convective phenomena were recorder. 326 soundings were collected in cases when no dangerous convective phenomena were observed.

Automation of the model running has been implemented in order to provide consistent processing of the set of soundings and to obtain output information in the required format.

Correction of the initial input data has been provided by setting dry adiabatic temperature gradient in the boundary layer and using Ippolitov's formula in order to define the height of the condensation level. Numerical tests show that after correction the model reproduces the convection development adequately in 91 % of cases, and the convection absence in 98 % of cases.

Three methods of machine learning: Support Vector Machine, Logistic Regression and Ridge Regression have been used for distinguishing the cases with dangerous convective phenomenon and without it. Only the thunderstorm is taken into account as a phenomenon at present stage of the research.

Feature selection for thunderstorm forecasting based on invariant statistical tests has been provided. Six features have been selected: the vertical component of the velocity, temperature excess inside the cloud, relative humidity, mixing ratio of water vapour, mixing ratio of water drops and thickness of the cloud.

Decision functions have been obtained for each method. Accuracy of the prediction amounts to 97.7 %, 98.6 % and 98.1 % for Support Vector Machine, Logistic Regression and Ridge Regression correspondingly.

The concept of multidimensional data base intended for distinguishing the types of the phenomena (thunderstorm, heavy rainfall and light rain) has been developed. Technical realisation is carried out using PHP API of the Jedox Palo BI Suite system produced by the German company Jedox AG.

Acknowledgment. This research was sponsored by the Russian Foundation for Basic Research under the projects: № 16-07-01113, № 16-07-00886, № 16-07-01111 and the Contract № 02. G25.31.0149 dated 01.12.2015 (Board of Education of Russia).

References

1. Petrov, D.A., Stankova, E.N.: Use of consolidation technology for meteorological data processing. In: Murgante, B., et al. (eds.) ICCSA 2014, Part I. LNCS, vol. 8579, pp. 440–451. Springer, Heidelberg (2014)
2. Petrov, D.A., Stankova, E.N.: Integrated information system for verification of the models of convective clouds. In: Gervasi, O., et al. (eds.) ICCSA 2015. LNCS, vol. 9158, pp. 321–330. Springer, Heidelberg (2015). doi:10.1007/978-3-319-21410-8_25
3. Stankova, E.N., Petrov, D.A.: Complex information system for organization of the input data of models of convective clouds. Appl. Math. Comput. Sci. Cont. Process. (3), 83–95 (2015) (in Russian). Vestnik of Saint-Petersburg University Series 10
4. Hastie, T., Tibshirani, R., Friedman, J.: The Elements of Statistical Learning, 2nd edn. Springer, Heidelberg (2009). http://statweb.stanford.edu/~tibs/ElemStatLearn/
5. Mitchell, T.: Machine Learning. Springer, Heidelberg (2009)
6. Raba, N.O. Stankova, E.N.: Research of influence of compensating descending flow on cloud's life cycle by means of 1.5-dimensional model with 2 cylinders. In: Proceedings of MGO, vol. 559, pp. 192–209 (2009) (in Russian)
7. Raba, N.O., Stankova, E.N., Ampilova, N.: On investigation of parallelization effectiveness with the help of multi-core processors. Procedia Comput. Sci. 1(1), 2757–2762 (2010)
8. Raba, N., Stankova, E.: On the possibilities of multi-core processor use for real-time forecast of dangerous convective phenomena. In: Taniar, D., Gervasi, O., Murgante, B., Pardede, E., Apduhan, B.O. (eds.) ICCSA 2010, Part II. LNCS, vol. 6017, pp. 130–138. Springer, Heidelberg (2010). ISBN: 978-3-642-12164-7

9. Raba, N.O., Stankova, E.N.: On the problem of numerical modeling of dangerous convective phenomena: possibilities of real-time forecast with the help of multi-core processors. In: Murgante, B., Gervasi, O., Iglesias, A., Taniar, D., Apduhan, B.O. (eds.) ICCSA 2011, Part V. LNCS, vol. 6786, pp. 633–642. Springer, Heidelberg (2011). ISSN: 0302-9743

10. Raba, N.O., Stankova, E.N.: On the effectiveness of using the GPU for numerical solution of stochastic collection equation. In: Murgante, B., Misra, S., Carlini, M., Torre, C.M., Nguyen, H.-Q., Taniar, D., Apduhan, B.O., Gervasi, O. (eds.) ICCSA 2013, Part V. LNCS, vol. 7975, pp. 248–258. Springer, Heidelberg (2013). doi:10.1007/978-3-642-39640-3_18

11. Date, C.J., Darwen, H.: Foundation for Future Database Systems: The Third Manifesto, 2nd edn., pp. 223–238. Addison Wesley Professional, Reading (2000)

12. Codd, E.F.: Providing OLAP (on-line analytical processing) to user-analysts: an IT mandate. Technical Report, E.F. Codd and Associates (1993)

13. Agrawal, R., Gupta, A., Sarawagi, S.: Modeling multi-dimensional databases. IBM Research Report, IBM Almaden Research Center, September 1995

14. Gyssens, M., Lakshmanan, L.V.S.: A foundation for multi-dimensional databases. Technical Report, Concordia University and University of Limburg, February 1997

15. Matveev, L.T.: Physics of Atmosphere. Saint-Petersburg, Hidrometeoizdat, p. 779 (in Russian)

16. Jedox. www.jedox.com/

Evaluation of the Financial Feasibility for Private Operators in Urban Redevelopment and Social Housing Investments

Pierluigi Morano[✉] and Francesco Tajani

Science Department of Civil Engineering and Architecture,
Polytechnic of Bari, Bari, Italy
pierluigi.morano@poliba.it, francescotajani@yahoo.it

Abstract. In this work an evaluation model to support Public Administration decisions in planning urban strategies that aim to involve private investors has been developed. The model allows to define the maximum amount of subsidized housing to be realized by the private investor and the administered selling price to be applied. This model has been developed translating in the field of urban planning, criteria and tools borrowed from the marginal economic theory. The results obtained by the application to a real case study confirm the potentialities and the user-friendly configuration of the model.

Keywords: Break-even analysis · Social housing · Urban redevelopment · Financial feasibility

1 Introduction

In Italy, the number of applicants for social housing has increased from 600,000 in 2008 to currently about 650,000, whereas the social housing production – related to an unchanged total expenditure for social housing – has suffered a slowdown since 2009 [11]. Indeed, Italy is the only State Member that has not registered an improvement in the social housing sector, due to the high number of units characterized by lack of basic amenities, further worsened by overcrowding issue. Despite new plans for the protection of the weaker members of society and the requalification of the existing housing patrimony [15], there are 2,5 million individuals unable to independently satisfy their housing needs [4].

In recent years, aiming to improve housing offers' methods and typologies for urban requalification [1, 2], there has been a development of several tools focused on involving, through different types of partnerships [3, 12, 13], private abilities and resources. Following the article 26 of L. 164/2014, an important directive has been issued, that aims to enhance unused public assets considering also social housing opportunities. According to the first subparagraph of the article 26, aiming to promote initiatives for enhancing unused public buildings to support economic and social development, a negotiated agreement between Public Administration and private investor is possible to consider - with regard to recovery actions of unused public buildings – as town planning amendment. In this way, the administrative procedure

© Springer International Publishing Switzerland 2016
O. Gervasi et al. (Eds.): ICCSA 2016, Part III, LNCS 9788, pp. 473–482, 2016.
DOI: 10.1007/978-3-319-42111-7_37

related to changes of functions for public buildings that need to be enhanced and at the same time to reduce operational time and risk, is possible to streamline. In particular, the aforementioned comma states that priority should be given to the recovery of public buildings intended, as a whole or partially, to develop new public housing units.

Although social housing production and urban renewal constitute both themes of primary importance in the current economic situation, hardly ever Public Administrations have appropriate skills to rationally set the best implementing modality [20–22]. As a consequence, the planned strategies almost always fail or are not duly taken into account by private investors, due to the wrong analysis of financial feasibility or even for the total absence of any kind of evaluation [14, 18].

2 Aims

With regard to the depicted scenario, the aim of this research is to develop an evaluation model to support Public Administration decisions in planning urban renewal initiatives involving private investors. The model allows to define the maximum amount of subsidized housing – as a percentage of the total housing gross floor surface that has been planned – to be realized by the private investor and the respective administered selling price, able to ensure the financial feasibility of the initiative. In this way, it is possible to facilitate the cooperation between collective instances of urban reorganization, social housing implementation and private needs for financial convenience [7, 9].

The model has been developed using the *Break-Even Analysis* (BEA). BEA is a decision support tool used for business planning to validate short-medium term choices. In this way, it is possible to obtain a flexible tool, user-friendly, implementable by collecting a few information that are easily traceable during the preliminary planning phases of the initiatives and usable in any territorial context.

In this paper, BEA is applied with an "instantaneous approach", without considering time effects. This means that costs/revenues of the procedure can be considered synchronously, with regard to the moment of the evaluation. Consequently, following this hypothesis, when BEA needs to be implemented with some financial items, influenced by time variable, they need to be evaluated as lump amounts. This is the case of interests on the capital borrowed from the credit institute to the private investor. This assumption is coherent with the aim of the work, that is the definition of a simple user-friendly tool. In this way the model can be used by inexperienced users, simply providing the amount of different variables involved within the process for a short-medium period.

The research is structured as follows. In Sect. 3 the basis of BEA are introduced using equations and assumptions. In Sect. 4 the evaluation model is presented. In Sect. 5 BEA is implemented to a real case concerning the urban renewal of an area in disuse, located in southern Italy. In Sect. 6 the conclusions of the work are drawn.

3   Outlines of BEA

In the international literature, BEA has been studied and applied since more than 50 years, especially in the Anglo-Saxon territory [5, 8, 16, 17].

In the evaluation of an investment, BEA considers only the monetary aspects of the initiatives in the *short term*, because it provides the order of magnitude of the variables examined over a period of a short duration [19]. Therefore, it operates by limiting the aspects of the monitored investment, and the analysis is focused: on the *total* costs (Ct), which are disjointed in the components of the *fixed* costs (Cf), that are cost items defined without considering the amount of the product to be realized (e.g. acquisition of land, its environmental remediation and restoration, the urbanization and the infrastructure for mobility, the recovery of existing buildings, the establishment of spaces and equipment of collective interest) and *variable* costs (Cv), that are cost items defined considering the amount of the product to be realized and sold within the initiative (e.g. energy costs, cost of raw materials directly used in the production, costs for the distribution and sale of the products, workers' salaries based on flexible contracts); on *total* revenues (Rt); on the quantity (q) of the goods or services that are expected to be produce and sell; on the financial feasibility of the initiative, computed in terms of *total* profit (Pt).

These are elements linked together in the mathematical relationship that expresses the total profit of the initiative:

$$Pt = Rt - Ct = Rt - (Cf + Cv) \tag{1}$$

According to microeconomic laws, the aforementioned parameters can vary depending on the amount of product (q) through a non-linear relation, coherently with the law of diminishing returns. For this reason, it is possible to introduce some assumptions to simplify BEA application: (i) costs and revenues are produced instantaneously, which means that the time dimension is not considered in the evaluation. In practice, it is as if the operator asserts expenses and realizes the value of the products and services at the same time; (ii) the total production costs have a linear trend. The variable costs can calculated as the product of the unit variable costs (Cvu) and the amount to be realized ($Cv = Cvu \cdot q$); (iii) the total revenues have equally a linear trend, so that they must be defined by the product of their unit price and the quantity to produce and sell ($Rt = pu \cdot q$).

Substituting into Eq. (1), the algebraic expressions of the variable costs and total revenues arising from the working hypothesis:

$$Pt = pu \cdot q - Cvu \cdot q - Cf \tag{2}$$

By determining the quantity (q) to be produced and sold and imposing the zero-total profit condition ($Pt = 0$), that – by definition – should be verified corresponding to the break-even point, it is possible to obtain:

$$Cf + (pu - Cvu) \cdot q = 0 \qquad (3)$$

By solving Eq. (3), it is possible to define q^*:

$$q^* = \frac{Cf}{pu - Cvu} \qquad (4)$$

This relation links the main financial variables of the investment and allows to calculate through a direct and rapid method the break-even quantity q^*, knowing fixed costs (Cf), selling price per unit (pu), variable production cost per unit (Cvu) of the initiative.

Within the fixed costs items it is also important to include the "normal" profit of the private investor. This profit is the expected compensation for the generic investor – in a specific area and for a specific typology of initiative – considering his activities of production's coordination and assumption of the risk investment. This means that q^* defines the minimum amount for the financial convenience, ensuring also the normal profit to the private investor. Amounts to be produced or sold that are bigger than the amount of q^*, will produce an extra-profit.

In order to check the feasibility of the initiative, the quantity q^* needs to respect all technical, normative and market restrictions. In fact, there will be a convenience for the private investor only if the break-even point is lower than the maximum threshold estimated. If it is possible to gather in the balance items of the initiative - in the form of measures of the fixed/variable cost and revenues - the financial "translations" of the restrictions, project choices, negotiated agreements for the solutions to be realized, and also considering the amount of public works to be realized by the private investor, the break-even analysis will be able to define the amount of q^* as building products to be realized and sold. This amount will ensure the balance between the several conveniences for the whole set of operators involved within the initiative.

4 The Model

With reference to urban renewal projects to be realized through the participation of the private investors, the proposed model has been developed to support territorial transformations for which Public Administration – taking into account also the financial convenience of the private – decides to maximize the percentage of subsidized housing to be realized by the private investor, that will sell the units respecting administered selling prices.

In fact, the share of social housing to be realized by the private investor reduces his total incomes. The selling price per unit of subsidized housing (p_{sh}) needs to be lower than the selling price per unit of housing for the free market (p_m). In Italy, criteria for the definition of administered prices are established by law, considering the selling prices related to the costs of construction of the housing units, and using a direct proportionality defined at regional level (L. No. 457/1978). However, it is not always verified that social housing prices are lower than housing prices in the free market: considering areas affected by depressed property market, the application of this principle could generate administrated selling prices that are incompatible with the local scenario. For this reason, the proposed model considers a multiplying coefficient (w), lower than 1, that compares administered prices with the free market, considering prices which normally are defined within the area of study, following the relation:

$$p_{sh} = w \cdot p_m \qquad 0 \le w < 1 \qquad (5)$$

At this point, starting from Eq. (2) it is possible to disaggregate the price per unit (pu) and variable cost per unit (Cvu), considering the different functions that contribute to define the project. Taking into account the Eq. (5) it is possible to write:

$$Pt = \left(\frac{p_m \cdot q_m + w \cdot p_m \cdot q_{sh} + p_c \cdot q_c}{q}\right) \cdot q - \left(\frac{Cvu_m \cdot q_m + Cvu_{sh} \cdot q_{sh} + Cvu_c \cdot q_c}{q}\right) \cdot q - Cf \qquad (6)$$

The meanings of the elements of Eqs. (5) and (6) are summarized in Table 1.

Table 1. Parameters of the model

Cf	fixed costs of the transformation [€]
Rt	total revenues of the transformation [€]
Pt	total profit (extra-profit) of the private investor [€]
q	total gross floor surface (GFS) of the project [m²]
q_m	GFS of housing sold in the free market [m²]
q_{sh}	GFS of subsidized housing [m²]
q_r	total residential GFS [m²]
q_c	GFS for not residential functions (e.g. commercial) [m²]
p_m	selling price per unit for housing in the free market [€/m²]
p_{sh}	selling price per unit for subsidized housing [€/m²]
p_c	selling price per unit for non residential functions [€/m²]
w	coefficient for the definition of the selling price per unit for subsidized housing
Cvu_m	variable cost per unit for housing in the free market [€/m²]
Cvu_{sh}	variable cost per unit for subsidized housing [€/m²]
Cvu_c	variable cost per unit for non residential functions [€/m²]

Considering also that the aim of the model is to define the amount of subsidized housing (q_{sh}) and the percentage (w) of deduction for the selling price per unit of housing in the free market able to nullify the total profit (Pt), Eq. (6) must be equal to zero:

$$P_m \cdot (q_m + w \cdot q_{sh}) + p_c \cdot q_c - Cvu_m \cdot q_m - Cvu_{sh} \cdot q_{sh} - Cvu_c \cdot q_c - Cf = 0 \qquad (7)$$

Specifying with q_r $(= q_m + q_{sh})$ the total gross floor area to be allocated to the housing units, it is possible to write:

$$p_m \cdot [q_r - (1 - w) \cdot q_{sh}] + p_c \cdot q_c - Cvu_m \cdot (q_r - q_{sh}) - Cvu_{sh} \cdot q_{sh} - Cvu_c \cdot q_c - Cf = 0 \quad (8)$$

From Eq. (8), isolating the amount of subsidized housing (q_{sh}) that zeros the total profit:

$$q_{sh} \frac{Cf + (Cvu_c - p_c) \cdot q_c + (Cvu_m - p_m) \cdot q_r}{[Cvu_m - Cvu_{sh} - (1 - w) \cdot p_m]} \quad (9)$$

Knowing dimensional data of the initiative, considering the distribution – defined through demand analysis - between the total gross floor area and the non residential functions, gathering registered fixed costs from the market, cost items that contribute to define variable cost per unit and selling prices for the different functions of the project's elements, Eq. (9) has two variables, i.e. q_{sh} e w. Actually, the variable cost per unit for the realization of subsidized housing (Cvu_{sh}), depends on the percentage of deduction (w) for the selling price per unit of housing in the free market: this cost item considers the normal profit per unit of the private investor, that can vary depending on the administered selling price per unit defined for subsidized housing.

If it is possible to prefigure several alternatives for the public and private actors involved within the urban renewal investment, Eq. (9) allows to define combinations of q_{sh} e w able to ensure the financial convenience of the initiative.

It is important to underpin how the empirical evidence shows that if w assumes values close to 1 – meaning the possibility to define social housing prices close to the ones of the free market – the amount of subsidized housing to be realized by the private investor increases; vice versa, if w assumes values close to 0 – meaning that the administrated price is really low, and the private can only gives the housing units for free to the Public Administration - the amount of subsidized housing to be realized by the private investor decrease considering the restriction of financial convenience.

5 Application of the Model

The illustrated model has been applied to an urban renewal investment of an area located in southern Italy. The intervention area is owned by the Public Administration, well served by infrastructures, extended for 11,200 m^2, located in an expansion area characterized by five levels buildings with commercial functions in the ground floors and residential units for the others.

The strong demand for affordable social housing expressed by local people induced the Public Administration to arrange a transformation project for the area. It consists of the realization of nine buildings with five levels above ground and one basement level to be realized by a private investor. In particular, a part of the housing accommodations should be sold in free market regime, another part with fixed price values, whereas ground floor commercial units and appurtenant basement parking should be sold on the free market.

Considering that the model borrows the BEA operative process, the organization of costs/revenues items within the financial balance in "fixed" and "variable" has been developed in Table 2.

Table 2. Organization of the private investor's balance items in "fixed" and "variable"

Fixed costs	
Land purchase	250,627 €
Taxes and notary's fees	27,569 €
Local planning fees for the commercial share	104,479 €
Normal profit of the investor (commercial)	962,050 €
Commercial construction costs	1,924,100 €
Realization of the green area	580,014 €
Technical and general expenses (commercial and green area)	200,329 €
Financial charges	100,322 €
Total	**4,149,490 €**
Variable unit costs	
Subsidized residential	
Local planning fees	13.80 €/m^2
Normal profit of the investor	$f(w)$
Technical and general expenses	72 €/m^2
Financial charges	32.04 €/m^2
Construction costs	900 €/m^2
Total	$f(w)$
Free market residential	
Local planning fees	45.80 €/m^2
Normal profit of the investor	440 €/m^2
Technical and general expenses	88 €/m^2
Financial charges	40.10 €/m^2
Construction costs	1,100 €/m^2
Total	**1,713.90 €/m^2**
Unit revenue	
Subsized residential sale	$f(w)$
Free market residential sale	**2,200 €/m^2**
Commercial sale	**2,500 €/m^2**

For the case study, within the fixed items there are land purchase, taxes and notary's fees, costs for the realization of the green area and costs items related to the realization of the commercial share (local planning and construction fees, normal profit of the private investor, construction costs).

Among *fixed* costs there are also technical and general expenditures related to the realization of the commercial share and of the green area. Furthermore, there are also finance charges related to the aforementioned items.

Within the *fixed* revenues there are the incomes generated by the sale of the commercial share, equal to 4,810,250 €. They have been evaluated applying to the surfaces with commercial purpose a price per unit defined through a market survey, equal to 2,500 €/m^2.

Among *variable costs* there are the sums related to the distribution of the total residential GFS in the subsidized share (q_{sh}) and in the share for the free market (q_m), local planning and construction fees for the subsidized housing and for the ones on the free market, construction costs, the amount of normal profit of the investor, technical and general expenses and finance charges related to the residential share. It is important to notice that the normal profit of the private investor related to the sale of subsidized residential GFS depends on w for the definition of the administered selling price per unit: this means that the realization cost per unit for subsidized housing units depends, in its turn, on the value that will be defined for the coefficient w. The variable cost per unit for residential share in the free market is equal to 1,713.90 €/m^2.

Within the *variable* revenues there are the incomes generated by the sale of subsidized housing GFS and GFS of the housing in the free market.

In Table 3, considering increases in the value of w equal to 0.10, the amount of subsidized housing (q_{sh}) defined through the Eq. (9) and the amount of housing in the free market (q_m) are reported. The latter is calculated through the difference between total residential GFS defined within the investment, equal to 10,450 m^2, and the amount of subsidized housing related to each w value considered. The value $w = 0.54$ – meaning an administered price per unit slightly higher than the half of the price per unit of the residential share in the free market – represents the maximum threshold for the case study. This is the scenario in which all the planned residential share can be intended for social housing.

Table 3. Outputs of the application of the model

w	q_{sh} [m^2]	q_m [m^2]
0.00	3,817	6,633
0.10	4,323	6,127
0.20	4,983	5,467
0.30	5,882	4,568
0.40	7,176	3,274
0.50	9,200	1,250
0.54	10,450	0.000

It is important to underpin how the empirical evidence shows that if w assumes values close to 1 – meaning the possibility to define social housing prices close to the ones of the free market – the amount of subsidized housing to be realized by the private investor increases; vice versa, if w assumes values close to 0 – meaning that the administrated price is really low, and the private can only gives the housing units for free to the Public Administration - the amount of subsidized housing to be realized by the private investor decreases considering the restriction of the financial convenience.

6 Conclusions

Urban requalification projects are the model's range of application. These kind of projects needs first of all a preview about housing to be sold with controlled price, because of the actual socio-economic conjuncture that allowed an increase of subjects unable to access to the free housing market. Moreover, with urban requalification projects are essential both the private investor's sources involvement and competences, even if he is interested to participate only in investment with verified restriction of financial feasibility.

The model is firstly composed of a procedure that borrow main logical features of BEA to later calibrate combinations of two variables, "price" and "share" of social housing, on total, that guarantee to the private investor the initiative's balance.

The logical and functional relations of the developed model allow to easily and mutually link the technical and financial variables of the initiative, underlining interconnections and critical aspects.

The application of the model to a tangible case has highlighted its adaptability to specific territorial conditions because of its simple structure, rationalizing the process of decision making [6].

The possibility to retrace operations legitimates the management of both private and public actors involved, with positive consequences about transparency and decision effectiveness.

Finally, stability and flexibility elements, whose introduction in the planning of the investments is allowed by the model, amortize unruliness and hesitations caused not only by property market changes [10] but also by the complex nature of urban requalification initiatives.

References

1. Attardi, R., Bonifazi, A., Torre, C.M.: Evaluating sustainability and democracy in the development of industrial port cities: some Italian cases. Sustainability 4(11), 3042–3065 (2012)
2. Attardi, R., De Rosa, F., Di Palma, M.: From visual features to shared future visions for Naples 2050. Appl. Spat. Anal. Policy 8(3), 249–271 (2015)
3. Calabrò, F., Della Spina, L.: The public-private partnerships in buildings regeneration: a model appraisal of the benefits and for land value capture. Adv. Mater. Res. 931, 555–559 (2014)
4. Censis. http://www.censis.it/1
5. Conine Jr., T.E.: A pedagogical note on cash break-even analysis. J. Bus. Finance Acc. 14(3), 437–441 (1987)
6. D'Alpaos, C.: The value of flexibility to switch between water supply sources. Appl. Math. Sci. 6(128), 6381–6401 (2012)

7. D'Alpaos, C., Marella, G.: Urban planning and option values. Appl. Math. Sci. **8**(157–160), 7845–7864 (2014)
8. Dean, J.: Cooperative research in cost-price relationships. Account. Rev. **14**(2), 181–184 (1969)
9. Del Giudice, V., De Paola, P., Torrieri, F.: An integrated choice model for the evaluation of urban sustainable renewal scenarios. Adv. Mater. Res. **1030–1032**, 2399–2406 (2014)
10. Del Giudice, V., Manganelli, B., De Paola, P.: Spline smoothing for estimating hedonic housing price models. In: Gervasi, O., Murgante, B., Misra, S., Gavrilova, M.L., Rocha, A. M.A.C., Torre, C., Taniar, D., Apduhan, B.O. (eds.) ICCSA 2015. LNCS, vol. 9157, pp. 210–219. Springer, Heidelberg (2015)
11. European Central Bank, Eurosystem, Statistical Data Warehouse. https://sdw.ecb.europa.eu/home.do
12. Gabrielli, L., Copiello, S.: Marginal costs and benefits in building energy retrofitting transaction. In: Hamburg International Conference on Sustainable Built Environment Strategies (SBE16), pp. 836–845. Zebau (2015)
13. Guarini, M.R., Battisti, F.: Evaluation and management of land-development processes based on the public-private partnership. Adv. Mater. Res. **869**, 154–161 (2013)
14. Guarini, M.R., Battisti, F.: Social housing and redevelopment of building complexes on brownfield sites: the financial sustainability of residential projects for vulnerable social groups. Adv. Mater. Res. **869**, 3–13 (2013)
15. Housing Europe, The state of housing in the EU 2015. http://www.housingeurope.eu/resource-468/the-state-of-housing-in-the-eu-2015
16. Ingraham, H.A.: Elementary presentation of volume, cost and profit relationships. Acc. Rev. **26**(3), 414–416 (1951)
17. Kee, R.C.: Implementing cost-volume profit analysis using an activity based costing system. Adv. Manag. Acc. **10**, 77–94 (2001)
18. Las Casas, G., Lombardo, S., Murgante, B., Pontrandolfi, P., Scorza, F.: Open data for territorial specialization assessment territorial specialization in attracting local development funds: an assessment procedure based on open data and open tools. TeMA. J. Land Use, Mobility Environ. 581–595 (2014)
19. Morano, P., Tajani, F.: Break Even Analysis for the financial verification of urban regeneration projects. Appl. Mech. Mater. **438**, 1830–1835 (2013)
20. Nesticò, A., Pipolo, O.: A protocol for sustainable building interventions: financial analysis and environmental effects. Int. J. Bus. Intell. Data Min. **10**(3), 199–212 (2015)
21. Rosato, P., Alberini, A., Zanatta, V., Breil, M.: Redeveloping derelict and underused historic city areas: evidence from a survey of real estate developers. J. Environ. Plann. Manage. **53**(2), 257–281 (2010)
22. Scorza, F., Casas, G.L.: Territorial specialization in attracting local development funds: an assessment procedure based on open data and open tools. In: Murgante, B., et al. (eds.) ICCSA 2014, Part II. LNCS, vol. 8580, pp. 750–757. Springer, Heidelberg (2014)

The ESPRESSO - Project – A European Approach for Smart City Standards

Jan-Philipp Exner[(✉)]

Department of CAD & Planning Methods in Urban Planning
and Architecture (CPE), University of Kaiserslautern,
Pfaffenbergstr. 95, 67663 Kaiserslautern, Germany
exner@rhrk.uni-kl.de

Abstract. Smart Cities are a trend topic in urban planning as well as in the ICT sector. The cities integrate physical, digital and human systems to deliver a sustainable, prosperous and inclusive future for its citizens. Many of these innovative solutions will be based on technological complexity, as well as the complexity of the various sectoral services involved within a Smart City, and require a system approach to standardization. Such an approach must promote the greatest possible reuse of existing open standards to accelerate the Smart City deployment. In an effort to tackle this issue, the Horizon 2020-project ESPRESSO (systEmic standardisation apPRoach to Empower Smart citieS and cOmmunities) will focus on the development of a conceptual Smart City Information Framework based on open standards within Europe. A further goal of ESPRESSO will be to envisage the impact of those technologies for urban planning and in societal terms. The partner cities will be engaged to analyze how their services can be improved and improved through large-scale use of standards. This will be done by analyzing, the downstream changes that the new scenarios enabled by large-scale interoperability and how they can be integrated in a future Smart City. Based on a detailed requirements-engineering campaign executed in close cooperation with cities, standardization organizations, administrative bodies, and private industry, the project will identify open standards matching the elicited requirements and will establish a baseline for interoperability between the various sectoral data sources and the Smart City enterprise application platform. In a comprehensive set of coordination, support and networking activities, the project will engage a very large number of stakeholders, such as Smart Cities (both existing and those with aspirations), European Standardization Organizations (ESOs), National Standardization Bodies (NSBs), Standards Development Organizations (SDOs), public administrations, industries, SMEs, and other institutions.

Keywords: Urban planning · Smart cities · Standards · ICT-Framework

1 Introduction

The transformation of urban areas over the next decade, including the way we live, work and use energy, transportation and other city resources and services will proceed with a significant change thanks to a range of innovative 'Smart City' solutions. A Smart City integrates physical, digital and human systems to deliver a sustainable,

© Springer International Publishing Switzerland 2016
O. Gervasi et al. (Eds.): ICCSA 2016, Part III, LNCS 9788, pp. 483–490, 2016.
DOI: 10.1007/978-3-319-42111-7_38

prosperous and inclusive future for its citizens. Many of these innovative solutions will be based on sophisticated information and communication technologies which are connected. However, technological complexity, as well as the complexity of the various sectoral services involved within a Smart City, are inducing compley requirements for a system approach to standardization. Such an approach must promote the greatest possible reuse of existing open standards to accelerate the Smart City deployment and exploit the enormous potential deriving from the use of disparate interoperable technologies and from reuse of interoperable applications and services among cities. To address this issue, ESPRESSO funded by the Horizon 2020- program by the European Commission as Coordination and Support action will focus on the concept of a European approach for the development of a conceptual Smart City Information Framework based on open standards.

2 Theoretical Framework

An intense discussion regarding the topic of Smart Cities is seen in urban planning though there is no sharp definition from a scientific point of view. An embracing explanation is a city, in which "ICT is merged with traditional infrastructures, coordinated and integrated using new digital technologies. These technologies establish the functions of the city and also provide ways in which citizen groups, governments, businesses, and various of agencies who have an interest in generating more efficient and equitable systems can interact in augmenting their understanding of the city and also providing essential engagement in the design and planning process" (Batty 2012, p. 492). The greatest potential in the use of networked and often centralized ICT solutions in urban areas lies in the context of increased efficiency through innovative technologies (less energy consumption, lower emissions, less CO_2 pollution, etc.). This is often promoted as a contribution for the urban quality of life. In addition to that, a common and open urban information platform to share the gathered data is very promising for companies and citizens with an ideal-typical vision of a central urban monitoring and simulation system in real-time. Due to this, the Smart City topic found its ways on the agendas of big corporations like IBM, Cisco Systems, Siemens, Accenture, Ferrovial and ABB. They are setting their sights on the urban market and are foreseeing a multi-billion dollar market (Ratti and Townsend 2011). Especially the connection to new trend topics such as the Internet of Things (IoT) and Big Data is seen as very promising for those companies. (Greenfield 2013; Kitchin 2014; Townsend 2013) It is considered as a big future business field in the ICT-sector for developing tools, which could improve the competitiveness and the quality of life for the citizens. Besides all of this technological potentials, it is important to understand, that, "a Smart City is something more than 'just' a digital or an intelligent city, where the attention is mainly drawn on the ICT components, as enabling connection and exchange of data and information within an urban environment" (Murgante and Borruso 2013, p. 630). In addition to that, considered from a holistic perspective, it is important also to understand a Smart City embracing from a technological, institutional and also social perspective (Exner 2015).

Considered from the technological perspective, the connection of smart cities & standards is obvious and they should foster interoperability within the various entities of a city. Multiple international studies emphasize that standards can increase productivity and innovation and "will provide the foundation for long term advances in the way software is built, bought and deployed". The 2011 update on the DIN study on "The Economic Benefits of Standardization" (Blind et al. 2011) quantifies the contribution of standards to the growth rate in each country. This rate is equivalent to 0.9 % in Germany, 0.8 % in France and Australia for example. ESPRESSO pursues an integrative approach that understands Smart Cities and Communities as a system of interlinked processes, components, workflows, legal and administrative constraints, or organizational guidelines. Data is provided by many sectoral and heterogeneous systems that need to be made interoperable in order to enable sustainable and economically powerful data integration and processing. ESPRESSO considers open standards as a prerequisite for any such system-oriented approach. It is the aim of the project to reflect all of these aspects in the structure of both its work packages and its consortium. The latter consists of standardization organizations, private industry (both large integrators and SMEs), and the cities as final users and customers. The consortium also links to strong user engagements, and research-oriented organizations to ensure embedding of standardization and reference architecture concepts in applied research studies in the field of urban planning within Europe. A further aim is to cross-fertilize and cooperate with on-going EU-projects in the field of Smart Cities and reuse existing networks and alliances to avoid replication of efforts.

3 Project Approach

Within the European Horizon 2020-project, the consortium tries to tackle these issues with 16 partners that provide an excellent combination of the necessary competences for achieving the project's objectives. ESPRESSO has an assembled and interdisciplinary team, consisting of Smart City cities, large-scale integrators, governmental owned organizations, SDOs and industry consortia, SME's, and applied research organizations. This group offers a unique combination of crosscutting skills and experiences suited to the concept of Smart City and Communities and standardization. In order to ensure social acceptance of developed solutions, ESPRESSO sets up a stakeholder communication network that ensures an early dialogue between standards development organizations, technology providers, and technology consumers (cities and citizens as end users) to avoid a mismatch between the design of technology solutions and cities' and citizens' needs. ESPRESSO ensures inclusion of end-users in terms of planning; design and knowledge transfer along the lifetime of the project. It is based on the assumption that cities plan to develop and provide a city platform on which most city applications and services will run. This platform will be the main IT backbone that will vertebrate many existing sectoral systems (e.g. energy efficient buildings, smart grid, intelligent transport systems, eHealth systems) and many new applications and systems specifically designed for Smart Cities and running on the city platform. Similar ambitions and efforts to develop Smart City platforms have been reported from especially Asia with China, Japan, Singapore and South Korea for example. In order to address these issues, the upcoming figure explains the approach of the project (Fig. 1).

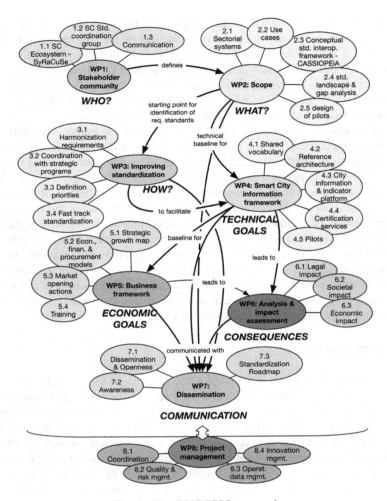

Fig. 1. The ESPRESSO approach

A lot of standardization efforts and results are already available in 'vertical' domains like eHealth, Building Energy Management Systems or Building Information Systems, Smart Grid, or Intelligent Transport Systems. These efforts should not be redone or even revised for the case of Smart City. The systems that are already available today provide most of what is required. Nevertheless, ESPRESSO acknowledges that in some areas (such as Internet of Things - IoT) standardization efforts are on-going and may extend the current standardization base relevant for Smart City. The direct involvement of Smart Cities and other relevant stakeholders from the beginning is to ensure reliable and real-world-need-driven development of requirements for open standards for Smart City with the goal to bring them to the market as efficiently as possible. It is also the aim to integrate business perspectives and relevant means, such as the development of a strategic growth map, which traces a long-term strategy for sighting of the effect of standardization in the Smart City domain. The upcoming work-packages show goals of the project:

- WP 1: First goal is to identify global stakeholders (industry, SDOs, etc.) and geographical clusters in the domain of Smart City (including "lighthouse" initiatives and their potential replicators and Smart City innovation zones) published as an online Interactive Atlas of Smart City and relevant stakeholders. To promote capacity building among the stakeholders in order to create a large eco-system (called the SmaCStak network), reaching out to hundreds of stakeholders, gathering a community, around ESPRESSO, made of local European and global stakeholders from both public and private sectors.
- WP 2: The second work-package helps to provide a consistent and shared definition of what a Smart City is today in terms of sectoral services and how this may evolve. Also mapping existing standards from various SDOs on top of the conceptual standards framework is part of this WP as well as to carry out a comprehensive SWOT analysis related to future standardization landscape. To define all aspects related to later interoperability, testbeds will be identified.
- WP 3: This work package uses the results from WP2 to identify standardization priorities based on a standardization criteria matrix and develops in close cooperation with CEN, CENELEC as well as international standardization organizations fast track recommendations that will be handed over to the corresponding SDOs for future consideration and implementation. In detail, WP3 has the objectives to analyze and document overlapping and subsequent harmonization potential of standards across different SDOs, Coordination requirements on new standards or components between European bodies CEN-CENELEC-ETSI,
- WP 4: Work package 4 has the objectives to satisfy the technical objectives within ESPRESSO, i.e. the development of a shared vocabulary, the definition of reference architecture, the city information and indicator platform, certification program, and execution of pilot-projects.
- WP 5: Work-package 5 aims to identify long-term strategic market implications of standardization in the Smart City domain brought by technological as well as societal evolution. It furthermore reflects new economic, financial and procurement models which can suit emerging Smart City scenarios and promotes a range of marketing opening actions targeted to standards in the domain of Smart City.
- WP 6: The goal of this work package is to understand and assess the legal, administrative, and societal impacts of Smart City platforms, applications, and workflows enabled by new information technology solutions.
  WP 7: The dissemination work-package aims to ensure maximum awareness of the projects activities and achievements through liaisons with other initiatives, through dissemination activities and through web-based publishing initiatives as well as a wide range of respective events including an embracing roadmap.
- WP 8: The last work package ensures that all the objectives of the project are met in line with the project schedule and with the highest quality standards, in terms of technical achievements as well as pilot and awareness activities.

The project at the moment is still in an early stage whereas the focus lies especially in developing the theoretical basis for the focus topics to share within the consortium. Furthermore, the reach-out activities will ensure, that the Smart City Stakeholder-Network (SmaCStak) will grow. This already founded network of Smart City experts will closely work with the consortium in order to achieve the best possible effect for the project aims.

4   Discussion

In regard of the project's aims within the ESPRESSO framework, the goal is to ensure interoperability of Smart City solutions, as well as avoidance of entry barriers or vendor lock-in through promoting common meta-data structures and interoperable (open) interfaces instead of proprietary ones. Considered from a technological perspective, there has to be a specific degree of standardization in the urban ICT networks. These have to be chosen so that they are not proprietary, prevent innovative Bottom-Up-solutions, and equally open to city administrations, companies and citizens. The required interoperability of services implies a certain degree of standardization in urban ICT networks. To build reliable and secure ICT structures, cooperations with the private sector are necessary. From an institutional point of perspective, the governmental requirements from a streamlining of processes are manifold and enhance the potential to implant standardization approaches. These issues will also be tackled with the close cooperation of our urban partners of the project as well as cooperation partners within the Smart City Stakeholder (SmaCStak)-Network. Therefore, it is important that they are not proprietary and prevent innovative bottom-up software solutions. The democratic legitimacy through full participation is essential in these standardization processes and thus part of a complex planning and city development process (Lojewski and Munziger 2013).

The dependencies between Smart Cities, standards and social aspects are far more complex, especially from an urban planning perspective. From a holistic perspective, the potentials of a networked ICT city are important, but also the dangers have to be taken in mind. Especially for public participation, there will be a complex potential, because, "design is a social process and not only a paternalistic process" (Klosterman 2008, p. 98). An important contribution to smarter cities could also lie in fostering innovation and creativity in the light of the knowledge-society. Most of the well known Smart Cities have concepts, which focus on optimization and efficiency, organized in a top-down manner and regarding the urban area simply as machine, which is controllable and adjustable though the social impacts of Smart Cities. The consortium is aware of this complex aspects and it is also the aim of the project to the develop the standardization approach in the light of these complex requirements. Though, the respective standardization approaches have to be considerate and will be part of the work for ESPRESSO, too. In order to guarantee social acceptance of developed solutions, the consortium sets up a stakeholder communication network that ensures an early dialogue between standards development organizations, technology providers, and technology consumers (cities and citizens as end users) to avoid a mismatch between the design of technology solutions and cities' and citizens' needs. ESPRESSO ensures inclusion of end-users in terms of planning; design and knowledge transfer along the lifetime of the project. Training material and sustainable platforms will further help to ensure social acceptance in the end. In addition, the exchange with comparable projects approaches also from an international level is very important and part of the project agenda.

Furthermore, there are barriers to the adoption of the ESPRESSO framework that will have to be overcome. As highlighted by several studies, the complexity of cities, in terms of stakeholders involved and processes represents one of the main barriers to adopting Smart City solutions. This complexity manifests itself across many areas of local government (policy, regulatory, governance, economic and organizational). The complexity of the ecosystem makes it difficult for city leaders and stakeholders to agree on the methodologies for implementing Smart City solutions. The provision of standards could reduce several barriers to Smart Cities implementation. On the other hand, standards needs to be made easily accessible to stakeholders in order to accelerate Smart City projects. Coordination and engagement with other standards bodies in order to coordinate and align relevant activities in the Smart City domain is needed. Another potential barrier for the adoption is the variation, at national level, in terms of contractual arrangements, legal implications, and national standards which may conflict with the practices of other nations particularity where these actions are not documented within a CEN or ISO standard. Some sectors, such as Smart Grids, have particularly high privacy and security concerns. ESPRESSO will consider open standards without doing implementation work for sensitive areas.

5   Conclusion

The project wants to show how standards can help smart cities and how the development could be fostered in Europe. First, there is the question of the understanding of the city and which kind of sectors are effected. From a technological perspective, the use of standards is very promising, from an institutional and social perspective, the issue is much more complicated because some aspects within the urban city are too complex and not suited for standardization purposes. So besides networks and capacity building, it is also the aim of the project to develop a common understanding which embraces potentials, but also constrains. To achieve the mentioned points, the knowledge and expertise exchange with other international standardization approaches in the field of smart cities are important in order to establish a positive influence to empower smart cities in Europe in a beneficial way for cities and citizens.

Acknowledgements. This project has been supported and received funding from the European Union's Horizon-2020 programme for research, technological development and demonstration under grant agreement No 691720. The author is grateful for the support of the Horizon 2020-consortium during the project work and represents the consortium for this publication. Further information can be found under: www.espresso-project.eu.

References

Batty, M.: Smart cities, big data. Environ. Plan. B: Plan. Des. **39**(2), 191–193 (2012)

Blind, K., Jungmittag, A., Mangelsdorf, A.: The economic benefits of standardization, pp. 1–24 (2011)

Exner, J.-P.: Smart cities–field of application for planning support systems in the 21st century? In: Ferreira, J., Goodspeed, R. (eds.) The 14th International Conference on Computers in Urban Planning and Urban Management (CUPUM), Boston, pp. 1–18 (2015)

Greenfield, A.: Against the smart city (The city is here to use) Do Projects (2013).http://www. amazon.de/Against-smart-city-here-you-ebook/dp/B00FHQ5DBS/ref=la_B001H6SA1C_1_ 1?s=books&ie=UTF8&qid=1392878639&sr=1-1

Kitchin, R.: The real-time city? Big data and smart urbanism. GeoJournal **79**(1), 1–14 (2014). http://doi.org/10.1007/s10708-013-9516-8

Klosterman, R.: A new tool for a new planning. In: Brail, R. (ed.) Planning Support Systems for Cities and Regions, pp. 85–99. Lincoln Institute of Land Policy (2008). http://www.whatifinc. biz/Resources/New_Tool.pdf

von Lojewski, H., Munziger, T.: Smart Cities und das Leitbild der europäischen Stadt. Städtetag Aktuell 9|13, pp. 10–11. Deutscher Städtetag (2013). http://www.staedtetag.de/imperia/md/ content/dst/veroeffentlichungen/dst_aktuell/staedtetag_aktuell_9_2013.pdf

Murgante, B., Borruso, G.: Cities and smartness: a critical analysis of opportunities and risks. In: Murgante, B., Misra, S., Carlini, M., Torre, C.M., Nguyen, H.-Q., Taniar, D., Apduhan, B.O., Gervasi, O. (eds.) ICCSA 2013, Part III. LNCS, vol. 7973, pp. 630–642. Springer, Heidelberg (2013)

Ratti, C., Townsend, A.: The social nexus. Sci. Am. **305**(3), 42–48 (2011)

Townsend, A.: Smart Cities. W.W. Norton & Company, New York (2013)

Geospatial Future Is Open: Lessons Learnt from Applications Based on Open Data

Branka Cuca[✉]

Department of Architecture, Built Environment and Construction Engineering,
Politecnico di Milano, Milan, Italy
branka.cuca@polimi.it

Abstract. Thirty five years ahead of the beginning of Open Data initiative, it is appropriate to evaluate the added value of such data and to reflect on how these can be exploited according to citizens' needs. This paper illustrates some first findings and lessons learnt during a research conducted on the use and re-use of geospatial Open Data by means of Virtual Hubs - an innovative method for brokering of geo-spatial information. Modern applications (APPs) related to topography, mapping of land use but also to historic maps and data of daily importance (including health facilities) are here examined so as to evaluate their thematic field of interest, the requirements of the end-users and the potential added value of Volunteered Geographic Information (VGI) in such context. The study finally evaluates the scale of impact and the level of awareness of developers towards a larger policy framework or their APPs.

Keywords: Open data · Geospatial information · Energic OD · VGI · Creative awareness · Intellectual data accessibility

1 Introduction and Background

Geo-spatial information is of crucial importance for all branches of sciences and research related to matters of topography, land use, earth observation technologies but also to territorial management, politics and geo-politics. Strong IT engines (e.g. Google, Bing etc.) have brought satellite imagery straight onto our screens and into our homes making us feel as true experts of all changes happening on the Earth's surface, from the latest volcano explosion or floods in the Balkans until destruction of the city of Palmira and the emergency situation of Syrian refugees. Phenomenon of Google Earth was in fact described by Goodchild as a "democratization of GIS" [1]. The reality is however much more complex, as the needs of end-users cannot be reduced to a simple display of data upon a click. Geo-spatial data offer in fact a framework for intelligent, shared and the right for interpretation of information. After Goodchild in [2], the activity of "geo-browsing" has become a major industry and it has introduced novel ways to explore and visualize overlaid information provided by the public and private sectors.

The right to information (RTI) movement and the call for open government data – OGD (including geospatial data) is based on a broader picture that started forming in the 80s with the requests of the civil society to make the governmental data accessible

© Springer International Publishing Switzerland 2016
O. Gervasi et al. (Eds.): ICCSA 2016, Part III, LNCS 9788, pp. 491–502, 2016.
DOI: 10.1007/978-3-319-42111-7_39

in machine readable formats. The research here illustrated is inserted in such a framework and it examines the specific cases developed upon the Virtual Hub (VH) concept established within the project "European NEtwork for Redistributing Geospatial Information to user Communities - Open Data (ENERGIC-OD)". The paper sets the scene in terms of intentions and possible impacts that innovative applications based on geospatial Open Data (geo OD) could have on the ongoing European policies and on OD initiative in general. The origins and the context given by RTI and OGD movements are firstly illustrated (Sect. 2) and the methodology for data collection on specific samples is explained (Sect. 3). The results of the Applications (APP) review is then illustrated and further discussed (Sect. 4). A discussion of impact of APPs based on Open geo Data is made (Sect. 5), with some first conclusions and lessons learnt on the use of VH systems for the distribution of such information.

2   Context

2.1   The Right to Information and Open Data Government

The right to information (RTI) movement and the call for open government data – OGD (including geospatial data) have started forming in the 1980s. These movements have the origin in the requests of the civil society to the governments regarding the data collected and held by public authorities (PAs) – such data should be made accessible in machine readable formats. The two movements seem very similar but they do contain some specific distinctions. Janssen in [3] makes a thorough study of similarities and differences between the two: she states that, while RTI is considered to be mostly "right-based" and rooted in civic society and democracy promotion activists, the OGD seems to be more "technology-driven" and promoted by techno-savvy and entrepreneurs. The drivers of the RTI movement seems to be based on the dual "right-obligation" flux where (1) public has the right to access the data and (2) the government has an obligation to actively disseminate the information of public interest. On the other side, four main drivers for OGD are identified as: (1) Transparency and accountability; (2) Participatory governance; (3) Innovation and economic growth and (4) Important internal value for public sector itself. While the factors such as transparency, accountability and public participation seems to be common to both, Janssen identifies one of the main gaps of OGD as the lack of "intellectual accessibility". Such feature could be defined as a justified fear that without appropriate skills and knowledge citizens will not be able to fully interpret and use of the "vast amounts of datasets thrown at them"; they will be hence incapable to fully access the information that can be obtained by the interpretation of such open data (OD). This paper focuses on geo-spatial OD and possibilities for their effective (re)distribution. The aim was to investigate how and if a closer relationship with application developers can influence a higher creative awareness and if such approach could contribute to an improvement of the "intellectual accessibility" by the citizens to Open geo Data.

2.2   The Policy Framework of Geospatial Open Data in Europe

There are currently several ongoing policies for access and use of public data in Europe. In 2003 an instrument called Directive on the re-use of public sector information (PSI Directive) has provided a common legal framework for a European market for government-held data i.e. public sector information [4]. The target set by the EC regarding this particular Directive is for overall gains of PSI to reach € 100 billion per year in EU by 2017, including new businesses development and efficiency gains in public sector services. Such context emphasizes in fact the aspect of economic growth and innovation that characterizes the overall OGD initiative. Member States were obliged to transpose Directive 2013/37/EU by 18 July 2015. As declared by the website of the European Commission [5], the EU Member States (MS) have implemented the PSI Directive in different ways. In detail, 13 have adopted specific PSI re-use measures (Belgium (https://ec.europa.eu/digital-agenda/en/news/implementation-psi-directive-belgium), Cyprus (https://ec.europa.eu/digital-agenda/en/news/implementation-psi-directive-cyprus), Germany (https://ec.europa.eu/digital-agenda/en/news/implementation-psi-directive-germany), Greece (https://ec.europa.eu/digital-agenda/en/news/implementation-psi-directive-greece), Hungary (https://ec.europa.eu/digital-agenda/en/news/implementation-psi-directive-hungary), Ireland (https://ec.europa.eu/digital-agenda/en/news/implementation-psi-directive-ireland), Italy (https://ec.europa.eu/digital-agenda/en/news/implementation-psi-directive-italy), Luxembourg (https://ec.europa.eu/digital-agenda/en/news/implementation-psi-directive-luxembourg), Malta (https://ec.europa.eu/digital-agenda/en/news/implementation-psi-directive-malta), Romania (https://ec.europa.eu/digital-agenda/en/news/implementation-psi-directive-romania), Spain (https://ec.europa.eu/digital-agenda/en/news/implementation-psi-directive-spain), Sweden (https://ec.europa.eu/digital-agenda/en/news/implementation-psi-directive-sweden), United Kingdom (https://ec.europa.eu/digital-agenda/en/news/implementation-psi-directive-uk)), 3 MS have used a combination of new measures specifically addressing re-use and legislation predating the Directive (Austria (https://ec.europa.eu/digital-agenda/en/news/implementation-psi-directive-austria), Denmark (https://ec.europa.eu/digital-agenda/en/news/implementation-psi-directive-denmark), Slovenia (https://ec.europa.eu/digital-agenda/en/news/implementation-psi-directive-slovenia)), while 9 MS have adapted their legislative framework for access to documents to include re-use of PSI (Bulgaria (https://ec.europa.eu/digital-agenda/en/news/implementation-psi-directive-bulgaria), Croatia (https://ec.europa.eu/digital-agenda/en/news/implementation-psi-directive-croatia), Czech Republic (https://ec.europa.eu/digital-agenda/en/news/implementation-psi-directive-czech-republic), Estonia (https://ec.europa.eu/digital-agenda/en/news/implementation-psi-directive-estonia), Finland (https://ec.europa.eu/digital-agenda/en/news/implementation-psi-directive-finland), France (https://ec.europa.eu/digital-agenda/en/news/implementation-psi-directive-france), Latvia (https://ec.europa.eu/digital-agenda/en/news/implementation-psi-directive-latvia), Lithuania (https://ec.europa.eu/digital-agenda/en/news/implementation-psi-directive-lithuania), Netherlands (https://ec.europa.eu/digital-agenda/en/news/implementation-psi-directive-netherlands), Poland (https://ec.europa.eu/digital-agenda/en/news/implementation-psi-directive-poland), Portugal (https://ec.europa.eu/

digital-agenda/en/news/implementation-psi-directive-portugal),   Slovak   Republic
(https://ec.europa.eu/digital-agenda/en/news/implementation-psi-directive-slovak-
republic)). Experts highlight the spatial data to be "one of the categories of PSI, of which
the exchange is particularly important" [6]. Furthermore, OD communication [7], one
of the main results of the Digital Agenda for Europe – DAE [8], emphasizes on intelligent
processing of information as essential factor for tackling the challenges of the contemporary society. Launched in the 2009, DAE is the first of the seven flagships initiatives
under the "Europe 2020" with an overall aim "to deliver sustainable economic and social
benefits from a digital single market based on fast and ultra-fast internet and interoperable applications". In particular, the Pillar VII "ICT-enabled benefits for EU society"
highlights the need of smart solutions for a vast number of users and for everyday
purposes. In such context the OD are considered to be crucial in addressing, environmental pressures, energy efficiency issues, land use and climate change, pollution and
traffic management and could have an important impact on more informed decision
making and policy creation in domains ranging from "decision support system for businesses, location based services and car navigation systems to weather forecasts and other
"apps" for our smartphones".

The use, interoperability and sharing geospatial information is inserted in such a
dynamic context with several other factors playing an important role. These factors
regard specific policy frameworks and ongoing international initiatives that largely
affect the way in which geospatial information is perceived, used and studied but also
the way in which information extracted can influence current and future decisions
regarding some of the main societal challenges. For example: the INSPIRE Directive
(Infrastructure for Spatial Information for Europe) as the first European (EU) framework
that enables sharing and re-use of publicly collected and held governmental data [9];
Copernicus program as the EU system for Earth Observation and Environmental monitoring and EU contribution to GEO and GEOSS; GEOSS – Global Earth Observation
System of Systems that fosters the connection and integration of geo OD for better
monitoring of environment and more informed decision making.

3   Methodology and Data Collection

In order to explore and observe the utilization of Open geospatial Data, the paper investigated the development of new innovative technological applications (hereafter APPs)
through a set of National Virtual Hubs (VH), specially developed for geo-spatial datasets. Such hubs exploit the concept of a broker i.e. an innovative method for facilitating
Open Data sharing and use based specific components (brokers) are dedicated to harmonize service interfaces, metadata and data models, enabling seamless discovery and
access to heterogeneous infrastructures and datasets. Such approach, already assessed
and adopted by a GEOSS Common Infrastructure [10], is expected to lower and possibly
remove the main barriers which hampers geo-information (GI) usage and exploitation
by application developers and end-users [11].

In order to access the added value of such systems, a series of requirements for APPs
have been studied in collaboration with APP developers. In addition to technological

requirements for application development (not treated by this paper), a series of non-technological aspects were here considered such as thematic fields of APPs and users to be addressed. The investigation methodology regarded two specific actions: (1) a close examination of the APP descriptions and a desktop research of related current initiatives and policies on Open geo Data and (2) a survey among the APP developers with specific queries. These will be explained in the following sections.

3.1 Desktop Research of Thematic Fields Influenced by Open Geo Data

A closer examination of these fields and of the first envisaged content of the APPs suggested that there could be a pattern connecting the thematic fields of applications with some specific European and International initiatives. In particular, the challenges established by Europe 2020 and deriving from the globally set millennium challenges, were considered as important starting point. European Innovation Partnerships (EIPs) were further chosen as the instruments designated to meet those challenges. GEO initiative was considered to examine the priorities of Geospatial information science with other economic sectors that might not traditionally exploit geo information to wider extent. Using the APPs as a "play-field", the goal was hence to explore, all the possible domains to which Geo information Sector and Geo Services could bring concrete benefits to the end-users to in the future. Domains investigated were:

- **Societal challenges.** "Intelligent processing of data is essential for addressing societal challenges", states Open Data communication [7]. The *informed decision making* is crucial for PAs to tackle the challenges faced by the contemporary society.
- **European Innovation Partnerships (EIPs),** a new approach under Innovation Union initiative aims to make the research and innovation more challenge-driven with focus on societal benefits and a rapid modernisation of the associated sectors.
- **GEO Societal Benefit Areas (SBAs)** has as main objective to provide earth observation and information products in support of societal needs, in a broad range of "Societal Benefit Areas" (SBAs): Agriculture; Biodiversity; Climate change; Disasters; Ecosystems; Energy; Health; Water and Weather (period 2010–2020).

Figure 1 illustrates a gamma of domains that could benefit and possibly even be measured. The use of geo-spatial information that keeps in mind the global scale of impact is considered to be an important international framework to local challenges, as further elaborated by the Geo science Report on resources for GEOSS [12].

3.2 Survey Mythology

This action was performed as a structured send-out survey that was distributed to all innovative APP developers and structured under two main sections:

- **Section 1 National/EU/International Policy - APP development framework:** Identification of the field of application and any relevant Directives or standards considered by the application and recommended on regional, country or EU level,

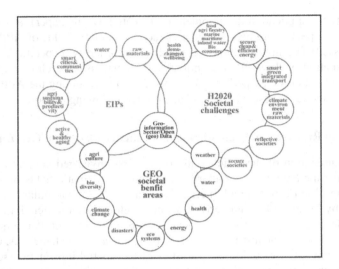

Fig. 1. Three main frameworks that could benefit by Open geoData explored

considering specific initiatives such as: *A. European Innovation Partnerships (EIPs);
B. EU Societal challenges; C. GEO SBAs.*

- **Section 2 User communities and final user requirements:** Identification of stakeholders (Professional Geomatics Communities; Volunteered Geo-information
communities; Open Data Networks; Small Medium Enterprises; PAs; Citizens for
example, elderly, youngsters, students, tourists) and the needs of the end-users to be
addressed.

APP developers were hence asked to perform three specific activities: (i) to identify
the field of expertise of their application; (ii) to identify possible relevant Directives,
regulations or other indications on regional/national/EU and international level; (iii) to
identify the most relevant keywords matching the APPs. Respectively these activities
had a purpose to: (i) identify thematic fields that could benefit from the concept of Virtual
Hubs; (ii) test the ground on developers' awareness of a larger policy framework around
GI sector and of EU Open Data initiative in particular; (iii) structure the first list of
taxonomies for the implementation of National VHs.

4   APP Framework and Results

The results here illustrated are based on the sample of ten APPs developed within the
project ENERGIC OD and based on VH concept. The results consider desktop research
and the survey conducted among developers. Due to respect of privacy towards APP
developers, all APPs are presented as anonymous.

4.1   A Brief Description of the APPs

A short description of ten examined APPs is here reported:

A1 will collect digitalized and georeferenced historic maps of the Zaragoza municipality from heterogeneous sources, enabling the user to browse, query and possibly download such historic information within a common spatial data framework.

A2, will enable location-aware communication between citizens and administration on infrastructure and services. This APP addresses a new market segment of users and use cases for open GI. The APP will aim to rely on the direct citizen participation.

A3 will propose to monitor the evolution of the coastline by using satellite imagery, public environmental GI data and GPS data provided by the users. This APP is dedicated to PAs and citizens who want to be informed of the status of the coastline.

A4 will generate noise maps of simulations and citizen-contributed noise data integrated by open traffic and open street map data. This novel platform will illustrate predicted noise levels in urban areas under dual aspect: (1) putting citizens at the center of a data acquisition and (2) using the community initiatives to engage the public.

A5, as a web based APP, will be able to give access to a directory of location-based healthcare services fed by remote web-services.

A6, foreseen as a web portal, will be used to raise awareness among actors involved in the production, processing and distribution of agricultural products of possible crop yield reductions caused by identified natural hazards events. Users will be provided with more reliable information about yield prediction and on possible damages caused by draught, floods and frosts. Such APP aims to prevent or reduce possible loses and hence to have a significant impact on the agriculture economy.

A7 will define the spatial distribution of areas with high deficits in the "biodiversity bird indicator" by analysing satellite and geospatial data within a GIS framework. This APP will aim to map areas with high deficits in habitat structure for every farmland bird included in the indicator and thus help to direct improvement measures.

A8 will collect in a common platform all information valuable for a sustainable territorial development, in particular with regards to landscape changes. By providing an spatial data infrastructure (SDI) of historical cadastral and topographical maps of Italy, integrated with panoramic views, such APP will facilitate and encourage the participation of experts, VGI communities and citizens.

A9, based the standards of OGC and OD the framework, will enable the search and structure of public knowledge of a particular place with a specific spatial extent.

A10 will develop an SDI to collect, transfer and visualise sensor data using standard OGC protocol or ad hoc bridges. A10 will collect these inexhaustible sources of data logged on the sensor or transmitted in almost real-time through web services.

4.2   Some Preliminary Results

The first results show that APP developers do not consider only technological feasibility and data availability but also the type of their potential end-users and their possible needs, as illustrated in the Table 1. The trend noticed is the overall tendency of developers to consider communities of data producers, data users and data *"produsers"* [13].

Table 1. Potential users/stakeholders of the new innovative APPs

	Potential users/ stake-holders	Profes-sional geomatic community	VGI com-munities	Open data net-works	SMEs	Public Authori-ties	Citi-zens
APPLICATION NUMBER	A1	YES	YES	YES	YES	YES	YES
	A2			YES			
	A3	YES	YES	YES		YES	YES
	A4	YES	YES	YES			YES
	A5	YES		YES	YES	YES	
	A6			YES	YES	YES	YES
	A7	YES	YES	YES		YES	YES
	A8	YES	YES	YES	YES	YES	YES
	A9			YES			YES
	A10	YES	YES	YES	YES		

Regarding the thematic fields of application, a summary of the inputs of the survey brought to some primary results that allowed to "cluster" different applications around similar topics. This first analysis suggests that APPs considered fall in two main categories i.e. (1) *"Land applications"* that consider a more geographical/landscape scale and (2) *"Urban applications"* that consider city/district scale. Both scales and the contributions of the applications could be of high relevance to the end-user communities and to citizens on the everyday bases. Several topics such Climate change, Agriculture, Health and Reflective Societies seem applicable for most APPs, revealing the main concerns or the main current focus of public attention. On one side, there is a strong awareness of the effects of climate change and the need for new services that could help to mitigate those changes, in particular in specific socially relevant fields – agriculture, food production and security, changes of landscapes and agri-landscapes, health services. Some other more specific apps underline the utility of open geo-information for a variety of services that cover landscape monitoring, biodiversity and ecosystems. Another significant category regards Smart Cities as many applications tackle the primary needs of citizens such as home search, wellbeing in urban areas and health emergency issues. Furthermore, most APPs refer to the Reflective Society challenge i.e. responding to the needs of more inclusive societies that use, benefit from and potentially voluntary contribute to ICT products and services. The situation seem to reflect the link between DAE Pillar VII and the societal challenges, indicating thus possible general creative awareness of developers.

The identification of the themes of application and potential user categories of APPs aimed here to provide a better insight into possible taxonomies for services based on geospatial OD. In [14] Nickerson et al. argue that in many disciplines the development of taxonomies follows an ad hoc approach and that a well-conceived method for developing taxonomies could be the basis for developing new taxonomies in Information

Systems so as to bring order to these complex areas and potentially lead to new directions of research. When it comes to spatial data, several examples discuss such requirements such as GEOSS taxonomy and Earth Observation [15]. An attempt to provide an insight of this sample of APPs was considered useful for common understanding of an appropriate terminology to be adopted by the VHs. This investigation will however need further research on a more critical number of APPs in order to provide major results on the issue.

4.3 Volunteered Geographic Information in APPs Using Open Geo Data

More than a half of the APP developers refer to VGI as an important data domain to be considered by their applications. In particular, such information is to be treated upon a double flux: on one side, developers think that VGI can effectively enrich an APP based on Open Data; on the other, developers are confident that VGI community can strongly benefit from an APP that integrates their information with other Open Data into an easy-to-use application that tackles specific domain of interest. To understand better the requirements of VGI communities and information generated on such premises, it is crucial to consider the type of VGI contributors and their motivation to provide and map information. Further to proposing a neologism of "produsers", in 2009 Coleman et al. have proposed to "distinguish contributors into five overlapping categories: Neophyte; Interested Amateur; Expert Amateur; Expert Professional and Expert Authority [13]". These categories have helped to further investigate and elaborate on the possible reasons that motivate volunteers to make a contribution. Several types of incentives were in fact defined and they range from *Altruism; Professional or Personal Interest; Intellectual Stimulation; Protection or enhancement of a personal investment; Social Reward; Enhanced Personal Reputation; Provides an Outlet for creative & independent self-expression* until the concept of *"Pride of Place"*.

Considering these aspects, a set of requirements, further illustrated in [16], is proposed in order to fully exploit the VH concept. Such approach should enable developers to consider the added value of VGI and to stimulate existing VGI communities on specific topics: (i) VHs should promote the increase of new VGI gathering different data sources; (ii) VGI generated by the APPs should bridge the gap of the current VGI in terms of lack of data and metadata standards; (iii) Easy and straight forward interaction with VGI should be guaranteed; (iv) Protocols, metadata standards and "smart" tools should be properly balanced to achieve the previously listed requirements; (v) VHs should connect current VGI relying on OGC standards and APIs; (vi) VHs should promote the publication of data and datasets in a standardized way and (vii) Metadata generation should be facilitated.

5   Discussion: Possible Impacts Geo OD on Current Policies

Regarding the non-technological requirements, the situation reflects the *"pull-and-push"* scenario elaborated in other economic sectors [17]. In this paper the "pull" of the market could be identified by information pull of public, volunteered and geo OD available for use and re-use, while the "push" refers to the contribution of the APPs to the existing communities and possibly to the change of the attitude of users towards the geospatial Open Data. For example, most of the developers represent SMEs that announce problems of lack of awareness of OD availability and their potentials for further APP development. PAs acknowledge the need of relevant information for more informed decision making and for more active involvement of citizens. PAs further emphasize that with such interactive approach, the needs of citizens might be more easily identified, categorized and possibly translated into future policies.

The potential impact of VH concept and APPs based on geo OD was explored with developers stimulated to consider the general frameworks of their APPs. Several relevant documents were identified as significant for this first sample of APPs, such as:

- Digital Agenda for Europe, Pillar VII ICT- enabled benefits for EU society;
- INSPIRE Directive;
- Common Agriculture Policy (CAP);
- Habitat Directive (Natura 2000);
- European Landscape Convention;
- European Convention on the Protection of the Archaeological Heritage;
- Environmental Noise Directive (END), Directive 2002/49/EC;
- Action Plan for Intelligent Transport System (ITS)
- Water framework Directive (WFD), Directive 2000/60/EC;
- Groundwater Framework Directive (GFD), 2006/118/EC.

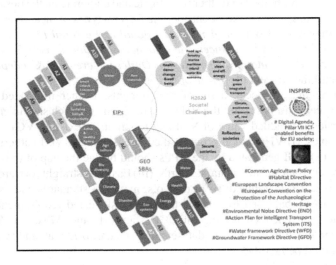

Fig. 2. Creative awareness at stage 2: thematic fields of ENERGIC OD APPs

Figure 2 shows APPs potential impact on the three main chosen frameworks. OD use was further observed in relation to the pull-push scenario and "awareness" in the creative process. Such "awareness" could work in two ways: (1) new products could rise awareness of users on certain important issues (e.g. environment, waste management, water scarcity etc.) and (2) the creation process could form a new generation of "creative thinkers", able to capture the needs of society and hence deliver better targeted products useful for specific challenges. Such approach was testes to examine a potential change in the "creative awareness" and/or the inclination to improve such skill that would enable the developers to perceive a "bigger picture" of their creative work.

6 Conclusions

The methodology applied was useful to understand what type of services based on geo OD can be beneficial to which thematic areas and if there is space for enlargement of this picture. Given that APP development is a highly creative process, the ability of understanding the larger framework of each APP was compared to the "creative awareness" i.e. an engine for innovation providing new marketable services based on geo OD and able to offer practical solutions for real social benefits. Lessons learnt highlight several interesting issues regarding Open geo-spatial Data:

- The scale of action for APP developed using Open geo Data can be both *landscape/ territorial* and *urban/city scale;*
- Socially relevant fields of main interest in this study are agriculture, food production and security, changes of landscapes/agri-landscapes, citizens wellbeing and health services and climate change;
- Advances in positioning technologies, Web mapping, cellular communications, smart phones, sensor Webs and wiki-based collaboration provides a more facilitated access to geospatial OD and location data: the role of VGI communities is to be considered as both information provider and pro-active user (*"producer"*);
- The policy framework around APPs is well perceived. Some further guidance is useful in order to fully translate such requests operationally into APP development;
- *"Creative awareness"*, that considers technological potentials and societal implications of APPs based on geo OD and VGI, could strongly contribute towards their effective use and improved *"intellectual accessibility"* of information by citizens.

On a more general note, the study confirms the need for systematic, easy and free of charge sharing of publicly collected geo-spatial data i.e. a full uptake of geo OD during the development of innovative services and APPs. Although predominantly technologically driven, OD initiative in Europe seems to have a serious potential to be beneficial to citizens on specific needs and in a wide gamma of thematic fields. Such approach could put focus on priorities of PAs, often subject to financial incentives, indicating novel trends for a better integration of VGI and geo OD. The study further confirms a burning need for (1) a more effective and immediate "translation" of facts (data) into information i.e. user-driven services with a tangible effect on the community; (2) a more user-friendly way for access and consumption of such information via easy-to-use technological devices (including VHs) and smart innovative APPs.

Acknowledgements. The research leading to the results of this paper is partially funded under the ICT Policy Support Programme (ICT PSP - CIP) (Grant Agreement no. 620400).

References

1. Butler, D.: Virtual globes: the web-wide world. Nature **439**, 776–778 (2006). doi: 10.1038/439776a
2. Goodchild, M.F.: Citizens as voluntary sensors: spatial data infrastructure in the world of Web 2.0. IJSDIR **2**, 24–32 (2007)
3. Janssen, K.: Open government data and right to information: opportunities and obstacles. J. Community Inform. **8**(2) (2012). http://ci-journal.net/index.php/ciej/article/view/952/954
4. Directive 2003/98/EC On the re-use of public sector information (2003)
5. European Commission (EC): Implementation of the Public Sector Information Directive (2015).   https://ec.europa.eu/digital-single-market/en/implementation-public-sector-information-directive-member-states. Accessed 25 June 2015
6. Janssen, K., Kuczerawy, A.: Increasing the availability of spatial data held by public sector bodies: some experiences and guidelines from the OneGeology-Europe Project. Int. J. Spat. Data Infrastruct. Res. **7**, 249–276 (2012)
7. EC, Communication from the Commission on Open data. An engine for innovation, growth and transparent governance [COM(2011)822] (2011)
8. EC, Communication from the Commission on "A Digital Agenda for Europe" [COM(2010) 245 final] (2010)
9. Directive 2007/2/EC: Establishing an Infrastructure for Spatial Information in the European Community (INSPIRE) (2007)
10. Nativi, S., Craglia, M., Pearlman, J.: Earth science infrastructures interoperability: the brokering approach. IEEE J. Sel. Top. Appl. Earth Obs. Remote Sens. **6**(3), 1118–1129 (2010). doi:10.1109/JSTARS.2013.2243113
11. Mazzetti, P., Latre, M.Á., Ernst, J., Brumana, R., Brauman, S., Nativi, S.: Virtual hubs for facilitating access to open data. Geophys. Res. Abstr. **17**, EGU2015-12080 (2015). EGU General Assembly 2015
12. Fellous, J., Béquignon, J.: Geo and Science. "Catalyzing Research and Development (R&D) Resources for GEOSS". https://www.earthobservations.org/index.php
13. Coleman, D.J., Georgiadou, Y., Labonte, J.: Volunteered geographic information: the nature and motivation of produsers. IJSDIR **4**, 332–358 (2009)
14. Nickerson, R.C., Varshney, U., Muntermann, J.: A method for taxonomy development and its application in information systems. EJIS **22**, 336–359 (2013). doi:10.1057/ejis.2012.26
15. EARSC: A Taxonomy for the EO services market: enhancing the perception and performance of the EO service industry (2015). https://www.earthobservations.org/documents/committees/stc/20100923_geo_and_science.pdf
16. ENERGIC OD project - Report D4.2 (2016). http://media.wix.com/ugd/4f5bdd_452025194e96422a859f03812d929496.pdf
17. Sesana, M.M., Cuca, B., Iannaccone, G., Brumana, R., Caccavelli, D., Gay, C.: Geomapping methodology for the GeoCluster Mapping Tool to assess deployment potential of technologies for energy efficiency in buildings. Sustainable Cities Soc. **17**, 22–34 (2015). doi: 10.1016/j.scs.2015.02.006

Definition of Luxury Dwellings Features for Regulatory Purposes and for Formation of Market Price

Maria Rosaria Guarini[1(✉)] and Anthea Chiovitti[2]

[1] Department of Architecture and Design (DIAP), Faculty of Architecture,
Sapienza University of Rome, Rome, Italy
mariarosaria.guarini@uniroma1.it

[2] Department of Architecture and Design (DIAP), Faculty of Architecture, Doctoral School
in Architecture and Construction (DRACO), Sapienza University of Rome, Rome, Italy
anthea.chiovitti@uniroma1.it

Abstract. In all the world, luxury dwellings represent a specific segment of residential Real Estate market. All Real Estate market operators/observers point out that luxury segment has very different features and trends from the residential market in general, especially in current and ongoing economic crisis.

The text shows the results of a Research which verify, with reference to a sample of 30 trades of luxury properties located in the historic center of Rome, presence and relevance of the features, listed by regulations, or that may affect the price formation.

Keywords: Luxury dwellings · Residential real estate market · Evaluation real estate market

1 Introduction and Aim of the Work

In all the world luxury dwellings stand out for having supply and demand features outside the ordinary residential real estate market, aimed mostly at satisfying a primary housing need. In a challenging business framework increasingly complex - marked by economic crisis and signs of diverging prospects for recovery - luxury dwellings are also often regarded as capital goods for a certain target buyers (Wealth-X and Sotheby's International Realty 2015; Liu and Yermack 2012) Some features of this specific market, in relation to characteristics of the places where these properties are located, can change, but they often have in common a significant share of international demand like location

The contribution is the result of the joint work of the two authors. In particularly, the paragraphs nos. 1, 2, 4.2 and 4.3 were written by M.R. Guarini; the paragraphs nos. 3, 4.1, 5 were written by A. Chiovitti who also worked on the process and return of gis data. The authors warmly thank the Real Estate agency that have made available the data relating to the examined properties, and Morgan de la Poer Horsley Beresfodr who collected and processed a part of these data during drafting of his of triennial degree thesis in "Gestione del Processo Edilizio - Project Management" (supervisor Prof. M.R. Guarini), Faculty of Architecture, Sapienza University of Rome.

© Springer International Publishing Switzerland 2016
O. Gervasi et al. (Eds.): ICCSA 2016, Part III, LNCS 9788, pp. 503–518, 2016.
DOI: 10.1007/978-3-319-42111-7_40

in prestigious areas characterized by environmental peculiarities or landscapes-characteristics, presence or proximity of commercial, leisure and high level cultural activities, presence of internal prestigious finishes (Christie's International Real Estate 2015a). The market luxury dwellings is constantly changing in relation to changes in lifestyle and possibility of faster movements which can modify prospects and financial investment preferences involving real estate (Christie's International Real Estate 2015b).

Also in Italy, the luxury dwellings represent in particular a specific segment of real estate market in which demand consists of parties with high income (with increasing numbers of foreigners) who generally seek real estate units that: have large square footage; are often not used as a single residential dwelling ("first home"), with specific prized features (special finishings and/or with recognizable exterior formal traits); and are located in "exclusive" settings, in terms of landscape and/or from the historical/artistic standpoint. Some regulatory provisions indicate what characteristics a building must possess in order to be considered "luxury" from the standpoint of construction, urban planning, and taxation. But the presence or absence of other characteristics may influence, positively or negatively, these dwellings' price on the market within the specific housing contexts in which they are set.

In Rome, in particular, the city's historical, artistic, and morphological features allow certain strategically located buildings, with privileged views or perspectives, to be enjoyed; similarly, the climate, mild for many months of the year, permits the almost continuous use of the spaces outside the dwelling and, lastly, these dwellings are often an integral part of that portion of the urban fabric with high historical and artistic value. Consequently, a real estate unit, even one not particularly large in size, situated in certain areas of the city, with an exterior prospect, preferably on a high floor, with a balcony, a loggia, and a terrace, has features that significantly augment its value, allowing it to easily acquire on the market the cachet of a luxury dwelling. In general, on the luxury residence market, we can see and typologically distinguish the real estate units as follows:

- those belonging to buildings of historical value and/or with unique characteristics of quality, generally located in prestigious urban settings (penthouses or flats)
- those located in stand-alone buildings, low, and with one or two residential units, marked essentially by the presence of a park and/or yard (villas, cottages);
- those consisting of a single building, ground up, located in a prestigious urban setting and/or with unique features of formal quality, even quite large in size. We may include in this set: Villas and palazzi with special features of both value and historical/artistic recognizability and of size, often in the historical fabric, taking up entire city blocks, that are placed in a non-ordinary market that is slightly active or inactive altogether; or buildings smaller in size, often integrated into the city blocks in the historical fabric ("palazzetto").

An objective of this text is to illustrate the results of a research work aimed at verifying, with reference to a significant sample of real estate sales located in Rome's historic centre, the presence and importance of the characteristics indicated by regulatory provisions, or that can influence how the price is formed.

In large European capitals, market for luxury dwellings has generally some similar characteristics, despite the cultural, climatic differences, historical, architectural the different States and cities that influence the specific supply and demand features (Christie's International Real Estate 2015a).

Rome certainly has some unique features, both in terms of the tangible and intangible values related to the presence of historical and artistic evidence in urban settlement, the shape of the city, the offer of high standard cultural and commercial activities. The identification of the characteristics affecting the price formation in the luxury dwellings, formulated on the basis of a sample referred to the city of Rome, it can be interesting to highlight the supply and demand special features for this specific market segment in the Italian capital, and to check the common aspects with other European and international contexts.

In this general context particular importance has to the legal definition of luxury property in Italy as well as defined by tax and planning/urban regulations.

Hereafter, Sect. 2 illustrates the various modes of defining luxury buildings in Italy; Sect. 3 briefly details the trends in the residential real estate market in general in Europe, Italy and in Rome (Sect. 3.1), and specifically in the luxury segment in Rome (Sect. 3.2); Sect. 4 discusses the purpose of the processing operations and the procedures for surveying and organizing the data (Sect. 4.1) used for the assessments that were made (Sect. 4.2) and the results obtained (Sect. 4.3); Sect. 5 states the conclusions.

2   Defining "Luxury Dwelling"

In Italy, the definition of "luxury dwellings" is established in the Ministerial Decree (Ministry of Infrastructure) of 02 August 1969 (hereinafter, referred to by its Italian initials "DM") on the "Characteristics of luxury dwellings." According to the DM, the following are to be considered "luxury":

1. "Dwellings built on areas zoned by adopted or approved urban planning instruments for "villas," "private park", or for constructions qualified by the aforementioned instruments as "luxury."
2. Dwellings built on areas for which adopted or approved urban planning instruments zone with a construction type of single-family homes, and with a specific requirement of lots of no less than 3000 m^2, excluding farm areas, even if residential construction is permitted.
3. Dwellings belonging buildings measuring more than 2,000 m^3 and are built on lots whose built volume is less than 25 m^3, void for full, for every 100 m^2 of area underneath the buildings.
4. Single-family dwellings with a swimming pool with an area of 80 m^2 or tennis court with drained foundation, covering an area of no less than 650 m^2.
5. Houses consisting of one or more rooms constituting a single master lodging with a total useful surface area exceeding 200 m^2 (excluding balconies, terraces, cellars, attics, stairways, and parking spaces) and having as appurtenance an uncovered area at least six times the covered area.

6. Individual real estate units having a total useful surface area exceeding 240 m^2 (excluding balconies, terraces, cellars, attics, stairways, and parking spaces).
7. Dwellings belonging to buildings or constituting buildings on areas at any rate zoned for residential construction, when the cost of the land covered and of the appurtenance exceeds by one and one half times the cost of the construction alone."

Moreover, the DM specifies that houses and individual real estate units that jointly have more than 4 characteristics out of those indicated in the table (Table 1) attached to the decree (hereinafter: DM criteria) are to be considered luxury dwellings.

Table 1. Characteristics jointly necessary (more than 4) to define the luxury buildings pursuant to the Ministerial Decree of 02 August 1969

Characteristics	Specification of characteristics
a) Area of the flat	Total useful surface area exceeding 160 m^2, with the calculation excluding terraces and balconies, cellars, attics, stairways, and parking space
b) Covered and uncovered terraces and balconies	When their total useful surface area exceeds 65 m^2 serving a single urban real estate unit
c) Lifts	When there is more than one lift for each stairway, each extra lift is counted as a characteristic if the stairway serves less than 7 above-ground storeys
d) Service stairway	When it is not required by laws or regulations, or imposed by accident- or fire-prevention needs
e) Goods or freight elevator	When serving fewer than 4 storeys.
f) Main stairway	a. with walls lined with prized materials for a height exceeding an average of 170 cm; b. with walls lined with materials worked in a prized manner.
g) Net clear height of the storey	Greater than 3.30 m, without prejudice to building regulations requiring greater minimum heights.
h) Entrance doors to flats from internal staircase	a. in prized solid and veneered wood; b. in carved, chiselled, or inlaid wood; c. with prized decorations, superimposed or stamped.
i) Internal window and door frames	As letters a), b), c) of characteristic h), even if with honeycomb support, if their total overall area exceeds 50% (fifty percent) of the total area.
l) Floors	Done for a total surface area exceeding 50% (fifty percent) of the total useful surface area of the flat: a. in prized material; b. with materials worked in a prized manner.
m) Walls	When for more than 30% (thirty percent) of their total surface area is: a. done with prized materials and handiwork; b. lined with fabrics or other prized materials.
n) Ceilings	If decorated and coffered, or decorated with plaster cast on site or hand-painted, excluding the small shapes of separation between walls and ceilings.
o) Swimming pool	Indoor or outdoor, below-ground, when it is at the service of a building or a complex of buildings comprising fewer than 15 real estate units.
p) Tennis court	When it is at the service of a building or a complex of buildings comprising fewer than 15 real estate units.

The DM makes no reference to cadastral categories. Only starting in 2014 for tax purposes are dwellings entered in one of the following cadastral categories to be considered luxury housing:

- A/1 (distinguished dwellings): real estate units belonging to buildings located in prestigious areas with characteristics of construction, technology, and finishings at a level exceeding that of residential-type buildings,
- A/8 (dwellings in villas): the term "villas" is to be understood as those buildings marked essentially by the presence of park and/or garden, built in urban areas zoned for these constructions or in prestigious areas with features of construction and finishings at a level exceeding the ordinary;
- A/9 (castles, palazzi of artistic and historic value): belonging to this category are eminent castles and palazzi that, for their structure, division of interior spaces, and built volumes are not comparable with the typical units of the other categories; they normally constitute a single real estate unit. Compatible with the category A/9 attribution is the presence of other, functionally independent units that may be recorded in the other categories.

In fact, since 01 January 2014, in the transfers of ownership of the (luxury) dwellings belonging to these cadastral categories, the tax benefits established for residential buildings belonging to other cadastral categories can no longer be applied (Nesticò and Galante 2015). In particular, the benefits related to the application of the VAT rate reduced to 4 % (art. 33 of Legislative Decree no. 175/2014 – Simplifications Decree) and, if the acquirer intends to access the first home benefit, the 2 % registration tax (pursuant to art. 26 of Legislative Decree no. 104/2013), cannot be used. A subsequent Circular (explaining Legislative Decree no. 175/2014) of Agenzia delle Entrate (Revenue Agency) (no. 31/E of 30 December 2014) specifies (point 24.2), to avoid a legislative *vulnus*, that the definition of "luxury dwelling" pursuant to the DM has no relevance any longer, also for the purposes of application of the 10 % VAT rate to transfers or to deeds establishing property rights with regard to:

- dwellings other than "first home"
- buildings or portions of building pursuant to art. 13 of law no. 408 of 02 July 1949 ("Tupini buildings"), under the conditions provided for by said art. no. 127-undecies).

Therefore, to access the aforementioned tax benefits, the definition of "luxury" dwelling, as intended under the DM, must be understood as obsolete. However, it appears to remain valid for the purposes of taxation. It must consequently be considered that the impossibility of accessing tax benefits may result in a different appreciation on the real estate market for luxury dwellings, with regard to productive-type factors (Morano et al. 2015).

In any event, regulatory indications appear to aim to take at least partial account of the fact that a dwelling's market price also reflects the appreciation expressed by demand for some elements and features of the sold property. Ordinarily, the following characteristics may trigger oscillations in market prices for homes (Forte and De Rossi 1979; Roscelli 2014):

- extrinsic/setting: in addition to the building's specific setting in the urban fabric, an expression of associated aspects of qualification: (i) infrastructural: proximity to the urban centre, accessibility to public services, accessibility to public transportation, (presence of basic commercial services); (ii) environmental: healthiness of the area, social context, lack of noise, low building density;

- intrinsic settings: associated with scenic quality or visibility, orientation, sunlight, light, ventilation, healthiness of the room;
- technological: connected with the level of the equipment and finishings present in the dwelling or in the common parts of the building and/or of the area where it is located: presence of an elevator, toilet facilities (and number thereof), quality of interior and exterior door and window frames, etc.
- productive: associated with tax exemption regimes, deductibility of maintenance expenses, limits on leasing, historical/artistic constraints or constraints consequent to leasing, maintenance conditions of common parts, type of heating and power consumption systems, presence of a park or common condominium services, etc.

Clearly, in addition to those indicated by the regulations, other aspects, too, may undoubtedly acquire particular and specific values for the so-called "luxury" properties, if considered with reference to the special features of the market setting in which the individual dwellings are placed. In addition to official sources (Agenzia delle Entrate, the real estate market observatory– OMI), it is also the commercial and productive operators that show the common features and distinctive traits that characterize luxury dwellings (see above), as well as the importance and particular nature of the market for housing of this kind. In fact, many of the most important real estate brokerages, held by major holding companies operating in the construction sector, have a division specialized in the international and domestic supply of integrated consulting services with regard to the sale and rental of buildings, private real estate, new constructions and top properties in luxury buildings.

Moreover, often, also to promote these specific areas of their expertise, they collect, process, and publish, generally every six months, data on the characteristics, dynamics, and trends in this market segment, referring from time to time specifically to "exclusive residences," to "prestige buildings," to "trophy buildings," but certainly always ascribable to the set of luxury dwellings. Of these, with regard to the purposes of the research illustrated in this text, which does not discuss "extraordinary" luxury buildings, the Observatory on "exclusive residences" (referred to hereunder by its Italian abbreviation ORE)[1] and the report on the "Prestigious buildings market" (RMIP)[2] may be cited.

In fact, the ORE, in specifying the criteria (set out differently for certain cities) that go towards delimiting the object of its scope of observation (Nomisma and Tirelli & Partners 2016), "considers as an *exclusive residence* a building that has:

---

[1] Since 2003, Tirelli & Partners has processed, in collaboration with Nomisma (and with Nuova Attici for surveying the data on the city of Rome), and published every six months, its monitoring of the trend in the most exclusive segment of the residential market. (http://www.tirelliandpartners.com/showPage.php?template=chisiamo&id=3).

[2] The analyses and surveys contained in the RMIP are based on the requests of the clients of the Santandrea (luxury house) agencies, which belong to the Gabetti Group, and are processed by its studies office.

1. a price per square metre, or overall price, greater than a given threshold;
2. a yearly rent per square metre, or total yearly rent, greater than a given threshold;
3. a location, out of the various urban areas, in those classified as prestigious, or in a so-called 'residual zone,' indicating those market situations that, although not situated in the identified areas, still possess the requirements of value 1 and 2."

The RMIP proposes, for certain Italian cities (Rome, Milan, Genoa, Florence, Naples, and Turin), a classification both of the features in greatest demand (single or double garage, terrace, scenic view, yard/park, triple living room, three or more baths, plus reception rooms, hobby room, spaces for service personnel, concierge/security services), and those features with lesser appeal (high asking price for the building's quality, building's less than optimum interior and exterior conditions, mezzanine and first story location, only bath being windowless, a surrounding setting that presents elements of decay, lack of balconies, lack of garage/parking space, proximity to noise sources, exposure to busy thoroughfares, high condominium expenses) for prestigious buildings (Sant'Andrea Luxury House 2016). In general, all real estate market operators/ observers agree in stressing that the segment of luxury residences presents, especially in the markets' enduring economic crisis, features and trends quite different from those characterizing the residential market in general (Tajani and Morano 2015; Nesticò and Pipolo 2015). At the same time, they state that certain characteristics count far less in the current phase than they did in the past: the floor area of the lodgings and, especially for used buildings, the presence of certain services and features that produce inevitable impacts on condominium expenses.

3 Trends in the Housing Market

3.1 In Europe, Italy and Rome

Rome's residential real estate market also shows the trend characterizing the current phase, marked on the one hand by signs of a slow recovery in transactions, and on the other by a bottoming out of prices, which stand today (2015), in nominal terms, below 2005 levels (Agenzia delle Entrate 2016; Agenzia delle Entrate 2015). In general, in Italy, the residential market from 2009 through 2014 saw a negative variation in prices on the order of −25 % to −30 %. In Italy, as in Europe, during the same period, the decline in the price for luxury flats did not exceed 20 % (Agenzia delle Entrate 2015; Knight Frank 2015). In this market segment, sales declined at a more contained rate (on the order of −25 %), while average times for selling a prestigious flat saw a significant increase. While in 2007 two months might suffice to make the sale, today (2015) it is hard to conclude a sale in under 8 months. Bucking this trend for the entire sector are prestigious areas, which recorded, in the first half of 2015, positive variations both for homes in excellent condition and for used ones.

3.2   The Market for Prestigious Dwellings in Rome

Substantially in line with the findings in other reports on the real estate market, the ORE (Nomisma and Tirelli & Partners 2016) shows that in Rome (since 2014), the residential market segment of prestige dwellings consisted of buildings marked by: (i) a purchase value of € 7,000 euro/m^2 or a total value exceeding € 1,000,000; (ii) rent exceeding € 200 euro/m^2 per year, or a total yearly fee exceeding € 30,000; (iii) location in one of the following areas: historic centre; Pinciano Veneto; Parioli - Salario – Trieste; Prati; Vigna Clara Camilluccia.

Market operators agree in stating that after sales held substantially steady in both halves of 2014, the first half 2015 recorded an increase in transactions in all the capital's prestigious areas (except for the Pinciano Veneto neighbourhood). Moreover, they record an improvement over the past in the quality level offered, especially due to the presence on the market of new real estate units that are high in quality (often connected with efficiency and energy savings requirements) and prestige meeting the expectations of demand at values that have become accessible. Foreign investors represent a significant share of the total demand, showing an increase in 2015 from a 10 % to 15 % share in the first half to 20–25 % in the second. As in Milan, in Rome smaller family size has also triggered, for luxury housing, a change in required floor area, which starting in the first half of 2014 stood at about 150 m at a price averaging around one million euros. Generally affected by this decease are professionals and executives, a significant portion of whom are Romans (first half of 2014: 15 % foreigners; 25 % non-Roman Italians) seeking homes for themselves and their families. In general, the luxury sector has recorded an increase in bargaining but not prices, which are gradually turning downward (Nomisma and Tirelli & Partners 2016; Sant'Andrea Luxury House 2016).

4   Direct Survey, Data Processing Purposes and Procedures

4.1   Purpose of the Direct Survey and Organization of Surveyed Data

A direct survey of some real estate agents dealing prevalently with the luxury property market supplied data referring to 30 properties sold in Rome in a period between the first quarter of 2013 and the third quarter of 2015:20 penthouses (abbreviated to "At" for *attico*) or flats (Ap, for *appartamento*), 5 ground-up stand-alone small buildings (P, for *palazzetto*), and 5 villas (V), all located in the neighbourhood of the historic centre (Fig. 1). As to the size and dynamics of this market segment in the examined housing context, these buildings may be considered a highly significant sample.

The survey collected data on the characteristics of the buildings referring both to the criteria indicated in the DM and to some of the factors that may ordinarily influence the formation of the dwellings' price. These data were reported in a database in which each property is identified with the its code: a letter for the type of property it belongs to (A, P, V) and a number marking the specific identifier of the individual real estate units for each type (A1, A2, A15; P1, P2, P5; V1, V2, …, V5). Moreover, each property's location was georeferenced using Qgis software (opensource) in the WGS 84 UTM Zone 32n reference system[3]. In this way, the querying of the database constructed with the data surveyed for each property made it possible to map the processing operations performed (Guarini et al. 2015; Song and Knaap 2003) as described in the following paragraphs.

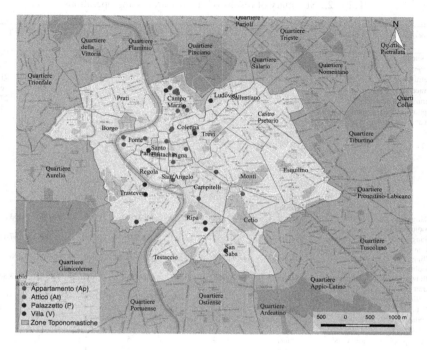

Fig. 1. Building in the survey, by location and type (Color figure online)

4.2   Data Processing: Objectives and Procedures

Table 2 sets out the summary results of the "basic" processing done on the surveyed data, broken down by category of properties (A, As, Ap, P, V). The table also sets out and calculates the values (total and average) for the entire reference sample, in order to be able to highlight the distinctive factors, among the various categories, within the overall universe of luxury housing taken into consideration.

Then, it was first checked that the sample of real estate units taken into consideration corresponded with the necessary requirements for being defined as luxury housing in accordance with the criteria dictated by the DM (Fig. 2) and by the regulations governing tax benefits in real estate sales. On the basis of these processing operations, it was then

---

[3] For the processing of the data in Qgis, use was made of the shapefiles available on ISTAT's website (www.): Territorial Bases 2011 - WGS 84 UTM Zone 32n, Rome, sub-municipal areas: Urban Planning Zones (*Quartieri*) and Topographic Zones (*Rioni*). The map support material was retrieved by the QuickMapServices Bing Satellite and Bing Map Ru services integrated into the Qgis software.

possible to determine the percentage of correspondence with these requirements for the set of cases being surveyed.

Table 2. Summary of results of the first processing operations

Properties	Unit of measurement	Processing operations	At	Ap	A	P	V	Total
Typology		total	17	3	20	5	5	30
Rione		number						
Useful area		average**	220	190	216	428	475	216
Terrace area	m²	average	70	-	70	55	46	64
Commercial area	m²	average**	283	245	277	536	620	277
Appurtenances and accessories	number/ type	total	6	3	9	6	8	23
		yard		1	1		5	6
		garage			-		3	3
		parking space	1		1			1
		garage stall	1		1			1
		cellar	2	2	4	1		5
		attic	1		1	1		2
		storage			-	2		2
		courtyard			-	2		2
		mansard	1		1			1
Asking price (absolute value)	euro	average**	5.770.588	2.900.000	5.340.000	6.120.000	11.125.000	6.463.333
Asking price (parametric)	commercial	average**	19.698	12.361	18.598	11.513	16.706	17.116
Selling price (absolute value)	euro	average**	4.764.706	2.516.667	4.427.500	4.970.000	7.700.000	5.090.000
Selling price (parametric)	commercial	average**	16.307	10.407	15.422	9.508	11.810	13.846
% variation between asking price and selling price			-17%	-15%	-17%	-18%	-28%	-19%
Sale time	month	number	12	13	12	15	17	13
Proposals received by real estate	number		3	4	3	5	3	3
State of property	renovated	number	4	1	5	1	3	9
	good	number	4	1	5	1		6
	needs renovation	number	9	1	10	3	2	15
Restriction	no	number	2	-	2	4	5	11
	yes	number	15	3	18	1		19
Views		number			-			-
Exposure	side	number	3	2	5	3	4	12
Ascensore	no		3		3	4	3	10
	yes		14	3	17	1	2	20
Entries to real estate unit	number	average*	1	1	1	1	3	2
Concierge services	no	number	6		6	5	5	16
	yes	number	11	3	14			14
Levels of real estate unit	number		1	1	1	4	3	2
Cadastral income rate	Euro	average*	7.701	3.262	7.035	7.497	14.846	8.414
Category cadastral	A1	number	5		5			5
	A2	number	12	2	14	3		17
	A3	number			0	1	1	2
	A4	number		1	1			1
	A7	number			0	1	4	5

** small stand-alone buildings and/or villas not included in the total

Lastly, for a set of 13 assessment criteria (ascribable to characteristics that can impact the properties' market price), attributing to each surveyed characteristic a value and a weight on a specifically constructed scale (Calabrò and Della Spina 2014), it was verified

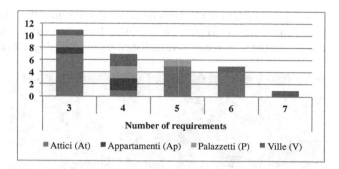

Fig. 2. Number of requirements consistent with the DM's criteria by types of properties (absolute value) (Color figure online)

whether the sample in consideration could be considered sufficiently homogeneous (total score obtained by each real estate unit taken into consideration) and therefore representative as a whole; on the other hand it was ascertained which of the characteristics taken into consideration recurred most frequently in the formation of the price. This makes it possible to highlight which of the considered criteria are actually more suited to reflect the current situation of the luxury real estate market in the historic centre of Rome.

4.3 Summary of Results of the Processing Operations Performed

The first basic processing operations (Table 2) show that the properties taken into consideration have an average selling price equal ($\approx$) to: for A), € 15,000/m² (At: € 16,000/m²; Ap, € 10,500/m²); price: min. € 10,000/m², max. € 28,800/m²; for P), to € 9,500/m²; price: min. € 8,400/m², max. € 12,100/m²; for V), € 12,000/m²; price: min. € 10,300/m², max. € 15,000/m² The gap between the selling price and the expected price, not considering villas and small stand-alone buildings, averages ≈-17 % (with variations ranging between a maximum value of 25 % and a minimum of 12 %). In fact, villas record far higher average gaps of about ≈-29 % (with variations ranging between a maximum value of 35 % and a minimum of 24 %), connected especially with the significant effort required for the ordinary and extraordinary maintenance of these properties, and the impossibility of enjoying tax credits.

The average selling time in the sample of properties taken into consideration equals 13 months. In this case as well, there are significant differences between the categories of properties observed and the interval between the minimum and maximum values. In particular, a variation in selling time was found that ranged: for At, between a minimum of 3 and a maximum of 18 months; for Ap, between a minimum of 11 and a maximum of 17 months; for P, between a minimum of 11 and a maximum of 24 months; for V, between a minimum of 12 and a maximum of 20 months. From all the standpoints taken into consideration, the sample observed with the direct survey presents, only for some of the types, some significant deviations from the average values indicated in the reports on the trend in this specific real estate market. With reference to the minimum requirements established by the DM, the verification that was made shows that more than 60 % of the real estate units that come under the Penthouses/Flats and small stand-alone buildings (*palazzetto*) types might not be considered luxury, because they do not satisfy at least 5 of the indicated characteristics. On the other hand, the villas that were sold corresponded more with the requirements established by the law (Figs. 2 and 3).

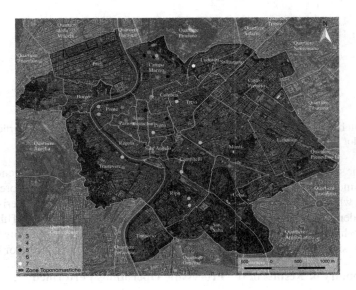

Fig. 3. Number of requirements consistent with the DM's criteria by types of properties (location) (Color figure online)

In detail the only criterion common to all the residential units taken into consideration is that referring to useful area greater than 160 m^2. Characteristic of a significant number of properties, albeit with different importance for the various types taken into consideration, are the criteria regarding: ceilings and floors in prized material, prized entrance doors (this term being be understood, in a manner more in line with the expectations of demand, as having an armoured door), and ceiling height exceeding 3.30 m. The presence of the terrace, however significant a factor it might be (present in 17/20 A, 2/5 P and 5/5 V), does not appear to be referable, as indicated in the DM, to a size greater than 65 m^2 (present only in 6 A). It must rather be observed that this factor takes on particular importance in relation to the view chat may be had from it. The presence of more than one lift per stairway, below-ground pool, or a tennis court do not appear to be distinctive characteristics. Certainly, these characteristics are featured more frequently in new buildings and/or buildings located in less central urban areas. On the other hand, in the historic centre of Rome, these characteristics are virtually impossible to find in old/antique buildings.

Also for the purposes of tax regulations, at the time of sale most of the properties were not registered in one of the cadastral categories allowing them to be considered luxury housing. In fact, only 5 real estate units belong to cadastral category A1; most (16) are registered in category A/2, Civil Dwellings (real estate units belonging to buildings with characteristics of construction, technology, and finishings at a level exceeding that of residential-type buildings). The remaining properties belong to other

Table 3. Criteria for allocating scores to the features considered

SCORES	5	4	3	2	1
FEATURES	**SCORES ATTRIBUTION**				
Proximity to the center	Internal Servian Wall	Inside Aurelian walls	Inside the ring road (tangenziale est)	Outside the ring road (tangenziale	Outside G.R.A.
Accessibility (proximity: distance in meters) in public/commercial	very high (<250 m)	high (250<m>500)	median (500<m>750)	low (750<m>1000)	very low (m>1000)
Accessibility (proximity) to public transport	very high (<250 m)	high (250<m>500)	median (500<m>750)	low (750<m>1000)	very low (m>1000)
Wholesomeness of the area	very high (especially green and protected areas)	high (mainly residential areas with medium and low urban density)	median (mixed urban areas of medium density)	low (areas with intense urban density and human activity)	very low (predominantly industrial areas)
Noise (in relation to the acoustic classification area of the township and soundproofing fixtures)	very high	high	median	low	very low
Panoramic views	view monument	panoramic view	view feature	Overlooking in the internal courtyard	Overlooking in little internal courtyard
Orientation/sunshine	south	south-west or south-east	East or North East	West; North-West	North
Interior fittings (level): depending on the presence two bathrooms, bathroom and kitchen renovated, number of valuable characteristics	very high	high	median	low	very low
Lift	present				not present
Fixtures (age/grade)	last generation (energy efficient)	replaced over the past 5 years (low energy efficiency)	replaced in the last 10 years (not energy-efficient)	old with change of glasses	outdated fixtures
taxation - cadastral income (euro)	high (€>13.000)	good (13.000<€>7.500)	median (7.500<€>3.500)	low (3.500<€>2.000)	very low (€<2000)
superintendence constraints	Building unconstrained		building with constraint		building and fronte with constraint
Energy consumption	energy class: A + or higher	energy class: A or B	energy class C or D	energy class: E or F	energy class: G or less

cadastral categories: A3 (3), 1 A4 (1) e A7 (5). With reference to the processing operations performed for the set of 13 assessment criteria that may be ascribed to the characteristics that may have an impact on the properties' market price (Table 3), we may
observe: (i) a substantial homogeneity in the scores attributed to each real estate unit),
both on the whole and considering the various types of properties (Fig. 4), also with
respect to the classification for their location (Fig. 5); (ii) a considerable level of significance for most of the features considered, both specifically of the individual categories,
and on the whole, compared with the almost total irrelevance of few others (Fig. 6). Both
are in fact suited to reflect the actual situation of the luxury real estate market in the
historic centre of Rome.

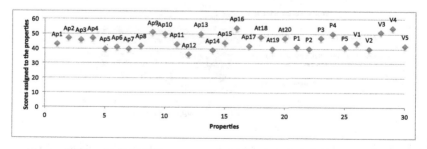

Fig. 4. Scores assigned to the properties for the features present

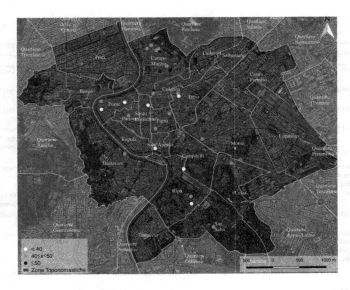

Fig. 5. Score for the features present in each property, by score classes for localization (Color figure online)

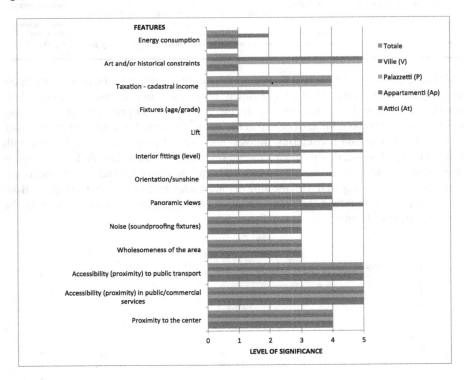

Fig. 6. Level of significance for the features considered (Color figure online)

5   Conclusion

The processing operations made it possible to underscore, for a significant sample of luxury properties located in Rome's historic centre, both correspondence with the regulatory-type requirements connected with urban planning and tax purposes (connected with the sale) and compliance with 13 characteristics that may have an impact on the properties' market price, which of the criteria taken into consideration are actually more suited to respond to the regulatory demands and to reflect the luxury real estate market in the considered territorial setting. The organization of the surveyed data in a database and in a georeferenced system (open source) permit even more detailed processing operations and allow them to be mapped.

What emerged in the research may be the basis for a new arrangement of regulatory provisions, more pertinent to the actual conditions of the supply and demand for the housing that currently (2015) constitutes this market segment.

References

Agenzia delle Entrate. Gli immobili in Italia 2015 (2016). http://www.agenziaentrate.gov.it/wps/content/nsilib/nsi/agenzia/agenzia+comunica/prodotti+editoriali/pubblicazioni+cartografia_catasto_mercato_immobiliare/immobili+in+italia/gli+immobili+in+italia+2015

Agenzia delle Entrate. Osservatorio sul mercato immobiliare - OMI. Rapporto immobiliare 2015. Il settore residenziale (2015). http://www.agenziaentrate.gov.it/wps/content/Nsilib/Nsi/Documentazione/ omi/Pubblicazioni/Rapporti+immobiliari+residenziali/

Calabrò, F., Della Spina, L.: The public-private partnerships in buildings regeneration: a model appraisal of the benefits and for land value capture. In: 5th International Engineering Conference 2014 (KKU-IENC 2014). Advanced Materials Research, vol. 931–932, pp. 555–559 (2014). doi:10.4028/www.scientific.net/AMR.931-932.555

Christie's International Real Estate. What price defines a "luxury" home in your market? Market insights. 2 July 2015 (2015a). http://luxurydefined.christiesrealestate.com/blog/market-insights/what-price-level-defines-a-luxury-home-in-your-market

Christie's International Real Estate. Luxury Defined 2015 white paper (2015b). http://luxurydefined.christiesrealestate.com/hubfs/CIRE_White_Paper_2015.pdf?pdf=luxury-defined

Forte, F., De Rossi, B.: Principi di economia ed estimo. Etas, Milano (1979)

Guarini, M.R., Locurcio, M., Battisti, F.: GIS-based multi-criteria decision analysis for the "highway in the sky". In: Gervasi, O., Murgante, B., Misra, S., Gavrilova, M.L., Rocha, A.M.A.C., Torre, C., Taniar, D., Apduhan, B.O. (eds.) ICCSA 2015. LNCS, vol. 9157, pp. 146–161. Springer, Heidelberg (2015). doi:10.1007/978-3-319-21470-2_11

Knight Frank. European cities review. Luxury residential market performance in europe's key cities 2014 (2015). http://content.knightfrank.com/research/635/documents/en/2014-2082.pdf

Liu, C., Yermack, D.: Where are the shareholders' mansions? CEOs' home purchases, stock sales, and subsequent company performance. In: Boubaker, S., Nguyen, B.D., Nguyen, D.K. (eds.) Corporate Governace, pp. 3–28. Springer, Heidelberg (2012). doi:10.1007/978-3-642-31579-4_1

Morano, P., Tajani, F., Locurcio, M.: Land use, economic welfare and property values. An analysis of the interdependencies of the real estate market with zonal and socio-economic variables in the municipalities of the Region of Puglia (Italy). Int. J. Agric. Environ. Inf. Syst. 6(4), 16–39 (2015). doi:10.4018/IJAEIS.2015100102

Nesticò, A., Galante, M.: An estimate model for the equalisation of real estate tax: a case study. Int. J. Bus. Intell. Data Min. 10(1), 19–32 (2015). doi:10.1504/IJBIDM.2015.069038

Nesticò, A., Pipolo, O.: A protocol for sustainable building interventions: financial analysis and environmental effects. Int. J. Bus. Intell. data Min. (IJBIDM)10(3), 199–212 (2015). doi: 10.1504/IJBIDM.2015.071325

Nomisma and Tirelli & Partners. Osservatorio sulle residenze esclusive. II Semestre 2015 (2016). http://www.tirelliandpartners.com/download/download_file_219369804.pdf

Roscelli, R. (ed.): Manuale di Estimo, Utet Università, De Agostini Scuola, Novara (2014)

Sant'Andrea Luxury House. Il mercato degli immobili di pregio settore residenziale H1 2015 (2016). http://www.immobilisantandrea.it/MEDIA/news/ufficio_studi/Immobili_di_Pregio_H1_2015.pdf

Song, Y., Knaap, G.J.: New urbanism and housing values: a disaggregate assessment. J. Urban Econ. 54(2), 218–238 (2003). doi:10.1016/S0094-1190(03)00059-7

Tajani, F., Morano, P.: An evaluation model of the financial feasibility of social housing in urban redevelopment. Property Manage. 33(2), 133–151 (2015). doi:10.1108/PM-02-2014-0007

Wealth-X and Sotheby's International Realty. The global luxury residential real estate report 2015 (2015). http://www.wealthx.com/wp-content/uploads/2015/02/Wealth-X-Sothebys-Global-Luxury-Residential-Real-Estate-Report-2015.pdf

GIS Infomobility for Travellers

Francesco Castelluccio[1], Gabriele D'Orso[1], Marco Migliore[1(✉)],
and Andrea Scianna[2]

[1] Department of Civil Environmental Aeronautics and Materials Engineering,
University of Palermo, Transport Research Group,
Viale delle Scienze Building 8, 90128 Palermo, Italy
{francesco.castelluccio,marco.migliore}@unipa.it
[2] ICAR-CNR, c/o GISLAB at Department of Architecture,
University of Palermo, Viale delle Scienze Building 8, 90128 Palermo, Italy
andrea.scianna@cnr.it

Abstract. Geographical Information Systems (GIS) are essential systems to support decisions on territorial and environmental aspects. But they not always have been properly used for this purpose. Only in recent years GIS have been getting better used for the planning, management and control of the territory. The application of GIS to the transport sector has become relevant both for management and decision-making in support of Public Administration (PA) and citizens. GIS are particularly useful for roads and routing graphs management capabilities as well as for searching the most suitable path. The results achieved in this research activity aimed to evaluate different road graphs, proprietary and free ones, in order to compare calculated distances and travel times.

Keywords: Webgis · Infomobility · Travel time · Road graph · Routing application

1 Introduction

The attention to Geographic Information Systems (GIS) [1] has developed recently, in order to support the decisions of both the Public Administration [4], (about the optimization of land use planning [9]) and citizens (as users of infomobility services [2]).

The main aim of this work is the construction of WebGIS applications that allow travelers using private transportation modes to plan their trips, allowing them to make use of infomobility services in a future perspective. They should ensure them a complete and updated overview of the transport supply, of travel times, of possible routes, of traffic and road conditions.

These systems could be extended to other modes of transport existing into a territory and with a more or less complex integration of real-time information regarding the possible congestions, the weather and the infrastructure conditions, in order to obtain an application as accurate as possible for a "real time routing" [11].

The effectiveness of available maps seems interesting with regard to the optimization of the user's travel time.

In order to have a large availability of routing applications, which are free and open to all potential users, it is particularly important to ensure the reliability of travel times

O. Gervasi et al. (Eds.): ICCSA 2016, Part III, LNCS 9788, pp. 519–529, 2016.
DOI: 10.1007/978-3-319-42111-7_41

of road links derived from OpenStreetMap (OSM), an open source project, which provides free map data. This opportunity, however, presents many risks associated with the verification of updated information, which should be supported by protocols with an high redundancy level. Therefore, it is interesting to verify that the travel times between two locations, derived from the use of the OSM graph, are reliable or at least comparable to those derived from research and specific studies, which include experimental investigations. So it is appropriate to compare three road graphs, the first acquired by the OSM, the second, owned by the Sicilian Region Administration, the third, owned by Google, in order to determine which graph is the most efficient if used within a WebGIS application, in searching for the shortest path between two locations.

2 Background

In 1959 Dijkstra Edsger Wybe [6] published a paper in which he described an algorithm for searching the shortest path between two nodes of a graph.

On this algorithm, more simple and efficient in respect to other ones existing, are based a plethora of computer application operating on graphs (ranging from routing of data over computer networks to applications on road graphs). Dijkstra algorithm is commonly used in GPS mobile navigators as well as WebGIS routing applications.

This algorithm has been successively improved; based on it other algorithms have been proposed like the A* [7].

Both Dijkstra's and A* algorithms have been implemented in GIS, WebGIS applications and navigation devices handling geographic data structured as graphs.

With the evolution of information technology and the advent of road navigators, many companies and public and private bodies started to produce and distribute different road graphs. Examples are Tele Atlas now acquired by Tomtom, Navtech and Google (initially user of Tele Atlas products) that now produces his own road graphs.

All these companies continuously update their databases on the roads, traffic, incidents and blocking situations; data acquisition is carried out using vehicles equipped with mobile mapping systems (MMSs); roads information is continuously updated using data received from satellite navigation systems and users smartphone connected to the Internet.

A graph is a mathematical structure consisting of nodes and links. Each link connects two nodes (also called vertices), while a node can be the endpoint of any number of links. By convention, often, the nodes are indicated by lower case letters, while the strings are indicated by the pair of endpoints enclosed in parentheses. It will thus, for example, the nodes i and j and the link (i, j). A graph is said weighed if quantitative characteristics are associated to each link, such as the transfer time. The links are classified according to road categories, through the number of lanes, the width, the slope and the windingness of the roadway, assigning different values of capacity and free flow speed for each road category. Road categories are typically: motorways, national roads, regional and provincial roads, municipal and urban roads. The cost function of a link is defined as the mathematical relation that binds the average cost of transport to the flow that affects the road and to the physical and functional characteristics of the connection represented by the link itself. In the scientific literature of the field [3], different cost functions applicable to urban and suburban roads are described.

The different companies producing road graphs adopted a data structure largely derived from the standard CEN TC278 and usually customized for their needs. Even for the Italian road cadastre there is a structure organized in hierarchical levels of progressive complexity adhering to the CEN TC278.

All the above products are proprietary commercial type.

In 2004, an open collaborative mapping project called OpenStreetMaps (OSM) has been started, aimed to creating road maps in open format. The project led to the creation of a world map of the roads and other related infrastructure, which today is actually a free and usable by all people. Obviously it is not guaranteed the continuity of the data upgrade given the voluntary collaborative nature of the project. The OSM road graph today covers enough the territory and can be effectively used for navigation on it.

The Sicilian Region in 2004 has commissioned the preparation of the Regional Transport Plan [10], for which it was specifically created a graph of the road network of regional interest.

3   The Road Graphs

The road graph acquired by OpenStreetMap (Fig. 1), being the result of field mapping using GPS/GNSS (Global Positioning System/Global Navigation Satellite System) receivers, consists of links that perfectly follow the topographic shape of the roads in the territory; the graph owned by the Sicilian Region Administration (Fig. 2), on the contrary, presents a lower level of detail, being constituted by straight and longer links (theoretical connections) than the first one. As for the speed, OpenStreetMap (OSM) assigns to each link a specific default speed, relatively to the type of road that the link represents; besides the road classification used for OSM does not reflect precisely the technical-functional classification and the administrative one proposed by the Italian laws. Regarding the graph provided by the Sicilian Region Administration, each link has a free-flow speed, depending on the orography of the territory and resulting from a transport modeling of the Sicilian road network.

Other characteristics are associated to each link such as the road category (motorway, national, regional, provincial and urban), the number of lanes, the road capacity. Each traffic generator or traffic attractor is interfaced with the road network via centroid connectors, which are abstract links connecting centroids to realistic access point on the physical network [8].

The third graph, is not provided in an open format by Google that only allows using its application, called "Maps", to find the shortest path between two locations.

4   Comparison of the Road Graphs

The software used to develop the experimentation was the PostgreSQL, a free and open source relational database software, which implements the SQL language to query and manage databases. The spatial extension PostGIS was used too in order to ensure that

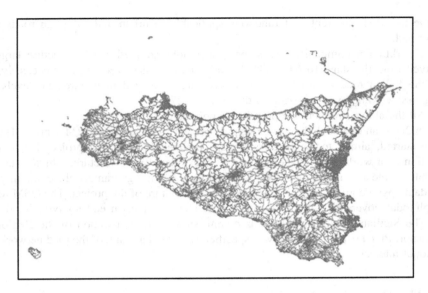

Fig. 1. The road graph acquired by OSM

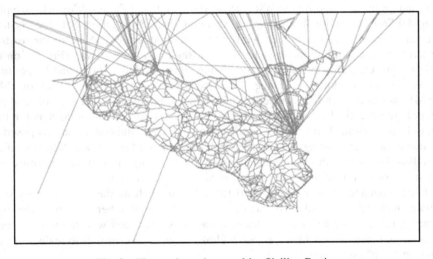

Fig. 2. The road graph owned by Sicilian Region

the PostgreSQL could support geo-referenced data, while the extension pgRouting was employed for applying routing algorithm. Through this extension it was possible to compare the road graphs by searching the shortest paths between different locations, applying each time the algorithms of Dijkstra and A* [5]. After the creation of three separate databases, obtained importing data regarding the two graphs and Google Maps, 15 Origin-Destination (OD) pairs common to all graphs were selected. The choice of the O-D couples took into account mainly the province of Ragusa, as study case, for the presence of important cultural and tourist attractions, such as Ragusa Ibla, Modica, Noto, and also for the presence of the strategic airport of Comiso and the maritime harbour of Pozzallo, by which Sicily is linked to Malta. For each OD pair the shortest path was calculated using SQL language, considering the travel time as a cost. The strings are such as the following:

```
SELECT * FROM shortest_path('
    SELECT id AS id,
           source,
           target,
           cost AS cost,
           reverse_cost AS reverse_cost
    FROM hh_2po_4pgr',
22284,
135280,
true,
true);
```

The free and open source software Quantum GIS has been used for graphical management of the graph and for showing results. For example the shortest paths between Pozzallo and Modica, identified by the algorithm of Dijkstra, are shown into Figs. 3, 4 and 5, respectively using the OSM road graph, the Sicilian Region graph, and Google Maps. While Fig. 6 shows an overlap of the three paths.

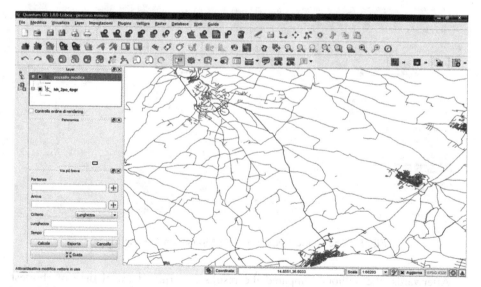

Fig. 3. The shortest path Pozzallo-Modica on the road graph acquired by OSM

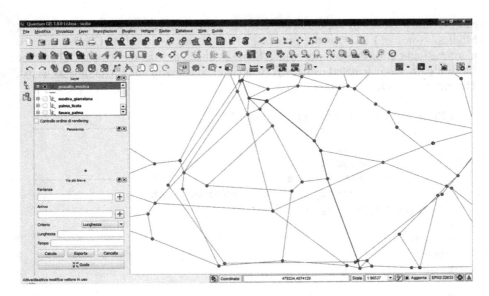

Fig. 4. The shortest path Pozzallo-Modica on the road graph provided by Sicilian Region

Fig. 5. The shortest path Pozzallo-Modica on Google Maps

Afterwards, the authors compared the results in terms of travel time of shortest paths on the three graphs taking into account the best approximation to the real situation. The real travel time was established by a on road campaign experimental

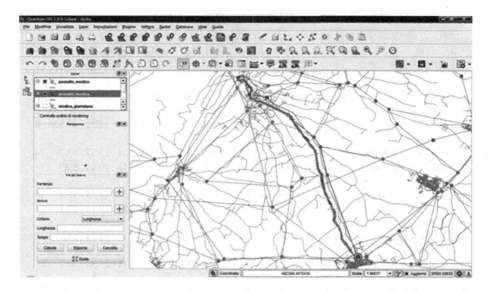

Fig. 6. Overlap of the three paths on the respective maps

measurements, in order to verify the accuracy of the results obtained from the analysis of the three graphs.

The measurements have been carried out during working days and with standard meteorological conditions. By analyzing the results in Table 1, Google Maps and Open Street Maps seem to be more accurate than the Sicilian Region graph into the determination of the distances, while travel time and speed are evaluated more correctly using the graph provided by Sicilian Region.

5 Analysis of Results

It can be observed that the graph owned by the Sicilian Region, which has a lower level of detail, is the most accurate in terms of determination of travel times, resulting into a mean error of 16 % with respect to the real measurement.

The graph acquired from OpenStreetMaps presents an error of 17 %, while Google Maps of 20 %.

It has been noted that Google Maps, in fact, almost always overestimates the travel time. Google Maps indicates travel times greater than the real one, being more imprecise than the other two, but well structured and coherent; the OSM and Sicilian Region graphs instead, overestimate the travel time in some cases and underestimate it in others, so they are not well structured as the first, as it is shown in Table 2.

This fact, in the case of the graph owned by Sicilian Region, may be due to the absence of a real modeling of the urban road network and, therefore, the speed in the links near the settlements is overestimated.

Table 1. Comparison of the lenght and of the travel time

Path	Origin	Destin.	Lenght [km]				Travel Time [min]			
			G	O	S	E.M.	G	O	S	E.M.
1	Via Ungaretti Pozzallo	Via Dente **Modica**	20	20.46	17.65	19.8	29	22.07	20.11	21
2	Via Ungaretti Pozzallo	Via Pace **Comiso**	52.4	47.63	50.66	46.3	50	45.3	44.62	40
3	Via Ungaretti Pozzallo	Via Odierna **Ragusa**	39.1	33.65	34	39.4	43	35.22	32.85	36
4	Via Odierna **Ragusa**	Via Pace **Comiso**	15.9	16.1	15.81	15.4	24	15.37	17.29	20
5	Via Pace **Comiso**	Via R. Settimo **Vittoria**	9.4	9.95	9.31	9.3	16	10.46	9.89	11
6	Via Dente **Modica**	Via Odierna **Ragusa**	15.6	14.81	12.83	15.3	24	14.53	15.4	24
7	Via Ungaretti Pozzallo	Via De Gama **Scicli**	21.6	25.11	18.74	25.3	29	27.14	22.13	24
8	Via R.Settimo **Vittoria**	Via Sassari **Gela**	34.3	34.51	34.15	35.8	41	31.34	34.81	35
9	Via Sassari **Gela**	Via della Salvia **Licata**	36.1	34.36	36.12	37	41	31.93	37.27	39
10	Via Ungaretti Pozzallo	Via Statale **Ispica**	9	7.81	8.31	8.5	11	7.33	9.98	10
11	Via Statale **Ispica**	Via Dente **Modica**	19.9	21.51	20.23	18.6	28	23.77	21.02	23
12	Via Odierna **Ragusa**	Via Archimede **Giarratana**	24	21.28	17.3	23.5	32	20.98	17.72	28
13	Via Cicchillo **Favara**	SS. Palma-Scalo **Palma di Montechiaro**	22.8	22.76	21.55	23	25	20.80	19.37	21
14	SS. Palma-Scalo **Palma di Montechiaro**	Via della Salvia **Licata**	18.4	18.44	18.22	19	16	18.41	17.06	18
15	Via Dente **Modica**	Via Archimede **Giarratana**	28.7	28.52	28.41	27.9	32	26.89	31.06	31

G: Google Map; O: Open Street Map; S: Sicilian Region; E.M. Experimental measurement

The paths going across different towns, such as the paths Pozzallo-Ragusa and Modica-Ragusa, which cross the town of Modica, or those which are for a large part within an urban center, such as the paths Ragusa-Comiso and Comiso-Vittoria, underestimated the travel time confirming that speeds are overestimated.

Lastly, it can be noted that for the windingness paths, such as Modica-Ragusa and Gela-Licata, Google Maps and the road graph owned by Sicilian Region calculate travel times closer to reality while OpenStreetMap makes errors.

So, although the graph owned by Sicilian Region does not follow with precision the planimetric layout of the streets, the speeds are assigned in this road graph in such a way as to take account of it. (Tables 3 and 4).

In the assignment of speed to links defined by OSM this factor was not considered, referring only to the type of road and other characteristics such as the number of lanes.

Table 2. Errors occurred in the determination of the travel time

Travel Times	Google Maps	OSM graph	Sicilian Region graph
Path	Square of the standard deviation		
1	64	1	1
2	100	28	21
3	49	1	10
4	16	21	7
5	25	0	1
6	0	90	74
7	25	10	3
8	36	13	0
9	4	50	3
10	1	7	0
11	25	1	4
12	16	49	106
13	16	0	3
14	4	0	1
15	1	17	0
RMSE (min)	**5.05**	**4.39**	**3.95**
Mean Error (%)	**20 %**	**17 %**	**16 %**

Table 3. Errors occurred in the determination of the lenght

Lenght	Google Maps	OSM graph	Sicilian Region graph
Path	Square of the standard deviation		
1	0.04	0.44	4.62
2	37.21	1.77	19.01
3	0.09	33.06	29.16
4	0.25	0.49	0.17
5	0.01	0.42	0.00
6	0.09	0.24	6.10
7	13.69	0.04	43.03
8	2.25	1.66	2.72
9	0.81	6.97	0.77
10	0.25	0.48	0.04
11	1.69	8.47	2.66
12	0.25	4.93	38.44
13	0.04	0.06	2.10
14	0.36	0.31	0.61
15	0.64	0.38	0.26
RMSE (km)	**1.96**	**2.00**	**3.16**
Mean Error (%)	**8 %**	**8 %**	**13 %**

6 Conclusion

This work is the basis for the creation of a WebGIS that allows route planning for all possible users of private transport systems, enabling users of an area to have a complete and updated overview of the transport supply, with the ability to calculate in advance the travel times of their movements and information about traffic and road conditions.

Table 4. Errors occurred in the determination of the average speed

Average Speed	Google Maps	OSM graph	Sicilian Region graph
Path	Square of the standard deviation		
1	230.80	0.90	15.30
2	43.16	40.50	1.76
3	123.40	69.58	12.72
4	41.60	277.21	75.07
5	239.55	40.29	33.11
6	0.56	524.70	137.76
7	344.49	59.87	154.78
8	124.91	22.07	6.30
9	16.76	58.42	1.50
10	3.64	167.16	1.08
11	34.56	33.33	85.07
12	28.70	110.27	67.58
13	120.87	0.00	1.08
14	32.11	10.47	0.56
15	0.04	92.87	0.78
RMSE (km/h)	**9.61**	**10.03**	**6.30**
Mean Error (%)	**17 %**	**18 %**	**11 %**

The system can further be extended to other modes of transport and integrated with real-time information concerning the movement (such as, for example, congestion, bad weather, accidents, road works), to the more correct determination of optimal and personalized route based to user requests and for more realistic traffic conditions.

The graph owned by the Sicilian Region Administration has the best approximation to the real measured travel times, because the best data on the real travel speeds, although with a significant error. It can be used in a WebGIS application that well satisfies the requirements of the users. However, being the graph acquired by OSM more accessible and free, it may be desirable also its future use in the creation of a WebGIS application designed to provide services in the routing and in the search for the shortest path between two locations.

However, the costs assigned to its links should be changed, equating them to the costs provided by the graph owned by the Region, which are more in line with reality.

References

1. Burrough, P.A.: Principles of geographical information systems for land resource assessment. Clarendon Press, Oxford (1986)
2. Burrough, P.A., McDonnel, R.: Principles of geographical information systems. Oxford University Press, Oxford (1998)
3. Cascetta, E.: Transportation systems engineering: theory and methods. Kluwer, Netherlands (2001)
4. Catalano, M., Migliore, M.: A Stackelberg-game approach to support the design of logistic terminals. J. Transp. Geogr. **41**, 63–73 (2014)
5. Cormen, T.H., Leiserson, C.E., Rivest, R.L.: Introduzione agli algoritmi. Jackson Libri, Italy (1999)
6. Dijkstra, E.W.: A note on two problems in connexion with graphs. Numer. Math. **1**(1), 269–271 (1959)
7. Hart, P.E., Nilsson, N.J., Raphael, B.: A formal basis for the heuristic determination of minimum cost paths. IEEE Trans. Syst. Sci. Cybern. **4**(2), 100–107 (1968)
8. Hensher, D.A., Button, K.J.: Handbook of Transport Modelling. Pergamon, Oxford (2000)
9. Migliore, M., Burgio, A.L., Di Giovanna, M.: Parking pricing for a sustainable transport system. Trans. Res. Procedia **3**, 403–412 (2014)
10. Sicilia, R.: Piano attuativo delle quattro modalità di trasporto: stradale, ferroviario, marittimo, aereo (2004)
11. Van Zeijl, N.: Including Practical Information in Route Planning Applications. Thesis Submitted in partial fulfillment of the Requirements of the University of Rotterdam for the Master in Operational Research and Quantitative Logistics. University of Rotterdam, Rotterdam (2013)

Resilience and Smartness of Coastal Regions. A Tool for Spatial Evaluation

Giampiero Lombardini[1] and Francesco Scorza[2(✉)]

[1] Department Science for Architecture, University of Genoa (I),
Stradone San'Agostino, 37, Genoa, Italy
g.lombardini@arch.unige.it
[2] School of Engineering, Laboratory of Regional and Urban Systems Engineering (LISUT),
University of Basilicata, Viale dell'Ateneo Lucano 10, 85100 Potenza, Italy
francesco.scorza@unibas.it

Abstract. The paper explores the potential of a method of analysis and evaluation in which the system of values are compared with the conditions of risk and degradation of landscape (due to environmental and man-made drivers) with the result of building maps of vulnerability, fragility and resilience.

A set of italian coastal regions and landscapes constitutes the case study and are interpreted as geographical spaces with high levels of sensitivity. The research objective is to build a spatial decision support system based on indicators, in which the relative parameters are measured to represent the resilient capacity of the system, taking into account indicators relating to the integration of technological system, system-law system, social system and economic system. The global indicator system provides the basis upon which to assess how much a territorial system is resilient and smart. The SDSS logical model allows to build different scenarios and to evaluate the effectiveness of policies and spatial strategies analyzed as intervention hypothesis.

Keywords: GIS · Resilience · Cultural landscapes · Indicators · Spatial decision · Spatial systems

1 Working on Resilience, Improving Sustainability

Among the consequences of actual rapid urban growth, projected to reach 6.4 billion by 2050, the increasing disaster vulnerability and exposure is one of the main relevant components. A global commitment is projected to tackle this issue and multi scale approaches are on the political debate involving civil protection agencies and governments. The results are fragmented in a wide numbers of local case studies and initiatives, sometimes accompaigned by a lack of documentation and so characterized by a weak transferability.

A renewed approach concerning Disaster Risks Management comes form the "Resilience Approach": we operatively refer to "Sendai Framework for Disaster Risk Reduction 2015–2030" [1].

© Springer International Publishing Switzerland 2016
O. Gervasi et al. (Eds.): ICCSA 2016, Part III, LNCS 9788, pp. 530–541, 2016.
DOI: 10.1007/978-3-319-42111-7_42

The Sendai Framework for Disaster Risk Reduction 2015–2030 (SFDRR) was adopted at the Third UN World Conference in Sendai, Japan, on March 18, 2015. It is the outcome of stakeholder consultations initiated in March 2012 and inter-governmental negotiations from July 2014 to March 2015, supported by the United Nations Office for Disaster Risk Reduction at the request of the UN General Assembly.

The SFDRR replaced the Hyogo Framework for Action (HFA) 2005-2015: Building the Resilience of Nations and Communities to Disasters (UN, 2005). The HFA was conceived to give further impetus to the global work under the International Framework for Action for the International Decade for Natural Disaster Reduction of 1989, and the Yokohama Strategy for a Safer World: Guidelines for Natural Disaster Prevention, Preparedness and Mitigation and its Plan of Action, adopted in 1994 and the International Strategy for Disaster Reduction of 1999.

Analyzing such framework global strategy it emerges an inclusive approach as leading tool for addressing four operative priorities:

- Priority 1: Understanding disaster risk
- Priority 2: Strengthening disaster risk governance to manage disaster risk
- Priority 3: Investing in disaster risk reduction for resilience
- Priority 4: Enhancing disaster preparedness for effective response and to "Build Back Better" in recovery, rehabilitation and reconstruction.

What is definitively clear is that the approach on disaster management shifted in anticipating actions (i.e. preparedness) based on a multistakeholders perspective. Involving local community in active process of empowerment (knowledge transfer), engagement (people and local groups commitment), and investments (from soft to hard scale) in reducing territorial vulnerability is main challenge to face future scenario.

If we focus on urban and territorial planning practices the comprehensive framework promoted by SFDRR could identify new tools and approaches to be included in the regulatory framework of territorial governance. Beyond traditional urban and territorial master plans the resilience approach needs to explore integrated territorial policies and projects for a concrete implementation. It means that more than a "resilience plan" we need to include resilience approach in each territorial transformation.

This represent a complex implementation framework especially if we consider current practices in managing everyday procedure in municipal technical office – especially where human resources and technical support systems were reduced to the minimum as a consequence of economic crisis, reduction of public funds or administrative reforms. Such dimension appears to be very far from SFDRR application but at the same time it is responsible of wide territorial extension, often characterized by rural space with high environmental values.

In other words if we believe that Metropolitan areas will rapidly develop tools for implementing SFDRR, what will happen in rural and marginal regions. That's the case of Italian coastal areas where seasonality in tourism and economy, depopulation of small town and progressive abandonment of agricultural tradition depict a fragile territorial framework highly vulnerable in terms of environmental risks and human pressure.

Probably a stronger commitment at global scale was affirmed in the domain of sustainability: we refer to Sustainable Development Goals (SDGs) (UN, 2015).

At the UN in New York the Open Working Group created by the UN General Assembly proposed a set of global Sustainable Development Goals (SDGs) which comprises 17 goals and 169 targets. We could say an "umbrella" statement already enlarged, compared to the first elaboration, oriented to include from generalized policies to specific intervention with the overall score of achieving sustainability, promoting sustainability, at least stressing the concept of sustainability in a viral perspective.

Some SDGs build on preceding Millennium Development Goals while others incorporate new ideas.

The current format of the proposed SDGs and their targets has laid a policy framework; however, without thorough expert and scientific follow up on their operationalisation the indicators may be ambiguous.

The debate on assessing sustainable growth is increasing: from the research of effective quantitative measure of sustainability [2] to the declaration of inapplicability in some cultural context [3].

Among others Uitto [4] affirm that sustainable development and disaster risk reduction are closely linked on many levels and the relationship cuts both ways. Disasters add often devastating costs to societies and communities in terms of financial losses, destroyed infrastructure and loss of life. They can set development back for years. Environmental destruction and lack of sustainable development exacerbate disaster risk and impact. Climate change is adding to the risk and uncertainty.

Those tree components: disaster risk reduction, Environmental sustainability, Climate change, define a new set of criteria to asses territorial transformation processes in order to achieve strategic goal defined by UN and accepted by governments world wide in a long term view.

The expectations regards an overall improvement of urban and territorial management processes based on an inclusive approach involving actively local communities and local groups in implementing every day actions.

Stake holders involvement brings benefits for policies implementation, such general statement was demonstrated in many application in differen sectors and domains. Among recent studies: Gillund et al. [5] in farming and agricultural application; Khalid et al. [6] in human right sustainable principles; Huang et al. in megalopolis development [7]; Aydın et al. concerning fiscal policies [8].

Such plurality of approaches and applications allow to assess the relevance of such global challenge. But which opportunities for resilience?

Sustainable growth implies durability in time of input-output resources cycle of a specific project. Such durability should consider as a priority the risk assessment according to a comprehensive and integrated multiparameter perspective. A relevant component of risks assessment regards environmental risks. This depends on territorial characteristics determined by the localization of the project. The preparedness to face risks represents the resilience degree of a specific action.

Such simplified chains represents a component of project cycle assessment but, starting from sustainability assessment, affirms that resilience represents a positive outcome of sustainable design for territorial transformation. A general improvement could come only from a distributed awareness in implementing such approach in several domains: i.e. urban regeneration, transport and infrastructure management, economic and industrial investments, agricultural practices, environmental preservation and exploitation etc.

2 The Landscape Dimension of Resilience: The Case of Coastal Landscapes

Generally, in most cases, information relating to the concept of resilience applications are used in the prevention and management of natural disasters [9]. In fact, the concept of the resilient capacity of a system seems to find its most fertile applicability with regards to shocks caused by extreme events. It can then be cause for debate whether it is possible and useful to adopt a similar paradigm for risk assessment and landscape degradation assessment. Certainly resilience is emerging in the scientific world as a paradigm widely used in different fields and with useful implications in all those matters in which the change is deep and rapid [10]. In the case of the landscape, conceptual models that adopt the resilience as logical structure seem to be useful because in the landscape we can observe intangible values that go beyond the physical and configurational space. These values take on the characteristics of elements of identity recognizable and recognized [11]. Identity values, however, must be preserved and protected, since these are configured as the main features of the landscape. On the other hand, the landscape itself is subjected, in the current historical phase, to a series of pressures and impacts (including environmental dynamics analysed in most of the studies on resilience, and, above all, the climate change) that, because their intensity, rarely has had place in the history [12]. The changes taking place in landscapes inherited from centuries of historical layers and have built their identity still recognized. These changes have place in a series of transformations that often, today, alter the cultural, symbolic as well as material meanings.

In the case of coastal landscapes, we are faced with situations in which the landscape is often not only the element of identity recognition by the communities, but also the key factor in the economic and social development of the region. A landscape and in particular a coastal system, in general, can be represented in three sub-critical systems: natural environment, human settlements and cultural heritage. Each of these sub-systems contribute to create a recognized expression of the sedimentation of the physical and symbolic signs accumulated during the historical process of regionalization [13, 14]. The coastal historical landscapes are often invested by change processes, which in some cases take on the size of real shocks. Climatic conditions, rapidly changing, along with an inheritance consisting of a chaotic and speculative urbanization and a progressive reduction of investments for the territory and its maintenance threaten the local landscape and with it the same economy of the communities. The double and parallel dynamics of urbanization (which is not only manifested in the land use) on the one hand and abandonment on the other, lead to a widespread deterioration of the conditions of vulnerability. In this context, a reading of the adaptive capacity of the regional context compared to the changes in place can be the basis for the construction of strategic scenarios project [15, 16] and evaluation frameworks for an assessment of the local conditions.

3   Resilience, Identity and Landscape

In terms of ecology and socio-ecology, the resilience can be understood as the ability of a system to acquire multiple equilibria: the resilience is presented as an intrinsic property of a system that passes from one equilibrium state to another without losing its internal fundamental structure, otherwise also defined in terms of "identity" [17]. Noting that particular type of complex systems such the urban regions and their landscapes, identify two main categories of failure: traumatic changes (shocks) and slow processes of change. While traumatic events (and their consequences and responses) have long been studied in the context of the disasters theories [18]. The second type of change (slow processes of mutation), which are observed in the systems in a gradual but steady transformation, can lead, albeit of longer times and for using of "molecular" transformations and punctual almost imperceptible or little impact if measured individually, also radical changes in the initial conditions. In these cases, the focus is not on the condition of balance and stability, but on the system's ability to adapt to change and to preserve its identity. This approach to the study of regional resilience [19], which can be called evolutionary, resumes conceptual models developed in the framework of studies on socio-ecological systems (SES).

As argued by Cumming [20], the main question is to define the identity of a system and at the same time the thresholds above which the identity of the system change. The question dealing with building indicators that are able to be used in the assessment of "active" conservation status of a system (and a landscape-system in the case considered here). The landscape, in particular, is also part of a larger system, which contribute to the environmental dimensions (eco-systemic), social dimensions (civic traditions, common knowledge, relations) and economic and technological perspectives [21], in a constantly changing dynamic way in which external factors (drivers) lead to more or less significant local modifications.

External conditions (external drivers: [22]), summarized in the sub-systems social context, economic dynamics, global natural phenomena (e.g.: climate change), culture, legal and regulatory environment, technology, lead to a state of pressure on local systems (and on landscapes in a specific way). Tangible and intangible components of the landscape, as well as their relationships, suffer the effects of these stressors and, depending on the local and puntual actions (or sometimes inactions), determined by the response prepared by the local system context (institutions, regulatory model framework, regional economies, local knowledge, technologies applicable), amend their initial conditions. Depending on the susceptibility to the change of various landscape components, these changes may lead either to an adaptation or to a loss (partial or total) of the preceding identity conditions.

To frame the landscape components within a single logical-conceptual model, the coastal landscape can be interpreted in three main dynamics of landscape values production [23]:

(a) Direct production of goods (through the landscape configuration, a local regional system produces goods and services: agricultural products, biodiversity, ecosystem services);

(b) Generation of values: production and reproduction of economic value, both public and private;

(c) Creation of identity values: permanence, persistence, transformation of collective recognition elements (landscape as a common good).

While the first two production dynamics of values, what is to be evaluated is the system's ability to maintain and possibly increase its potential production, in the case of identity values, what is decisive is the ability to preserve the conditions through the landscape-system can materialize the persistence in time of these values (Fig. 1).

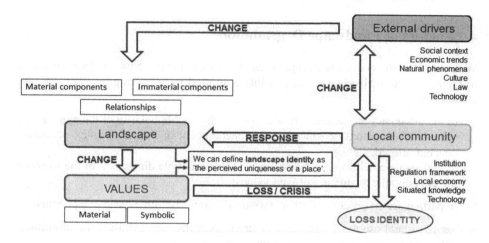

Fig. 1. Landscape as a socio-ecological system

The components of a coastal landscape can be represented by:

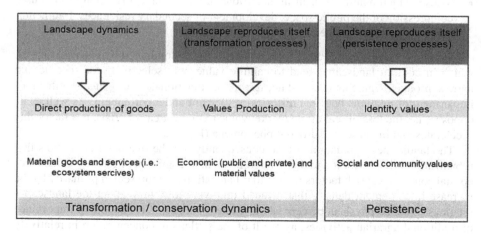

Fig. 2. Landscape values production dynamics

- the physical conditions and forms of the region (geomorphology, hydrology natural land forms, coastal landforms),
- biodiversity,
- land use and land cover
- the settlement morphologies
- the consistency and quality of the cultural heritage,
- the infrastructure system (not only roads but also the set of works and artifacts that allow a settled community to govern the life cycles (such as artificial water system, networks technological, etc.) (Fig. 2).

4 Factors of Landscape Degradation

With regard to the coastal landscape, it can be assumed that the two main phenomena triggered by external drivers (and possibly amplified by local fragility: [24, 25] are attributable:

1. to the urbanization categories (understood in the double polarity of the dynamic an spatial differentiated densification vs abandonment and then complex processes concentration/depletion)
2. to real estate conditions (to be understood in the double dimension of the over- or under economic evaluation of land values).

In particular, the process of urbanization may comprise three basic components:

- Intensive land use that results in pressure respect the convertibility of the building soils, and a general over-use of territorial assets;
- Abandonment in many rural areas: the dynamics of neglect of agricultural farms are increasingly serious factors of crisis, causing dangerous erosion or a messy (without regulation) return of nature (re-naturalization), which in turn further increases vulnerability factors (hydrological, fire risk, usability of forest resources, etc.)
- Misuse, which manifests itself in distortions in the spatial distribution of land uses (compression of the public space, development of secondary residences, fragmentation of service facilities, etc.), activities and functions.

Similarly the distorting effects of a housing market dominated by external factors, which in cultural landscapes tend to capture value only selectively for specific and narrow product segments (real estate, territorial, functional), can generate effects of impairment of relations between historical settlement and environment. As well as its opposite, i.e. the loss of real estate market value to many areas of crisis, is a factor that accelerates and increases the dropout phenomena (Fig. 3).

The landscape degradation is then consequently, on the one hand the effects that urbanization brings with it and that compromise the recognition of identity values. For coastal contexts, the risk factors related to urbanization are represented primarily by the increase in settlement density that, even if relatively low, may affect the landscapes relations systems (physical, perceptual and functional: [26, 27]. The loss of importance of traditional agrarian activities, as result of the settlement concentration in relatively

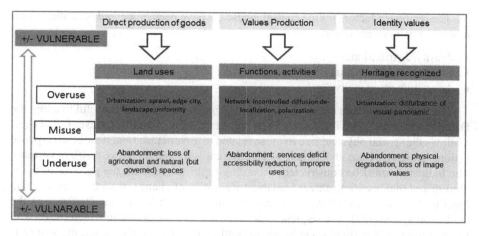

Fig. 3. Landscape dynamics between overuse and abandonment.

few areas with the highest use value and competition over land use to the benefit of "quality" features, implies a drastic reduction of ordinary land maintenance activities in situations in which the urban character of the settlement prevail over those agricultural and to a radical abandonment of a number of oversight activities in areas a time suited to extensive agricultural activities (areas of cultivated forest, the system of the meadows-grazing). The abandonment of the outermost regions, in turn, leads to a profound change in land cover [28], with an impressive of the forest advanced. A forest which does not already goes to colonize natural areas, but areas already populated and transformed. Such ungoverned naturalization phenomenon creates serious ecological and environmental stability problems, with a cumulative impact on the conditions of risk and a progressive landscape degradation. Finally, another phenomenon linked to processes of urbanization that affect the potential degradation of landscape consists of the phenomena of misuse of the land and the artifacts result by history stratifications, that led to the consolidation in landscapes highest cultural value.

5 Indicators for Assessing the Resilience of Coastal Landscapes

The assumption at the basis of process of building a framework for analysis of the landscape resilience is that all complex systems reflect, to a certain level of analysis, the fundamental structure and physical principles of our universe [29]. Having similar behaviours, we can represent the same processes in a logical structure that have similar behaviours in different fields, hollowing out a number of useful concepts for the definition of resilience and resiliency threshold (or vulnerability). In this sense, we can define those processes that concur to define the ability of a system to maintain their identity, namely to show a resilient behaviour: the main appear to be those of connection (network behaviour, strength of the relations, hierarchy, organization), fragmentation, heterogeneity, redundancy. The resilience of a territory conditions can therefore be represented, in the first instance, by these four processes.

The resilience factors for the coastal landscape have been processed according to the following pairs of elements:

- density/distribution;
- Connection/decoupling.
- diversity/homogeneity,

If the aim is to assess the resilience through indicators, it may be useful to adopt the DPSIR model (i.e.: Driving forces, Pressure, State, Impact and Response: [30, 31], as it allows the spatial physical size to those related to governance (and hence the response actions, which here are understood as adaptive capacity and therefore resilient system). The indicators must be constructed, according to this logic, with the purpose of interpreting and making some extent readable trajectories of change of landscapes, holding together the dynamics arising from external drivers with local characteristics of the landscape-system [32]. The sequence, the intensity, and the period in which the driving forces acting on a particular landscape depend on the characteristics and the susceptibility of the latter, in turn determined by the course of history, from the geographical characteristics and the barriers imposed by beliefs and local practices (cultural, legal, etc.: Lambin et al. [33]). The external drivers modify the landscape according to various sizes, extensions, frequencies and speed, leading to the creation of a fragmented landscape, composed of a mosaic of different natural and manmade systems at different levels of modification and use [34].

The landscape-environmental knowledge elements are compared to the structural and spatial elements resulting from human activities that generate impacts and effects (indicators of driving forces and impacts) on the local system. The latter in turn can be articulated in two models: the first concerns the representation of the territorial status conditions (status indicators) and the second the representation within a logical assessment model of the elements-processes which define vulnerability and resilience of the territorial system, and in particular of the landscape system (pressure indicators and response).

In the first phase (status indicators), the reconstruction of the environmental features and settlement of the coastal landscape is built by the construction of an heritage atlas which was subsequently amended as synthetic vision. The indicators are intended as measurement and evaluation tools that can measure not only static but dynamic in place according to the concept of transition (e.g. Kaligarič and Ivajnšič [28] using a transition matrix based on neural networks and Markov chains). What the indicator must represent is the historical dynamics of landscape change phenomena, highlighting the transition from one state to another of different areas. The themes through which to interpret the changes in the landscape-environmental conditions are related to soils (defined as the quality of the land and as a systematic set of man-made accommodation, mainly for the purpose of agricultural management, for example, terraces, works of drainage and water systems management, excavations and filling, shaping of slopes), the settlement morphology, to the road network [35] and the land cover. The indicators that represent the system transitions and therefore should then form the basis for assessing the resilience of landscapes bring back the reading of the state of landscape components to an interpretation of their transition through retention cycles, re-use, intensification of use, transformation. This dynamic representation allows to consider the elements and factors on which action to promote those system changes that can ensure the adaptability of the system landscape (Fig. 4).

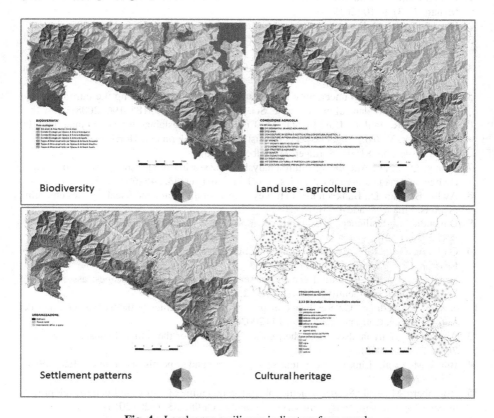

Fig. 4. Landscape resilience indicators framework

The GIS-based system assessment model, in this perspective, assumes the character of a system of knowledge based on an unitary ontology [36] that allows the representation of the basic components of landscapes (geomorphology, hydrology and coastal dynamics, biodiversity, land use and agriculture, settlement patterns, cultural heritage, activities, functions and infrastructure) through indicators measuring levels of fragility, vulnerability and resilience depending on the size of resilience before mentioned: access, distribution, diversity/homogeneity, productivity, memory) refers both to the same components in their integrated and cumulative reading [37]. The dual dynamic of thickening and thinning of artefacts, values, customs (a result of urbanization, which is spatially selective and produce different effects from place to place) is the dynamic that crossed with the state of landscape components provide elements of fragility evaluation of the territory and exposure to the risk of cultural values.

References

1. United Nations: Sendai framework for disaster risk reduction 2015–2030. Aust. J. Emerg. Manag. **30**(3), 9–10 (2015)
2. Hak, T., Sustainability indicators: a scientific assessment. J. Ind. Ecol. **12** (2008)
3. El-Zein, A., et al.: Health and ecological sustainability in the Arab world: a matter of survival. Lancet **383**(9915), 458–476 (2014)
4. Uitto, J.I., Shaw, R.: Sustainable Development and Disaster Risk Reduction. Springer Japan, Tokyo (2015)
5. Gillund, F., et al.: Do uncertainty analyses reveal uncertainties? using the introduction of DNA vaccines to aquaculture as a case. Sci. Total Environ. **407**, 185–196 (2008)
6. Md Khalid, R., Jalil, F., Bin Mokhtar, M.: Environmental sustainability as a human right. In: Mauerhofer, V. (ed.) Legal Aspects of Sustainable Development: Horizontal and Sectorial Policy Issues, pp. 81–94. Springer, Heidelberg (2016)
7. Huang, Q., et al.: Alternative future analysis for assessing the potential impact of climate change on urban landscape dynamics. Sci. Total Environ. **532**, 48–60 (2015)
8. Aydin, B., Igan, D.: Bank lending in turkey: effects of monetary and fiscal policies emerging markets. Financ. Trade **48**(5), 78 (2012)
9. Davoudi, S.: Resilience: a bridging concept or a dead end? Plan. Theory Pract. **13**(2), 299–307 (2012)
10. Gallopín, G.C.: Linkages between vulnerability, resilience, and adaptive capacity. Glob. Environ. Change **16**(3), 293–303 (2006)
11. Tengberg, A., et al.: Cultural ecosystem services provided by landscapes: assessment of heritage values and identity. Ecosyst. Serv. **2**, 14–26 (2012)
12. Kinzig, A., Redman, C.: Resilience of past landscapes: resilience theory, society, and the Longue Duree. Conserv. Ecol. **7**(1), 14 (2003)
13. Antrop, M.: Why landscapes of the past are important for the future. Landsc. Urban Plan. **70**(1), 21–34 (2005)
14. Bürgi, M., et al.: Linking ecosystem services with landscape history. Landsc. Ecol. **30**(1), 11–20 (2015)
15. Rosenberg, M., et al.: Scenario methodology for modelling of future landscape developments as basis for assessing ecosystem services. Landsc. Online **33**, 1–20 (2014)
16. Ramos, I.L.: Exploratory landscape scenarios in the formulation of 'landscape quality objectives'. Futures **42**(7), 682–692 (2010)
17. Berkes, F., Colding, J., Folke, C.: Navigating Social-Ecological Systems: Building Resilience for Complexity and Change. Cambridge University Press, Cambrige (2003)
18. Carpenter, S., et al.: Catastrophic shifts in ecosystems. Nature **413**(11), 591–596 (2001)
19. Folke, C., et al.: Resilience thinking: integrating resilience, adaptability and transformability. Ecol. Soc. **15**(4), 20 (2010)
20. Cumming, G.: Spatial resilience: integrating landscape ecology, resilience, and sustainability. Landsc. Ecol. **26**(7), 899–909 (2011)
21. Bieling, C., Plieninger, T.: Resilience and the Cultural Landscape. Understanding and Managing Change in Human-Shaped Environments. Cambridge University Press, Cambridge (2012)
22. Bürgi, M., et al.: Driving forces of landscape change. Current and new directions. Landsc. Ecol. **19**(8), 857–868 (2004)
23. Stephenson, J.: The cultural values model: an integrated approach to values in landscapes. Landsc. Urban Plan. **84**(2), 127–139 (2008)

24. Hersperger, A.M., Bürgi, M.: Going beyond landscape change description: quantifying the importance of driving forces of landscape change in a central europe case study. Land Use Policy **26**(3), 640–648 (2009)
25. Schneeberger, N., et al.: Driving forces and rates of landscape change as a promising combination for landscape change research—an application on the Northern Fringe of the Swiss Alps. Land Use Policy **24**(2), 349–361 (2007)
26. Hadley, D.: Land use and the coastal zone. Land Use Policy **26**, 198–203 (2009)
27. Fontana, V., et al.: Comparing land-use alternatives: using the ecosystem services concept to define a multi-criteria decision analysis. Ecol. Econ. **93**, 128–136 (2013)
28. Kaligarič, M., Ivajnšič, D.: Vanishing landscape of the "Classic" Karst: changed landscape identity and projections for the future. Landsc. Urban Plan. **132**, 148–158 (2014)
29. Cumming, G.S.: Spatial Resilience in Social-Ecological Systems. Springer, Dordrecht (2011)
30. Cassatella, C., Peano, A.: Landscape Indicators: Assessing and Monitoring Landscape Quality. Springer Netherlands, Dordrecht (2011)
31. Vallega, A.: Indicatori per il paesaggio. Franco Angeli, Milano (2008)
32. Bertolo, L.S., et al.: Identifying change trajectories and evolutive phases on coastal landscapes. Case study: São Sebastião Island, Brazil. Landsc. Urban Plan. **106**(1), 115–123 (2012)
33. Lambin, E.F., et al.: The causes of land-use and land-cover change: moving beyond the myths. Glob. Environ. Change **11**, 261–269 (2001)
34. Antrop, M.: Landscape change and the urbanization process in Europe. Landsc. Urban Plan. **67**, 9–26 (2003)
35. Serra, M., Pinho, P.: Dynamics of periurban spatial structures: investigating differentiated patterns of change on Oporto's urban fringe. Environ. Plan. B: Plan. Des. **38**(2), 359–382 (2011)
36. Lombardini, G.: Dealing with resilience conceptualization. Formal ontologies as a tool for implementation of intelligent geographic information systems. TEMA J. Land Use, Mobil. Envorin. 633–644 (2014)
37. Pavlickova, K., Vyskupova, M.: A method proposal for cumulative environmental impact assessment based on the landscape vulnerability evaluation. Environ. Impact Assess. Rev. **50**, 74–84 (2015)

Political Challenges, Best Practices and Recommendations for Energy Sustainable Municipalities

Valentin Grecu[1]([✉]) and Silviu Nate[2]

[1] Department of Industrial Engineering and Management,
Lucian Blaga University of Sibiu, Sibiu, Romania
valentin.grecu@ulbsibiu.ro
[2] Department of International Relations, Political Science and Security Studies,
Lucian Blaga University of Sibiu, Sibiu, Romania
silviu.nate@ulbsibiu.ro

Abstract. Integrated system solutions, together with improved management of municipal infrastructure and environments, are essential to address energy challenges. A number of aspects should be examined and developed, including eco-cycle models demonstrating integrated solutions for energy, waste and water, integrated land use and transportation, ecosystems planning, sustainable building design, and strategies to reduce air pollution. It is imperative to apply the principles of sustainable development and support smart solutions today if we are not to compromise the ability of future generations to meet their needs tomorrow. We should think of urban systems as living organisms and find synergies between urban processes in a mutually beneficial union. Human beings are social beings, and urban environments provide a wide range and choice of social, educational, cultural and economic opportunities. This paper presents the challenges related to sustainable development and proposes solutions for municipalities that aim to become sustainable.

Keywords: Sustainability · Political challenges · Sustainable municipalities

1 Setting the Context

Earth has a limited capacity to meet the growing demand for natural resources of the socio-economic system and to absorb the destructive effects of their use. Climate change, erosion and desertification, soil, water and air pollution, reduction of the area of tropical forest systems and of wetlands, extinction or endangering a large number of species of terrestrial or aquatic plants and animals, accelerated depletion of non-renewable natural resources have begun to have measurable negative effects on socio-economic development and quality of life of people in vast areas of the planet (Parmenter et al. 2007).

The international community decided to address environmental issues through global collective actions, which it sought to define and implement through appropriate international framework. The international framework for action is formed in a dynamic evolution, including binding legal measures such as treaties or conventions or non-binding measures, in the form of declarations, resolutions, or sets of guidelines and policy recommendations, institutional measures and sustainable funding mechanisms.

© Springer International Publishing Switzerland 2016
O. Gervasi et al. (Eds.): ICCSA 2016, Part III, LNCS 9788, pp. 542–551, 2016.
DOI: 10.1007/978-3-319-42111-7_43

2   Integration of Sustainable Development in EU Policies

Since 1997 through its inclusion in the Maastricht Treaty, sustainability has become a political objective of the European Union. In 2001, the Gothenburg European Council adopted the Sustainable Development Strategy of the European Union, which was added an external dimension in Barcelona, in 2002 (Grecu 2014).

The European Commission launched a review of the Strategy, in 2005, making a critical assessment of the progress made since 2001, pointing out a number of courses of action to follow next. Some unsustainable trends were identified, with negative effects on the environment, which could affect the future development of the European Union, namely climate change, threats to public health, poverty and social exclusion, depletion of natural resources and biodiversity erosion.

In June 2005, following this assessment, the Heads of State and Government of EU countries have adopted a declaration on sustainable development guidelines incorporating the revised Lisbon Agenda for growth and job creation as an essential component of overarching objective of sustainable development (Measuring Progress towards a more Sustainable Europe 2005).

The European Commission proposed on 13 December 2005, a review of the Gothenburg Strategy in 2001. Following this, the renewed Sustainable Development Strategy for an enlarged Europe was adopted on 9 June 2006. General objective of this document is carrying out actions to enable the EU to achieve continuous improvement in quality of life for present and future generations through the creation of sustainable communities able to manage and use resources efficiently and to the potential of eco-innovation and social economy to ensure prosperity, environmental protection and social cohesion (NSDSR 2008).

For this purpose, there identified four key objectives:

- Environmental protection through measures to decouple economic growth from negative environmental impacts;
- Equity and social cohesion, respecting fundamental rights, cultural diversity, equal opportunities and combating discrimination of any kind;
- Economic prosperity by promoting knowledge, innovation and competitiveness to ensure high standards of living and abundant and well paid jobs;
- Meeting the EU's international responsibilities through the promotion of democratic institutions in service of peace, security and freedom and the principles and practices of sustainable development worldwide.

The key objectives of EU policies are summarized in Fig. 1:

EU Strategy sets the following guiding principles to ensure balanced integration and correlation of economic, ecological and socio-cultural aspects of sustainable development (NSDSR 2008):

- Promotion and protection of fundamental human rights;
- Solidarity within generations and between generations;
- Fostering an open and democratic society;
- Informing and involving citizens in decision making;

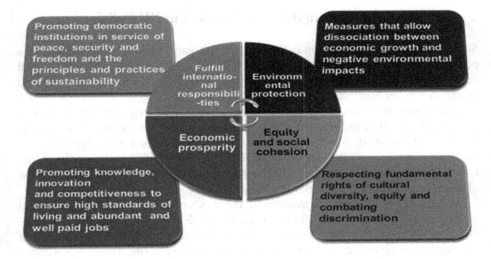

Fig. 1. The key objectives of EU policies

- Involvement of business and social partners;
- Policy consistency and quality of governance at local, regional, national and global levels;
- Integration of economic, social and environmental impact assessments and consultation with stakeholders (stakeholders);
- Use of modern knowledge to ensure economic efficiency and investment;
- Applying the precautionary principle for uncertain scientific information;
- Apply the "polluter pays".

The substance of the EU Strategy focuses on a number of crucial challenges 7 and 2 cross-cutting policy areas. Many of the targets agreed in the EU are set in numerical or percentage, with strict deadlines for implementation, which are mandatory for all Member States.

3   Growing Role of Sustainable Development in EU Policy Making

In recent years the EU has clearly demonstrated its commitment to sustainable development and to successfully integrate this dimension in many policy areas. EU policies on climate change and energy are evidence of the impact that sustainable development strategy has had on the political agenda. The EU has started to integrate the sustainable development in many other policy areas.

EU Agenda regarding better regulation contributes to the integration of policy objectives and improves the effectiveness of policy decisions in terms of costs. This was achieved through the simplification of EU legislation and reducing administrative burden (Wilkinson et al. 2005; Radaelli and Meuwese 2009). It should also be particularly stressed the role of impact assessment. Impact assessment system used by the Commission contributes to sustainable development by assessing the potential economic, social and environmental impacts of new legislation or new policy proposals through an integrated approach (Tamborra 2003; Abaza et al. 2004; Jacob and Volkery 2004).

Commission's Renewed Social Agenda presented an integrated and holistic approach, and several policy initiatives aimed at various policy areas. The Renewed Social Agenda stressed the importance the Commission attaches to sustainable development of our societies and welfare announced targets "beyond GDP" (COM 2009).

Guidelines on employment are part of the European strategy for employment and provide a framework for developing and implementing measures consistent with the objectives of sustainable development strategy (Ball 2001). For example, EU structural funds used to support the efforts of Member States to achieve a low carbon economy and carbon dioxide in terms of efficient use of resources.

Continuing to develop an economy with low emissions of carbon dioxide will also be essential to lead the EU to reverse. Corporate social responsibility (CSR) is an opportunity for businesses to combine the economic, social and environmental objectives. If the European businesses take CSR stronger commitment for sustainable development, this will enhance Europe's capacity of sustainable development (Utting 2000). Commission and several EU member states have intensified efforts to promote the adoption of the concept of CSR, focusing on dialogue between stakeholders (Orbie and Babarinde 2008).

The EU has made progress in integrating sustainable development strategy agenda in its foreign policies, e.g. sustainable development impact assessments undertaken in the context of free trade agreements and training activities on climate change (Persson 2009).

Member States are also developing innovative solutions that are relevant to sustainable development agenda. Using virtual carbon prices to reflect the social costs of emissions of greenhouse gases (GHG) in the evaluation of policy options is becoming increasingly common. France has committed to invest neutral in terms of carbon dioxide emissions with funds for cohesion policy 2007–2013. Another example is the allocation of "budget" of carbon dioxide different departments (United Kingdom of Great Britain and Northern Ireland). There has been progress in terms of carbon accounting in enterprises (Kolk et al. 2008). The results of tests conducted in several Member States have shown that using smart meters can lead to lower energy consumption by up to 10 %. Some Member States have created new systems of energy audit, including financial support, which contributed significantly to reducing energy consumption in construction. Also, the initiative "Grenelle de l'Environnement" in France brought together government, business and civil society in high-level debates on new measures for sustainable development.

4   The Role of Governments in Framing for a Sustainable Future

Raivio (2011) underlines that norms and regulations issued by the government or multinational authorities, like the European Union, "are needed in many instances, not only to protect and guide consumers but also to discourage unsustainable production practices." Other authors show that government intervention through policymaking is an important factor as it can to some extent facilitate transitions toward more sustainable directions by correcting market failures (Lovio et al. 2011; Rotmans et al. 2001; Unruh 2000).

Kivimaa and Mickwitz (2011) emphasize the importance of framing in policy strategies. Framing, they say, "has become a key concept in the social sciences, because how an issue is framed largely determines what should be done." Frames define problems, diagnose causes, make moral interpretations, and suggest action (Entman, 1993). Other authors suggest that framing in the context of policy domains can contribute to the social shaping of technological options (Jorgensen et al. 2009; Klein and Kleinman 2002) by favouring 'optimal' technological solutions or by presenting the context for certain technological or system solutions.

The role of government intervention is often to "support those options that are politically desirable but cannot succeed in the market without government intervention" (Kivimaa and Mickwitz 2011). Developing new technologies takes much longer than reframing existing ones. However, reframing doesn't lead inevitably to promoting new technologies in the strategies, but rather changes the meaning given to existing, dominant technologies (Kivimaa and Mickwitz 2011).

Hence, sustainable development has started to feature on the Romanian Government's development strategy (Romanian Ministry of Environment and Sustainable Development 2008) and the government, aligning its policies with the EU, calls for a transition towards a sustainable society. Governments increasingly need to share these responsibilities with the private sector (Keijzers 2002), and therefore there is increasing pressure on the private sector to change existing models of production and consumption.

5   Political Challenges and Best Practices/Case Studies Analysis

It is extremely important to interact and benefit in order to develop urban areas in a way that saves resources, drives sustainable growth, enhances human capital and alleviates poverty.

Unlocking the synergies between urban systems opens up a wealth of benefits – environmental, social and economic. Encouraging multi and trans-disciplinary cooperation between stakeholders like communities, municipalities, regional and national governments, institutes and universities, civil society organizations and private companies, can achieve remarkable results in their cultural context by working together/ networking across boundaries to create better cities, to identify practical and integrated systems solutions.

Developing visions, scenarios, strategies and solutions for sustainable urban development and renewables applications could be a new focus for energy policy in EU.

Urban resilience represents the durable ability of municipalities to take action, while continuing to transform and develop. In a time of amplified preoccupations on ecological and social structure, comprehending how to strengthen such systems is indispensable. "The ability of a social or ecological system to absorb disturbances while retaining the same basic structure and ways of functioning, the capacity for self-organization, and the capacity to adapt to stress and change" (Francis and Beckera 2014).

Urban resilience to the impacts of climate change includes health and safety aspects, food security and energy provision. It must be related to the larger circumstance of the adjacent neighbourhood (local, national and global interdependencies), including the provision of vital resources as water, food, energy resources, labour, capital and functional markets.

Ecological footprint is defined as the area of productive land required to supply energy, food and other resources. The ecological footprint of rich countries increased by 400 % during the 20th Century. During the same period, the land available globally for food production fell by 25 % relative to that available in 1900, from 6 to 1.8 ha per capita, due to the population increase (WWF 2010).

The footprint concept reveals an important reality: due to high population densities, the rapid rise in per capita energy and resource consumption, and growing dependence on trade, the ecological locations of human settlements no longer coincide with their geographical locations. Cities and industrial regions are dependent for survival and growth on a vast and increasingly global hinterland of ecologically productive landscapes.

The extensive use of cars in urban areas generates noise and air pollution. Even though vehicles can be improved technically does not solve the problems of space utilization and congestion. However, integrated public transport systems can diminish such problems and increase mobility.

Monitoring energy consumption and evaluating the socio-economic context and potential should be the first step for shaping future plans. Independent effective public lightning is an alternative to reduce the community energy costs.

Urban areas need a partnership with nature in order to recycle their waste products. Through composting and waste management cities could produce inputs for farming, gardening and energy production. Small green areas in cities can maintain high biodiversity, especially wetlands.

Socio-cultural and socio-economic aspects of urban life are essential for human well-being and sustainable urban development. Good urban governance, development planning and design need to identify and prioritize measures to improve living standards and conditions for the poor and for vulnerable groups such as the elderly, disabled, young children and minority groups. Improved access to education, health services and income generating opportunities are essential for more equitable social and economic development.

Safety and security are also critical, and appropriate urban structures and well-designed, well-lit public spaces enhance surveillance, safety and social interaction.

UN-Habitat defines governance as: "The sum of the many ways individuals and institutions, public and private, plan and manage the common affairs of the city. It is a continuing process through which the conflicting or diverse interests may be accommodated and cooperative action be taken. It includes formal institutions as well as informal arrangements and the social capital of citizens".

Urban governance is significant, as it engages the administration of cities, and the financial, technical, organizational, and human resources necessary for sustainable urban development. The World Bank defines good governance as: "Predictable, open and enlightened policy making, a bureaucracy imbued with a professional ethos acting for the public good, the rule of law, transparent processes, and a strong civil society participating in public affairs".

Increased local self-governance requires the strengthening of institutional capacity and technical support, improved public communication and participation, professional auditing and a national system for monitoring municipal performance. Decentralization and mixed models (semi-autonomous) play a major role for mutual community achievements. Subsidies and aid offer to improve the energy management of public buildings could be a political key issue for local decision-makers.

An integrated approach to urban development will identify contrary policies and consider alternative development scenarios at an early stage in planning processes, and involve stakeholders in balancing different objectives to achieve solutions that best combine ecological, socio-cultural, economic and spatial considerations.

Awareness raising is the starting point for participation and the quality of urban planning processes and their outcomes is significantly affected by the degree of openness and participation. For example, both producers and consumers need to be informed about environmental problems and engaged in solutions.

Local authorities can engage stakeholders and communities in a participatory process to: strengthen local democracy by actively involving more people in local processes, increase transparency and effectiveness, achieve a better understanding of citizens' priorities, use citizens' knowledge as a planning resource, inform citizens about the purpose and services of the municipality, increase participation in local elections, increase citizens' contributions to local development.

There are different methods of communicating and establishing dialogue, depending on the purpose and stage in the planning process, including: Dissemination of information – especially in the initial stages, to increase awareness; Consultation – when proposals are presented and stakeholders are invited to respond; Participation – on site needs' assessments and formulating solutions; Mobilization – of stakeholders to participate in planning or implementation.

Promoting economic sustainability seeks to stimulate interaction between the business sector, academia and the public sector. The energy used by buildings differs dramatically, depending on their design and function. Minimizing the energy needs of buildings is a key factor for energy efficiency.

Good design can make buildings more energy efficient decreasing their environmental footprint and saving costs. Appropriate design of the structure and choice of materials and technical systems, especially for heating and cooling, are key aspects.

The aim should be to reduce the use of energy that is piped, wired and trucked into cities. Globally, buildings account for more than 40 % of total energy use and some 15 % of global CO_2 emissions (Cioca et al. 2015).

A good political strategy based on achieved results will strengthen transmission and distribution efficiency; reduce dependency on single sources; promote decentralized systems.

Digestion of biodegradable waste (e.g. wastewater sludge) can produce gas but waste collection creates a high transportation load. Waste utilities, private actors and traffic planners need to plan how to make waste transportation efficient (Torretta et al. 2015).

6   Conclusions

In order to articulate the energy strategy and policy at the European level, an analysis of communities' energy development potential, as well as the feasibility and urban design components must be financed first through EU programs. In this sense, there is a need of a methodological framework at the European level, based on variables and predetermined development indicators for monitoring and evaluation of energy projects. Ideally, it should encourage greater autonomy of the municipalities and regional administrations, consequently leading to a synergic compatibility of local development needs with their strategic aspirations, without a noticeable interference with multiple bureaucratic and hierarchical obstacles at the national level. Thus, a redistribution of prerogatives towards local governments and citizens through an adjustment of the national legislation would facilitate a proper implementation of energy projects.

Creating conditions for the development of standard implementations procedures at the European level, debates focused on the convergence of local projects and their integration into the national and European level should become the main objective during the implementation of local energy projects. This step becomes more effective once the growth and development of social infrastructure for energy is stimulated through identifying a co-interest between the citizens, the public sector, academic and business environment. It is recommended to identify a clear set of steps to access the benefits and responsibilities/liabilities of the actors involved, both providers and recipients of energy. The development of a manual of good practices for the development of energy efficient communities should ensure the convergence of future strategies, policies, tools, procedures and results at the EU level.

In order to provide short-term (1–2 years) energy security and rational exploitation of the environment, the European Commission, the European Parliament and national governments must be engaged in an effort to ensure de-bureaucratization and decentralization of energy policies in order to support specific implementation of models/local projects for future sustainable bio-communities.

Policy recommendations are decision path premises for the development of a project or an energy efficient community. The policy recommendations were developed by interpreting elements of renewable energy policy of the last decade at the local, national and European levels.

By identifying the local needs, all the actors involved gain a certain degree of importance. They are no longer just simple beneficiaries of developed policies – they become suppliers of human security. Therefore, the connection to European values and ensuring a citizen-cantered transfer of good practices will ensure mutually beneficial interdependence between the existing resources and developed processes/implementation mechanisms – decision making output and legitimacy.

Changing the mentality of stakeholders in order to develop sustainable communities remains the main global policy challenge.

References

Abaza, H., Bisset, R., Saddler, B.: Environmental Impact Assessment and Strategic Environmental Assessment: Towards an Integrated Approach. United Nations Environmental Programme, Geneva (2004). http://www.unep.ch/etu/publications/textONUbr.pdf. Accessed 4 Dec 2010

Ball, S.: The European employment strategy: the will but not the way? Ind. Law J. **30**(4), 353–374 (2001)

Cioca, L.I., Ivascu, L., Rada, E.C., Torretta, V., Ionescu, G.: sustainable development and technological impact on CO2 reducing conditions in Romania. Sustainability **7**(2), 1637–1650 (2015)

Entman, R.: Framing: toward clarification of a fractured paradigm. J. Commun. **43**, 51–58 (1993)

Francis, R., Bekera, B.: A metric and frameworks for resilience analysis of engineered and infrastructure systems. Reliab. Eng. Syst. Safety **121**, 90–103 (2014)

Grecu, V.: Contributions to Sustainability in Universities. LAP-LAMBERT Academic Publishing, Saarbrücken (2014). ISBN 978-3-659-63666-0. 564 pagini

COM: Integrating Sustainable Development in EU Policies: 2009 Report on the Sustainable Development Strategy of the European Union [original title: Integrarea dezvoltarii durabile în politicile UE: raport de analiza pe anul 2009 a Strategieide dezvoltare durabila a Uniunii Europene] - Bruxelles, 24 July 2009 (2009). http://eurlex.europa.eu/LexUriServ/LexUriServ.do?uri=COM:2009:0400:FIN:RO:PDF

Jacob, K., Volkery, A.: Institutions and instruments for government self-regulation: environmental policy integration in a cross-country perspective. J. Comp. Policy Anal. Res. Pract. **6**(3), 291–309 (2004)

Jorgensen, M.S., Jorgensen, U., Clausen, C.: The social shaping approach to technology foresight. Futures **41**, 80–86 (2009)

Keijzers, G.: The sustainable enterprise. J. Clean. Prod. **10**(4), 349–359 (2002)

Kivimaa, P., Mickwitz, P.: Public policy as part of transforming energy systems: framing bioenergy in Finnish energy policy. J. Clean. Prod. **19**(16), 1812–1821 (2011)

Klein, H.K., Kleinman, D.L.: The social construction of technology: structural considerations. Sci. Technol. Hum. Values **27**, 28–52 (2002)

Kolk, A., Levy, D., Pinkse, J.: Corporate responses in an emerging climate regime: the institutionalization and commensuration of carbon disclosure. Eur. Account. Rev. **17**(4), 721–747 (2008)

Lovio, R., Mickwitz, P., Heiskanen, E.: Path dependence, path creation and creative destruction in the evolution of energy systems. In: Wüstenhagen, R., Wuebker, R. (eds.) The Handbook of Research on Energy Entrepreneurship. pp. 274–301. Edward Elgar Publishing, Cheltenham (2011)

Measuring Progress towards a more Sustainable Europe. Office for Official Publications of the European Communities, Luxembourg (2005). http://epp.eurostat.ec.europa.eu/cache/ITY_OFFPUB/KS-68-05-551/EN/KS-68-05-551-EN.PDF. Accessed 5 Dec 2010

National Sustainable Development Strategy of Romania (NSDSR) (2008). http://www.sdnp.ro/documents/national_strategy_for_sustainable_development/SNDD_2008_EN.pdf. Accessed 9 June 2010

Orbie, J., Babarinde, O.: The social dimension of globalization and EU development policy: promoting core labour standards and corporate social responsibility. J. Eur. Integr. **30**(3), 459–477 (2008)

Parmenter, R., Cain, S., Lipp, J.: Assessing the full cost of energy in Nova Scotia: a GPI atlantic approach. In: Proceedings of the Second International Conference on Gross National Happiness, Rethinking Development, St. Francis Xavier University, Antigonish, Nova Scotia, Canada, 20–24 June 2005, pp 47–63 (2007)

Persson, Å. : Mainstreaming climate change adaptation into official development assistance: a case of international policy integration. EPIGOV Papers, (36) (2008)

Radaelli, C.M., Meuwese, A.C.: Better regulation in Europe: between public management and regulatory reform. Public Adm. **87**, 639–654 (2009)

Raivio, K.: Sustainability as an educational agenda. J. Clean. Prod. **19**(16), 1906–1907 (2011)

Rotmans, J., Kemp, R., Van Asselt, M.: More evolution than revolution: transition management in public policy. Foresight J. Future Stud. Strateg. Thinking Policy 3, 15-31 (2001). World Wide Fund for Nature - WWF. (2010, 01 17). Ecological Footprint. Accessed from WWF Global: http://wwf.panda.org/about_our_earth/teacher_resources/webfieldtrips/ecological_balance/eco_footprint/

Tamborra, M.: Developing tools for sustainability impact assessment: the role of socioeconomic research in the EU. Paper presented at the EDIAIS Conference, Manchester, 24–25 November 2003 (2003). http://www.enterprise-impact.org.uk/pdf/Tamborra.pdf. Accessed 4 Dec 2010

Torretta, V., Rada, E.C., Ragazzi, M., Trulli, E., Istrate, I.A., Cioca, L.I.: Treatment and disposal of tyres: two EU approaches. A review. Waste Manag. **45**, 152–160 (2015)

Unruh, G.C.: Understanding carbon lock-in. Energy Policy **28**, 817–830 (2000)

Utting, P.: Business responsibility for sustainable development. Occasional Paper No. 2. UNRISD, Geneva (2000)

Wilkinson, D., Monkhouse, C., Herodes, M., Farmer, A.: For Better or For Worse? The EU's "Better Regulation" Agenda and the Environment. Institute for European Environmental Policy, London (2005)

Sustainable Urban Regeneration Policy Making: Inclusive Participation Practice

Piergiuseppe Pontrandolfi[1] and Francesco Scorza[2(✉)]

[1] DICEM, University of Basilicata,
Via Lazazzera, 75100 Matera, Italy
`piergiuseppe.pontrandolfi@unibas.it`
[2] School of Engineering, Laboratory of Urban and Regional Systems Engineering,
University of Basilicata, 10, Viale dell'Ateneo Lucano, 85100 Potenza, Italy
`francesco.scorza@unibas.it`

Abstract. In physical and economic planning at both urban and regional scale, the role of participation is a key element of the planning process. Participation is often linked to organization forms (we call it "structures of participation") that find a heterogeneous applications in the Urban Center model. This work analyses the first results of Project CAST (Active Citizenship for Sustainable Development of Territory), especially those related to the neighbourhood of Poggio Tre Galli in Potenza (Italy), where the test of a traditional/technological participatory approach has allowed the development of urban regeneration scenarios characterized by an inclusive approach "Citizens centred". This is an operative contribution in terms of Inclusive Smart Planning, and so of evolutionary 2.0 approaches oriented to an inclusive and participative urban management through ICT tools. A singular feature of the ICT platform developed during the project CAST is the integration of management tools and streaming analysis of the main social networks with a SDI. The experience, described both in quantitative terms and as a strategic design of urban regeneration at neighbourhood scale, shows from one hand the request of bottom-up contributions, especially from institutions (i.e. Municipality), on the other hand, the need to test effective solutions to balance the commitment to manage and configure complex information systems according to quality results. Research perspectives look at the definition of web-assisted procedures for the participation in urban and territorial government choices that can reinforce bottom-up practices such as DSS, starting from ICT tools tested during the project CAST.

Keywords: Participation 2.0 · Urban regeneration · Virtual urban center

1 Introduction

The process addressed to the identification of development models and urban regeneration passes through the implementation of advanced participative processes in which actors' needs can find spaces and technical studies aimed at identifying sustainable strategies for urban development (cfr. [1]).

© Springer International Publishing Switzerland 2016
O. Gervasi et al. (Eds.): ICCSA 2016, Part III, LNCS 9788, pp. 552–560, 2016.
DOI: 10.1007/978-3-319-42111-7_44

In this sense, the project CAST (Active Citizenship for Sustainable Development of Territory) is a suitable case study in which a participatory process - supported by advanced ICT tools [2] - has generated a great interest of the local community and - consequently - of the public administration about the opportunity of defining urban regeneration strategies in a particular and limited context, possible candidates for urban agenda 2014–2020 funding. From a procedural point of view, it is important to underline this strict "bottom-up" contribution represents an element of value in a context in which public administrations live a moment of crisis of means and tools [3, 4] to investigate, develop and encourage the active involvement of citizens.

This paper presents an operative contribution in terms of Inclusive Smart Planning, and so of evolutionary 2.0 approaches (in comparison with Murgante et al. described in 2011 [5, 6]) oriented to an inclusive and participative urban management through ICT tools.

The project considers an interdisciplinary approach in the design and implementation of participative processes looking at qualified case studies described in literature [7–9]. Experts of several branches of knowledge (especially urban planning, architecture and sociology) will interact with local communities to test innovative forms of e-governance, through ICT and web-based tools too.

In this work, after the presentation of the implementation context and of the main methodological references, we describe the strategic framework deriving from the participative process in general. Conclusions refer to later developments and analysis of methods and procedures adopted in the organization of "virtual urban centres".

2 The Implementation Context of the Project

Territories involved in the project are: the metropolitan area of Potenza and the town of Matera. In particular, the latter town will develop important initiatives regarding innovation and urban regeneration in consequence of its appointment as "European Capital of Culture 2019".

Within the Basilicata Region, such territories are beneficiary of the programming action of Structural Funds for sustainable development of cities in OP ERDF 2014-2020. This research provides contributions to the development of operative scenarios that include outcomes of social innovation actions (empowerment) in the territory [10].

The proposal comes from the idea of promoting a wider participation process to define choices and economic - territorial planning interventions within a renewed local development strategy in Basilicata, in order to face clear elements of social and economic weakness that do not fully promote the relaunch of endogenous development processes in urban and territorial area, despite the availability of significant natural, cultural, economic and social resources (cfr. [11]).

The main topics are territorial balance, social inclusion, the organization of basic services according to principles of equity, efficiency and the proper use of resources to promote the main guidelines for a sustainable urban project that can benefit from available financial resources and EU Funds of the next years.

While in Matera the project goal is to examine in depth a city development vision with the European Capital of Culture 2019 perspective, in Potenza the question is to enable an overall process of regeneration and improvement of the urban fabric, mainly in the peripheral areas in a singular context: potential disarray of the municipal administration and programming of significant extraordinary resources coming from EU Structural Funds.

3   Technologies and Participation: An Integrated Open-Source Solution

The technological component of the project is based on an on-line ICT platform, which integrates open-source tools and frameworks to develop an integrated technological infrastructure allowing a high level of interaction among users compared with participation dimensions previously described.

The project of this technological infrastructure comes from some experiences recently developed in this domain; among these the project "Cilentolabscape" (cfr. [6, 12]) seems to be representative: it proposes an integrated platform, which combines traditional functions for data displaying and innovative social interaction tools.

The system combines CMS features (Content Management System), a Geo-portal to get territorial information, advanced systems for the management of online polls and votes, OGC services for data sharing according to OPEN DATA standards, the integration of social networks and the management of spatial social alerts for participation and collaborative mapping [13].

The system is formed by the following parts: a technological platform - CMS & SDI, a Geo-portal, widgets, app POI builder, a mobile app for web access.

The integrated technological platform: main parts

- The design shows a Spatial Data Infrastructure (SDI), formed by a central database (PostgreSQL/PostGis), a geospatial application server (GeoServer) and a content management system (Joomla).
- The Spatial Data Infrastructure is the basis for the provision and use of all of mapping services. It is composed by two different lines of reasoning, depending on data typology (vector or raster).
- The Geo-portal will organize spatial information, produced by participatory process.
- Social Mapping & Social Alert

Nowadays social networks represent the most important data bank in the world. As far as Twitter, a social network that provides a "short message" service with a localization tool, everyday there are about 400 million of tweets, which are made available by the system through API.

A tweet is an innovative form of geospatial information (it comes from different mobile devices through GPS or GSM) and multimedia information (images or texts organized in a synthetic and significant way).

Web-Gis and social network integration improves interaction between users and the system, increasing the quality of the participatory process thought for the project C.A.S.T.

The key element is the opportunity to indicate localized instances referring to punctual emergencies, which can be categorized on the basis of topics discussed on the platform.

4 The Methodological Approach: LFA

In terms of methodology, the experience was conducted according to the Logical Framework Approach procedure.

References are traceable in the research and application of LISUT [14–17] that considers LFA as a structured and useful methodological reference for a rational approach to the planning process. The LFA is characterized by a phase of analysis that defines problems and operative purposes for the individuation of strategies (options), through stakeholders' involvement supported by technical analysis and implementation context assessments. It is followed by a synthesis phase, in which the Logical Framework Matrix defines the operative form of the program structure connecting purposes, results, actions and resources according to a rigorous logic linked to "objectively verifiable" indicators.

The Logical Framework (LF) is useful to structure the logic of plan activities in order to facilitate its assessment at different stages of the Project Cycle Management.

Las Casas and Scorza [18] proposed in 2009 a version of the LFA aiming at clarifying pertinence and relevance of policy choices according to the context of implementation,

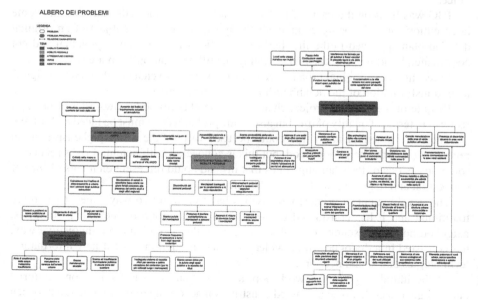

Fig. 1. Problem Tree – "Poggio Tre Galli" – Potenza (Color figure online)

looking for a place/context based policy, besides the principles of effectiveness and efficiency.

Aune [19] warns us of the danger: "Form over substance". As for aids to enterprises, the "form" of LFA often replaces the "substance". In fact, in the widest applications of LFA, the "compiling of matrix" beyond the utility levels required by the project [20] can represent the victory of form over substance. Coleman [21] argues that LFA approach is an "aid to think" rather than a set of procedures.

Especially the phase of analysis of LFA has been developed during project CAST. Starting from technical elaborations and urban assessments, the analysis of results obtained through online surveys and streaming analysis of social network, some planned meetings, problem tree and objective tree have been identified (Fig. 1).

5   Participatory Planning Workshop in "Poggio Tre Galli"

In the context of project CAST, activities and experiences have been developed in Potenza to promote forms of active citizenship in city government. Activities that Associations intend to develop and strengthen in the coming months.

Especially, we want to underline the experience of a Participatory Planning Workshop about urban regeneration in the west neighbourhoods of the city (Poggio Tre Galli and G area), that has involved citizens, municipal administration representatives, cultural associations and city volunteers.

This urban area comprises the neighbourhood of Poggio Tre Galli (P3G), the G area and the area of the so-called "Study-Centre" where many institutes for higher education and primary schools are located, as well as several buildings destined to house regional offices.

P3G was built on the basis of an "area plan" between the '70s and' 80s and some interventions are still being completed; Study-Centre area, the subject of an original detailed plan approved in the first half of the '70s, was subjected to punctual and not coordinated building projects, and nowadays it still has non-built-up areas.

In recent years, original urban forecasts were subject to changes that led to a significant transformation of the neighbourhood urban structure.

Despite of the original project changes, the neighbourhood shows a lot of positive features in terms of quality of life, but also some problems that must be faced and solved.

There are potential areas that could be used to promote a comprehensive urban regeneration of the neighbourhood, but this is connected with potential risks related to some provisions of the Planning Rules that instead could represent elements of further congestion and degradation of the neighbourhood.

The development of workshop activities has followed two ways: the first is more traditional and it has considered the work of a small group of individuals (experts, neighbourhood representatives, and associations); the second, a more innovative way, has consisted in a dialogue with a wider audience through the use of new information technologies and social networks.

At the initial stage of the workshop, a great attention was focused on territorial knowledge as a collective and shared construction according to the interaction between different actors. The assessment process aimed at defining the most important problems of the neighbourhood came from discussions of the early phase of knowledge. The sharing of problems started the discussion and definition of objectives on which credible intervention strategies could be organized, considering the integration of sectorial issues.

Strategies and actions have been shown and described in a final meeting to provide a further contribution to discussion and to submit proposals to the municipal administration, indicating possible priorities.

The urban regeneration strategy, based on technical assessments and proposals emerged during the participatory workshop, is based on organizational and material interventions characterized by feasibility, synergies and as sustainable as possible under an environmental point of view. All the proposed interventions were defined consistently with the shared objectives whose achievement should be tackled effectively and efficiently to the demands and needs coming from citizens and users living in the neighbourhood.

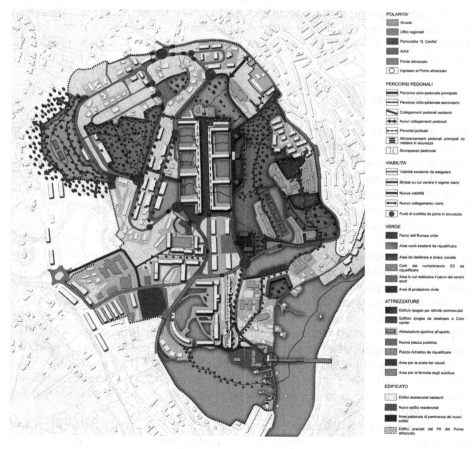

Fig. 2. Stretegic urban development design (Color figure online)

The overall vision for the proposal and intervention strategies developed in the participatory workshop refers to an urban regeneration project based on the promotion of sustainable mobility and the development of pedestrian mobility and an effective construction of a network system of green infrastructure, services and open spaces for the community of citizens the fundamental cornerstones for the promotion of a sustainable city. The proposal is to develop an urban structure based on green, pedestrian zones and integrated services.

The following figure shows the general pattern identified through the participatory process indicating the main point and linear proposed interventions and classification of retail areas.

The actions proposed refer to development and mitigation strategies of vehicular traffic mobility, the promotion of slow mobility (especially walking and cycling, also enhancing and recovering space and the various paths), the increase of the budget of green infrastructure, promotion of urban agriculture, the physical and functional connection between the different scope urban parts (especially among the Study Centre area and the Poggio 3 Galli district), the increase and the best use of the total allocation of spaces and equipment for recreation and leisure (even with the recovery of existing equipment), the increase of the availability of equipment and spaces for public use, the redevelopment of areas for parking and parking to connect with pedestrian traffic (Fig. 2).

6 Conclusions

The overall experience of the CAST project, proposed here as a preliminary assessment of the results of the implemented procedure, follows the path of inclusive practices of urban regeneration. The context of the trial and described the main results describe a methodological and operational framework (with reference to the implemented ICT tools and knowledge and design studies) that effectively has been a growing interest from the community to contribute to a collaborative design process it starts from the bottom and that crosses the city administration attention.

The project proposal, developed mainly in strategic terms, is a bottom-up contribution to the administration in support of the Urban Agenda Planning processes 2014–2020 and in this sense is an element that demonstrates the utility derived from the integration between programming processes and actors that operate at different levels.

The contribution of ICT tools has represented an added value in the stages of involvement and listening to citizens and local actors [14]. The result is an evolved form of participation in the urban development programming, tested and validated by subsequent tests and applications, it can be an important aid to local government processes and regeneration of the city. The e-participation forms can indeed make a significant contribution, particularly in the context where the participation is struggling to become established and where there is a strong inertia of public decision-makers in recognizing effective utility.

In these terms it strengthens the proposal of the CAST project to the municipal administration of the power project for the construction of a "virtual urban center" qualified organization to promote widespread forms of participation and aimed to investigate and experiment with a new approach to city planning and of the territory based on the principles of equity, efficiency and conservation of resources [16, 22].

References

1. Luisi, D.: Dinamiche inclusive e costruzione dell'agency nelle politiche pubbliche partecipate. ARCHIVIO DI STUDI URBANI E REGIONALI (2016). Knapp, S., Coors, V.: The use of eParticipation systems in public participation: the VEPs example. In: Coors, V. et al. (eds) Urban and Regional Data Management, pp. 93–104. Taylor and Francis, London (2008)
2. Lanza, V., Prosperi, D.: Collaborative E-Governance: Describing and Pre-Calibrating the Digital Milieu in Urban and Regional Planning. Taylor and Francis, London (2009)
3. Cutini, V., Rusci, S.: Ai tempi della crisi. il mercato immobiliare e le influenze sulla pianificazione. ARCHIVIO DI STUDI URBANI E REGIONALI (2016)
4. Manganelli, B., Pontrandolfi, P., Azzato, A., Murgante, B.: Urban residential land value analysis: the case of Potenza. In: Murgante, B., Misra, S., Carlini, M., Torre, C.M., Nguyen, H.-Q., Taniar, D., Apduhan, B.O., Gervasi, O. (eds.) ICCSA 2013, Part IV. LNCS, vol. 7974, pp. 304–314. Springer, Heidelberg (2013)
5. Murgante, B., Tilio, L., Lanza, V., Scorza, F.: Using participative GIS and e-tools for involving citizens of Marmo Platano-Melandro area in European programming activities. J. Balkan Near Eastern Stud. 13(1), 97–115 (2011)
6. Attardi, R., Cerreta, M., Franciosa, A., Gravagnuolo, A.: Valuing cultural landscape services: a multidimensional and multi-group SDSS for scenario simulations. In: Murgante, B., et al. (eds.) ICCSA 2014, Part III. LNCS, vol. 8581, pp. 398–413. Springer, Heidelberg (2014)
7. Ave, G.: Play it again Turin. Analisi del piano strategico di Torino come strumento di pianificazione della rigenerazione urbana. In: Martinelli, F. (ed.) La pianificazione strategica in Italia e in Europa: Metodologie ed esiti a confronto, pp. 35–67. Franco Angeli, Milan (2005)
8. Gibelli, M.C.: Vivibilità e nuova urbanità nelle politiche e nei progetti di rigenerazione urbana. In: Boniburini, L. (a cura di) Alla ricerca della città vivibile, Alinea, Firenze, pp. 75–90 (2009)
9. Tedesco, C.: Una politica europea per la città?: l'implementazione di urban a Bari, Bristol, Londra e Roma, vol. 104. FrancoAngeli (2005)
10. Pontrandolfi, P.: Contenuti strategici della pianificazione ed esperienze di partecipazione ai processi decisionali. Territorio della Ricerca su Insediamenti e Ambiente. Rivista internazionale di cultura urbanistica 3(6), 115–126 (2010)
11. Pontrandolfi, P.: La pianificazione del territorio extra-urbano di Potenza. Aestimum (30) (2009)
12. Cerreta, M., Inglese, P., Manzi, M.L.: A multi-methodological decision-making process for cultural landscapes evaluation: the green lucania project. Procedia-Soc. Behav. Sci. 216, 578–590 (2016)
13. Scorza, F., Pontrandolfi, P.: Citizen participation and technologies: the CAST architecture. In: Gervasi, O., Murgante, B., Misra, S., Gavrilova, M.L., Rocha, A.M.A.C., Torre, C., Taniar, D., Apduhan, B.O. (eds.) ICCSA 2015. LNCS, vol. 9156, pp. 747–755. Springer, Heidelberg (2015)

14. Las Casas, G., Murgante, B., Scorza, F.: Regional local development strategies benefiting from open data and open tools and an outlook on the renewable energy sources contribution. In: Papa, R., Fistola, R. (eds.) Smart Energy in the Smart City, pp. 275–290. Springer, Heidelberg (2016)
15. Las Casas, G.B.: Processo di decisione e processo di Piano. In: Clemente, F. (a cura di) Pianificazione del Territorio e sistema informativo. FrancoAngeli, Milano (1984)
16. Las Casas, G.B.: L'etica della Razionalità. Urbanistica e Informazioni, vol 144, (1995). ISSN 0392-5005
17. Las Casas, G.B., Sansone, A.: Un approccio rinnovato alla razionalità nel piano. In: Depilano, G. (a cura di) Politiche e strumenti per il recupero urbano, EdicomEdizioni, Monfalcone (2004)
18. Las Casas, G.B., Scorza, F.: Un approccio "context-based" e "valutazione integrata" per il futuro della programmazione operativa regionale in Europa". In: Lo Sviluppo Territoriale Nell'economia Della Conoscenza: Teorie, Attori Strategie, Collana Scienze Regionali, 41 (2009)
19. Aune, J.B.: Logical framework approach. Dev. Methods Approaches **214** (2000)
20. Bakewell, O., Garbutt, A.: The Use and Abuse of the Logical Framework Approach, vol. 27. Stockholm, Swedish International Development Cooperation Agency (Sida) (2005)
21. Coleman, G.: Logical framework approach to the monitoring and evaluation of agricultural and rural development projects. Proj. Appraisal **2**, 251–259 (1987)
22. Wilson, A.: New roles for urban models: planning for the long term. Reg. Stud. Reg. Sci. **3**(1), 48–57 (2016). doi:10.1080/21681376.2015.1109474

GI2NK Geographic Information: Need to Know Towards a More Demand-Driven Geospatial Workforce Education/Training System

Mauro Salvemini[3], Laura Berardi[3], Monica Sebillo[2],
Giuliana Vitiello[2], Sergio Farruggia[1], and Beniamino Murgante[4(✉)]

[1] AMFM GIS Italia, Via Ugo Ojetti 427, 00137 Roma, Italy
sergio.farruggia@gmail.com
[2] University of Salerno, Via Giovanni Paolo II, 132, 84084 Fisciano, SA, Italy
{msebillo,gvitiello}@unisa.it
[3] Sapienza University of Rome, Piazza Borghese, 9, 00186 Roma, Italy
{mauro.salvemini,laura.berardi}@uniroma1.it
[4] University of Basilicata, 10, Viale dell'Ateneo Lucano, 85100 Potenza, Italy
beniamino.murgante@unibas.it

Abstract. The paper presents GI-N2K (Geographic information: Need to Know), a European project aiming to improve the way in which future GI professionals are prepared for the labour market. Its main goal is twofold: updating the existing Body of Knowledge on the basis of the new technological developments and the European perspective, and realizing advanced tools to define curriculum, training opportunities and courses. This project focused on foundational research into the creation of a transformational, dynamic environment for pedagogy, knowledge construction, discourse, collaboration, and research in the domain of Geographic Information Science and Technology (GIS&T). After an initial integrated analysis of the demand for and supply of geospatial education and training, the revision of the BoK and the design of the Virtual Laboratory for the BoK (VirLaBok) are currently under investigation. In particular, the Consortium is now involved into the recognition of a proper revision strategy, in terms of content and its usability through an e-platform. The goal of this paper is to discuss preliminary results of this phase and illustrate problems due to a possible overlapping among different knowledge areas. Finally, a prototyping version of the ontology underlying the VirLaBok is presented.

Keywords: GIS · Body of knowledge · GIS job market · Ontology · Geospatial education · Geospatial professionals

1 Introduction

The geospatial industry is a rapidly growing industry and involves high value/high tech jobs, innovative services and fast evolving technologies. In the European context, the need to prepare Europe's GI S&T workforce competently to answer to the requirements

O. Gervasi et al. (Eds.): ICCSA 2016, Part III, LNCS 9788, pp. 561–572, 2016.
DOI: 10.1007/978-3-319-42111-7_45

of the rapidly evolving, innovative European knowledge society is driven by the objectives of several European strategies and policies, like Europe 2020, Horizon 2020, the Digital Agenda for Europe, the Smart Cities initiative, the INSPIRE directive and the European Location Framework and many other initiatives. While in 2000 the Pira study [1] estimated the economic value added by geographic public sector information to the economy in 1999 at € 36 billion, the US Department of Labor's High Growth Industry Profile – Geospatial Technology report came to the conclusion that the geospatial market is "growing at an annual rate of almost 35 %, with the commercial subsection of the market expanding at the rate of 100 % each year." The same Department identified geographic information system technology "as one of the three most important and evolving fields, along with nanotechnology and biotechnology".

On the basis of such assumptions, it is clear how important it is to build professional profiles based on the analysis of current demand in relation to required and existing knowledge in the field of GI. The goal is to create an easy integration of industry experts in the labour market.

To address this challenge, GI S & T BoK [2] (GI Science and Technology Body of Knowledge) realized for the USA University GIS Consortium in 2006 has been considered as reference. GI S & T BoK is the tool and the repository containing data, instructions and information on how to develop knowledge in the GI sector, considering demand and supply at the same time.

The aim of BoK is giving greater coherence and effectiveness to the academic education offered by American universities, aimed at satisfying users' demand (government agencies, enterprises and NGOs). This version of BoK was able to combine, for a long time, quality of academic educational offer with demands from the labour market.

Nevertheless, in 2014 the document appeared outdated, considering the latest conceptual and technological breakthroughs in the field of GI from academic and professional sectors.

In order to answer this demand, an European project aiming to improve the way in which future GI professionals are prepared for the labour market, has been proposed. The European project Geographic information: Need to Know GI-N2K (http://www.gi-n2k.eu) has been proposed and financed.

2   Geographic Information: Need to Know (GI-N2K)

The project was commissioned by the European Union (Lifelong Learning Program of the Education, Audiovisual and Cultural Executive Agency of the European Commission) to a consortium of 31 partners from 25 countries (Austria, Belgium, Bulgaria, Czech Republic, Denmark, Estonia, Finland, France, Germany, Iceland, Greece, Hungary, Ireland, Italy, Lithuania, Macedonia, Netherlands, Poland, Portugal, Romania, Slovakia, Slovenia Spain, Sweden, UK), its objective being an improved version of the GIS&T BoK, including tools to use and maintain it.

The project started in October 2013 and it will finish in October 2016.

The consortium has been carefully composed bringing together the demanders for and suppliers of the revised BoK (and tools) through the involvement of academic, private and public partners in the domain of GI S&T. An important strength of the consortium is the involvement of networks and associations in many countries, e.g. Italy (AM/FM Italia), the Netherlands (Geonovum and Stichting Arbeidsmarkt Nederland) and Belgium (AGORIA).

The involvement of these partners ensures that the project development and results are endorsed and applied by the wide GI community not only during the project period but also beyond the project period.

The project will be developed by the full partners - mainly composed of European Academic Institutions including the European academic network AGILE. The testing and validation of the project outcomes will also involve the associated partners – mainly composed of demanders for a revised BoK who then can help with the further dissemination of the results when successfully applied.

The network of networks idea behind GI-N2K is illustrated in Fig. 1. The revised BoK and tools will be mainly developed by the full partners of the project (1st circle). The project outcomes will be tested, validated and distributed by the associated partners of the project (2nd circle).

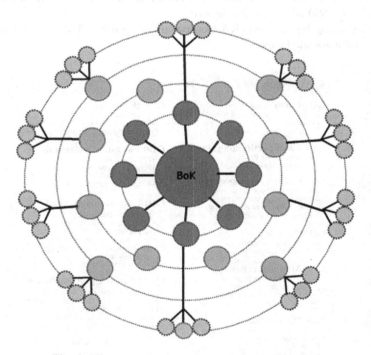

Fig. 1. The network of networks idea behind GI-N2K

The 3rd circle of involved entities consists of representatives of different GI communities like Open Geospatial Consortium Europe, University Consortium for Geographic Information Science and the Joint Research Centre of the EC, who are involved in the project as members of the advisory board.

Each of these circles consists of several partners that are participating and/or managing a broad network of relevant stakeholders (4th circle). The ambition of the project is to strengthen existing relationships and build relationships to new stakeholders in order to broaden and strengthen the network of GI S&T stakeholders.

The European project Geographic information: Need to Know GI-N2K is based on the methodology of the Body of Knowledge (BoK) which is basically the agreed ontology of a specific professional domain. GI S&T sector is linked with a lot of disciplines and knowledge areas consequently it is fundamental to develop a BoK be able to cover and describe all disciplines and knowledge areas.

In particular, it was shown that a major cause is a significant mismatch between training in the field of GI S & T and current demands from the labour market.

A lot of postgraduate courses on GIS strand are mainly based on geography topics, as well as many computer courses do not provide the necessary geographical knowledge. To address this important issue, the proposed solution is the design and the realization of a support system to the formation of figures with geospatial skills.

The first version of the GI S & T BoK was based on a hierarchical organization including 10 Knowledge Areas (KA), 73 units (26 defined as "core unit"), 329 topics, and more than 1600 educational objectives.

As an example, Fig. 2 shows KA Geospatial Data (GD) with its twelve units (GD1 GD12 ÷) and associated topics.

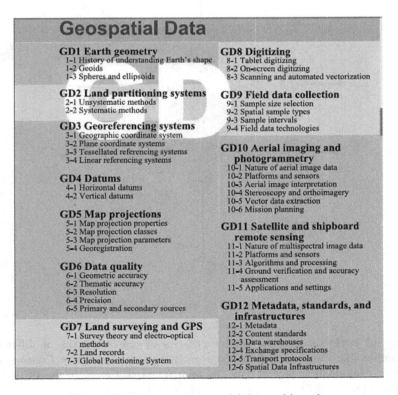

Fig. 2. Knowledge area geospatial data and its units

Subsequently, each KA is briefly introduced and each unit is detailed in terms of topics and learning objectives.

Table 1 (row 1) shows the initial description of GD, while educational objectives of a single topic GD3-1 (geographic coordinate system) are shown in Table 1 (row 2).

Table 1. KA geospatial data and GD3 georeferencing systems with the topic GD3-1.

Knowledge area: Geospatial Data (GD)	Geospatial data represent measurements of the locations and attributes of phenomena at or near Earth's surface. Information is data made meaningful in the context of a question or problem. Information is rendered from data by analytical methods. Information quality and value depends to a large extent on the quality and currency of data (though historical data are valuable for many applications). Geospatial data may have spatial, temporal, and attribute (descriptive) components, as well as associated metadata. Data may be acquired from primary or secondary data sources. Examples of primary data sources include surveying, remote sensing (including aerial and satellite imaging), the global positioning system (GPS), work logs (e.g., police traffic crash reports), environmental monitoring stations, and field surveys. Secondary geospatial or geospatial-temporal data can be acquired by digitizing and scanning analog maps, as well as from other sources, such as governmental agencies. [...]
GD3- Georeferencing systems (core unit) GD3-1 (Geographic coordinate system)	Distinguish between various latitude definitions (e.g., geocentric, geodetic, astronomic latitudes) Explain the angular measurements represented by latitude and longitude coordinates Locate on a globe the positions represented by latitude and longitude coordinates Write an algorithm that converts geographic coordinates from decimal degrees (DD) to degrees, minutes, seconds (DMS) format Calculate the latitude and longitude coordinates of a given location on the map using the coordinate grid ticks in the collar of a topographic map and the appropriate interpolation formula Mathematically express the relationship between Cartesian coordinates and polar coordinates Calculate the uncertainty of a ground position defined by latitude and longitude coordinates specified in decimal degrees to a given number of decimal places Use GIS software and base data encoded as geographic coordinates to geocode a list of address-referenced locations

Since 2006, BoK has been used for several initiatives. Each initiative highlighted the potential of this instrument in the process of creation and exploration of GI S & T BoK content [3–5].

GI-N2K will incorporate results and findings of the offer and demand analysis and will take advantage of the network that was created in VESTA-GIS. The overall aim of VESTA-GIS (www.vesta-gis.eu), was to pool knowledge in the GIS domain (technology, applications), to share experience and foster innovation (new approaches) in vocational training by bringing together experts, organizations and users of GI.

The project is divided into 8 work packages:

- WP1 – Analysis of demand and supply
- WP2 – Revision of the Body of Knowledge
- WP3 – The Virtual Lab for the BoK: VirLaBok
- WP4 – Testing & Validation
- WP5 – Quality assurance
- WP6 – Dissemination
- WP7 – Exploitation and Sustainability
- WP8 - Management

WP1, WP2, WP3 and WP4 are the implementation work packages of the project, in which the project outcomes are prepared, developed, tested and validated. The activities in these work packages deal with different methods:

- In WP1 survey research and in-depth interviews will be used to identify the knowledge areas (KAs), units, topics and concepts that should be included in the revised version of the GI S&T BoK. In addition, an analysis of the current supply of GI S&T education and training will be made based on the approaches and results of previous initiatives.
- WP2 focuses on the examination and improvement of the existing GI S&T BoK. After the definition of the structure of the BoK, i.e. the knowledge areas that will be part of the BoK, the structure, concepts, usability, relevance, and up-to-dateness of the content will be further developed in different expert teams, in which key experts of a specific knowledge area are represented.
- WP3 involves the design and development of an online BoK repository supported by several tools for updating and using the BoK. In order to gain insight in the requirements with regard to the functionalities of the toolsets and the characteristics of the data repositories, a survey and interviews will be conducted and a workshop will be organized. An additional workshop will be organized to demonstrate and test the repository and toolset.
- In WP4 the new BoK and the developed tools will be applied on several real world cases, in order to gain insight on how the output of WP2 and WP3 can be modified and improved. The selection of use cases will be based on the results of a survey and interviews. The selected set of use cases will be presented and discussed during a series of workshops. Integrated tests of different cases will also take place in workshops organized in different countries.

3   The Results of the BoK Evaluation Process

WP1 was completed in May 2014. WP1 results include:

- an overview of awareness of the GIS&T BoK and its use among respondents of the surveys in at least the 25 European countries that participate in the project;
- a comparison of employer's demands with the supply of GI teaching;
- an overview of subjects to supplement GIS&T BoK with.

At the end of these activities, an interesting report has been prepared (www.gi-n2k.eu/surveys-results/) where results of questionnaires and interviews, later on used to carry out the review of BoK, were presented and discussed. It is based on reports on the outcomes of surveys about the Demand side [6] and the Supply side [7], that were held early in 2014 as part of this work package. Generally, the evaluation has been focused on:

- awareness and use of GIS T BoK by expert communities,
- teaching gap between demand and supply of skills,
- content gap: incompleteness of GIS & T BoK.

From the demand point of view, the analysis highlighted three fundamental requests. It is important to shift the focus from the acquisition of primary data to the management of large amounts of spatial data. The lack of skills in programming and application development should be taken into account. And finally it is important to increase the role of the Web in all its forms, divulgative, support for the education, training, etc.

Regarding the importance of the content of BoK compared to the request of skills, the different KAs were evaluated in a different way.

The highest average rating was achieved by KA Geospatial Data, followed by Cartography and Visualization and Design Aspects. At the lower level there is Geocomputation KA (Fig. 3). In a consistent way also the units within the same KA were evaluated generally with different ratings. For example, while the basic operations such as measurement of geometrical properties and execution of queries were considered "very important" in Analytical Methods KA, advanced methods such as spatial regression or mathematical optimization were evaluated "less relevant" (Fig. 4).

Consequently, the Geocomputation KA, which is entirely composed of these advanced units, obtained generally the lowest rating (Fig. 5).

Finally, other KA, like GIS & T and Society, have achieved a significant number of "somehow relevant" which indicates that GIS & T is still mostly seen as a technical discipline (Fig. 6).

Regarding general trends and needs, the most popular keywords in GIS & T domain, in terms of trends, needs (educational objectives) or deficiencies are GIS, Data and Analysis.

In general, the expected outcomes are mainly oriented to web and mobile applications, to the management of big and open data, and programming (Fig. 7).

The completely missing concepts in BoK, generally, are related to the development of applications (E.g., API, GeoJSON, python, javascript), to WebGIS (e.g., HTML5, smartphone, Wireless, GPRS, RESTful, semantic web), to SDI - Spatial Data

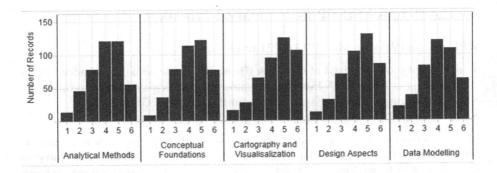

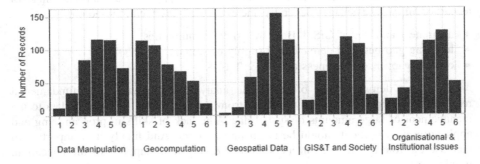

Fig. 3. The ratings by European GI professionals of the overall relevance of individual GIS&T BoK knowledge areas on a scale of 1-6 (Wallentin et al. [6]).

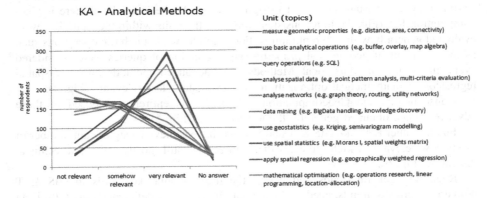

Fig. 4. Analytical methods - how relevant have the following competences been in your professional work in the last year? (Color figure online) (Wallentin et al. [6]).

Infrastructure (e.g., INSPIRE, Harmonization, GeoPortal) to data acquisition (e.g., UAVs, VGI, crowdsourcing), and other topics such as big data, augmented reality and standards for 3D modeling such as CityGML.

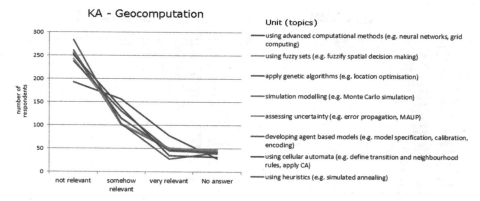

Fig. 5. How relevant have the following competences been in your professional work in the last year? - geocomputation units are seen very relevant in professional work only by every tenth respondent (Color figure online) (Wallentin et al. [6]).

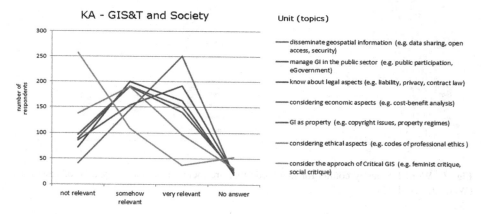

Fig. 6. GIS&T and society - how relevant have the following competences been in your professional work in the last year? (Color figure online) (Wallentin et al. [6]).

Finally, some interesting comments from GI experts regarding future competencies expected by professionals were collected.

In summary, there is broad consensus on the need for qualified staff in the field of sensors and mobile applications, as well as a relevant aspect is the integration of huge masses of data and the use of NoSQL databases.

Furthermore, it is always stronger the tendency to distinguish two different types of professional roles: works technically oriented to the realization of GI services, and work-oriented projects that require a deeper understanding of the fundamental concepts and specific domain requirements.

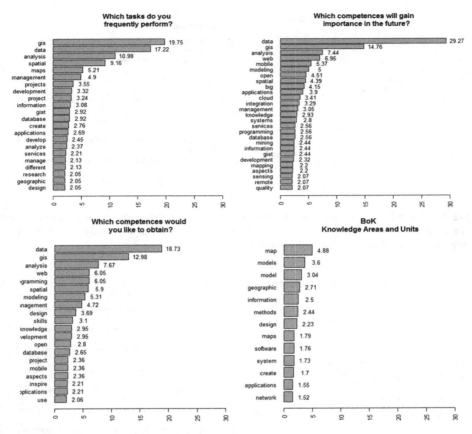

Fig. 7. Word frequency counts for keywords that are mentioned at least by 2 % of responses (Wallentin et al. [6]).

4 An Ontology for the Revisited BoK

In May 2015, a two-days workshop was held in Lisbon with the aim of defining a revision strategy for BoK. The consortium partners have agreed on a revision method which provides to add to the 10 KA of the original BoK an additional KA concerning the latest technological developments, completely absent in the 2006 version of BoK.

Simultaneously, 11 groups of experts were defined, from the consortium partners and from other experts in specific fields.

In this first phase, each group will work on different subjects and units in order to decide which ones have become obsolete in the meantime, which ones will have to be modified, and finally which concepts will be introduced as new topics or units.

Meanwhile, a first transposition of the content of BoK into an ontological structure (Fig. 8) has been carried out. Each KA is hierarchically connected to the corresponding unit, and then to the topics.

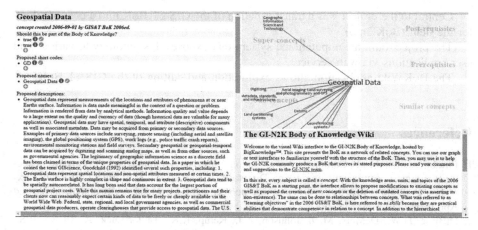

Fig. 8. The ontological structure of the BoK.

The results of upcoming activities of WP2, will enrich the ontology with the semantic relationships between useful concepts. This aspect is important to ensure interoperability between them.

In particular, a network of concepts linked by different types of relations, Hierarchies, Super-concepts & Sub-concepts, Similarity, Dependency, etc., will be built. Moreover, each concept will have attributes (properties) - e.g. a name and a description.

With the conclusion of WP 2 it will be possible to have an updated version of BoK to test with real use case, to arrive at the definition of a body of knowledge within the GI actually adhering to the needs of the labour market, able to support the definition of profiles that have the right methodological and technological response to the sector demands.

5 Conclusions

The current supply of geospatial professionals is inadequate and the geospatial workers appear to be inadequately prepared to answer to the challenges and opportunities of this field.

For the identification of the specific knowledge areas that a professional needs to master for proficiency and success in its field or profession, it is proposed to use the methodology of the Body of Knowledge (BoK) which is basically the agreed ontology of a specific professional domain (reference framework).

The goal of GI-N2K is to build upon this discussion and an advanced European-authentic and dynamic GI S&T BoK.

This project focused on foundational research into the creation of a transformational, dynamic environment for pedagogy, knowledge construction, discourse, collaboration, and research in the domain of Geographic Information Science and Technology (GIS&T).

Central to the project was a transformation of the GIS&T BoK into a core ontology for the field (BoKOnto).

The project built a computational system that exposes this ontology to various end user applications via web services.

The BoKWiki application allows exploration and editing of the GIS&T BoK.

References

1. Pira International Ltd.: Commercial exploitation of Europe's public sector information - executive summary, European Commission Directorate General for the Information Society Luxembourg (2000). ftp://ftp.cordis.europa.eu/pub/econtent/docs/2000_1558_en.pdf. Accessed Apr 2016
2. DiBiase, D.W., DeMers, M.N., Johnson, A.J., Kemp, K.K., Taylor-Luck, A., Plewe, B.S., Wentz, E.A. (eds.): Geographic Information Science and Technology Body of Knowledge, 1st edn. University Consortium for Geographic Information Science and Association of American Geographers, Washington, D.C. (2006)
3. Hossain, I., Reinhardt, W.: Curriculum design based on the UC GI S&T body of knowledge supported by a software tool. In: Proceedings of the 8th European GIS Education Seminar-EUGISES 'GIS-Education: Where Are the Boundaries?' Leuven, BE, 6–9 September 2012, pp. 19–26 (2012)
4. Painho, M., Curvelo, P.: BoK e-Tool prototype an ontological-based approach to the exploration of geographic information science and technology body of knowledge. In: Online Proceedings of the 6th European GIS Education Seminar-EUGISES 2008, Cirencester, UK, pp. 1–8 (2008)
5. Rip, F.: GI-education: the impact of EduMapping. In: Proceedings of the Workshop 'GIS-Education in a Changing Academic Environment-LeGIO', Leuven, BE, 18 November 2011, pp. 3–12 (2011)
6. Wallentin, G., Hofer, B., Traun, C.: GI-N2K. Analysis of the demand for geospatial education and training (Task 1.1), Project Deliverable 1.1.1. University of Salzburg, Interfaculty Department of Geoinformatics – Z_GIS (2014)
7. Rip, F.I., van Lammeren, R.J.A., Bergsma, A.R.: GI-N2K supply survey. An exploratory survey about teaching geo-information in 28 European countries. Wageningen University CGI report 2014-2. Wageningen, The Netherlands (2014)

Author Index

Printed in the United States
By Bookmasters